Lecture Notes in Computer Science 1234
Edited by G. Goos, J. Hartmanis and J. van Leeuwen

Advisory Board: W. Brauer D. Gries J. Stoer

Springer
Berlin
Heidelberg
New York
Barcelona
Budapest
Hong Kong
London
Milan
Paris
Santa Clara
Singapore
Tokyo

Sergei Adian   Anil Nerode (Eds.)

Logical Foundations of Computer Science

4th International Symposium, LFCS'97
Yaroslavl, Russia, July 6-12, 1997
Proceedings

 Springer

Series Editors

Gerhard Goos, Karlsruhe University, Germany

Juris Hartmanis, Cornell University, NY, USA

Jan van Leeuwen, Utrecht University, The Netherlands

Volume Editors

Sergei Adian
Russian Academy of Sciences, Steklov Mathematical Institute
8 Goubkina Street, Moscow 117966, Russia
E-mail: adian@genesis.mi.ras.ru

Anil Nerode
Mathematical Sciences Institute, Cornell University
Ithaca, NY 14853, USA
E-mail: anil@math.cornell.edu

Cataloging-in-Publication data applied for

Die Deutsche Bibliothek - CIP-Einheitsaufnahme

Logical foundations of computer science : 4th international
symposium ; proceedings / LFCS '97, Yaroslavl, Russia, July 6 - 12,
1997. Sergei Adian , Anil Nerode (ed.). - Berlin ; Heidelberg ; New
York ; Barcelona ; Budapest ; Hong Kong ; London ; Milan ; Paris ;
Santa Clara ; Singapore ; Tokyo : Springer, 1997
   (Lecture notes in computer science ; Vol. 1234)
   ISBN 3-540-63045-7

CR Subject Classification (1991): F.4, F.3

ISSN 0302-9743
ISBN 3-540-63045-7 Springer-Verlag Berlin Heidelberg New York

© Springer-Verlag Berlin Heidelberg 1997
Printed in Germany

Typesetting: Camera-ready by author
SPIN 10548775   06/3142 – 5 4 3 2 1 0   Printed on acid-free paper

Preface

This volume consists of refereed papers accepted for the symposium "Logical Foundations of Computer Science '97, Logic at Yaroslavl". This is the fourth symposium in the LFCS series. The first, Logic at Botik, took place in 1989. The second, Logic at Tver, took place in Sokol, near Tver, in 1992. The third, Logic at St Petersburg, took place in 1994.

The changing panorama of the post-Cold War world has created unprecedented opportunities for the development of ties between the countries of the former Soviet Union and the West. All scientists can profit from new opportunities, new contacts, and new scientific challenges. The LFCS series is one of these exciting opportunities. Conceived as a forum for the exchange of ideas and as a meeting place for scientists from the former Soviet Union and the West, LFCS attracts scientists devoted to logic in computer science. Equally, it is a forum for the presentation of challenges to logic presented by computer science. The interaction of logic and computer science is an especially fertile ground for new ideas. We hope that future symposia will contribute equally to both science and better mutual understanding.

We would like to express our gratitude to the members of the international program committee who took part in the reviewing process, producing a quality scientific program, and to the Yaroslavl organizing committee, which provided splendid support. We acknowledge the financial support of the Russian Foundation for Basic Research, the Mathematical Sciences Institute of Cornell University, the ARO under the MURI program "Integrated Approach to Intelligent Systems" – grant number DAA H04-96-1-0341, the Steklov Institute in Moscow, and Yaroslavl University.

This volume could not have been produced without the assistance of S. Artemov, L.D. Beklemishev, J. Davoren, D. Drake, V. Marek, A. Nogin, E. Nogina, and A. Walz.

March 1997

S.I. Adian, Moscow, Russia
A. Nerode, Ithaca NY, USA

Program Committee:

S.I. Adian (Moscow, co-chair)
A. Nerode (Ithaca, co-chair)
S.N. Artemov (Moscow and Ithaca)
A. Blass (Ann Arbor)
V.G. Durnev (Yaroslavl)
E. Engeler (Zurich)
S.S. Goncharov (Novosibirsk)
G. Gottlob (Vienna)
J. Jaffar (Yorktown Heights)
D. Leivant (Bloomington)
G. Longo (Paris)
A. Macintyre (Oxford)
J. Makowsky (Haifa)
V. Marek (Lexington)
Yu.V. Matiyasevich (St. Petersburg)
J.-E. Pin (Paris)
A.A. Razborov (Moscow)
J. Remmel (San Diego)
H. Schwichtenberg (Munich)
M.A. Taitslin (Tver)
P. Vitanyi (Amsterdam)

Organizing Committee:

V.G. Durnev (Chair, Yaroslavl)
S.I. Adian (Moscow)
L.D. Beklemishev (Moscow)
V.A. Bondarenko (Yaroslavl)
G.S. Mironov (Yaroslavl)
M.R. Pentus (Moscow)
V.A. Sokolov (Yaroslavl)
I.G. Zetkin (Yaroslavl)
O.V. Zetkina (Yaroslavl)

Table of Contents

Topological Semantics for Hybrid Systems

Sergei Artemov *, Jennifer Davoren ** and Anil Nerode ***

Mathematical Sciences Institute
Cornell University
Ithaca, NY 14850, U.S.A.
{artemov,jennifer,anil}@math.cornell.edu

1 Introduction

Hybrid systems are interacting networks of real time digital programs, implemented as finite automata, and continuous physical systems, usually referred to as plants. The purpose of the real time digital programs is to control the behavior of the plants. Our purpose is to introduce a topological semantics for hybrid systems so that proofs of digital program correctness imply that behavior is correct in the presence of fluctuations in critical parameters such as observation of plant state. This is not provided by existing languages for non-linear plants.

The theory of hybrid systems is advancing rapidly. For background and terminology on hybrid systems see the Kohn-Nerode papers: [NK93a], [NK93b], [KNRY95], [KNR95], [GKNR96], [KJN96], [KNR96a], [KNR96b], [KRNJ96], [KRN97]. Also consult the volumes from the hybrid systems workshop series in which some of these papers appear, and the proceedings of recent IEEE control theory conferences: International Symposium on Intelligent Control, Conference on Decision and Control (CDC), and Symposium on Computer-Aided Control System Design (CACSD).

The purpose of logics for hybrid systems is to give an interpreted language for reasoning about hybrid systems. They may encode the system performance specification for the plants and the control programs of the hybrid system, provide a deductive means to extract digital control programs from performance specifications plus the mathematical simulation model for the plant, and provide deductive semantical tools to verify that a proposed digital control program forces the hybrid system to meet its performance specification. In the physical world, there are always unmodelled dynamics. A ubiquitous source of unmodelled dynamics is the inevitable deviation of the physical parameters or measurements from those used to construct the model. This is our main concern here. We want a language in which validity of a program correctness statement

---

 * Supported in part by the Russian Foundation for Basic Research, grant No. 96-01-01135, and by INTAS grant No. 94-2412.
** American Association of University Women (AAUW) Educational Foundation 1996-97 International Fellow.
*** Research supported by ARO under the MURI program "Integrated Approach to Intelligent Systems", grant number DAA H04-96-1-0341.

implies that the digital control program works when small error are present in specified parameters.

Usually logics of hybrid systems, whether based on temporal logics or on automata languages or otherwise, come with a natural classical semantics in which basic variables range over the state spaces associated with the hybrid dynamical system. These state spaces include spaces of plant states, control values, sensor readings and disturbances.

When automata languages are used, there is an explicit or implicit reduction from the original continuous dynamical systems state spaces to strictly finite versions of these spaces that the digital program can deal with in digital terms. This is achieved by partitioning each state space involved into finitely many partition sets, and letting variables previously ranging over a state space now run over the finitely many partition sets of that space instead. The plant is represented by a non-deterministic input-output automaton. The digital control automaton is usually extracted to control this finite reduction of the plant, entirely within the world of discrete finitary mathematics. The digital control programs designed, extracted, and verified within such automata languages treat plant states in the same partition set as indistinguishable in deciding what control orders to send to the actuators.

The physical plants whose states are being controlled, the sensors, and the actuators, know nothing about the partitioning process and simply follow their evolutionary equations subject to controls, disturbances and other fluctuations. In such an automaton language one tries to verify that the control automaton satisfies the performance specification assuming that the plant automaton has been represented in some way. To obtain this representation, one discretizes states, time of sensing and time of control, and then hopefully takes the care to go through epsilon-delta analysis, done entirely outside the formal automaton language, to figure out possible transitions from partition to partition. But often the plant automata representation is such that there is no controller that could achieve the given performance specification. In this case, one seeks a better plant representation by taking finer partitions in the hope of finding a better digital controller. But this may be impossible too, and the performance specification may have to be weakened. When one succeeds, the automata often have millions of states.

2 Motivation

The dynamical systems state spaces associated with a hybrid system come equipped with natural topologies and continuous operations. Their behavior, especially relative to fluctuations, needs to be analyzed by mathematical tools such as analysis, calculus of variations, geometry of manifolds, Lie algebras, symbolic methods in differential equations, numerical analysis, etc. We want to be sure that proving a program correct within the language forces the hybrid system to satisfy its specification in the face of the small inevitable uncertainties in hybrid system parameters. The plant state partition handed to the control program is often not the plant state partition of the true state because the true state is close to the boundary between two partitions and small fluctuations in sensor readings and analog to digital conversion make the observed state's partition different from the true state's partition.

There are two ways a language can be altered to yield correctness proofs that prove sufficiently small changes do not take one outside the performance specification. The first is to extend the syntactical expressiveness of the language. The second is to change the semantics, axioms, and rules of inference so that proving a program correct has a different meaning, which ensures that small fluctuations do not harm the program satisfying its performance specification. The first we think leads to unwieldy formal systems which might as well be set theory. We propose the second. Thus we wish to associate a topological semantics with the hybrid system with associated valid axioms and rules of inference such that correctness proofs preserve validity in the semantics, and validity in the semantics implies that small changes in prescribed system parameters leave the controlled hybrid system within its performance specification.

Here is the origin of the topological semantics we use. The only access a digital control automaton has to current information about plant state is by analog to digital conversion of a sensor of plant state. The digital conversion of the sensor reading fires one of a finite number of input symbols for the control automaton since a real time digital system has only a finite number of input letters. The control orders that the control automaton can send to actuators are among a finite number of possibilities for the same reason. The control automaton is finite state for the same reason.

We argue as a physical realizability requirement that arbitrarily small changes in input sensor readings should not change the input letter fired. We interpret this as meaning that each input letter is fired by all states in an open set. We assume that every plant state fires some input letter, and may fire many. The control automaton has no information about plant state other than the set of input letters it fires. We think of each input letter as denoting the open set of states that fire it. This is the origin of our models. The open sets firing the various letters generate a finite subtopology of the plant state topology, a finite distributive lattice of open sets. If one takes all the letters fired by a given state and intersects the denoted open set, one gets the join irreducibles of the topology as a lattice. Every open set of the finite topology is a join of join irreducibles. We can add, without changing anything theoretically, input letters so that for each join irreducible there is an input letter fired exactly by states in that join irreducible. In fact, the join irreducibles might as well be taken as the input alphabet. Define two states as observationally indistinguishable with respect to the finite automaton if they fire exactly the same input letters. The equivalence class of a state is the join irreducible it is in, minus the union of all smaller join irreducibles. The equivalence classes of observational indistinguishable states form a finite partition of the states. These are exactly the atoms of the Boolean algebra of subsets of the domain generated by the finite topology. A confusing point at first is that the partition sets are not open; they are the difference of an open and a closed set. Note that the finite topology cannot be recovered from knowledge of the partition, which is the origin of our difficulties in expressing behavior with respect to fluctuations. The finite topology is usually non-Hausdorff. The set of partition sets is the quotient of the state space under observational equivalence, and is a finite T_0 space, usually not discrete. We view the control automaton as seeing, about the state space, only the partition set containing the current state. This is a member of the quotient space.

In the Kohn-Nerode theory ([NK93a], [NK93b], [KJN96]), finite topologies on the state space and on the control space are extracted along with the control program. In order to avoid the explosion entailed in writing out the transition matrix for a plant automaton (they have often millions of states), Kohn-Nerode use a Prolog program as an implicit representation of the plant automaton. Its possible premises are the input letters for the control automaton. They are atomic statements which fire when the sensor reports that the plant is in the open set that fires that input letter. We say that the atomic statement denotes the open set that fires it. This is the origin of our topological models of Prolog and first order logic for hybrid systems. The Prolog control program may be thought of as having a predicate $P(x, y, z)$ which holds when x is a plant state partition set, y is one of finitely many control values that the automaton could issue at the beginning of a control interval, z is the partition set of a plant state that could result at the end of a control interval by the plant evolving from a state in the x partition with control value y at the beginning of a control interval. The program also has a predicate $G(x, z)$ of partition sets of states which holds when x is the partition set of the initial state, z is in the partition set of the end state of the control interval, and z is an acceptable local goal. The Prolog deduction mechanism instantiates y and z when x has been instantiated by the sensor reading so as to satisfy both $P(x, y, z)$ and $G(x, z)$. The instantiation of y is the control that the control automaton will impose at the beginning of the control interval. The Prolog answer substitution mechanism thus is used to select a particular control, giving a deterministic automaton based "cross-section" of the plant description $P(x, y, z)$. That is, it gives a control policy telling what control y to use in initial state x with goal Q, based on the partition sets. Where does the local goal Q come from in Kohn-Nerode? They use dynamic programming on manifolds on which the plant state evolves to reduce global goals to local ones, and automata theory and approximation to determine the finite topology on the plant space and the finite automaton. The automaton and the denotations of its input letters can be read off from finite topologies that come up in the Prolog Program extraction process (see the appendix to [NK93a]). A number of readers have thought that the finite topology associated with input letters was to characterize sensors. Instead it is a requirement on the accuracy needed by sensors in order that the control automaton provably work correctly when subjected to small variations in sense data. The sizes of the variations that can be tolerated also come out of the extraction process.

We can express the topological semantics of the Kohn-Nerode Prolog programs as follows. These are Horn clause programs, so only universal quantifiers are involved, and one is deducing answer substitutions since this is where control laws come up. Due to the fact that the partitions of state spaces induced by observational equivalence are finite in number, we can regard ourselves as working over a finite (multi-sorted) domain, each sort a finite partition of some state space. One sort will be a state space partition, another a control space partition or a finite set of controls, etc. Thus the original dynamical system is replaced by a finite simulacrum, as in all automata based approaches. But the partition sets carry the induced T_0 topologies under observational equivalences; they are a topological structure. So we have a finite quotient structure with T_0 topologies on everything, which are generally not discrete topologies. We have the predicates on these partition sets induced by the predicates on the original state space, the ones expressed in the Prolog program. Since in this instance the domain is finite and every partition set can be given a name, for simplicity we can look at the ground version of the Prolog program. In practice, one does not do this because it is computationally inappropriate. But here the explanation is simpler. The extraction procedure gives an open set denotation to the premises and conclusions of each ground clause, in such a way that being a state in the intersection of the open sets denoted means classically satisfying the premises, and any such not only satisfies the conclusion, but is in the open set denoted by the conclusion. The finite topology is then generated by the open sets in the denotations of clauses.

In fact, we are not limited to Horn clauses. Take predicate logic based on "and", "or", "implies", "falsehood", and the quantifiers. Interpret quantifiers as unions and intersections of open sets over the finite domain, "or" and "and" similarly. One gets by induction an open set denotation for all sentences in the finite topology. If one introduces a product space with infinitely many copies of each finite partition space, one for each variable ranging over a sort in the language of first order logic, one gets a natural space to define satisfaction so that every formula denotes an open set, and valid formulas denote the whole space. This is a substantial refinement of the Tarski-Rasiowa-Sikorski topological semantics ([Ta38],[RS63]; see also [Da86], [Du77]). In their work, there is no topology on the underlying domain of the model. Rather, the domain is just a set carrying functions and relations, and atomic sentences are assigned open sets in a topology in the same way they are assigned values in a Heyting algebra (pseudo-Boolean algebra). In our topological semantics, an atomic sentence $R(a)$ must denote an open set containing the point denoted by a; the domain of the model must be itself a topological space. The finite topologies coming from the Prolog program are all subtopologies of an original topology on the state space, and our semantics captures the notion of finite approximation of truth. We express program correctness as validity in this model; that is, the program correctness assertion denotes the whole space. Statements of performance specification can involve quantifiers in any way.

What axioms and rules of inference should we use? We have postponed mentioning Intuitionistic logic till now, because that logic was designed for constructive mathematics based on a constructive theology. The origin here is purely topological. The answer is: the rules and axioms of Intuitionistic first order logic. If one starts out with premises denoting the whole space, and uses these axioms and rules, one will end up with conclusions denoting the whole space. Note there is no extension of first order logic, rather only a restriction to Intuitionistic rules of inference and axioms. An inductive examination of the definition of denotation familiar to Intuitionistic logicians reveals exactly why, with quantified statements, the condition that sufficiently small changes in sensor state do not move one out of the performance specification is satisfied by program correctness statements valid in this interpretation. We nowhere syntactically mention small changes. The constructive character of Intuitionistic proofs shows that by following them in detail one can compute from the denotations of atomic statements how big fluctuations can be, and still keep within the performance specification. These are sufficient bounds, not necessary ones. They represent epsilon-delta arguments. We are interpreting a Prolog program over a finite domain consisting of partition sets and relations between them. If we interpret the finite topology itself as a partially ordered set under inclusion, and when an atomic statement denotes an open set, we say that the open set *forces* the atomic statement, we get a Kripke model. We then find that topological validity is validity in that Kripke model, which is finite and decidable by a P-space complete procedure ([St79]). Thus the Prolog program, and its specification, are in a small decidable theory of a finite topological automaton. We intend to develop topological automata in a future publication.

3 Conclusion

This procedure for strengthening correctness to correctness relative to small changes in any interesting parameters can be extended to apply to any hybrid systems language, with slight changes in primitive relations. The key idea is that the physical systems are dynamical systems equipped with topologies on state spaces, that finite automata have to work relative to finite space quotient topologies, and that one needs to respect observational equivalence to get correctness proofs that allow small variations, which means using Intuitionistic deductions.

References

[BD74]   Raymond Balbes and Philip Dwinger, *Distributive Lattices* (University of Missouri Press, Columbia, 1974).

[Br93]   Michael Branicky, "Topology of Hybrid Systems", *Proceedings of the 32nd IEEE Conference on Decision and Control* (IEEE Computer Society Press, Los Alamitos, 1993), Vol. 3, 2309-2314.

[Da86]   Dirk van Dalen, "Intuitionistic Logic", in D. Gabbbay and F. Guenthner (eds.), *Handbook of Philosophical Logic, Volume III: Alternatives to Classical Logic* (D. Reidel, Dordrecht, 1986), 225-339.

[Du77]   Michael Dummett, *Elements of Intuitionism* (Clarendon Press, Oxford, 1977).

[FS79]   Michael P. Fourman and Dana S. Scott, "Sheaves and Logic", in M. P. Fourman, C. J. Mulvey and D. S. Scott (eds.), *Applications of Sheaves*, Proceedings of the Research Symposium on Applications of Sheaf Theory to Logic, Algebra and Analysis, Durham, England, July 9-21, 1977; Springer Lecture Notes in Mathematics **753** (Springer-Verlag, Berlin 1979), 302-401.

[GKNR96] Xiaolin Ge, Wolf Kohn, Anil Nerode, and Jeffrey B. Remmel, "Hybrid Systems: Chattering Approximations to Relaxed Controls", in R. Alur, T. A. Henzinger, E. D. Sontag (eds.), *Hybrid Systems III*, Lecture Notes in Computer Science **1066** (Springer-Verlag, Berlin, 1996), 76-100.

[KJN96]   Wolf Kohn, John James and Anil Nerode, "The Declarative Approach to the Design of Robust Control Systems", *Proceedings of the IEEE Symposium on Computer-Aided Control System Design 1996 (CACSD '96)*, 26-31.

[KJNRC96] Wolf Kohn, John James, Anil Nerode, Jeffrey B. Remmel and Benjamin Cummings, "A New Approach to Generating Finite-State Control Programs for Hybrid Systems", *Proceedings of the 13th Triennial World Congress of the International Federation of Automatic Control (IFAC) 1996* (Pergamon Press, 1996), 461-466.

[KNR95]   Wolf Kohn, Anil Nerode and Jeffrey B. Remmel, "Hybrid Systems as Finsler Manifolds: Finite State Control as Approximation to Connections", in P. Antsaklis, W. Kohn, A. Nerode and S. Sastry (eds.), *Hybrid Systems II*, Lecture Notes in Computer Science **999** (Springer-Verlag, Berlin, 1995), 294-321.

[KNR96a]   Wolf Kohn, Anil Nerode and Jeffrey B. Remmel, "Continualization: A Hybrid Systems Control Technique for Computing", *Proceedings of 1996 IMACS Conference on Computation Engineering in Systems Applications (CEAS '96)*.

[KNR96b]   Wolf Kohn, Anil Nerode and Jeffrey B. Remmel, "Feedback Derivations: Near Optimal Controls for Hybrid Systems", *Proceedings of the 1996 IMACS conference on Computation Engineering in Systems Applications (CESA '96)*, 517-521.

[KNRY95]   Wolf Kohn, Anil Nerode, Jeffrey B. Remmel and Alexander Yakhnis, "Viability in Hybrid Systems", *Theoretical Computer Science* **138** (1995) 141-168.

[KRN97]   Wolf Kohn, Jeffrey B. Remmel, Anil Nerode, "Scalable Data and Sensor Fusion via Multiple Agent Hybrid Systems", *IEEE Transactions on Automatic Control*, special issue on Hybrid Systems, to appear 1997.

[KRNJ96]   Wolf Kohn, Jeffrey B. Remmel, Anil Nerode, John James, "Multiple Agent Hybrid Control for Manufacturing Systems", *Proceedings of the 11th IEEE International Symposium on Intelligent Control* (IEEE, New York 1996), 348-353.

[Ll87]   John W. Lloyd, *Foundations of Logic Programming*, 2nd edition (Springer-Verlag, Berlin 1987).

[Ne59]   Anil Nerode, "Some Stone Spaces and Recursion Theory", *Duke Mathematical Journal* **26** (1959) 397-406.

[Ne90]   Anil Nerode, "Some Lectures on Intuitionistic Logic", in P. Odifreddi (ed.), *Logic and Computer Science*, Lecture Notes in Mathematics **1429** (Springer-Verlag, Berlin, 1990), 12-59.

[NK93a]   Anil Nerode and Wolf Kohn, "Models for Hybrid Systems: Automata, Topologies, Controllability, Observability", in R. Grossman, A. Nerode, A. Ravn and H. Rischel (eds.), *Hybrid Systems*, Springer Lecture Notes in Computer Science **736** (Springer-Verlag, Berlin,1993), 297-316.

[NK93b]   Anil Nerode and Wolf Kohn, "Multiple Agent Hybrid Control Architecture", in R. Grossman, A. Nerode, A. Ravn and H. Rischel (eds.), *Hybrid Systems*, Springer Lecture Notes in Computer Science **736** (Springer-Verlag, Berlin,1993), 317-356.

[RS63]   Helena Rasiowa and Roman Sikorski, *The Mathematics of Metamathematics*, Polska Akademia Nauk Monografie Matematyczne, Tom **41** (Państwowe Wydawnictwo Naukowe, Warsaw, 1963).

[St79]   Richard Statman, "Intuitionistic Propositional Logic is Polynomial Space Complete", *Theoretical Computer Science* **9** (1979) 67-72.

[Ta38]   Alfred Tarski, "Der Aussagenkalkül und die Topologie", *Fundamenta Mathematicae* **31** (1938) 103-134. Reprinted (and translated by J. H. Woodger) as "Sentential Calculus and Topology", in A. Tarski, *Logic, Semantics, Metamathematics* (Oxford University Press, 1956), 421-454.

The Concurrency Complexity for the Horn Fragment of Linear Logic

Sergey M. Dudakov

Dept. of CS, Tver St. Univ., 33 Zheljabova str., Tver, Russia, 170000

Abstract. The provability problem for the Horn fragment of linear logic is NP-complete [4, 1]. In this work we investigate various definitions of concurrency proposed in [2] and establish the complexity of the provability problem and the problem of concurrency recognition. The notion of k-maximal concurrency is introduced which guarantees polynomial time provability. Theorems on hierarchy and on complexity of recognition of the property are proved.

1 Introduction

The problem of provability for the Horn fragment of linear logic was proposed in [5]. Kanovich in [4] proved that the problem is *NP*-complete. Another simple proof of this fact was presented in [1]. So it is interesting to find such natural subclasses of the problem which can be solved in polynomial time. In [2] two subclasses were defined based on the idea of parallel computations: the class of superconcurrent multisets of Horn implications **SUC** and the class of concurrent multisets of Horn implications **CC**. The former class allows to apply implications in any order, the second one provides nondeterministic computations which are "maximally" concurrent. Some properties of classes **SUC** and **CC** were established and it was proven that the problem of recognition if a multiset of implications Γ belongs to **CC** is *co-NP*-hard.

This work continues the investigations of [2]. A new class of strong concurrent multisets of implications **STC** is defined in Section 2 with the polynomial time provability problem. The complexity of provability and recognition problems for all three classes established in Section 3. In particular, the bound of [2] for **CC** is refined: the recognition problem for this class is Π_2^P-complete. If the size of alphabet is bounded by a constant then the problem belongs to Δ_2^P. For classes **SUC** and **CC** recognition problems are *co-NP*-complete. In Section 4 a series of classes $\mathbf{MC}_k$, $k = 1, 2, 3, \ldots$ is introduced. Parameter k in the name of class denotes the number of simultaneously applied implications. The provability problem for every class $\mathbf{MC}_k$ is solvable in polynomial time. It is shown that classes $\mathbf{MC}_k$ forms a hierarchy between classes **STC** and **CC**. The recognition problem for $\mathbf{MC}_k$ is *NP*-complete. Class **MC** which is the union of all classes $\mathbf{MC}_k$ has *NP*-complete provability problem. But for its subclasses which use alphabets of bounded size the provability problem is solvable in polynomial time.

Due to space limitations only short sketches of the proofs are included.

We consider rather broad classes of DDBs. Therefore, not surprisingly, the complexity of the properties of $\forall\forall$- and $\forall\exists$-stability in these classes sometimes turns out to be very high. However, in real applications the intensional description of system behavior is composed from a variety of independent small modules. For such DDBs our results guarantee quite tractable upper complexity bounds. Very efficient CAD systems can be created on the ground of the proposed concepts, which can support an interactive experimental analysis of behavior of complex discrete dynamic systems in broad classes of applications.

References

1. Apt, K.R., Blair, H. and Walker A., *Towards a theory of declarative knowledge.* in: J. Minker (ed.) *Foundations of deductive databases and logic programming.* Morgan Kaufman Pub., Los Altos, 89-148, 1988.
2. Chandra, A.K., Kozen, D.C., Stockmeyer, L.J., *Alternation.* J. Ass. Comput. Mach., v.28, n.1, 114-133, 1981.
3. Dayal, U., Hanson,E., and Widom, J., *Active database systems.* In W. Kim (ed.) *Modern Database Systems.* 436-456, Addison Wesley, 1995.
4. Dekhtyar, M.I., Dikovsky, A.Ja., *Dynamic Deductive Data Bases with Steady Behavior.* In L. Sterling (ed.) *Proc. of the 12th International Conf. on Logic Programming,* The MIT Press, 183-197, 1995.
5. Dekhtyar, M.I., Dikovsky, A.Ja., *On Homeostatic Behavior of Dynamic Deductive Data Bases.* In: D. Bjorner, M.Broy, I.Pottosin (eds.) *Proc. 2nd Int. A.P.Ershov Memorial Conference "Perspectives of System Informatics".* Lecture Notes in Computer Science. 1996, Vol. 1181, 420-432.
6. Eiter, T., Gottlob, G., *On the complexity of propositional knowledge base revision, updates, and counterfactuals.* Artificial Intelligence, vol. 57, 227-270, 1992.
7. Gelfond, M., Lifschitz, V., *The stable semantics for logic programs.* In: R.Kovalsky and K.Bowen (eds.) *Proc. of the 5th Intern. Symp. on Logic Programming.* 1070-1080, Cambridge, MA, 1988, MIT Press.
8. Gottlob, G., Moercotte, G., Subrahmanian, V.S., *The PARK semantics for Active Databases.* In *Proc. of EDBT'96.* Avignon, France, 1996.
9. Halfeld Ferrari Alves, M., Laurent, D., Spyratos, N. *Update rules in Datalog programs.* Rapport de Recherche n. 1024, 01 / 1996, Université de Paris-Sud, Centre d'Orsay, LRI.
10. Katsuno, H., Mendelzon, A. O., *Propositional knowledge base revision and minimal change.* Artificial Intelligence, vol. 52, 253-294, 1991.
11. Marek, V.W., Truszcińsky, M. *Revision programming, database updates and integrity constraints.* In: *International Conference on Data Base theory,* ICDT, LNCS n. 893, 368-382, 1995.
12. Przymusinski, T.C., Turner, H., *Update by Means of Inference Rules.* In: V.W.Marek, A.Nerode, M.Truszczyński (eds.) *Logic Programming and Nonmonotonic Reasoning,* Proc. of the Third Int. Conf. LPNMR'95, Lexington, KY, USA, 166-174, 1995.

2 Abstract type systems

In this section we introduce the notion of an abstract type system, which provides a convenient framework for the presentation and comparison of different notions of type system.

Definition 1. An *abstract type system* (ATS) is a triple $(V, T, \vdash)$ where

1. V is a set of *variables*;
2. T is a set of *terms* with $V \subseteq T$;
3. $C = (V \times T)^*$ is the set of *contexts*.[4] The empty context is denoted by $\langle\rangle$; The domain of a context Γ is $\mathrm{dom}(\Gamma) = \{x \mid \exists t \text{ s.t. } x : t \in \Gamma\}$.
4. $J = C \times T \times T$ is the set of *judgements*;
5. $\vdash \subseteq J$ is the *derivability* relation;

satisfying the following closure property

$$((\Gamma, x : B, \Gamma'), M, A) \in \vdash \quad \Rightarrow \quad ((\Gamma, x : B), x, B) \in \vdash$$

In the sequel, we write $\Gamma \vdash M : A$ for $(\Gamma, M, A) \in \vdash$ and $\vdash M : A$ for $(\langle\rangle, M, A) \in \vdash$.

The derivability relation may be used to define legal terms.

Definition 2. Let $(V, T, \vdash)$ be an ATS.

1. A judgement (Γ, M, A) is *legal* if $\Gamma \vdash M : A$.
2. A context Γ is *legal* if $\Gamma \vdash M : A$ for some M and A.
3. A term M is *legal* if $\Gamma \vdash M : A$ or $\Gamma \vdash A : M$ for some Γ and A.

Morphisms of ATSs provide an important tool to compare type systems. Here we let an ATS-morphism be a map between the underlying sets of terms which preserves derivability.

Definition 3. Let $(V, T, \vdash)$ and $(V', T', \vdash')$ be ATSs. A map $h : T \to T'$ is naturally lifted to maps $h^C : C \to C'$ on contexts and $h^J : J \to J'$ on judgements:

$$h^C(\langle\rangle) = \langle\rangle$$
$$h^C(\Gamma, x : A) = h^C(\Gamma), h(x) : h(A)$$

$$h^J(\Gamma, M, A) = (h^C(\Gamma), h(M), h(A))$$

The map $h : T \to T'$ is an ATS-*morphism* if

1. for every $x \in V$, $h(x) \in V'$;
2. for every $j \in J$, $j \in \vdash$ implies $h^J(j) \in \vdash'$.

---

[4] S^* denotes the set of finite lists over S.

Domain-Free Pure Type Systems

Gilles Barthe[1] & Morten Heine Sørensen[2]

[1] Centrum voor Wiskunde en Informatica (CWI)
PO Box 94079, 1090 GB Amsterdam, The Netherlands, gilles@cwi.nl

[2] Department of Computer Science University of Copenhagen (DIKU)
Universitetsparken 1, DK-2100 Copenhagen Ø, Denmark, rambo@diku.dk

Abstract. Pure type systems feature domain-specified λ-abstractions $\lambda x : A.M$. We present a variant of pure type systems, which we call domain-free pure type systems, with domain-free λ-abstractions $\lambda x.M$. We study the basic properties of domain-free pure type systems and establish their formal relationship with pure type systems.

1   Introduction

Typed versions of the λ-calculus were introduced independently by Church [6] and Curry [7]. In Church's system abstractions have *domains*, i.e. are of the form $\lambda x{:}A.t$, whereas in Curry's system abstractions have no domain, i.e. are of the form $\lambda x.t$. Over the years, many type systems have appeared, the majority of which use domain-specified abstractions. Barendregt and others give an abstract, unifying view of type systems with domain-specified abstractions in terms of the notion of *pure type system*—see e.g. [2, 3, 8, 9]. In this paper, we consider for every pure type system a domain-free version in which abstractions have no domain. We call such systems *domain-free pure type systems*.[3] The main technical contribution of the paper—expressed in Theorem 26—states, under suitable hypotheses, a connection between a pure type system and its corresponding domain-free pure type system. Domain-free pure type systems and Theorem 26 have proved useful for defining continuation-passing style translations for pure type systems [4], for proving strong normalisation from weak normalisation of pure type systems [15], and for studying classical pure type systems [5].

Contents of the paper. In Section 2 we introduce abstract type systems. These are used in Section 3 to present the notion of pure type system and domain-free pure type system. In Section 4 we develop some basic properties of domain-free pure type systems, and in Section 5 we relate pure type systems to domain-free pure type systems. In Section 6 we compare domain-free pure type systems with the type assignment systems of [1]. In Section 7 we discuss type checking issues. We conclude in Section 8.

---

[3] These systems were informally suggested in [10].

The map $h : T \to T'$ is an ATS-*reflection* if moreover, for every $j' \in \vdash'$, there exists $j \in \vdash$ s.t. $h^{\mathcal{J}}(j) = j'$.

Examples of reflections, apart from the ones considered in this paper, can be found in [1, 3, 8]. Reflections are closed under composition but need not be injective and might not have an inverse. The following observation will be useful in Section 6.

Lemma 4. *Let* $(V_1, T_1, \vdash_1)$, $(V_2, T_2, \vdash_2)$ *and* $(V_3, T_3, \vdash_3)$ *be ATSs. Moreover let* $h : T_1 \to T_2$, $h' : T_2 \to T_3$ *and* $h'' : T_1 \to T_3$ *be ATSs morphisms s.t.* $h' \circ h(M) = h''(M)$ *for every legal* M. *If* h'' *is an ATS-reflection, then* h' *is an ATS-reflection.*

3   Pure type systems and domain-free pure type systems

In this section we present pure type systems and domain-free pure type systems. This approach is inspired by [14].

Definition 5.

1. A *specification* is a triple $(\mathcal{S}, \mathcal{A}, \mathcal{R})$ where
   (a) $\mathcal{S}$ is a set of *sorts*;
   (b) $\mathcal{A} \subseteq \mathcal{S} \times \mathcal{S}$ is a set of *axioms*;
   (c) $\mathcal{R} \subseteq \mathcal{S} \times \mathcal{S} \times \mathcal{S}$ is a set of *rules*.
2. A specification $(\mathcal{S}, \mathcal{A}, \mathcal{R})$ is *functional* if for every $s_1, s_2, s_2', s_3, s_3' \in \mathcal{S}$,
   (a) $(s_1, s_2) \in \mathcal{A}$, $(s_1, s_2') \in \mathcal{A}$ $\Rightarrow$ $s_2 \equiv s_2'$
   (b) $(s_1, s_2, s_3) \in \mathcal{R}$, $(s_1, s_2, s_3') \in \mathcal{R}$ $\Rightarrow$ $s_3 \equiv s_3'$
3. $s \in \mathcal{S}$ is a *top-sort* if there is no $s' \in \mathcal{S}$ s.t. $(s, s') \in \mathcal{A}$. The set of top-sorts is denoted by $\mathcal{S}^\mathsf{T}$.

In the rest of this section, we let $\mathbf{S} = (\mathcal{S}, \mathcal{A}, \mathcal{R})$ denote a fixed specification and V denote a fixed countably infinite set of variables.

Definition 6.

1. The set $\mathcal{T}$ of PTS-*pseudo-terms* is given by the abstract syntax:

$$\mathcal{T} = V \mid \mathcal{S} \mid \mathcal{T}\mathcal{T} \mid \lambda V : \mathcal{T}.\mathcal{T} \mid \varPi V : \mathcal{T}.\mathcal{T}$$

2. β-*reduction* $\to_\beta$ is defined as the compatible closure of the contraction

$$(\lambda x{:}A.M)\, N \to_\beta M[x := N]$$

   where $\bullet[\bullet := \bullet]$ is the standard substitution operator.
3. β-*equality* $=_\beta$ is the reflexive, transitive, symmetric closure of $\to_\beta$.
4. The PTS *derivability* relation $\vdash$ is given by the rules of Table 1.

Every specification $\mathbf{S}$ induces an ATS $\lambda\mathbf{S}$ with $\mathcal{T}$ as the set of terms and $\vdash$ as the derivability relation. Such an ATS is called a *pure type system*, or a *PTS*.

(axiom)	$\langle\rangle \vdash s_1 : s_2$	if $(s_1, s_2) \in \mathcal{A}$
(start)	$\dfrac{\Gamma \vdash A : s}{\Gamma, x : A \vdash x : A}$	if $x \notin \mathrm{dom}(\Gamma)$
(weakening)	$\dfrac{\Gamma \vdash A : B \quad \Gamma \vdash C : s}{\Gamma, x : C \vdash A : B}$	if $x \notin \mathrm{dom}(\Gamma)$
(product)	$\dfrac{\Gamma \vdash A : s_1 \quad \Gamma, x : A \vdash B : s_2}{\Gamma \vdash (\Pi x{:}A.\,B) : s_3}$	if $(s_1, s_2, s_3) \in \mathcal{R}$
(application)	$\dfrac{\Gamma \vdash F : (\Pi x{:}A.\,B) \quad \Gamma \vdash a : A}{\Gamma \vdash F\,a : B[x := a]}$	
(abstraction)	$\dfrac{\Gamma, x : A \vdash b : B \quad \Gamma \vdash (\Pi x{:}A.\,B) : s}{\Gamma \vdash \lambda x{:}A.b : \Pi x{:}A.\,B}$	
(conversion)	$\dfrac{\Gamma \vdash A : B \quad \Gamma \vdash B' : s}{\Gamma \vdash A : B'}$	if $B =_\beta B'$

Table 1. PURE TYPE SYSTEMS

Definition 7.

1. The set $\mathcal{L}$ of DFPTS-*pseudo-terms* is given by the abstract syntax:

$$\mathcal{L} = V \mid \mathcal{S} \mid \mathcal{L}\mathcal{L} \mid \lambda V.\mathcal{L} \mid \Pi V : \mathcal{L}.\mathcal{L}$$

2. *β-reduction* $\to_\beta$ is defined as the compatible closure of the contraction

$$(\lambda x.M)\, N \to_\beta M\{x := N\}$$

   where $\bullet\{\bullet := \bullet\}$ is the obvious substitution operator.
3. *β-equality* $=_\beta$ is the reflexive, transitive, symmetric closure of $\to_\beta$.
4. The DFPTS *derivability* relation $\vdash$ is given by the rules of Table 2.

Every specification **S** induces an ATS λ**S** with $\mathcal{L}$ as the set of terms and $\vdash$ as the derivability relation. Such an ATS is called a *domain-free pure type system*, or a *DFPTS*. The two most significant DFPTSs that appear in the literature are Curry's version of the simply typed λ-calculus $\lambda{\to}$ and Martin-Löf's Logical Framework λP. Further examples of DFPTSs are provided by the remaining specifications of the cube [2, 3], as defined below.

Definition 8. Let $\mathcal{S} = \{*, \Box\}$ and $\mathcal{A} = \{(* : \Box)\}$. The *cube*-specifications are

$$
\begin{aligned}
\rightarrow &= (\mathcal{S}, \mathcal{A}, \{(*, *)\}) & P &= (\mathcal{S}, \mathcal{A}, \{(*, *), (*, \Box)\}) \\
2 &= (\mathcal{S}, \mathcal{A}, \{(*, *), (\Box, *)\}) & P2 &= (\mathcal{S}, \mathcal{A}, \{(*, *), (\Box, *), (*, \Box)\}) \\
\underline{\omega} &= (\mathcal{S}, \mathcal{A}, \{(*, *), (\Box, \Box)\}) & P\underline{\omega} &= (\mathcal{S}, \mathcal{A}, \{(*, *), (\Box, \Box), (*, \Box)\}) \\
\omega &= (\mathcal{S}, \mathcal{A}, \{(*, *), (\Box, *), (\Box, \Box)\}) & P\omega &= (\mathcal{S}, \mathcal{A}, \{(*, *), (\Box, *), (\Box, \Box), (*, \Box)\})
\end{aligned}
$$

where (s_1, s_2) denotes (s_1, s_2, s_2). The λ-*cube* consists of the eight PTSs λS, where S is one of the cube-specifications. Similarly, the $\underline{\lambda}$-*cube* consists the eight DFPTSs $\underline{\lambda}$S, where S is one of the cube-specifications.

(axiom)	$\langle\rangle \vdash s_1 : s_2$	if $(s_1, s_2) \in \mathcal{A}$
(start)	$\dfrac{\Gamma \vdash A : s}{\Gamma, x : A \vdash x : A}$	if $x \notin \mathrm{dom}(\Gamma)$
(weakening)	$\dfrac{\Gamma \vdash A : B \quad \Gamma \vdash C : s}{\Gamma, x : C \vdash A : B}$	if $x \notin \mathrm{dom}(\Gamma)$
(product)	$\dfrac{\Gamma \vdash A : s_1 \quad \Gamma, x : A \vdash B : s_2}{\Gamma \vdash (\Pi x{:}A.\, B) : s_3}$	if $(s_1, s_2, s_3) \in \mathcal{R}$
(application)	$\dfrac{\Gamma \vdash F : (\Pi x{:}A.\, B) \quad \Gamma \vdash a : A}{\Gamma \vdash F\, a : B\{x := a\}}$	
(abstraction)	$\dfrac{\Gamma, x : A \vdash b : B \quad \Gamma \vdash (\Pi x{:}A.\, B) : s}{\Gamma \vdash \lambda x.b : \Pi x{:}A.\, B}$	
(conversion)	$\dfrac{\Gamma \vdash A : B \quad \Gamma \vdash B' : s}{\Gamma \vdash A : B'}$	if $B =_{\underline{\beta}} B'$

Table 2. DOMAIN-FREE PURE TYPE SYSTEMS

Definition 9. Let (R, T) be $(\beta, \mathcal{T})$ or $(\underline{\beta}, \mathcal{L})$.

1. $\twoheadrightarrow_R$ is the reflexive transitive closure of $\rightarrow_R$;
2. $M \in \mathrm{NF}_R \Leftrightarrow$ there is no $N \in T$ s.t. $M \rightarrow_R N$;
3. $M \in \mathrm{SN}_R \Leftrightarrow$ there is no infinite sequence $M_0 \rightarrow_R M_1 \rightarrow_R M_2 \rightarrow_R \cdots$;
4. $M \in \mathrm{WN}_R \Leftrightarrow$ there is $N \in \mathrm{NF}_R$ s.t. $M \twoheadrightarrow_R N$.

Elements of NF_R, SN_R, WN_R are *R-normal forms*, *R-strongly normalizing*, and *R-weakly normalizing*, respectively.

Definition 10. A specification **S** has *normalizing (resp. strongly normalizing)* PTS-*types* if $M \in WN_\beta$ (resp. $M \in SN_\beta$) for every legal judgement $\Gamma \vdash M : s$ with $s \in \mathcal{S}$.

4 Properties of domain-free pure type systems

In this section, we state some basic facts about DFPTSs. We follow the structure of [3, Section 5.2]. The proofs are simply inductions, similar to those for the corresponding results for PTSs and are therefore omitted for brevity. Throughout the section, **S** denotes a fixed specification $(\mathcal{S}, \mathcal{A}, \mathcal{R})$.

Proposition 11 (Church-Rosser). $\rightarrow_\beta$ *is confluent on* $\mathcal{L}$.

Proof. By the technique of Tait and Martin-Löf. □

Lemma 12 (Substitution). *Assume* $\Gamma, x : A, \Delta \vdash B : C$ *and* $\Gamma \vdash a : A$. *Then also* $\Gamma, \Delta\{x := a\} \vdash A\{x := a\} : B\{x := a\}$.

Lemma 13 (Thinning). *If* $\Gamma \vdash A : B$ *and* $\Delta \supseteq \Gamma$ *is legal then* $\Delta \vdash A : B$.

Lemma 14 (Generation).

1. $\Gamma \vdash s : C \Rightarrow \exists (s, s') \in \mathcal{A} . C =_\beta s'$;
2. $\Gamma \vdash x : C \Rightarrow \exists s \in \mathcal{S}, D \in \mathcal{L}. C =_\beta D, x : D \in \Gamma, \Gamma \vdash D : s$;
3. $\Gamma \vdash \lambda x.b : C \Rightarrow \exists s \in \mathcal{S}, A, B \in \mathcal{L}. C =_\beta \Pi x:A.B, \Gamma, x : A \vdash b : B, \Gamma \vdash \Pi x:A.B : s$;
4. $\Gamma \vdash \Pi x:A.B : C \Rightarrow \exists (s_1, s_2, s_3) \in \mathcal{R}. C =_\beta s_3, \Gamma \vdash A : s_1, \Gamma, x : A \vdash B : s_2$;
5. $\Gamma \vdash F a : C \Rightarrow \exists x \in V, A, B \in \mathcal{L}. C =_\beta B\{x := a\}, \Gamma \vdash F : \Pi x:A.B, \Gamma \vdash a : A$.

Lemma 15 (Correctness of types). *If* $\Gamma \vdash A : B$ *then either* $B \in \mathcal{S}^\mathsf{T}$ *or* $\exists s \in \mathcal{S}. \Gamma \vdash B : s$.

Theorem 16 (Subject Reduction). *If* $\Gamma \vdash A : B$ *and* $A \rightarrow_\beta A'$ *then* $\Gamma \vdash A' : B$.

One important difference between PTSs and DFPTSs is that even in the simply typed case, the domain-free system does not satisfy Uniqueness of Types.

Lemma 17 (Failure of Uniqueness of Types). *In* $\lambda\rightarrow$, *there exists a term* M *and a context* Γ *s.t.* $\Gamma \vdash M : C$ *and* $\Gamma \vdash M : C'$ *with* $C \neq_\beta C'$.

Proof. Take $\Gamma \equiv A : *, B : *$ and $M \equiv \lambda x.x$. Then $\Gamma \vdash M : A \rightarrow A$ and $\Gamma \vdash M : B \rightarrow B$. □

The failure of Uniqueness of Types is not catastrophic per se but is often accompanied by the loss of Decidability of Type Checking—see Section 7. Interestingly, the Classification Property still holds under some mild condition.

Definition 18. S is *classifiable* if it is functional and for all $s_1, s_3, s, s' \in \mathcal{S}$,

1. $\Gamma \vdash A : s \;\; \wedge \;\; \Gamma \vdash A' : s' \;\; \wedge \;\; A =_\beta A' \;\;\Rightarrow\;\; s \equiv s'$
2. $(s, s_1) \in \mathcal{A} \;\; \wedge \;\; (s', s_1) \in \mathcal{A} \;\;\Rightarrow\;\; s \equiv s'$
3. $(s_1, s, s_3) \in \mathcal{R} \;\; \wedge \;\; (s_1, s', s_3) \in \mathcal{R} \;\;\Rightarrow\;\; s \equiv s'$

For example, the cube-specifications are classifiable.

The Classification Lemma is proved for a variant of DFPTSs with sorted variables. See [8] for a variant of PTSs based on sorted variables.

Proposition 19. *Assume that* S *is classifiable. For every sorts* $s \not\equiv s'$,

$$\mathsf{Term}^s \cap \mathsf{Term}^{s'} = \emptyset$$
$$\mathsf{Type}^s \cap \mathsf{Type}^{s'} = \emptyset$$

where

$$\mathsf{Type}^s = \{M \in \mathcal{L} \mid \Gamma \vdash M : s \text{ for some context } \Gamma\}$$
$$\mathsf{Term}^s = \{M \in \mathcal{L} \mid \Gamma \vdash M : A \text{ for some context } \Gamma \text{ and } A \in \mathsf{Type}^s\}$$

5 Reflection and applications

In this section we study the relation between PTSs and DFPTSs. Throughout the section, S denotes a fixed specification.

Every PTS-pseudo-term induces a DFPTS-pseudo-term by erasing the domains of abstractions. This erasing function is used by Geuvers [8] to study PTSs with $\beta\eta$-conversion.

Definition 20. The *erasure* map $|.| : \mathcal{T} \to \mathcal{L}$ is defined as follows:

$$|x| = x$$
$$|s| = s$$
$$|t\ u| = |t|\ |u|$$
$$|\lambda x : A.t| = \lambda x.|t|$$
$$|\Pi x : A.B| = \Pi x : |A|.|B|$$

Erasure preserves reduction, equality and typing:

Proposition 21.

1. *If* $M \to_\beta N$ *then* $|M| \twoheadrightarrow_\beta |N|$;
2. *If* $M =_\beta N$ *then* $|M| =_\beta |N|$;
3. *If* $\Gamma \vdash M : A$ *then* $|\Gamma| \vdash |M| : |A|$.

Proof. First prove by induction on M that

$$|M[x := N]| \equiv |M|\{x := |N|\} \tag{*}$$

Then prove (1) using $(*)$ by induction on the structure of M, (2) by induction on the derivation of $M =_\beta N$ and (3) by induction on the derivation of $\Gamma \vdash M : A$, using $(*)$ and 2. $\qquad\square$

Corollary 22. $|.|$ *is an* ATS-*morphism from* λS *to* $\underline{\lambda} S$.

The main result of this section is that, under suitable conditions, $|.|$ is an ATS-reflection. This generalizes [3, Proposition 3.2.15] where a similar result is proved for simply typed λ-calculus—see also Section 6. Before proving the main result, we start with three preliminary lemmas.

The first lemma is about the relation between $\rightarrow_\beta$ and $\rightarrow_{\underline{\beta}}$.

Lemma 23 [8].

1. If $|A| \rightarrow_{\underline{\beta}} F$ then there is B such that $A \rightarrow_\beta B$ and $|B| \equiv F$;
2. If $|A| =_{\underline{\beta}} s$ then $A =_\beta s$.

The second lemma establishes the fundamental property of erasure.

Lemma 24. *Assume* $M, M' \in \mathrm{NF}_\beta$, $|M| \equiv |M'|$, $\Gamma \vdash M : A$, *and* $\Gamma \vdash M' : A'$.

1. If $A \equiv A'$ then $M \equiv M'$.
2. If $A \equiv s$, $A' \equiv s'$ then $M \equiv M'$.

The third lemma provides the necssary machinery to prove the main result.

Lemma 25. *Let* S *be a specification with normalizing* PTS-*types.*

1. If $\Gamma \vdash M : s$, $\Gamma \vdash N : s'$, $|M| =_{\underline{\beta}} |N|$, then $M =_\beta N$.
2. If Γ_1, Γ_2 are legal and $|\Gamma_1| =_{\underline{\beta}} |\Gamma_2|$ then $\Gamma_1 =_\beta \Gamma_2$.
3. If $\Gamma \vdash P : M$, $\Gamma \vdash N : s'$, $|M| =_{\underline{\beta}} |N|$, then $M =_\beta N$.

Theorem 26 (Main result). *Let* S *be a functional specification with normalizing* PTS-*types. Then* $|.|$ *is an* ATS-*reflection.*

Proof. By induction on the structure of derivations, using the above lemmas. $\qquad\square$

We conclude this section with some applications of Theorem 26. These applications are used in [4, 5, 15]. First we examine how normalization is reflected.

Definition 27. Let (ϕ, T) be $(\lambda, \mathcal{T})$ or $(\underline{\lambda}, \mathcal{L})$. Assume $X \subseteq T$. We write $\phi S \models X$ if every legal ϕS-term t belongs to X.

We have:

Proposition 28. *Let* S *be a functional specification.*

1. $\lambda S \models SN_\beta$ *implies* $\underline{\lambda}S \models SN_\beta$.
2. $\lambda S \models WN_\beta$ *implies* $\underline{\lambda}S \models WN_\beta$.
3. *If* S *has strongly normalizing* $\overline{PTS}$-*types, then* $\underline{\lambda}S \models SN_\beta$ *implies* $\lambda S \models SN_\beta$.
4. *If* S *has normalizing* PTS-*types, then* $\underline{\lambda}S \models WN_\beta$ *implies* $\lambda S \models WN_\beta$.

Proof. (1): By Thm. 26 and Lem. 23. (2): By Thm. 26, Prop. 21(1), and by noting that the erasure of a β-normal form is a β-normal form. (3) and (4): By Prop. 21(3) and some elementary reasoning on reductions. $\square$

This result is useful in work on the Barendregt-Geuvers-Klop conjecture which states that for every specification S, $\lambda S \models WN_\beta$ implies $\lambda S \models SN_\beta$—see [15, 4]. Moreover, it implies strong normalisation for the $\underline{\lambda}$-cube:

Proposition 29. $\underline{\lambda}S \models SN_\beta$ *for every cube-specification* S.

Proof. By Proposition 28(1) and strong normalization of the λ-cube. $\square$

As an application of Theorem 26, we can also infer consistency of a DFPTS from the corresponding PTS:

Proposition 30. *For every cube-specification, there is no* $M \in \mathcal{L}$ *s.t.* $x : * \vdash M : x$.

Proof. By Theorem 26 and consistency of the λ-cube. $\square$

6 Comparison with type assignment systems

There are two ways to perceive Curry's version of simply typed λ-calculus:

1. terms in Curry's system are those of the untyped λ-calculus;
2. terms in Curry's system are those of Church's system with domain-free abstractions.

View (2) leads to the notion of domain-free pure type system studied in this paper whereas view (1) leads to the notion of *type assignment system*, or TAS. In recent work [1], van Bakel, Liquori, Ronchi della Roncha and Urzyczyn define for each cube-specification S a TAS $\overline{\lambda}S$. These systems, the $\overline{\lambda}$-cube, include simple types $\overline{\lambda}{\rightarrow}$ introduced by Curry [7], second-order types $\overline{\lambda}2$ introduced by Leivant [12] and higher-order types $\overline{\lambda}\omega$ introduced by Giannini and Ronchi della Rocca [11].

An important aspect of the $\overline{\lambda}$-cube is the separation of the set $\mathcal{U}$ of pseudo-terms into three different syntactic categories: terms, constructors and kinds. Terms use domain-free abstractions whereas constructors use domain-specified abstractions. As a consequence, the erasure map[5] E is an ATS-reflection from λS to $\overline{\lambda}S$ only for the specifications which do not combine polymorphism and

---

[5] E is defined as a map on legal PTS-terms but can of course be extended to a map on PTS-pseudo-terms, e.g. by taking $E(M) = *$ if M is not legal.

dependent types, i.e. $\rightarrow$, 2, ω, P, $\underline{\omega}$ and $P\underline{\omega}$. For the remaining specifications $P2$ and $P\omega$, E is simply an ATS-morphism.

Note that $\underline{\lambda}\mathbf{S}$ and $\overline{\lambda}\mathbf{S}$ are identical for $\mathbf{S} = \rightarrow$ but diverge for the other cube-specifications. Indeed, consider the $\lambda 2$-term $\lambda\alpha{:}{*}.\lambda x{:}\alpha.x$ of type $\forall\alpha{:}{*}.\alpha \rightarrow \alpha$. The corresponding term in $\overline{\lambda}2$ is $E(M) \equiv \lambda x.x$ whereas the corresponding domain-free term in $\underline{\lambda}2$ is $|M| \equiv \lambda\alpha.\lambda x.x$.

We now turn to the relation between TASs and DFPTSs. Define the erasure map $E' : \mathcal{U} \rightarrow \mathcal{L}$ which removes the domains of constructor abstractions.

Proposition 31. *If $\mathbf{S}$ is a cube-specification without polymorphism (i.e. $\mathbf{S}$ is $\rightarrow$, P, $\underline{\omega}$ or $P\underline{\omega}$), then E' is an ATS-reflection from $\overline{\lambda}\mathbf{S}$ to $\underline{\lambda}\mathbf{S}$.*

Proof. Prove by induction on the structure of derivations that E' is an ATS-morphism and that $|M| = E'(E\,M)$ for every legal PTS-term M. Conclude by Theorem 26 and Lemma 4. □

Note that E' is not an ATS-morphism from $\overline{\lambda}2$ to $\underline{\lambda}2$: consider the term M given above —a similar remark applies to ω, $P2$ and $P\omega$. However, one may define an erasure map $F : \mathcal{L} \rightarrow \mathcal{U}$ which removes type abstractions and type applications.[6]

Proposition 32. *F is an ATS-reflection.*

Proof. Prove by induction on the structure of derivations that F is an ATS-morphism and that $E(M) = F(|M|)$ for every legal PTS-term M. Conclude from the fact that E is an ATS-reflection from $\lambda 2$ to $\overline{\lambda}2$ and Lemma 4. □

7   Type checking

The *type checking problem* (TC) for a specification $\mathbf{S}$ consists in deciding whether a given DFPTS-judgement (Γ, M, A) is legal. In this section we briefly discuss some results related to type checking.

Proposition 33. *Given an arbitrary specification $\mathbf{S}$, it is decidable whether a DFPTS-judgement in β-normal is derivable.*

People working on the ALF system claim that this limited form of decidability is sufficient in practice because of the presence of definitions—see e.g. [13].

Proposition 34. *TC is decidable for $\underline{\lambda}\rightarrow$ but undecidable for $\underline{\lambda}P$.*

It would be interesting to know whether TC is decidable for $\underline{\lambda}2$ and $\underline{\lambda}\omega$.

---

[6] It requires that we work with sorted variables, as suggested at the end of Section 4.

8 Conclusion

We have introduced the notion of a domain-free pure type system, developed its basic properties and established its exact relationship with the notion of pure type system. Thus far, DFPTSs have proved useful in several applications [4, 5, 15]. Moreover, —variants of—DFPTSs are used in the implementation of proof-assistant systems based on type theory [13]. It is also possible that DFPTSs will play a role in the design of programming languages; for example, one may envisage a dependently typed extension of ML based on λP. Finally, we would like to point out that DFPTSs sometimes provide a more natural basis than PTSs for extended type systems. For example, the extension of pure type systems with classical operators [5] yields better results when the abstractions are domain-free.

References

1. S. van Bakel, L. Liquori, S. Ronchi della Roncha, and P. Urzyczyn. Comparing cubes. In A. Nerode and Y.N. Matiyasevich, editors, *Proceedings of LFCS'94*, volume 813 of *Lecture Notes in Computer Science*, pages 353–365. Springer-Verlag, 1994. An extended version is to appear in *Annals of Pure and Applied Logic*.
2. H. Barendregt. Introduction to Generalised Type Systems. *J. Functional Programming*, 1(2):125–154, April 1991.
3. H. Barendregt. Lambda calculi with types. In S. Abramsky, D. M. Gabbay, and T.S.E. Maibaum, editors, *Handbook of Logic in Computer Science*, pages 117–309. Oxford Science Publications, 1992. Volume 2.
4. G. Barthe, J. Hatcliff, and M.H. Sørensen. CPS-translation and applications: the cube and beyond. In O. Danvy, editor, *Proc. of the 2nd ACM SIGPLAN Workshop on Continuations*, number NS-96-13 in BRICS Notes, pages 4:1–31, 1996.
5. G. Barthe, J. Hatcliff, and M.H. Sørensen. Classical pure type systems. In *Proceedings of MFPS'97*, Electronic Notes in Theoretical Computer Science, 1997.
6. A. Church. A formulation of the simple theory of types. *Journal of Symbolic Logic*, 5:56–68, 1940.
7. H. Curry. Functionality in combinatory logic. *Proceedings of the National Academy of Science USA*, 20:584–590, 1934.
8. H. Geuvers. *Logics and type systems*. PhD thesis, University of Nijmegen, 1993.
9. H. Geuvers and M.-J. Nederhof. A modular proof of strong normalisation for the Calculus of Constructions. *Journal of Functional Programming*, 1:155–189, 1991.
10. H. Geuvers and B. Werner. On the Church-Rosser property for expressive type systems and its consequence for their metatheoretic study. In *Proceedings of LICS'94*, pages 320–329. IEEE Computer Society Press, 1994.
11. P. Giannini and S. Ronchi Della Rocca. Characterization of typings in polymorphic type discipline. In *Proceedings of LICS'88*, pages 61–70. IEEE Computer Society Press, 1988.
12. D. Leivant. Polymorphic type inference. In *Proceedings of POPL'83*, pages 88–98. ACM Press, 1983.
13. L. Magnusson. *The implementation of ALF: a proof editor based on Martin-Löf's monomorphic type theory with explicit substitution*. PhD thesis, Department of Computer Science, Chalmers University, 1994.

14. P. Severi. *Normalisation in lambda calculus and its relation to type inference.* PhD thesis, Department of Computer Science, Technical University of Eindhoven, 1996.
15. M.H. Sørensen. Strong normalization from weak normalization in typed λ-calculi. *Information and Computation*, 133(1):35–71, 1997.

Generic Queries Over Quasi-*o*-minimal Domains

Oleg V. Belegradek[1] *, Alexei P. Stolboushkin[2] **, and Michael A. Taitslin[3] ***

[1] Department of Mathematics
Kemerovo State University
Kemerovo, Russia 650043
beleg@kaskad.kemerovo.su
[2] Fourth Dimension Software
555 Twin Dolphin Dr., Redwood City, CA 94404
and UCLA Mathematics Dept., Los Angeles, CA 90095
aps@math.ucla.edu
[3] Department of Computer Science
Tver State University
Tver, Russia 170000
mat@tversu.ac.ru

Abstract. We consider relational databases organized over an ordered domain with some additional relations—a typical example is the ordered domain of rational numbers together with the ternary relation + of addition. In the focus of our study are the first order (FO) queries that are invariant under order-preserving "permutations" — such queries are called order-generic. It has recently been discovered that for some domains order-generic FO queries fail to express more than pure order queries. For example, every order-generic FO query over rational numbers with + can be rewritten without +. For some other domains, however, this is not the case.

We establish a very general condition on a domain, the so-called *Isolation Property*, that ensures collapse of order-generic FO queries to pure order queries over this domain. We further distinguish one broad class of domains satisfying the Isolation Property, the so-called *quasi-o-minimal* domains. This class includes all *o*-minimal domains, but also the ordered group of integer numbers and the ordered semigroup of natural numbers, and some other domains.

Recent results by Stolboushkin and Taitslin imply that all the results of the present paper continue to hold over the so-called finitely representable states. We show, however, that these results cannot be transfered to arbitrary infinite states.

* A part of this work had been done while Oleg Belegradek was visiting the Fields Institute for Research in Mathematical Sciences in Toronto (January–March, 1997).
** This work of Alexei P. Stolboushkin has been partially supported by NSF Grant CCR 9403809.
*** A part of this research was carried out while M. A. Taitslin was visiting UCLA (partially supported by NSF Grant CCR 9403809), DIMACS and Princeton (partially supported by a grant from Princeton University). The work of the author was partially supported by the Russian Foundation of Basic Research (project code: 96-01-00086).

1 Introduction

We consider relational databases organized over an infinite domain, for example, over integer or rational numbers. First order (FO) queries expressible over such a domain are *generic* (see [CH80]), meaning that they are preservable under arbitrary permutations of the domain.

The expressive power of the pure FO with respect to generic queries is, however, severely limited—a classical example is inexpressibility of the parity query asserting that the cardinality of a finite relation in the database scheme is even. One of the ways to try to enhance the expressive power of the query language is by allowing certain *domain functions/relations*, or *givens* to be used in the queries. The simplest example is the relation < of linear order. These givens are considered to be a part of the domain—rather than of the database scheme—and to have a fixed meaning. Throwing in such givens does obviously increase the expressive power of FO, but what is often not obvious is whether any new *generic* queries become expressible.

Yu. Gurevich [G90] showed that the language FO(<) of first order logic with a relation of linear order does indeed express more generic queries than FO, although the parity continues to be inexpressible. The natural question has been, whether allowing certain other givens, in specific situations, enhances the expressive power of generic FO(<) even more. And while in some situations the answer is trivially affirmative—for example, allowing + and × over integer (or rational) numbers makes it possible to express all computable queries—in others the question may be hard.

Let **N**, **Z**, **Q** and **R** be the sets of all natural, integer, rational and real numbers, respectively. Practically, the most interesting cases have been:

1. $(\mathbf{Q}, <, +)$ and $(\mathbf{R}, <, +)$
2. $(\mathbf{Z}, <, +)$ and $(\mathbf{N}, <, +)$
3. $(\mathbf{R}, <, +, \times)$

Paradaens et al. [PVV95] showed that (1) does not extend over the power of FO(<) w.r.t. generic queries. In [ST96], Stolboushkin and Taitslin proved a more general result on *recursive* translation of generic FO(<, +) queries into equivalent FO(<) queries over every ordered divisible Abelian group. Finally, Benedikt et al. [BDLW96] showed collapse of generic FO(<, Ω) to FO(<) over o-minimal domains—a broad class of domains that includes, for example, (1,3), but not (2).

Actually, all the collapse results above have been shown in a stronger form—collapse of *order-generic* FO(<, Ω) to FO(<) for specific domains. A query is called *order-generic* if it is preservable under arbitrary order-automorphisms of the domain. Observe that this type of collapse trivially implies collapse of generic queries to FO(<). In the technical part of the paper, we will simply refer to the order-generic queries as "generic".

For some domains, however, notably for $(\mathbf{N}, <)$, the notion of order-genericity does not make much sense, because there are no nontrivial order-automorphisms.

Therefore, we are actually going to consider *local order-genericity* that only requires preservation under local order-isomorphisms of the active domain. Observe that for $(\mathbf{Q}, <)$ or $(\mathbf{R}, <)$, as well as for several other domains, these two notions coincide.

In this paper, we use the technique of special models to establish a very general condition on a domain, the so-called *Isolation Property*, that ensures collapse of locally order-generic FO queries over this domain to pure order queries. Although the Isolation Property can be directly used in classifying domains, proving this condition for a specific domain may be a bit technical. However, we identify a broad class of domains—the so-called *quasi o-minimal domains*—that all satisfy the Isolation Property. Examples of quasi o-minimal domains include all the o-minimal domains, but also $(\mathbf{Z}, <, +)$, $(\mathbf{N}, <, +)$, and the ordered real numbers with the distinguished subset of rational numbers. These our results are discussed in Section 3.

Recent results by Stolboushkin and Taitslin [ST96] imply that all the results in the present paper continue to hold over the so-called finitely representable states. We show in Section 4, however, that these results cannot be transfered further to arbitrary infinite states.

2 Preliminaries

A structure of a relational signature L is a non-empty set with a mapping that assigns to every relational symbol in L a relation of the same arity over the set. Let U be an infinite structure over the signature L. This structure is called the *universe*. In this paper, we always consider ordered universes. This means that L includes a binary relational symbol $<$ whose interpretation in U satisfies the axioms of linear order. Let us denote $L_0 = \{<\}$, and $\Omega = L \setminus L_0$.

Databases operating over U use non-signature relational symbols as well. A *database scheme* is a finite collection of relational symbols of fixed arities. A *database state* (over U) is an assignment to these relational symbols of concrete relations of corresponding arities over U. These relations are called *database relations*. A database state is *finite* iff all the relations are finite. The set of all elements of the universe those occur in some tuple in some relation of a database state s is called the active domain of s; we denote it by $AD(s)$. We fix a database scheme SC and denote $L_0^+ = L_0 \bigcup SC$, and $L^+ = L \bigcup SC$.

A *database query* can formally be defined as a mapping that takes in a database state (of a fixed d.b. scheme), and produces a new relation, of a fixed arity, over U. Thus, every query has an arity. Specifically, queries of arity 0 are called *Boolean*. A Boolean query defines a mapping from d.b. states to $\{0, 1\}$, or, in other words, subsets of all possible d.b. states of a given d.b. scheme.

Queries are formulated using *query languages*, the simplest being the language of first-order logic FO. Formulas (queries) of this language use $=$, as well as the relational symbols of the signature and of the database scheme. Thus, a d.b. state essentially defines a structure of a larger signature with U as the domain; then a formula with n free variables defines an n-ary relation over U.

Generally, an FO query may yield an infinite answer even in a finite d.b. state. [KKR90] introduced the notion of *finitely representable* database state as a database state where every relation corresponding to a relation name from SC is defined (independently on the others) by a quantifier-free formula using $=, <$, and constants for the elements of U.

We consider two languages for querying. Queries of the first one are FO formulas of the signature L_0^+ — we call them restricted. Queries of the second language are FO formulas of the signature L^+ — we call them extended.

It is easy to see that the restricted queries are *generic,* in the sense that they are preserved under order-preserving permutations of U.[4] In other words, if $\phi : U \to U$ is an automorphism of $\langle U, < \rangle$, and a restricted query Q transforms a database state s into a relation R, then Q transforms $\phi(s)$ into $\phi(R)$; in other words, $Q(\phi(s)) = \phi(Q(s))$. The problem with extended queries is, they may be not generic.

We will also use a stronger notion of *locally generic* query. A k-ary query Q is said to be *locally generic over finite states* if $\bar{a} \in Q(s)$ iff $\phi(\bar{a}) \in Q(\phi(s))$, for any partial $<$-isomorphism $\phi : X \to U$ with $X \subseteq U$, for any finite state s over X, and for any k-tuple $\bar{a}$ in X.

For any finite representation σ over a subset X of U and for any partial $<$-isomorphism $\phi : X \to U$, a finite representation $\phi(\sigma)$ can be naturally defined, by replacing any parameter a that occur in σ with the parameter $\phi(a)$. So, for finitely representable states, the notion of local genericity can be defined as follows. A k-ary query Q is said to be *locally generic over finitely representable states* if $\bar{a} \in Q(\sigma)$ iff $\phi(\bar{a}) \in Q(\phi(\sigma))$, for any partial $<$-isomorphism $\phi : X \to U$ with $X \subseteq U$, for any finite representation σ over X, and for any k-tuple $\bar{a}$ in X. Here we denote by $Q(\sigma)$ the state into which the query Q transforms the state finitely represented by σ.

Of course, every locally generic query is generic. Conversely, for some domains, every generic query is locally generic. A sufficient condition for this is the so-called *double transitivity* of the domain: a domain is called doubly transitive if for any $a_1 < b_1$ and $a_2 < b_2$ in the domain, there exists a $<$-automorphism of the domain mapping a_1 to a_2, and b_1 to b_2. For instance, real and rational numbers are doubly transitive, while integer numbers are not. The Boolean query *'there are even and odd numbers in P'* is an example of a query which is generic but not locally generic over finite states over $\mathbf{Z}$.

3 Collapse of extended locally generic queries

In this section, we pursue collapse results over finite states. However, all the results can be transfered to finitely representable states by directly applying Theorem 4.8 from [ST96].

For convenience, consider database schemes that contain not only relation symbols, but also finitely many constant symbols. A database state over a uni-

---

[4] As discussed in Introduction, the term "generic" is sometimes understood in a more restrictive sense.

verse U for such a scheme is a mapping that assigns to any relation symbol in the scheme a relation on U of the corresponding arity, and to any constant symbol in the scheme an element in U. In this case the active domain of a database state is defined to be the union of the active domain of the relational part of the state and the set of values of all constants of the scheme. For a relational database scheme SC, denote by SC_k the scheme $SC \cup \{c_1, \ldots, c_k\}$, where the c_i are new constant symbols.

Clearly, two k-ary SC-queries $\phi(x_1, \ldots, x_k)$ and $\psi(x_1, \ldots, x_k)$ are equivalent over finite states over a universe U if the Boolean SC_k-queries $\phi(c_1, \ldots, c_k)$ and $\psi(c_1, \ldots, c_k)$ are equivalent over finite states over U.

The notions of genericity and local genericity for SC_k-queries are defined exactly the same way as for SC-queries. Clearly, a k-ary SC-query $\phi(x_1, \ldots, x_k)$ is generic (locally generic) over U iff the Boolean SC_k-query $\phi(c_1, \ldots, c_k)$ is generic (locally generic) over U.

Our ultimate goal is to prove that, under certain conditions on the universe U, any locally generic extended query is equivalent over finite states over U to a restricted query. Hence, it suffices to prove such a result for Boolean queries (for database schemes with constant symbols).

For a class K of structures of an arbitrary signature L (in symbols, L-structures), the first order L-theory of K (in symbols, $\mathrm{Th}(K)$) is the set of all the first order formulas ϕ of signature L (in symbols, L-formulas) such that ϕ is a sentence (has no free variables) and ϕ holds in M for any $M \in K$. Two L-structures M amd N are called *elementarily equivalent* (in symbols, $M \equiv N$), if ϕ holds in M iff ϕ holds in N, for any L-sentence ϕ. $M \equiv N$ iff $\mathrm{Th}(M) = \mathrm{Th}(N)$.

Let ρ be a database scheme $\{R_1, \ldots, R_n, c_1 \ldots, c_k\}$. We denote $L \cup \rho$ by $L(\rho)$. A ρ-state s over an L-structure W is said to be *pseudo-finite in* W if (W, s) is a model of the first order $L(\rho)$-theory $F(W, \rho)$ of all (W, r), where r is a finite ρ-state over W.

For a first order $L(\rho)$-sentence ψ and $m < \omega$, there is a first order L-sentence ψ_m such that, for any L-structure V, the sentence ψ_m holds in V iff ψ holds for all ρ-states over V, whose active domain has cardinality at most m. Thus, $\psi \in F(W, \rho)$ iff $\{\psi_m : m < \omega\} \subseteq \mathrm{Th}(W)$. It follows that $W \equiv V$ implies $F(V, \rho) = F(W, \rho)$. An L-theory is a set of first order L-sentences. An L-theory T is complete if it is $\mathrm{Th}(M)$ for an L-structure M. Any such an M is a model of T. For a complete L-theory T, the first order $L(\rho)$-theory $F(T, \rho)$ is well-defined to be $F(W, \rho)$, where W is an arbitrary model of T.

Let ρ' be a disjoint copy $\{R'_1, \ldots, R'_n, c'_1 \ldots, c'_k\}$ of ρ. For an $L(\rho)$-sentence θ denote by $\theta(\rho')$ its $L(\rho')$-copy, that is, the result of replacement of every occurrence of R_i and c_j in θ with R'_i and c'_j, respectively. Let $\bar{\rho} = \rho \cup \rho'$.

We say that a Boolean extended ρ-query ϕ is *generic for pseudo-finite states in* V if the following holds: for any $\bar{\rho}$-state (r, r'), if it is pseudo-finite in V and r can be transformed to r' by an L_0-automorphism of V, then $\phi(\rho)$ holds in (V, r) iff $\phi(\rho')$ holds in (V, r').

As we will use the technique of the so-called special models, we summarize its basic definitions and facts (see [CK90] for detail).

For a structure M of an arbitrary signature L and a subset A of M, denote by $L(A)$ the signature obtained by adjoining to L names for the elements of A. We do not normally distinguish between elements of A and their names.

We say that M is an *elementary substructure* of N (in symbols, $M \preceq N$ or $N \succeq M$), if M is a substructure of N, and ϕ holds in M iff ϕ holds in N, for any $L(M)$-sentence ϕ.

A set p of first order $L(A)$-formulas with one free variable x is said to be a *type* over A in M if every finite subset $\{\phi_1(x), \ldots, \phi_k(x)\}$ of p is realized in M $((\exists x)(\phi_1(x)\& \ldots \&\phi_k(x))$ holds in $M)$, and, for every $L(A)$-formula $\phi(x)$, either $\phi \in p$ or $\neg\phi \in p$. We say that a subset q of p *isolates* p if p is the only type over A in M containing q.

Let A be a subset of M. For any $N \succeq M$ and $a \in N$, the set of all $L(A)$-formulas $\phi(x)$ such that $\phi(a)$ holds in N forms a type over A in M; denote it by $\mathrm{tp}(a/A)$. For any type p over A in M, there are $N \succeq M$ and $a \in N$ such that $p = \mathrm{tp}(a/A)$. We denote $\mathrm{tp}(A)$ the set of all $L(A)$-sentences which hold in M.

For a cardinal λ, a structure M is said to be λ-*saturated* if any type p over any its subset A of power $< \lambda$ is realized in M; that is, $p = \mathrm{tp}(a/A)$, for some $a \in M$. For any infinite $\lambda \geq |L|$, every two elementarily equivalent λ-saturated structures of power λ are isomorphic.

An L-structure M of power $\aleph_\alpha$ is said to be *special* if $M = \bigcup\{M_\beta : \beta < \alpha\}$, where $M_\beta \preceq M_\gamma \preceq M$, for $\beta < \gamma < \alpha$, and each M_β is $\aleph_{\beta+1}$-saturated. Every two elementarily equivalent special structures of the same power are isomorphic. For any infinite L-structure M and any cardinal $\lambda > |L|, |M|$ with $\lambda^* = \lambda$, there exists a special $N \succeq M$ of power λ. Here $\aleph_\alpha^*$ is defined to be $\sum_{\beta<\alpha} 2^{\aleph_\beta}$. It is easy to construct cardinals λ with $\lambda^* = \lambda$ of arbitrarily large cofinality.

A linearly ordered subset I of a structure M is said to be an *indiscernible sequence* in M if $\theta(a_1, \ldots, a_n)$ holds in M iff $\theta(a'_1, \ldots, a'_n)$ holds in M, for every first order L-formula $\theta(x_1, \ldots, x_n)$ and every $a_1, \ldots, a_n, a'_1, \ldots, a'_n$ in I with $a_1 < \ldots < a_n$ and $a'_1 < \ldots < a'_n$.

Theorem 1. *For any countable universe U and any Boolean extended ρ-query ϕ the following conditions are equivalent:*

1. *there is a restricted ρ-query ψ which is equivalent to ϕ over finite database states over U*
2. *ϕ is generic for pseudo-finite states over V, for all $V \equiv U$*
3. *for every uncountable power κ with $\kappa = \kappa^*$, the query ϕ is generic over pseudo-finite states over the special model $V \equiv U$ with $|V| = \kappa$*

Let I be a set in a universe V. We say that a Boolean extended ρ-query ϕ is *locally generic over pseudo-finite states over I in V* if the following holds: if an $\bar{\rho}$-state (r, r') over I is pseudo-finite in V and r can be transformed into r' by a partial L_0-isomorphism in V then $\phi(\rho)$ holds in (V, r) iff $\phi(\rho')$ holds in (V, r').

Theorem 2. *Let an extended Boolean ρ-query ϕ be locally generic for finite states over U. Suppose, for some uncountable κ with $\kappa = \kappa^*$, there is a special model $V \equiv U$ of power κ such that, for any infinite indiscernible sequence I in*

V, the query ϕ is locally generic over pseudo-finite states over I in V. Then ϕ is equivalent over finite states over U to a restricted ρ-query.

We say that a complete theory T has the *Isolation Property*, if there is a cardinal λ such that, for any pseudo-finite set A and any element a in a model of T, there is $A_0 \subseteq A$ with $|A_0| < \lambda$ such that $tp(a/A_0)$ isolates $tp(a/A)$.

Theorem 3. *Suppose the theory of a universe U has the Isolation Property. Let an extended query ϕ be locally generic over finite states over U. Then ϕ is equivalent over finite states over U to a restricted query.*

Proof. It suffices to prove that ϕ satisfies the condition of Theorem 2. Let λ be the cardinal from the definition of the Isolation Property. Let κ be an uncountable cardinal with $\kappa^* = \kappa$ and $cf(\kappa) > \lambda$. Let $V \equiv U$ be a special model of power κ; then V is λ-saturated. Let I be an infinite L-indiscernible sequence in V. Suppose $\bar{\rho}$-state (r, r') over I is pseudo-finite in V and r can be transformed to r' by a partial L_0-isomorphism g in V, whose domain is the active domain of r and whose range is the active domain of r'. We need to show that $\phi(\rho)$ holds in (V, r) iff $\phi(\rho')$ holds in (V, r'). In fact, using a Fraïssé-Ehrenfeucht game, we will show that g is an $L(\rho)$-elementary map from (V, r) to (V, r'). Due to the L-indiscernibility of I, the map g is an L-elementary map, and, in particular, a partial $L(\rho)$-isomorphism. Therefore, to complete the proof of the theorem, it suffices to prove the following lemmas:

Lemma 4. *The active domain of any pseudo-finite database state is a pseudo-finite set.*

Lemma 5. *Let A be a pseudo-finite set in V, and $a \in V$. Then $A \cup \{a\}$ is a pseudo-finite set.*

Lemma 6. *Let A and B be pseudo-finite sets in V, and $a \in V$. Let $h : A \to B$ be an L-elementary map. Then there is $b \in V$ such that $h \cup \{(a, b)\}$ is an L-elementary map.*

Proof of Lemma 4. Consider the database scheme $\tau = \{P\}$, where P is a unary relation name. For any $L(\tau)$-sentence γ, there is an $L(\rho)$-sentence γ^* such that $(V, s) \models \gamma^*$ iff $(V, AD(s)) \models \gamma$, for any ρ-state s. Suppose a state s is pseudo-finite in V, and $\gamma \in F(V, \tau)$. Since the active domain of any finite state is finite, we have $(V, r) \models \gamma^*$ for all finite ρ-states r. So $(V, s) \models \gamma^*$, and hence $(V, AD(s)) \models \gamma$.

Proof of Lemma 5. Consider the database scheme $\tau = \{P\}$, where P is a unary relation name. Let $\theta \in F(V, \tau)$. Let $\theta^*(x)$ be the result of replacement of every occurrence of $P(y)$ in θ with $P(y) \vee y = x$, where x is a new variable. Then $\forall x \theta^*(x)$ belongs to $F(V, \tau)$ and so holds in (V, A). Hence θ holds in $(V, A \cup \{a\})$. Thus, $A \cup \{a\}$ is pseudo-finite.

Proof of Lemma 6. Choose $A_0 \subseteq A$ with $|A_0| < \lambda$ such that $p_0 = \text{tp}(a/A_0)$ isolates $p = \text{tp}(a/A)$. Since the map h is elementary, $h(p)$ is a type over B, and $h(p_0)$ isolates $h(p)$. (For a set $q(x)$ of formulas over A we denote by $h(q)$ the set $\{\theta(x, h(\bar{c})) : \theta(x, \bar{c}) \in q\}$.) As V is λ-saturated, there is $b \in V$ realizing $h(p_0)$ and hence $h(p)$. So $h \cup \{(a, b)\}$ is an L-elementary map.

Theorem 7 below gives a broad class of theories with the Isolation Property.

A complete L-theory T is said to be *o-minimal* if in every its model any definable set is a finite union of singletons and open intervals.

We call a complete L-theory T *quasi-o-minimal* iff there exists T', an extension of T by definitions in a language L', such that any L'-formula $\theta(x, \bar{y})$ is T'-equivalent to a disjunction of formulas of the form $\phi(x) \wedge \psi(\bar{y}) \wedge \rho(x, \bar{y})$, where $\rho(x, \bar{y})$ has one of the following forms, for some L'-terms t and t' in the variables $\bar{y}$:

$$x = x, \quad x = t, \quad x < t, \quad t < x, \quad t < x < t'.$$

Every *o-minimal* theory is quasi-*o*-minimal. Clearly, in every model of a quasi-*o*-minimal theory any definable set is a finite union of singletons and sets of the form $I \cap D$, where I is an open interval and D is a set definable without parameters.

There exist quasi-*o*-minimal theories which are not *o*-minimal. The simplest example is the theory T of dense ordered sets without endpoints with a distinguished subset which is dense and codense in the universe. It can be easily shown that T is the theory of the structure $(\mathbf{R}, <, \mathbf{Q})$. The theory is not *o*-minimal because the distinguished subset is not a finite union of singletons and open intervals. Standard arguments show that T admits quantifier elimination; obviously, we can take T as T' from the definition of quasi-*o*-minimality.

Another example of a quasi-*o*-minimal theory is the theory T of $(\mathbf{Z}, <, +)$. Indeed, by Presburger's Theorem, the definitional expansion of the model by the constants 0, 1 and the unary predicates 'n divides x', for all positive integers n, admits quantifier elimination. For a positive integer n, define the function $f_n(x)$ by the condition $0 \leq x - n f_n(x) < n$. Since 'n divides x' iff $n f_n(x) = x$, and $nx = t$ iff $x = f_n(t)$, and $nx < t$ iff $x < f_n(t)$, and $nx > t$ iff $x > f_n(t)$, the theory T' of the definitional expansion of $(\mathbf{Z}, <, +)$ by the constants 0, 1 and all the functions $f_n(x)$ satisfies the condition of the definition of quasi-*o*-minimality. However, the theory of $(\mathbf{Z}, <, +)$ is not *o*-minimal because the definable subsets $n\mathbf{Z}$ are not finite unions of singletons and open intervals.

Similarly, the theory of $(\mathbf{N}, <, +)$ is quasi-*o*-minimal but not *o*-minimal.

Theorem 7. *Any quasi-o-minimal theory has the Isolation Property.*

Proof. Let A be a pseudo-finite set in a model V of a quasi-*o*-minimal theory T, and $a \in V$. We will show that there is $A_0 \subseteq A$ with $|A_0| \leq |T|$ such that $\text{tp}(a/A_0)$ isolates $\text{tp}(a/A)$.

Let T' be the extension by definitions of T from the definition of quasi-*o*-minimality. Let $\eta(x, \bar{y})$ be an L-formula. It is T'-equivalent to a disjunction

$\theta(x, \bar{y})$ of L'-formulas of the form $\phi(x) \wedge \psi(\bar{y}) \wedge \rho(x, \bar{y})$, where $\rho(x, \bar{y})$ has one of the following forms, for some L'-terms t and t' in the variables $\bar{y}$:

$$x = x, \quad x = t, \quad x < t, \quad t < x, \quad t < x < t'.$$

Denote by S_θ the finite set of all L'-terms t, t' involved in ρ in some of the disjuncts of θ. Let F be a finite subset of V. Then, for every $d \in V$, in the finite subset of V

$$\{t(\bar{c}) : t \in S_\theta, \quad \bar{c} \in F, \quad t(\bar{c}) \leq d\}$$

there is a maximal element $m_\theta(F, d)$, provided the subset is not empty. Similarly, for every $d \in V$, in the finite subset of V

$$\{t(\bar{c}) : t \in S_\theta, \quad \bar{c} \in F, \quad t(\bar{c}) \geq d\}$$

there is a minimal element $m^\theta(F, d)$, provided the subset is not empty.

As the set A is pseudo-finite in V, the same holds for A instead of F. Denote $m_\theta = m_\theta(A, a)$, and $m^\theta = m^\theta(A, a)$.

Let m_θ be $t_\theta(\bar{c}_\theta)$, where $t_\theta \in S_\theta$, $\bar{c}_\theta \in A$, when m_θ is defined; otherwise we write $m_\theta = -\infty$. Let m^θ be $t^\theta(\bar{c}^\theta)$, where $t^\theta \in S_\theta$, $\bar{c}^\theta \in A$, when m^θ is defined; otherwise we write $m^\theta = \infty$. Clearly, $m_\theta = m^\theta$ iff $a = t(\bar{c})$ for some $t(\bar{y}) \in S_\theta$ and $\bar{c} \in A$. If $m_\theta \neq m^\theta$ then $m_\theta < a < m^\theta$.

Suppose $\eta(x, \bar{c})$ is in $\mathrm{tp}(a/A)$. Since $\eta(a, \bar{c})$ holds in V, one of the disjuncts $\phi(a) \wedge \psi(\bar{c}) \wedge \rho(a, \bar{c})$ of $\theta(a, \bar{c})$ holds; we have $\phi(x) \in \mathrm{tp}(a/\emptyset)$ and $\psi(\bar{y}) \in \mathrm{tp}(A)$.

If $m_\theta = m^\theta$ then $a = t(\bar{c})$, for some $t(\bar{y}) \in S_\theta$ and $\bar{c} \in A$. In this case $\mathrm{tp}(a/\bar{c})$ isolates $\mathrm{tp}(a/A)$, and we can take $\bar{c}$ as A_0.

Now suppose $m_\theta \neq m^\theta$. We will show that

$$\mathrm{tp}(A) \cup \mathrm{tp}(a/\emptyset) \vdash m_\theta < x < m^\theta \rightarrow \eta(x, \bar{c}).$$

It suffices to prove that

$$\mathrm{tp}(A) \vdash m_\theta < x < m^\theta \rightarrow \rho(x, \bar{c}).$$

The case when $\rho(x, \bar{y})$ is $x = x$ is trivial. The formula $\rho(x, \bar{y})$ cannot be of the form $x = t(\bar{y})$: otherwise $a = t(\bar{c})$, and we would have $m_\theta = m^\theta$.

Suppose $\rho(x, \bar{y})$ is of the form $t(\bar{y}) < x < t'(\bar{y})$; so we have $t(\bar{c}) < a < t'(\bar{c})$. Therefore the formulas $t(\bar{c}) \leq t_\theta(\bar{c}_\theta))$ and $t^\theta(\bar{c}^\theta) \leq t'(\bar{c})$ belong to $\mathrm{tp}(A)$. Then $\mathrm{tp}(A)$ together with $m_\theta < x < m^\theta$ implies $t(\bar{c}) < x < t'(\bar{c})$. The cases when $\rho(x, \bar{y})$ is of the form $x < t(\bar{y})$ or $t(\bar{y}) < x$ can be considered similarly.

So the set A_0 of all c_θ, c^θ satisfies the desired conditions.

4 Impossibility of translation over arbitrary states

The goal of this section is to compare restricted and generic extended queries from the viewpoint of their expressive power over *all possible states*, whether finitely representable or not. We show that, in general, extended generic querying is more expressive than restricted, even in very simple situations. Our results in this section substantially generalize Theorem 3.1 in [ST96] by extending the result proved there for rational numbers, to a broad class of domains.

But first, let us observe that the results of Section 3, specifically, Theorem 3, cannot be proved for generic, but not locally generic, queries (we remind the reader that for many domains these two notions coincide).

Theorem 8. *There is an an extended Boolean query Q over $(\mathbf{Z}, +, <)$, the ordered group of integer numbers, such that*

1. *Q is generic over all database states; in particular, Q is generic over all finite states;*
2. *Q is not equivalent, over finite database states, to a restricted query; in particular, Q is not equivalent, over all database states, to a restricted query.*

Note that Q constructed is obviously not locally generic. Moreover, it has been shown in Section 3 that *every* FO extended query, which is locally generic over finite states over $(\mathbf{Z}, <, +)$, is equivalent, for finite database states, to an FO restricted query. A similar result has been proved for $(\mathbf{Q}, <, +)$, the ordered group of rational numbers.

By the way, the mentioned result from Section 3 concerning $\mathbf{Z}$ has a curious corollary: *the query '$|P|$ is even' cannot be expressed as an extended FO query for finite database states over $(\mathbf{Z}, <, +)$, as opposed to the query 'there are even and odd numbers in P'.* Indeed, the query '$|P|$ is even' is obviously locally generic, even over all database states, and essentially the same arguments, as in the proof of Theorem 8, show that the query is not equivalent, for finite database states, to an FO extended query over $(\mathbf{Z}, <, +)$. Note, for contrast, that the query '$|P|$ is finite' can be expressed as a restricted FO query over $(\mathbf{Z}, <, +)$, because a set of integers is finite iff it is bounded.

It is natural to ask whether Theorem 8 holds for $\mathbf{Q}$ instead of $\mathbf{Z}$. In this situation, in contrast to the case of $(\mathbf{Z}, <, +)$, the notions of genericity and local genericity coincide. However, for $(\mathbf{Q}, <, +)$, we will give an example of a restricted query which is generic over all database states, but not equivalent, over all database states, to a restricted one. That example draws a line between finite and finitely representable database states, on one side, and essentially infinite states, on the other.

In fact, we will prove a more general result:

Theorem 9. *Extended querying is more expressive than restricted with respect to generic Boolean queries over any divisible Archimedean ordered Abelian group not isomorphic to the ordered group of reals.*

Classical examples of divisible Archimedean ordered Abelian groups are the ordered groups of rational and real numbers. It is known that, up to isomorphism, Archimedean ordered groups are exactly the subgroups of the ordered group of reals. We don't know whether the result of Theorem 9 holds for the ordered group of reals.

To prove Theorem 9, it suffices to prove the following two results.

Theorem 10. *Let $(A, +, <)$ be an Archimedean ordered Abelian group. Then the finiteness of database states over A is expressible by an extended FO query.*

Theorem 11. *Let A be a set of reals containing $\mathbf{Q}$. Then the finiteness of database states over A is expressible by a restricted FO query iff $A = \mathbf{R}$.*

Note, for contrast, that in $(\mathbf{Z}, <)$ the finiteness is expressible by a restricted FO query because a set of integers is finite iff it is bounded.

5 Conclusion

Further work needs to be done for integer numbers. For instance, the authors are under impression that an effective translation algorithm for locally generic queries over $(\mathbf{Z}, <, +)$ can be extracted from a Fraïssé-Ehrenfeucht-style proof of decidability of Presburger Arithmetic. Then, how much higher than $+$ in $(\mathbf{Z}, <)$ can we go? Obviously, $+, \times$ make locally generic extended queries impossible to express as pure order ones because, for example, the query "the cardinality of P is even" is locally generic and expressible in the FO extended language but not in the restricted language. We **conjecture** that if, in an extension of $(\mathbf{Z}, <, +)$, locally generic extended queries express more than restricted, the first order theory of the domain is undecidable. Stolboushkin and Taitslin [ST95] discussed irrelevance of undecidable domains to databases.

Acknowledgments

We are grateful to the late Paris Kanellakis, and also to Michael Benedikt, Yuri Gurevich, Leonid Libkin, Jianwen Su, and Jan Van den Bussche for helpful comments and suggestions.

References

[BDLW96] M. Benedikt, G. Dong, L. Libkin, and L. Wong. Relational expressive power of constraint query languages. In *Proc. 15th ACM Symp. on Principles of Database Systems*, pages 5–16, 1996.

[BL96] M. Benedikt and L. Libkin. On the structure of queries in constraint query languages. In *Proc. 11th IEEE Symp. on Logic in Computer Science*, 1996. to appear.

[CH80]   A. Chandra and D. Harel. Computable queries for relational databases. *Journal of Computer and System Sciences*, 21:156–178, 1980.

[CK90]   C.C. Chang and H.J. Keisler. *Model Theory*. North Holland, 3rd edition, 1990.

[G90]   Yu. Gurevich. Private communication, 1990.

[KKR90]   P.C. Kanellakis, G.M. Kuper, and P.Z. Revesz. Constraint query languages. In *Proc. 9th ACM Symp. on Principles of Database Systems*, pages 299–313, 1990.

[OV95]   M. Otto and J. Van den Bussche. First-order queries on databases embedded in an infinite structure. Manuscript, 1995.

[PVV95]   J. Paradaens, J. Van den Bussche, and D. Van Gucht. First-order queries on finite structures over reals. In *Proc. 10th IEEE Symp. on Logic in Computer Science*, pages 79–87. IEEE Computer Society Press, 1995.

[ST95]   A.P. Stolboushkin and M.A. Taitslin. Finite queries do not have effective syntax. In *Proc. 14th ACM Symp. on Principles of Database Systems*, pages 277–285, 1995.

[ST96]   A.P. Stolboushkin and M.A. Taitslin. Linear vs. order constraint queries over rational databases. In *Proc. 15th ACM Symp. on Principles of Database Systems*, pages 17–27, 1996.

Towards Computing Distances Between Programs via Scott Domains*

Michael A. Bukatin[1] and Joshua S. Scott[2]

[1] Department of Computer Science, Brandeis University, Waltham, MA 02254, USA;
bukatin@cs.brandeis.edu; http://www.cs.brandeis.edu/~bukatin/papers.html;
Parametric Technology Corp., 128 Technology Drive, Waltham, MA 02154, USA
[2] Department of Mathematics, Northeastern University, 360 Huntington Avenue,
Boston, MA 02115, USA; josh@neu.edu

Abstract. This paper introduces an approach to defining and computing distances between programs via *continuous generalized distance functions* $\rho : A \times A \to D$, where A and D are directed complete partial orders with the induced Scott topology, A is a semantic domain, and D is a domain representing distances (usually, some version of interval numbers). A continuous distance function ρ can define a T_0 topology on a nontrivial domain A only if the axiom $\exists 0 \in D.\forall x \in A.\rho(x,x) = 0$ does not hold. Hence, the notion of *relaxed metric* is introduced for domains — the axiom $\rho(x,x) = 0$ is eliminated, but the axiom $\rho(x,y) = \rho(y,x)$ and a version of the triangle inequality tailored for the domain D remain.

The paper constructs continuous relaxed metrics yielding the Scott topology for all continuous Scott domains with countable bases. This construction is closely related to partial metrics of Matthews and valuation spaces of O'Neill, but it describes a wider class of domains in a more intuitive way from the computational point of view.

1 Introduction

In this paper we presume that the methods of denotational semantics allow us to obtain adequate descriptions of program behavior (e.g., see [10]). The term *domain* in this paper denotes a directed complete partial order (*dcpo*) equipped with the Scott topology.

The traditional paradigm of denotational semantics states that all data types should be represented by domains and all computable functions should be represented by Scott continuous functions between domains. For the purposes of this paper all *continuous* functions are Scott continuous.

Consider the typical setting in denotational semantics — a syntactic domain of programs, P, a semantic domain of meanings, A, and a continuous semantic function, $[\![\,]\!] : P \to A$. The syntactic domain P (called a syntactic lattice in [10]) represents a data type of program parse trees, but we say colloquially that programs belong to P.

* Partially supported by NSF Grant CCR-9216185 and Office of Naval Research Grant ONR N00014-93-1-1015.

Assume that we have a domain representing distances, D, and a continuous generalized distance function, $\rho : A \times A \rightarrow D$. Assume that we can construct a *generalized metric topology*, $\mathcal{T}[\rho]$, on A via ρ. It would be reasonable to say that ρ reflects computational properties of A, if $\mathcal{T}[\rho]$ is the Scott topology on A.

Then $\rho([\![p_1]\!], [\![p_2]\!])$ would yield a computationally meaningful distance between programs p_1 and p_2. The continuous function ρ cannot possess all properties of ordinary metrics because we want $\mathcal{T}[\rho]$ to be non-Hausdorff.

1.1 Axiom $\rho(x, x) = 0$ Cannot Hold

Recall that a non-empty partially ordered set (poset), $(S, \sqsubseteq)$, is *directed* if $\forall x, y \in S.\ \exists z \in S.\ x \sqsubseteq z, y \sqsubseteq z$. A poset, $(A, \sqsubseteq)$, is a *dcpo* if it has the least element, $\bot$, and for any directed $S \subseteq A$, A contains $\sqcup S$ — the least upper bound of S. A set $U \subseteq A$ is *Scott open* if $\forall x, y \in A.\ x \in U, x \sqsubseteq y \Rightarrow y \in U$ and for any directed poset $S \subseteq A$, $\sqcup S \in U \Rightarrow \exists s \in S.\ s \in U$. The Scott open subsets of a dcpo form the *Scott topology*.

Consider dcpo's $(A, \sqsubseteq_A)$ and $(B, \sqsubseteq_B)$ with the respective Scott topologies. $f : A \rightarrow B$ is (Scott) continuous iff it is monotonic ($x \sqsubseteq_A y \Rightarrow f(x) \sqsubseteq_B f(y)$) and for any directed poset $S \subseteq A$, $f(\sqcup_A S) = \sqcup_B \{f(s) \mid s \in S\}$.

Assume that there is an element $0 \in D$ representing the ordinary numerical 0. Let us show that $\forall x.\ \rho(x, x) = 0$ cannot be true under reasonable assumptions. We will see later that all other properties of ordinary metrics can be preserved at least for continuous Scott domains with countable bases (Sect. 5).

It seems reasonable to assume that any reasonable construction of $\mathcal{T}[\rho]$ for any generalized distance function $\rho : A \times A \rightarrow D$, should satisfy the following axiom, regardless of whether the distance space D is a domain, or whether ρ is continuous:

Axiom 1. *For all $x, y \in A$, $\rho(x, y) = \rho(y, x) = 0$ implies that x and y share the same system of open sets, i.e. for all open sets $U \in \mathcal{T}[\rho]$, $x \in U$ iff $y \in U$.*

We assume this axiom for the rest of the paper. A topology is called T_0, if different elements do not share the systems of open set.

Corollary 2. *If there are $x, y \in A$, such that $x \neq y$ and $\rho(x, y) = \rho(y, x) = 0$, then $\mathcal{T}[\rho]$ is not a T_0 topology.*

Let us return to our main case, where D is a domain and ρ is a continuous function.

Lemma 3. *Assume that there are at least two elements $x, y \in A$, such that $x \sqsubseteq_A y$. Assume that $\rho : A \times A \rightarrow D$ is a continuous function. If $\rho(x, x) = \rho(y, y) = d \in D$, then $\rho(x, y) = \rho(y, x) = d$*

Proof. The continuity of ρ implies its monotonicity with respect to the both of its arguments. Then $x \sqsubseteq_A y$ implies $d = \rho(x, x) \sqsubseteq_D \rho(x, y) \sqsubseteq_D \rho(y, y) = d$. This yields $\rho(x, y) = d$, and, similarly, $\rho(y, x) = d$. □

Then we can obtain the following simple, but important result.

Theorem 4. *Assume that there are at least two elements $x, y \in A$, such that $x \sqsubseteq_A y$. Assume that $\rho : A \times A \to D$ is a continuous generalized distance function and $\mathcal{T}[\rho]$ is a T_0 topology. Then the double equality $\rho(x, x) = \rho(y, y) = 0$ does not hold.*

Proof. By Lemma 3, $\rho(x, x) = \rho(y, y) = 0$ would imply $\rho(x, y) = \rho(y, x) = 0$. Then, by Corollary 2, $\mathcal{T}[\rho]$ would not be T_0, contradicting our assumptions. $\square$

The topologies used in domain theory are usually T_0; in particular, the Scott topology is T_0. This justifies studying continuous generalized metrics ρ, such that $\rho(x, x) = 0$ is false for some x, more closely.

1.2 Intuition behind $\rho(x, x) \neq 0$

There are compelling intuitive reasons not to expect $\rho(x, x) = 0$, when x is not a maximal element of A. The computational intuition behind $\rho(x, y)$ is that the elements in question are actually x' and y', $x \sqsubseteq_A x', y \sqsubseteq_A y'$, but not all information is usually known about them. The correctness condition $\rho(x, y) \sqsubseteq_D \rho(x', y')$ is provided by the monotonicity of ρ.

In particular, even if $x = y$, this only means that we know the same information about x' and y', but this does not mean that $x' = y'$. Consider $x' \neq y'$, such that $x \sqsubseteq_A x'$ and $x \sqsubseteq_A y'$. Then $\rho(x, x) \sqsubseteq_D \rho(x', y')$ and $\rho(x, x) \sqsubseteq_D \rho(y', x')$, and at least one of $\rho(x', y')$ and $\rho(y', x')$ is non-zero, if we want ρ to yield a T_0 topology (we do not assume symmetry yet).

Example 1. Here is an important example — a continuous generalized distance on the domain of interval numbers R^I — $\rho : R^I \times R^I \to R^I$ (See Sect. 2 for the definition of R^I). Consider intervals $[a, b]$ and $[c, d]$ and set $S = \{|x' - y'| \mid a \leq x' \leq b, c \leq y' \leq d\}$. Define $\rho([a, b], [c, d]) = [\min S, \max S]$. In particular, $\rho([a, b], [a, b]) = [0, b - a] \neq [0, 0]$, and $\rho([a, a], [b, b]) = [|a - b|, |a - b|]$.

1.3 Related Work: Quasi-Metrics

Quasi-metrics [9] and Kopperman-Flagg generalized distances [4] are asymmetric generalized distances. They satisfy axiom $\rho(x, x) = 0$ and yield Scott topology for various classes of domains via a construction satisfying Axiom 1.

Theorem 4 means that if one wishes to represent the distance spaces via domains, these asymmetric distances cannot be made continuous unless their nature is changed substantially.

The practice of representing all computable functions via continuous functions between domains suggests that quasi-metrics cannot, in general, be computed (see Sect. 7).

1.4 Related Work: Partial Metrics

Historically, partial metrics are the first generalized distances on domains for which axiom $\rho(x, x) = 0$ does not hold. They were introduced by Matthews [7, 6] and further investigated by Vickers [11] and O'Neill [8].

Partial metrics satisfy a number of additional axioms in lieu of $\rho(x, x) = 0$ (see Sect. 4). Matthews and Vickers state that $\rho(x, x) \neq 0$ is caused by the fact, that x expresses a partially defined object. The most essential component of the central construction in our paper is a partial metric (Sect. 5).

1.5 Our Contribution

We build partial metrics yielding Scott topologies for a wider class of domains that was known before (Sect. 5, Theorem 7). This class — all continuous Scott domains with countable bases — is sufficiently big to solve interesting domain equations and to define denotational semantics of at least sequential deterministic programming languages [10].

We introduce the notion of *relaxed metric* (Sect. 2), which maintains the intuitively clear requirement to reject axiom $\rho(x, x) = 0$, but does not impose the specific axioms of partial metrics. We believe that the applicability of these specific axioms is more limited (see Sect. 8.1).

We introduce the idea that a space of distances should be thought of as a data type in the context of denotational semantics and thus, should be represented by a domain. We also introduce the requirement that distance functions should be computable and thus, Scott continuous (the use of *continuous valuations* in [8] should be considered as a step in this direction).

These considerations lead to an understanding that relaxed metrics should map pairs of partial elements to *upper estimates* of some "ideal" distances, where the *distance domain of upper estimates*, R^-, is equipped with a dual informational order: $\sqsubseteq_{R^-} = \geq$. We also consider lower estimates of "ideal" distances, thus, introducing the *distance domain of interval numbers*, R^I. Continuous lower estimates are useful during actual computations of distances (Sect. 7) and for defining and computing an induced metric structure on the space of total elements (Theorem 8).

There is a comparison in [7] between partial metrics and alternative generalized distance structures such as quasi-metrics and weighted metrics [5]. We provide what we believe to be the strongest argument in favor of partial metrics so far — among all those alternatives only partial metrics can be thought of as Scott continuous, computable functions.

2 Relaxed Metrics

Consider distance domains in greater detail. It is conventional to think about distances as non-negative real numbers. When it comes to considering approximate information about reals, it is conventional to use some kind of *interval numbers*.

We follow both conventions in this text. The distance domain consists of pairs $< a, b >$ (also denoted as $[a, b]$) of non-negative reals ($+\infty$ included), such that $a \leq b$. We denote this domain as R^I. $[a, b] \sqsubseteq_{R^I} [c, d]$ iff $a \leq c$ and $d \leq b$.

We can also think about R^I as a subset of $R^+ \times R^-$, where $\sqsubseteq_{R^+} = \leq$, $\sqsubseteq_{R^-} = \geq$, and both R^+ and R^- consist of non-negative reals and $+\infty$. We call R^+ a *domain of lower bounds*, and R^- — a *domain of upper bounds*. Thus a distance function $\rho : A \times A \to R^I$ can be thought of as a pair of distance functions $< l, u >$, $l : A \times A \to R^+$, $u : A \times A \to R^-$.

We think about $l(x, y)$ and $u(x, y)$ as, respectively, lower and upper bounds of some "ideal" distance $\sigma(x, y)$. We do not try to formalize the "ideal" distances, but we refer to them to motivate our axioms. There are good reasons to impose the triangle inequality, $u(x, z) \leq u(x, y) + u(y, z)$. Assume that for our "ideal" distance, the triangle inequality, $\sigma(x, z) \leq \sigma(x, y) + \sigma(y, z)$, holds. If $u(x, z) > u(x, y) + u(y, z)$, then $u(x, y) + u(y, z)$ gives a better upper estimate for $\sigma(x, z)$ than $u(x, z)$. This means that unless $u(x, z) \leq u(x, y) + u(y, z)$, u could be easily improved and, hence, would be very imperfect.

This kind of reasoning is not valid for $l(x, z)$. In fact, there are reasonable situations, when $l(x, z) \neq 0$, but $l(x, y) = l(y, z) = 0$. E.g., consider Example 1 and take $x = [2, 2], y = [2, 3], z = [3, 3]$.

Also only u plays a role in the subsequent definition of the relaxed metric topology, and the most important results remain true even if we take $l(x, y) = 0$. In the last case we sometimes take $D = R^-$ instead of $D = R^I$ making the distance domain look more like ordinary numbers (it is important to remember, that $\sqsubseteq_{R^-} = \geq$ and, hence, 0 is the largest element of R^-).

We also impose the symmetry axiom on the function ρ. The motivation here is that we presume our "ideal" distance to be symmetric, hence, we should be able to make symmetric upper and lower estimates.

We state a definition summarizing the discourse above:

Definition 5. A symmetric function $u : A \times A \to R^-$ is called a *relaxed metric* when it satisfies the triangle inequality. A symmetric function $\rho : A \times A \to R^I$ is called a *relaxed metric* when its upper part u is a relaxed metric.

3 Relaxed Metric Topology

An *open ball* with a center $x \in A$ and a real radius ϵ is defined as $B_{x, \epsilon} = \{y \in A \mid u(x, y) < \epsilon\}$. Notice that only upper bounds are used in this definition — the ball only includes those points y, about which we are *sure* that they are not too far from x.

We should formulate the notion of a relaxed metric open set more carefully than for ordinary metrics, because it is now possible to have a ball of a non-zero positive radius, which does not contain its own center.

Definition 6. A subset U of A is *relaxed metric open* if for any point $x \in U$, there is an $\epsilon > \rho(x, x)$ such that $B_{x, \epsilon} \subseteq U$.

It is easy to show that for a continuous relaxed metric on a dcpo all relaxed metric open sets are Scott open and form a topology.

4 Partial Metrics

The distances p with $p(x,x) \neq 0$ were first introduced by Matthews [7, 6]. They are known as *partial metrics* and obey the following axioms:

1. $x = y$ iff $p(x,x) = p(x,y) = p(y,y)$.
2. $p(x,x) \leq p(x,y)$.
3. $p(x,y) = p(y,x)$.
4. $p(x,z) \leq p(x,y) + p(y,z) - p(y,y)$.

The last axiom (due to Vickers [11]) implies the ordinary triangle inequality, since the distances are non-negative. O'Neill found it useful to introduce negative distances in [8], but this is avoided in the present paper.

Whenever partial metrics are used to describe a partially ordered domain, a stronger form of the first two axioms is used: If $x \sqsubseteq y$ then $\rho(x,x) = \rho(x,y)$, otherwise $\rho(x,x) < \rho(x,y)$. We include the stronger form in the definition of partial metrics for the purposes of this paper.

5 Central Construction

Here we construct continuous relaxed metrics yielding the Scott topology for all continuous Scott domains with countable bases. Our construction closely resembles one by O'Neill [8]. We also use valuations, but we consider continuous valuations on the powerset of the basis instead of continuous valuations on the domain itself. This allows us to handle a wider class of domains.

We define continuous Scott domains in the spirit of [3]. Consider a dcpo $(A, \sqsubseteq)$. We say that $x \ll y$ (x is *way below* y) if for any directed set $S \subseteq A$, $y \sqsubseteq \sqcup S \Rightarrow \exists s \in S. \, x \sqsubseteq s$. An element x, such that $x \ll x$, is called *compact*. We say that A is *bounded complete* if $\forall B \subseteq A. \, (\exists a \in A. \, \forall b \in B. b \sqsubseteq a) \Rightarrow \sqcup_A B \in A$.

Consider a set $K \subseteq A$. Assume that $\perp_A \in K$. We say that a bounded complete dcpo A is a *continuous Scott domain* with *basis* K, if for any $a \in A$, the set $K_a = \{k \in K \mid k \ll a\}$ is directed and $a = \sqcup K_a$. We call elements of K *basic* elements.

Enumerate elements of $K \setminus \{\perp_A\}$: $k_1, ..., k_i, ...$ Associate weights with all basic elements: $w(\perp_A) = 0$, and let $w(k_i)$ form a converging sequence of strictly positive weights. For convenience we agree that the sum of weights of all basic elements equals 1. For example, one might wish to consider $w(k_i) = 2^{-i}$ or $w(k_i) = \epsilon(1+\epsilon)^{-i}, \epsilon > 0$. Then for any $K_0 \subseteq K$, the weight of set K_0, $W(K_0) = \sum_{k \in K_0} w(k_0)$, is well defined and belongs to $[0,1]$.

We have several versions of function ρ. For most purposes it is enough to consider $u(x,y) = 1 - W(K_x \cap K_y)$ and $l(x,y) = 0$. Sometimes it is useful to consider a better lower bound function $l(x,y) = W(I_x \cup I_y)$, where $I_x = \{k \in K \mid k \sqcup x$ does not exist$\}$ for the computational purposes (see Sect. 7).

Theorem 7. *The function u is a partial metric. The function ρ is a continuous relaxed metric. The relaxed metric topology coincides with the Scott topology.*

If, in addition, we would like the next theorem to hold, we have to consider a different version of ρ with $u(x,y) = 1 - W(K_x \cap K_y) - W(I_x \cap I_y)$ and $l(x,y) = W(K_x \cap I_y) + W(K_y \cap I_x)$. The previous theorem still holds.

We introduce the notion of a *regular basis*. A set of maximal elements in A is denoted as $Total(A)$. We say that the basis K is *regular* if $\forall k \in K, x \in Total(A)$. $k \sqsubseteq x \Rightarrow k \ll x$. In particular, if K consists of compact elements, thus making A an algebraic Scott domain, K is regular.

Theorem 8. *Let K be a regular basis in A. Then for all x and y from $Total(A)$, $l(x,y) = u(x,y)$. Consider $\mu : Total(A) \times Total(A) \to \mathbb{R}$, $\mu(x,y) = l(x,y) = u(x,y)$. Then $(Total(A), \mu)$ is a metric space, and its metric topology is the subspace topology induced by the Scott topology on A.*

6 Proofs of Theorems 7 and 8

Here we prove relatively difficult parts of these theorems for the case when $u(x,y) = 1 - W(K_x \cap K_y) - W(I_x \cap I_y)$ and $l(x,y) = W(K_x \cap I_y) + W(K_y \cap I_x)$. Lemma 10 is needed for Theorem 8, and other lemmas are needed for Theorem 7.

Lemma 9 (Correctness of lower bounds). $l(x,y) \le u(x,y)$.

Proof. Using $K_x \cap I_x = \emptyset$ we can rewrite u and l. $u(x,y) = 1 - W(K_x \cap K_y) - W(I_x \cap I_y) = W(\overline{U})$, where $U = (K_x \cap K_y) \cup (I_x \cap I_y)$. $l(x,y) = W(V)$, where $V = (K_x \cap I_y) \cup (K_y \cap I_x)$.

We want to show that $V \subseteq \overline{U}$, for which it is enough to show that $V \cap U = \emptyset$. We show that $(K_x \cap I_y) \cap U = \emptyset$. Then by symmetry the same will be true for $K_y \cap I_x$, and hence for V.

$(K_x \cap I_y) \cap U = (K_x \cap I_y \cap K_x \cap K_y) \cup (K_x \cap I_y \cap I_x \cap I_y)$. But $K_x \cap I_y \cap K_x \cap K_y \subseteq I_y \cap K_y = \emptyset$. Similarly, $K_x \cap I_y \cap I_x \cap I_y = \emptyset$. $\qquad\square$

Lemma 10. *If K is a regular basis and $x, y \in Total(A)$, then $l(x,y) = u(x,y)$.*

Proof. Using the notations of the previous proof we want to show that $\overline{U} \subseteq V$.

Let us show first that $K_x \cup I_x = K_y \cup I_y = K$. Consider $k \in K$. Since $x \in Total(A)$, if $k \notin I_x$, then $k \sqsubseteq x$. Now from the regularity of K we obtain $k \ll x$ and $k \in K_x$. Same for y.

Now, if $k \notin U$, then $k \notin K_x$ or $k \notin K_y$. Because of the symmetry it is enough to consider $k \notin K_x$. Then $k \in I_x$. Then, using $k \notin U$ once again, $k \notin I_y$. Then $k \in K_y$. So we have $k \in K_y \cap I_x \subseteq V$. $\qquad\square$

Lemma 11 (Vickers-Matthews triangle inequality for upper bounds). $u(x,z) \le u(x,y) + u(y,z) - u(y,y)$.

Proof. We want to show $1 - W(K_x \cap K_z) - W(I_x \cap I_z) \leq 1 - W(K_x \cap K_y) - W(I_x \cap I_y) + 1 - W(K_y \cap K_z) - W(I_y \cap I_z) - 1 + W(K_y) + W(I_y)$. This is equivalent to $W(K_x \cap K_y) + W(K_y \cap K_z) + W(I_x \cap I_y) + W(I_y \cap I_z) \leq W(K_y) + W(I_y) + W(K_x \cap K_z) + W(I_x \cap I_z)$.

Notice that $W(K_x \cap K_y) + W(K_y \cap K_z) = W(K_x \cap K_y \cap K_z) + W(K_y \cap (K_x \cup K_z))$, and the similar formula holds for I's.

Then the result follows from the following simple facts: $W(K_x \cap K_y \cap K_z) \leq W(K_x \cap K_z)$, $W(K_y \cap (K_x \cup K_z)) \leq W(K_y)$, and the similar inequalities for I's. $\qquad \square$

Lemma 12. *Function $\rho : A \times A \to R^I$ is continuous.*

Proof. Monotonicity of ρ is trivial.

Consider a directed set $B \subseteq A$ and some $z \in A$. We have to show that $\rho(z, \sqcup B) = \sqcup_{R^I} \{\rho(z, x) \mid x \in B\}$, which is equivalent to $u(z, \sqcup B) = \inf\{u(z, x) \mid x \in B\}$ and $l(z, \sqcup B) = \sup\{l(z, x) \mid x \in B\}$.

Rewriting this, we want to show that $W(K_z \cap K_{\sqcup B}) + W(I_z \cap I_{\sqcup B}) = \sup\{W(K_z \cap K_x) + W(I_z \cap I_x) \mid x \in B\}$ and $W(K_z \cap I_{\sqcup B}) + W(I_z \cap K_{\sqcup B}) = \sup\{W(K_z \cap I_x) + W(I_z \cap K_x) \mid x \in B\}$. Monotonicity considerations trivially yield both "$\geq$" inequalities, so it is enough to show "$\leq$" inequalities. In fact, we will show that for any sets $C \subseteq A$ and $D \subseteq A$, $W(C \cap K_{\sqcup B}) + W(D \cap I_{\sqcup B}) \leq \sup\{W(C \cap K_x) + W(D \cap I_x) \mid x \in B\}$ holds.

It is easy to show that $K_{\sqcup B} = \cup\{K_x \mid x \in B\}$ by showing first that the set $\cup\{K_x \mid x \in B\}$ is directed and $\sqcup B = \sqcup(\cup\{K_x \mid x \in B\})$. Let us prove that $I_{\sqcup B} = \cup\{I_x \mid x \in B\}$. "$\supseteq$" is trivial. Let us prove "$\subseteq$". Assume that $k \notin \cup\{I_x \mid x \in B\}$, i.e. $\forall x \in B$ $k \sqcup x$ exists. It is easy to see that because B is a directed set, $\{k \sqcup x \mid x \in B\}$ is also directed. Then $k \sqsubseteq \sqcup\{k \sqcup x \mid x \in B\} \sqsupseteq \sqcup B$, implying existence of $k \sqcup (\sqcup B)$ and, hence, $k \notin I_{\sqcup B}$.

Now consider enumerations of the countable or finite sets $K_{\sqcup B}$ and $I_{\sqcup B}$: $k_1, ..., k_n, ...$ and $k'_1, ..., k'_n, ...$, respectively. Define tail sums $S_n = w(k_n) + w(k_{n+1}) + ...$ and $S'_n = w(k'_n) + w(k'_{n+1}) + ...$. Observe that (S_n) and (S'_n) converge to 0.

Pick for every k_n some $x_n \in B$ such that $k_n \in K_{x_n}$. Pick for every k'_n some $x'_n \in B$ such that $k'_n \in I_{x'_n}$. Then using the directness of B, we can for any n pick such $y_n \in B$, that $x_1, ..., x_n, x'_1, ..., x'_n \sqsubseteq y_n$. Then $k_1, ..., k_n \in K_{y_n}$ and $k'_1, ..., k'_n \in I_{y_n}$. It is easy to see that $(W(C \cap K_{\sqcup B}) + W(D \cap I_{\sqcup B})) - (W(C \cap K_{y_n}) + W(D \cap I_{y_n})) = W(C \cap K_{\sqcup B}) - W(C \cap K_{y_n}) + W(D \cap I_{\sqcup B}) - W(D \cap I_{y_n}) < S_n + S'_n$, which can be made as small as we like. $\qquad \square$

Lemma 13. *If $B \subseteq A$ is Scott open then it is relaxed metric open.*

Proof. Consider $x \in B$. Because B is Scott open there is a basic element $k \in B$ such that $k \ll x$. We must find $\epsilon > 0$ such that $x \in B_{x,\epsilon} \subseteq B$. Let $\epsilon = u(x, x) + w(k)/2$. Clearly $x \in B_{x,\epsilon}$. We claim that $B_{x,\epsilon} \subseteq \{y \mid y \gg k\} \subseteq B$. Assume, by contradiction, that $y \gg k$ is false. Then $k \notin K_y$ and $k \in K_x$. Then $W(K_x \cap K_y) + W(I_x \cap I_y) + w(k) \leq W(K_x) + W(I_x)$. Then $u(x, y) - w(k) = 1 - W(K_x \cap K_y) - W(I_x \cap I_y) - w(k) \geq 1 - W(K_x) - W(I_x) = u(x, x)$. Therefore $u(x, y) \geq u(x, x) + w(k)$ and thus $y \notin B_{x,\epsilon}$. $\qquad \square$

7 Computability and Continuity

We do not have space for a proper discussion on effective structures on domains (see [1] for some of it). We have to ask the reader to take on faith that whenever one expects element x to be computable, it is reasonable to impose the requirement that K_x and I_x must be recursively enumerable, but one almost never should expect them to be recursive.

The actual computation of $\rho(x, y)$ goes as follows. Start with $[0, 1]$ and go along the recursive enumerations of K_x, K_y, I_x, and I_y. Whenever we discover that some k occurs in both K_x and K_y, or in both I_x and I_y, subtract $w(k)$ from the upper boundary. Whenever we discover that some k occurs in both K_x and I_y, or in both I_x and K_y, add $w(k)$ to the lower boundary. If this process continues long enough, $[l(x, y), u(x, y)]$ is approximated as well as desired.

However, there is no general way to compute a better lower estimate for $u(x, y)$ than $l(x, y)$, or to compute a better upper estimate for $l(x, y)$ than $u(x, y)$. Consequently, there is no general way to determine how close is the convergence process to the actual values of $l(x, y)$ and $u(x, y)$, except that we know that $u(x, y)$ is not less than currently computed lower bound, and $l(x, y)$ is not greater than currently computed upper bound. Of course, for large x and y this knowledge might provide a lot of information, and if the basis of our domain is regular, for total elements x and y this knowledge provides us with precise estimates — i.e. if the basis is regular, then the resulting classical metric on $Total(A)$ can be nicely computed.

The computational situation is very different with regard to quasi-metrics. Consider $u(x, y) = 1 - W(K_x \cap K_y)$ and $d(x, y) = u(x, y) - u(x, x) = W(K_x \setminus K_y)$. This is a quasi-metric in the style of [9, 4], and it yields a Scott topology [1]. However, as discussed in [1], typically $K_x \setminus K_y$ is not recursive. Moreover, one should not expect $K_x \setminus K_y$ or its complement to be recursively enumerable. This precludes us from building a generally applicable method computing $d(x, y)$ and illustrates that it is computationally incorrect to subtract one upper bound from another.

8 Some Open Issues in the Theory of Relaxed Metrics

8.1 Should the Axioms of Partial Metrics Hold?

Consider relaxed metric ρ and its upper part u. Should we expect function u to satisfy the axioms of partial metrics? Example 1 describes a natural relaxed metric on interval numbers, where u is not a partial metric. Function u gives a better upper estimate of the "ideal" distance between interval numbers, than the partial metric $p([a, b], [c, d]) = \max(b, d) - \min(a, c)$ described in [7]. For example, $u([0, 2], [1, 1]) = 1$, which is what one expects — if we know that one of the numbers is somewhere between 0 and 2, and another number equals 1, then we know that the distance between them is no greater than 1. However, $p([0, 2], [1, 1]) = p([0, 2], [0, 2]) = 2$.

Now we describe two situations when the axioms of partial metrics are justified. Consider again function $d(x, y) = u(x, y) - u(x, x)$ from Sect. 7. Vickers notes in [11] that the triangle inequality $d(x, z) \leq d(x, y) + d(y, z)$ is equivalent to $u(x, z) \leq u(x, y) + u(y, z) - u(y, y)$, and $d(x, y) \geq 0$ is equivalent to $u(x, x) \leq u(x, y)$. This means that function u is a partial metric if and only if function d is a quasi-metric.

Another justification comes from the consideration of the proof of Lemma 11. Whenever the upper part $u(x, y)$ of a relaxed metric is based on *common information* shared by x and y yielding a *negative contribution* to the distance (we subtract the weight of common information from the universal distance 1 in this paper), both $u(x, z) \leq u(x, y) + u(y, z) - u(y, y)$ and $u(x, x) \leq u(x, y)$ should hold.

However, to specify function u in Example 1, we use information about x and y, which cannot be thought of as common information shared by x and y. In such case we still expect a relaxed metric ρ, but its upper part u does not have to satisfy the axioms of partial metrics.

8.2 Potential Relationship with Measure Theory

Important applications of domains to measure theory have emerged recently [2]. The construction developed in this paper suggests that fruitful applications in the opposite direction are possible. For example, in order to compute upper bound $u(x, y)$ we measure common information shared by x and y and subtract it from the universal distance 1. Our way to measure this information is fairly primitive — it is based on assigning a converging system of weights to a countable set of basic points and adding weights of basic points belonging to the sets of interest. This is similar to measuring areas on the plane via assigning a converging system of finite weights to points with rational coordinates and adding weights of such points belonging to the sets of interest. We should be able to do better, at least for domains without many compact elements.

9 Conclusion

Let us briefly state where we stand with regard to the applications to programs. We are able to introduce relaxed metrics on a class of domains sufficiently large for practical applications in the spirit of [10].

For example, consider $X = [\![$while B do S$]\!]$, and the sequence of programs, $P_1 = $ loop; ...; $P_N = $ if B then S;P_{N-1}else skip endif; .... Define $X_N = [\![P_N]\!]$. Typically $X_{N-1} \sqsubseteq X_N$ and $X = \sqcup X_N$ hold. We agreed that the distances between programs will be distances between their meanings. Assume that $M \leq N$. Then regardless of specific weights, $u(P_M, P_M) = u(P_M, P_N) = u(P_M, P)$, also $u(P_M, P_M) \geq u(P_N, P_N) \geq u(P, P)$, and $u(P, P) = \inf u(P_N, P_N)$. Of course, none of these distances has to be zero.

However, we do not yet know how to build relaxed distances so that not only nice convergence properties are true, but also that distances between particular pairs of programs "look right" — a notion, which is more difficult to formalize, than convergence. Also, we compute these distances via recursive enumeration now, and a more efficient scheme is needed.

Acknowledgements

The authors thank Michael Alekhnovich, Will Clinger, Ross Viselman, Steve Matthews, Mitch Wand, and especially Bob Flagg and Simon O'Neill for helpful discussions, and Steve Vickers for his notes and remarks.

References

1. Bukatin M.A., Scott J.S. Towards Computing Distances between Programs via Domains: a Symmetric Continuous Generalized Metric for Scott Topology on Continuous Scott Domains with Countable Bases. Available via URL http://www.cs.brandeis.edu/~bukatin/dist_new.ps.gz, December 1996.
2. Edalat A. Domain theory and integration. *Theoretical Computer Science*, **151** (1995) 163–193.
3. Hoofman R. Continuous information systems. *Information and Computation*, **105** (1993) 42–71.
4. Kopperman R.D., Flagg R.C. The asymmetric topology of computer science. In S. Brooks et al., eds., Mathematical Foundations of Programming Semantics, *Lecture Notes in Computer Science*, **802**, 544–553, Springer, 1993.
5. Kunzi H.P.A., Vajner V. Weighted quasi-metrics. In S. Andima et al., eds., Proc. 8th Summer Conference on General Topology and Applications, *Annals of the New York Academy of Sciences*, **728**, 64–77, New York, 1994.
6. Matthews S.G. An extensional treatment of lazy data flow deadlock. *Theoretical Computer Science*, **151** (1995), 195–205.
7. Matthews S.G. Partial metric topology. In S. Andima et al., eds., Proc. 8th Summer Conference on General Topology and Applications, *Annals of the New York Academy of Sciences*, **728**, 183–197, New York, 1994.
8. O'Neill S.J. Partial metrics, valuations and domain theory. In S. Andima et al., eds., Proc. 11th Summer Conference on General Topology and Applications, *Annals of the New York Academy of Sciences*, **806**, 304–315, New York, 1997.
9. Smyth M.B. Quasi-uniformities: reconciling domains and metric spaces. In M. Main et al., eds., Mathematical Foundations of Programming Language Semantics, *Lecture Notes in Computer Science*, **298**, 236–253, Springer, 1988.
10. Stoy J.E. *Denotational Semantics: The Scott-Strachey Approach to Programming Language Semantics*. MIT Press, Cambridge, Massachusetts, 1977.
11. Vickers S. Matthews Metrics. Unpublished notes, Imperial College, UK, 1987.

A Safe Recursion Scheme for Exponential Time

Peter Clote*

Institut für Informatik, Ludwig-Maximilians-Universität München, Oettingenstraße 67, D-80538 München, GERMANY. clote@informatik.uni-muenchen.de

Abstract. Using a function algebra characterization of exponential time due to Monien [5], in the style of Bellantoni-Cook [2], we characterize exponential time functions of linear growth via a *safe* course-of-values recursion scheme.

1 Introduction

In 1991 [2], S. Bellantoni and S.A. Cook characterized the class FP of polynomial time computable functions as the smallest class of functions containing certain initial functions, and closed under *safe* composition and *safe* recursion on notation. In 1965, A. Cobham had earlier characterized FP in a similar manner using composition and *bounded* recursion on notation, where f is defined by bounded recursion on notation from g, h_0, h_1, k, if

$$f(0, \mathbf{y}) = g(\mathbf{y})$$
$$f(2x, \mathbf{y}) = h_0(x, \mathbf{y}, f(x, \mathbf{y})), \quad \text{if } x \neq 0$$
$$f(2x + 1, \mathbf{y}) = h_1(x, \mathbf{y}, f(x, \mathbf{y})),$$

provided that $f(x, \mathbf{y}) \leq k(x, \mathbf{y})$. In addition to removing an initial function required by Cobham for polynomial growth rate, the novelty of the Bellantoni-Cook construction was to remove the bounding requirement $f(x, \mathbf{y}) \leq k(x, \mathbf{y})$ by imposing a syntactic restriction on the manner in which variables can be used in the recursion — variables are either *normal* or *safe*, whereby only normal variables can be used as the principal variable x in a definition by recursion in $f(x, \mathbf{y})$.

Since publication of the Bellantoni-Cook characterization of polynomial time, many function complexity classes have been similarly characterized by *safe* recursion schemes without a bounding requirement: linear space, levels of the polynomial time hierarchy, logarithmic space, NC (S. Bellantoni[2] [1]; alternating logarithmic time and polylogarithmic time (S. Bloch [3]), and other classes. The purpose of this note is to similarly characterize the class of exponential time computable functions of linear growth.

* Research supported in part by NSF CCR-9408090, US-Czech Science and Technology Grant 93-025 and by the Volkswagen Stiftung.
  Key words and phrases: exponential time, safe recursion, function algebra
[2] Linear space was characterized in a similar manner as well by W. Handley and independently D. Leivant.)

2 Preliminaries

Following [2], variables are designated as *normal* or *safe* and distinguished by their position in a function $f(x_1, \ldots, x_n; y_1, \ldots, y_m)$ — variables x_i occurring to the left of the semi-colon are normal, whereas the variables y_i occurring to the right are called safe. Often we will write normal variables as x, y, z, etc. and safe variables as a, b, c, etc.

Following Bellantoni [1], define the following initial functions.

1. (0-ary constant function) 0, or more formally $0(\ ;\ )$.

2. (projections) $I_j^{n,m}(x_1, \ldots, x_n; a_1, \ldots, a_m) = \begin{cases} x_j & \text{if } 1 \le j \le n \\ a_{j-n} & \text{if } n < j \le n + m \end{cases}$

3. (successor) $S(; a) = a + 1$

4. (predecessor) $Pr(; a) = \begin{cases} a - 1 & \text{if } a > 0 \\ 0 & \text{else} \end{cases}$

5. (conditional) $C(; a, b, c) = \begin{cases} b & \text{if } a = 0 \\ c & \text{else.} \end{cases}$

We designate the collection of all projection functions by I, and define the function algebra $[f, g, h : \mathcal{O}_1, \mathcal{O}_2]$ as the smallest class of functions containing the initial functions f, g, h and closed under the operations $\mathcal{O}_1$ and $\mathcal{O}_2$. If $\mathcal{F}$ is a function algebra, then by $\mathcal{F}_*$ we denote the class of predicates whose characteristic function belongs to $\mathcal{F}$.

Definition 1 Bellantoni [1]. The function f is defined by safe composition (SCOMP) from $g, u_1, \ldots, u_n, v_1, \ldots, v_m$ if

$$f(\mathbf{x}; \mathbf{a}) = g(u_1(\mathbf{x};), \ldots, u_n(\mathbf{x};); v_1(\mathbf{x}; \mathbf{a}), \ldots, v_m(\mathbf{x}; \mathbf{a})).$$

The function f is defined by safe recursion (SR) from functions g, h if

$$f(0, \mathbf{y}; \mathbf{a}) = g(\mathbf{y}; \mathbf{a})$$
$$f(x + 1, \mathbf{y}; \mathbf{a}) = h(x, \mathbf{y}; \mathbf{a}, f(x, \mathbf{y}; \mathbf{a})).$$

It is important to note that in safe composition, no function having safe variables may be substituted into a normal position of another function. Note that if $f(x; y)$ is defined, then SCOMP allows us to define

$$g(x, y;) = f(I_1^{2,0}(x, y;); I_2^{2,0}(x, y;)) = f(x; y).$$

Thus normal variables can be moved into safe positions, but not vice-versa. As well, note that in defining $f(x + 1, \mathbf{y}; \mathbf{a})$, the previous value $f(x, \mathbf{y}; \mathbf{a})$ must be placed into a safe position in the function h. The intuition is that a safe variable cannot be used as the principal variable of an induction, and by constraining composition as above, the role played by a safe variable in a function is limited to composition.

By $([0, I, S, Pr, C : \text{SCOMP}, \text{SR}] \cap \text{NORMAL})$ we denote the class of functions $f(\mathbf{x};)$ having only normal variables, such that f belongs to the function algebra $[0, I, S, Pr, C : \text{SCOMP}, \text{SR}]$.

Theorem 2 Bellantoni [1]. *The class $\mathcal{E}^2$ of linear space computable functions equals*

$$([0, I, S, Pr, C : \text{SCOMP}, \text{SR}] \cap \text{NORMAL}).$$

Define the function

$$m_0(0; x) = x$$
$$m_0(y + 1; x) = Pr(; m_0(y_1; x))$$
$$M(x, y;) = m_0(I_2^{2,0}(x, y;); I_1^{2,0}(x, y;))$$

Thus $M(x, y;) = x \dot{-} y$, where modified difference (sometimes called *monus*) of x and y is 0 if $x - y$ is negative, else $x - y$. The recursion scheme investigated in this paper is defined as follows.

Definition 3. The function f is defined from g, h, r by safe value recursion[3] (SVR) if

$$f(0, \mathbf{y}; \mathbf{a}) = g(\mathbf{y}; \mathbf{a})$$
$$f(x + 1, \mathbf{y}; \mathbf{a}) = h(x, \mathbf{y}; \mathbf{a}, f(x, \mathbf{y}; \mathbf{a}), f(M(x, r(x, \mathbf{y};);), \mathbf{y}; \mathbf{a})).$$

This recursion scheme is the safe version of a scheme investigated by Monien [5] (see [6] or [4] for a presentation). Our use of the function $M(x, r(x, \mathbf{y};);)$ is to ensure that this argument is at most x.

3 Main result

Definition 4. Let $\mathcal{F}$ denote the class of functions $f(\mathbf{x})$ computable in exponential time $ETIME = TIME(2^{O(n)})$ on a multitape Turing machine, which are of linear growth rate, i.e.

$$|(f(x_1, \ldots, x_n)| \le c \cdot \max(|x_1|, \ldots, |x_n|) + d$$

for constants c, d independent of $\mathbf{x}$.

The main result is the following.

Theorem 5. $\mathcal{F}$ *equals*

$$([0, I, S, Pr, C : \text{SCOMP}, \text{SR}, \text{VSR}] \cap \text{NORMAL}).$$

Corollary 6. *ETIME equals*

$$([0, I, S, Pr, C : \text{SCOMP}, \text{SR}, \text{VSR}] \cap \text{NORMAL}).$$

---

[3] This is a course-of-values recursion on two values.

The corollary is immediate from the theorem. The remainder of the paper consists of the proof of the theorem.

Consider the direction from left to right. Suppose that $L \in ETIME$, and that the Turing machine $M = (Q, \Sigma, \Gamma, \delta, q_0, B, q_a, q_r)$ accepts L in exponential time, where

1. $q_0, q_a, q_r \in Q$ are respectively the initial, accepting and rejecting states;
2. $B \notin \Sigma$ is the blank symbol, Σ is the input tape alphabet, and Γ is the work tape alphabet;
3. δ is the transition function for M.

For simplicity, assume that M has only one tape, and that the transition function

$$\delta : (Q - \{q_a, q_r\}) \times (\Sigma \cup \{B\}) \rightarrow Q \times (\Sigma \cup \{B\}) \times \{-1, 0, 1\}.$$

W.l.o.g. we take $Q, \Sigma, \Gamma \subset \mathbf{N}$, $B \in \mathbf{N}$, and that $q_0 = 0$. Since M is fixed, using case statements, we can define a simple bijection τ_3 mapping $Q \times (\Sigma \cup \{B\}) \times \{-1, 0, 1\}$ onto a finite set $A = \{0, \ldots, k_M\}$ of integers. Denote the inverse projection functions corresponding to τ_3 by Π_1^3, Π_2^3, and Π_3^3. Clearly the transition function δ, as well as τ_3 and its inverse projections can be defined by safe composition from $C(; a, b, c), Pr(; a), S(; a), 0(;)$, and we have

$$\tau_3(; \Pi_1^3(; d), \Pi_2^3(; d), \Pi_3^3(; d)) = d$$

for all $d \leq k_M$. We may assume that δ satisfies

$$\delta(; q, \sigma) = \tau_3(; newstate(; q, \sigma), newsymbol(; q, \sigma), direction(; q, \sigma))$$

for suitable functions $newstate, newsymbol, direction$ defined by

$$newstate(; q, \sigma) = \Pi_1^3(\delta(; q, \sigma))$$
$$newsymbol(; q, \sigma) = \Pi_2^3(\delta(; q, \sigma))$$
$$direction(; q, \sigma) = \Pi_3^3(\delta(; q, \sigma)).$$

The initial configuration of M on input $x = x_1 \cdots x_n$ is as follows:

$$\boxed{B} \boxed{x_1} \ldots \boxed{x_n} \boxed{B}$$

where M starts in state q_0 on the leftmost tape square. W.l.o.g. we may assume that the head movements of M on any input are of the following form (if not, then an equivalent Turing machine M' with these head movements can be given which runs in time $2 \cdot \sum_{i=0}^{2^{cn}} i \leq 2^{dn}$).

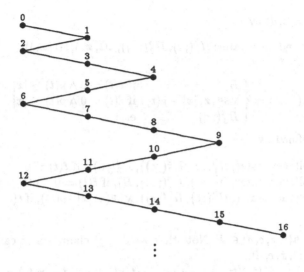

$$\vdots$$

The idea is to describe a function for the arithmetization of the computation of M on input x. To this end, define $g(n) = 2 \cdot \sum_{i \leq n} i$, and $\nu(t) = \max n \leq t[g(n) \leq t]$.

Note that $\nu(t)$ can alternately be defined as $\min n \leq t[g(n+1) > t]$. Define $p(t) = \min(t - g(\nu(t)), g(\nu(t)+1) - t)$ and note that $p(t)$ gives the position of M's head at time t. Define

$$\ell(t) = \begin{cases} 0 & \text{if } (\forall i < t)[p(i) \neq p(t)] \\ 1 + \max i < t[p(i) = p(t)] & \text{else} \end{cases}$$

and note that $\ell(t)$ is 0 if at time t, M's head is on a tape square never visited before, while otherwise $\ell(t)$ is the most recent time $t' < t$ for which M's head visited the same tape square.

Using SVR, we will define

$$f(t, x;) = \tau_3(; s(t, x;), a(t, x;), b(t, x;)) \tag{1}$$

where $s(t, x;)$ is the state of M at time t <u>before</u> invoking δ, and $a(t, x;)$ is the symbol under M's head at time t <u>before</u> invoking δ, and $b(t, x;)$ is the symbol which M writes upon invoking δ at time t ; i.e.

$$b(t, x;) = \text{newsymbol}(; s(t, x;), a(t, x;)).$$

To show that functions $a, b \in \mathcal{F}$, we introduce some auxiliary functions. Define $\text{MSP}(x, y) = \lfloor x/2^y \rfloor$, $|x| = \lceil \log_2 x + 1 \rceil$ (length of binary representation of x), and recall that p (position), ℓ (last time), and $M(x, y;) = x \dot- y$ have already been defined. Functions $\ell, p, |x|, \text{MSP}, \dot-$ and relation $\leq$ are representable in $\mathcal{F}$ via functions having only normal variables (by applying Bellantoni's theorem, since these functions are in linear space).

Define $h(t, x; c, d)$ by

$$\tau_3(; \text{newstate}(; \Pi_1^3(; c), \Pi_2^3(; c)), a(t, x;), b(t, x;))$$

where

$$a(t, x;) = \begin{cases} B & \text{if } \ell(t) = 0 \wedge p(t) > |x| \\ \text{MSP}(x, |x| - p(t)) & \text{if } \ell(t) = 0 \wedge p(t) \le |x| \\ \Pi_3^3(; d) & \text{else} \end{cases}$$

and $b(t, x;)$ defined by

1. $\text{newsymbol}(; \text{newstate}(; \Pi_1^3(; c), \Pi_2^3(; c)), \Pi_3^3(; d))$, if $\ell(t) \ne 0$
2. $\text{newsymbol}(; \text{newstate}(; \Pi_1^3(; c), \Pi_2^3(; c)), B)$, if $\ell(t) = 0 \wedge p(t) > |x|$
3. $\text{newsymbol}(; \text{newstate}(; \Pi_1^3(; c), \Pi_2^3(; c)), \text{MSP}(x, |x| \dot- p(t)))$, if $\ell(t) = 0 \wedge p(t) \le |x|$

It follows that $h(t, x; c, d) \in \mathcal{F}$. Note that we do <u>not</u> claim that h can be written in the form $h(; t, x, c, d)$.

Define $r(t;) = M(t, \ell(t;);)$, so that $M(t, r(t;);) = t \dot- r(t;) = \ell(t;)$ Using SVR define

$$f(0, x;) = \tau_3(; 0, B, \text{nextsymbol}(; 0, B))$$
$$f(t + 1, x;) = h(t, x; f(t, x;), f(M(t, r(t;);), x;)) = h(t, x; f(t, x;), f(\ell(t;), x;))$$

By induction on t, it is easy to verify that $f(t, x;)$ satisfies Equation (1). We may suppose that the Turing machine M terminates its computation in $q(x) \simeq 2^{c \cdot |x|}$ steps, where q is a polynomial. Polynomials are clearly computable in linear space, so by Bellantoni's theorem $q \in \mathcal{F}$. It follows that M accepts input x iff

$$q_a = \Pi_1^3(; f(q(x;), x;)).$$

Using a simple case statement, we obtain that the characteristic function of L belongs to $\mathcal{F}$.

Now if $g \in$ ETIME and $|g(x)| = O(|x|)$, then using standard techniques one can define g in $\mathcal{F}$. This concludes the direction from left to right.

Now consider the direction from right to left. We show by induction on term formation in the function algebra $\mathcal{F}$ that every function in $\mathcal{F}$ has linear growth rate and is computable in $ETIME$. The following lemma is a trivial modification of [1].

Lemma 7. *For each $f(\mathbf{x}; \mathbf{a}) \in \mathcal{F}$ there is a monotone increasing polynomial p_f such that for all $\mathbf{x}, \mathbf{a}$*

$$f(\mathbf{x}; \mathbf{a}) \le p_f(\mathbf{x}) + \max(\mathbf{a}).$$

Proof. By induction on term formation in algebra $\mathcal{F}$. If f is any of $0, I_i^{n,m}, S, Pr, C$, then $f(\mathbf{x}; \mathbf{a}) \le \sum_{i=1}^{n} x_i + \max(\mathbf{a})$. Suppose that f is defined by SCOMP from $g, u_1, \ldots, u_m, v_1, \ldots, v_n$

$$f(\mathbf{x}; \mathbf{a}) = g(u_1(\mathbf{x};), \ldots, u_m(\mathbf{x};); v_1(\mathbf{x}; \mathbf{a}), \ldots, v_n(\mathbf{x}; \mathbf{a})).$$

By the induction hypothesis, assume that there exist polynomials p_g, p_{u_i}, p_{v_i} such that

$$g(y_1, \ldots, y_m; b_1, \ldots, b_n) \leq p_g(y_1, \ldots, y_m) + \max(b_1, \ldots, b_n)$$

$$u_i(\mathbf{x};) \leq p_{u_i}(\mathbf{x})$$
$$v_i(\mathbf{x}; \mathbf{a}) \leq p_{v_i}(\mathbf{x}) + \max(\mathbf{a}).$$

Then

$$
\begin{aligned}
f(\mathbf{x}; \mathbf{a}) &\leq p_g(u_1(\mathbf{x};), \ldots, u_m(\mathbf{x};)) + \max(v_1(\mathbf{x}; \mathbf{a}), \ldots, v_n(\mathbf{x}; \mathbf{a})) \\
&\leq p_g(p_{u_1}(\mathbf{x}), \ldots, p_{u_m}(\mathbf{x})) + \max_i(p_{v_i}(\mathbf{x}) + \max(\mathbf{a})) \\
&\leq p_g(p_{u_1}(\mathbf{x}), \ldots, p_{u_m}(\mathbf{x})) + p_{v_1}(\mathbf{x}) + \ldots + p_{v_n}(\mathbf{x}) + \max(\mathbf{a}).
\end{aligned}
$$

If we define

$$p_f(\mathbf{x}) = p_g(p_{u_1}(\mathbf{x}), \ldots, p_{u_m}(\mathbf{x})) + \sum_{i=1}^{n} p_{v_i}(\mathbf{x})$$

then we have

$$f(\mathbf{x}; \mathbf{a}) \leq p_f(\mathbf{x}) + \max(\mathbf{a}).$$

Now suppose that f is defined by SVR from g, h, r so that

$$f(0, \mathbf{x}; \mathbf{a}) = g(\mathbf{x}; \mathbf{a})$$

$$f(t + 1, \mathbf{x}; \mathbf{a}) = h(t, \mathbf{x}; \mathbf{a}, f(t, \mathbf{x}; \mathbf{a}), f(M(t, r(t, \mathbf{x};)), \mathbf{x}; \mathbf{a})).$$

By the induction hypothesis, assume there exist monotone increasing polynomials p_g, p_h such that

$$g(\mathbf{x}; \mathbf{a}) \leq p_g(\mathbf{x}) + \max(\mathbf{a})$$
$$h(t, \mathbf{x}; a, b, c) \leq p_h(t, \mathbf{x}) + \max(a, b, c).$$

Define

$$p_f(t, \mathbf{x}) = t \cdot p_h(t, \mathbf{x}) + p_g(\mathbf{x}).$$

We show by induction on t that

$$f(t, \mathbf{x}; \mathbf{a}) \leq p_f(t, \mathbf{x}) + \max(\mathbf{a}).$$

Now

$$f(0, \mathbf{x}; \mathbf{a}) = g(\mathbf{x}; \mathbf{a}) \leq p_g(\mathbf{x}) + \max(\mathbf{a}) = p_f(0, \mathbf{x}) + \max(\mathbf{a})$$

so the base of the induction is established. Assume that

$$f(t, \mathbf{x}; \mathbf{a}) \leq p_f(t, \mathbf{x}) + \max(\mathbf{a}).$$

Then

$$f(t+1, \mathbf{x}; \mathbf{a}) = h(t, \mathbf{x}; \mathbf{a}, f(t, \mathbf{x}; \mathbf{a}), f(M(t, r(t, \mathbf{x};);), \mathbf{x}; \mathbf{a}))$$
$$\leq p_h(t, \mathbf{x}) + \max(\mathbf{a}, f(t, \mathbf{x}; \mathbf{a}), f(M(t, r(t, \mathbf{x};);), \mathbf{x}; \mathbf{a}))$$
$$\leq p_h(t, \mathbf{x}) +$$
$$\quad \max(\mathbf{a}, p_f(t, \mathbf{x}) + \max(\mathbf{a}), p_f(M(t, r(t, \mathbf{x};);), \mathbf{x}) + \max(\mathbf{a}))$$
$$\leq p_h(t, \mathbf{x}) + \max(\mathbf{a}) + \max(p_f(t, \mathbf{x}), p_f(M(t, r(t, \mathbf{x};);), \mathbf{x}))$$
$$\leq p_h(t, \mathbf{x}) + \max(\mathbf{a}) + p_f(t, \mathbf{x}) \qquad \text{by monotonicity of } p_f$$
$$\leq p_h(t, \mathbf{x}) + \max(\mathbf{a}) + (t \cdot p_h(t, \mathbf{x}) + p_g(\mathbf{x}))$$
$$\leq (t+1) \cdot p_h(t, \mathbf{x}) + p_g(\mathbf{x}) + \max(\mathbf{a})$$
$$\leq p_f(t, \mathbf{x}) + \max(\mathbf{a}).$$

This establishes the claim when f is defined by SVR from g, h, r. In the case where f is defined by SR from g, h

$$f(t+1, \mathbf{x}; \mathbf{a}) = h(t, \mathbf{x}; \mathbf{a}, f(t, \mathbf{x}; \mathbf{a}))$$

an identical argument establishes that

$$f(t, \mathbf{x}; \mathbf{a}) \leq t \cdot p_h(t, \mathbf{x}) + p_g(\mathbf{x}) + \max(\mathbf{a})$$
$$= p_f(t, \mathbf{x}) + \max(\mathbf{a}).$$

By induction on term formation in the algebra $\mathcal{F}$, the lemma now follows.

Proposition 8. *If* $f(\mathbf{x}; \mathbf{a}) \in \mathcal{F}$ *then* $f \in$ ETIME *and*

$$|f(\mathbf{x}; \mathbf{a})| = O(\max_i(|x_i|), \max_j(|a_j|)).$$

Proof. ¿From the preceding lemma, $f(\mathbf{x}; \mathbf{a}) \leq c \cdot \max(\mathbf{x})^k + \max(\mathbf{a}) + d$, for appropriate k, c, d, so the bound on $|f(\mathbf{x}; \mathbf{a})|$ follows immediately.

We show by induction on term formation that if $f(\mathbf{x}; \mathbf{a}) \in \mathcal{F}$ then $f \in$ ETIME. This is clearly true if f is any of $0, I_i^{n,m}, S, Pr, C$. For functions $g, u, \ldots, u_m, v_1, \ldots, v_n$ of linear growth rate which belong to ETIME, it is clear that

$$f(\mathbf{x}; \mathbf{a}) = g(u_1(\mathbf{x};), \ldots, u_m(\mathbf{x};); v_1(\mathbf{x}; \mathbf{a}), \ldots, v_n(\mathbf{x}; \mathbf{a}))$$

belongs to ETIME. If f is defined from g, h, r by SVR, where g, h, r belong to ETIME and are of linear growth rate, then by standard techniques the function

$$F(t, \mathbf{x}, \mathbf{a}) := \langle f(0, \mathbf{x}; \mathbf{a}), \ldots, f(t-1, \mathbf{x}; \mathbf{a}) \rangle$$

encoding the sequence of f's previous values can be computed in time $t \cdot 2^{c \cdot \max(|x_i|, |a_j|)} + d$ which is bounded by $2^{c' \cdot \max(|t|, |x_i|, |a_j|)} + d'$. The proposition now follows.

This concludes the proof of the theorem.

We end with the following open question whether there is a safe characterization of the class of *rudimentary* (coined *constructive arithmetic* by Smullyan) functions.

Definition 9. Let COMP denote the scheme of ordinary composition, and BMIN to denote the scheme of bounded minimization. The class RF of rudimentary functions is $[0, I, s, +, \times; \text{COMP}, \text{BMIN}]$. An equivalent definition of RF is to replace bounded recursion by bounded minimization in Grzegorczyk's class $\mathcal{E}^2$. (See [4] for more information about such function algebras.)

Definition 10 S. Bellantoni. The function f is defined by *safe minimization* (SMIN) from the function g, denoted $f(\mathbf{x}; \mathbf{b}) = s_1(\mu a[g(\mathbf{x}; a, \mathbf{b}) \bmod 2 = 0)])$, if

$$f(\mathbf{x}; \mathbf{b}) = \begin{cases} s_1(\min\{a : g(\mathbf{x}; a, \mathbf{b}) \equiv 0 \bmod 2\}), & \text{if such exists,} \\ 0 & \text{else.} \end{cases}$$

Question 11. Can one characterize RF by $[0, I, S, Pr, C; \text{SCOMP}, \text{SMIN}]$?

4 Acknowledgements

I would like to thank Karl-Heinz Niggl, and especially Thorsten Altenkirch for comments on this paper. Thorsten Altenkirch pointed out that the pairing function could be defined by cases, since the function values are bounded in the case at hand. This observation is crucial to our proof. Finally, thanks to Irmgard Mignani for typing assistance.

References

1. S. Bellantoni. Predicative recursion and computational complexity. Technical Report 264/92, University of Toronto, Computer Science Department, September 1992. 164 pages.
2. S. Bellantoni and S. Cook. A new recursion-theoretic characterization of the polytime functions. *Computational Complexity*, 2:97–110, 1992.
3. S. Bloch. Function-algebraic characterizations of log and polylog parallel time. *Computational Complexity*, 4(2):175–205, 1994.
4. P. Clote. Computation models and function algebras. In E. Griffor, editor, *Handbook of Recursion Theory*. North Holland, in preparation.
5. B. Monien. A recursive and grammatical characterization of exponential time languages. *Theoretical Computer Science*, 3:61–74, 1977.
6. K. Wagner and G. Wechsung. *Computational Complexity*. Reidel Publishing Co., 1986.

Finite Model Theory, Universal Algebra and Graph Grammars

Bruno Courcelle

Bordeaux-1 University, LaBRI 351 Cours de la Liberation, 33405 Talence, France
courcell@labri.u-bordeaux.fr
http://dept-info.labri.u-bordeaux.fr/~courcell/courcell.html

There are two main ways to specify a set of finite graphs. The first one consists in using a characteristic property, the second consists in describing graphs as combinations of smaller graphs of the same set or of auxilliary ones. In order to prove general results (comparisons of defining techniques, complexity of membership algorithms), one need formal definitions. We shall consider the following possibilities.

Logic: it can be used to express graph properties, because a graph can be defined quite naturally as a logical structure (by a graph, we mean an isomorphism class of "concrete" graphs; no logical formula can distinguish two isomorphic concrete graphs). The set of finite models of a given formula (written with the appropriate logical symbols) is the set defined by this formula. The classical logical languages, first-order logic, monadic second-order logic, second-order logic (and their sub-hierarchies based on restrictions of quantifications) yield hierarchies of graph properties. These hierarchies are linked to complexity classes, and some conjectures of finite model theory are equivalent to hard conjectures in complexity theory.

Universal Algebra: it is the appropriate tool for handling hierachical descriptions of graphs. If we equip the set of (finite) graphs with operations that, given two graphs produce a larger one by gluing them or connecting them in some precise way, then we make the set of graphs into an algebra. The so-called *equational subsets* of this algebra, i.e., the least solutions of certain systems of equations that define sets by mutual recursion (with the help of set union and extensions to sets of the gluing operations) provide the second main formal method of description of sets of graphs. The equational sets of graphs extend the classical context-free languages (which can be viewed as the equational subsets of free monoids). They form the basis of the theory of *graph grammars*.

Universal Algebra provides us also with the notion of a recognizable set, i.e., a set which is the union of classes of a congruence having finitely many classes. The *recognizable sets of graphs* extend the regular languages. (There are still other definitions of classes of graphs, based on rewriting techniques, that can be also formulated in general algebraic terms but they are outside the scope of this presentation.)

Among the logical languages useful to describe sets of graphs, and in view of applications to the theory of graph grammars, we think that *monadic second-order logic* and some close variants of it are of special interest. The reason is that the sets of graphs defined by monadic second-order formulas (for example the set of planar graphs) have a remarkable algebraic structure: *they are recognizable.* Many interesting results follow from this fact. Apart from them, it is interesting because it relates two very different ways of viewing graphs: as logical stuctures on the one hand and as algebraic objects on the other. What is behind is nothing but a classical theorem in first-order logic by Feferman and Vaught extended by Shelah to monadic second-order logic. The conceptual framework is thus clear.

In the lecture, we shall review the following notions: graph operations, equational and recognizable sets, monadic second-order expressibility of graph properties using quantification over sets of edges. We shall review the main results relating three classes of graphs (equational, recognizable, and monadic second-order definable sets), and their applications to graph grammars and graph algorithms. We shall then discuss the extension to graphs of the theorem by Elgot and Büchi stating that a set of words is recognizable if and only if it is monadic second-order definable. It is conjectured that if a set of graphs of *bounded treewidth* is recognizable, then it is definable, and this conjecture has been proved in some special cases. Other results and conjectures regarding monadic second-order logic without edge quantifications will be presented.

References

1. S. Arnborg, J. Lagergren, D. Seese. Easy problems for tree decomposable graphs. *J. of Algorithms* 12:308-340, 1991.
2. B. Courcelle. The monadic second-order logic of graphs I: Recognizable sets of finite graphs. *Information and Computation* 85:12-75, 1990
3. B. Courcelle. The monadic second-order logic of graphs V: On closing the gap beween definability and recognizability. *Theoret. Comput. Sci.* 80:153-202, 1991
4. B. Courcelle. The monadic second-order logic of graphs VI: On several representations of graphs by relational structures *Discrete Applied Mathematics* 54:117-149, 1994
5. B. Courcelle. The monadic second order logic of graphs X : Linear orders. *Theoret. Comput. Sci.* 160:87-143, 1996.
6. B. Courcelle. Basic notions of universal algebra; applications to formal language theory and graph grammars. *Theoret. Comput. Sci.* 163:1-54, 1996
7. B. Courcelle. The expression of graph properties and graph transformations in monadic-second-order logic. In *Handbook of graph transformations, volume 1: Foundations*, G.Rozenberg ed., 1997, in press.
8. B. Courcelle. On the expression of graph properties in some fragments of monadic-second-order logic. In *Descriptive complexity and finite models*, N. Immerman and Ph. Kolaitis eds., Contemporary Mathematics, AMS, 1997, in press.
9. B. Courcelle, M. Mosbah. Monadic second-order evaluations on tree-decomposable graphs. *Theoret. Comput. Sci.* 109:49-82, 1993
10. R. Fagin. Generalized first-order spectra and polynomial time recognizable sets. In *Complexity of Computation* R. Karp ed., SIAM-AMS proceedings 7:43-73, 1974
11. R. Fagin. Finite model theory - a personal perspective. *Theoret. Comput. Sci.* 116:3-31, 1993

12. H. Gaifman. On local and non local properties. *Logic Colloquium'81*, J. Stern ed., North-Holland, pp. 105-135, 1982
13. N. Immerman. Languages that capture complexity classes. *SIAM J. Comput.* 16:760-778, 1987
14. D. Kaller. Definability equals recognizability of partial 3-trees. newblock Proceedings of WG'96, Como, Italy, *Lecture Notes in Computer Science*. 1997, to appear.
15. V. Kabanets. Recognizability equals definability for partial k-paths, 1996, Submitted
16. J. Makowsky Model theory and computer science: an appetizer. Chapter I.6 of *Handbook of Logic in Computer Science, Vol. 1*, S. Abramsky et al. eds., Oxford University Press, 1992.
17. J. Makowsky, Y. Pnueli, Arity vs. Alternation in second-order logic. *Logical Foundations of Computer Science*, A. Nerode, Y. Matiyasevich eds., Lecture Notes in Computer Science 813:240-252, 1994
18. D. Seese The structure of the models of decidable monadic theories of graphs. *Ann. Pure Applied Logic* 53:169-195, 1991
19. L. Stockmeyer The polynomial-time hierarchy. *Theoret. Comput. Sci.* 3:1-22, 1977

Complexity of Query Answering in Logic Databases with Complex Values

Evgeny Dantsin[1]* and Andrei Voronkov[2]**

[1] Steklov Institute of Mathematics at St.Petersburg
Fontanka 27, St.Petersburg 191011, Russia
[2] Computing Science Department, Uppsala University
Box 311, S-751 05 Uppsala, Sweden

Abstract. This paper studies the complexity of the query answering problem in logic databases with complex values. Logic databases are represented as Horn clause logic programs; complex values are described in a data model based on equational theories. As examples of complex values, we consider trees, bags and finite sets. We give a natural sufficient condition under which the query answering problem for non-recursive programs with complex values is in NEXP. In particular, logic programs with trees, bags and finite sets satisfy this condition. We also show that the query answering problem for non-recursive range restricted logic programs is NEXP-hard. Thereby, the query answering problem for logic databases with trees, bags and sets turns out to be NEXP-complete.

1 Introduction

The relational model of data deals with simple values, namely tuples consisting of atomic components. Various generalizations and formalisms have been proposed to handle more complex values like nested tuples, tuples of sets, etc. For example, Kuper [14] introduced logic databases with sets. A thorough discussion of such formalisms and their expressive power may be found in Abiteboul and Beeri [2]. Our paper studies complexity aspects of databases with complex values.

There are three main kinds of complexity in databases: the *data complexity*, the *program complexity* (both defined in Vardi [22]) and the *combined complexity* (see e.g. the survey [6]). The data complexity means that data (represented by the facts of a logic programs) are fixed, while a query (represented by the rules of a logic program and a goal) is variable. The program complexity (sometimes called the *expression complexity*) means that data are variable, while a query is fixed. The combined complexity means that both data and query are variable.

By the *query answering complexity* we mean the combined complexity of checking whether a given goal succeeds with respect to a given logic program. Imagine a database user asking a query. In the case of complex values, an answer

---

* Supported by grants from the Swedish Royal Academy of Sciences, the Swedish Institute and RFBR/INTAS.
** Supported by a TFR grant.

can be too large or even infinite. The query answering complexity measures how long the user should wait until the fact of the mere existence of a positive answer is established.

The query answering problem can be naturally formalized as the decision problem for fragments of first-order logic corresponding to Horn clause logic programs. Decidability and complexity of fragments of predicate calculus have been actively studied (see e.g. the recent monograph [5]). As for logic programming, the decidability and complexity have been basically considered in the context of non-monotonic semantics of propositional programs or programs without function symbols. Surprisingly, very little is known about the complexity of query languages for complex values.

This paper contains some complexity bounds for logic programs with complex values, including trees, bags and sets. Such results are useful for the design of future database query languages able to manage complex values. To the best of our knowledge, this is the first analysis of the complexity for databases with complex values. So far, the complexity was mainly analyzed for constraint satisfiability.

This paper is structured as follows. In Section 2 we introduce our data model and logic programs with complex values. We also give examples of complex values. In Section 3 we prove NEXP-hardness of the query answering problem for non-recursive range restricted logic programs. In Section 4 we consider complex values described by theories satisfying some very general assumption, called NP-solvability. We show that the query answering problem for logic programs with such complex values is in NEXP. In Section 5 we use NP-solvability for trees, bags and sets to obtain NEXP-completeness for a number of classes of non-recursive databases with complex values.

2 Preliminaries

2.1 Data model

In order to define what we mean by *complex values*, we shall introduce the corresponding data model. Since we consider the logical model of computation, databases are represented as logic programs. In logic programs, values are represented by ground terms, and two values are equal when the corresponding terms coincide. In our data model, values are also represented by ground terms, but the interpretation of equality can be arbitrary. To formalize this, we use the notion of *Herbrand interpretations with respect to equational theories*. Thus, our notion of logic databases with complex values is more general than *well-behaved equational logic programs* by Gallier and Raatz [10] but in some sense less general than that of constraint logic programs or constraint databases (see Jaffar and Maher [11]). The generalization of our results to constraint logic programs can also be done, but the notion of data becomes less transparent.

Herbrand interpretations. Let Σ be a signature, i.e. a set of function and predicate symbols containing at least one constant; constants are considered to be function symbols of arity 0. Terms, atoms, substitutions, instances, variants etc. are defined in the standard way (see, for example, Apt [3]). The *Herbrand universe* U_Σ for Σ is the set of all *ground terms over* Σ, that is terms without variables. The *Herbrand base* B_Σ for Σ is the set of all ground atoms of Σ. A *Herbrand interpretation for* Σ is any subset of B_Σ (all atoms in this set are considered to be true, all other ground atoms are considered to be false).

Equational theories. By an *equation* over Σ we mean an atom $s = t$, where the equality symbol $=$ does *not* belong to Σ. An equation is said to be *ground* if s and t are ground. An *equational theory* E over Σ is a set of ground equations closed under the logical consequence relation, i.e. such that

1. E contains all ground equations of the form $t = t$;
2. if E contains $s = t$ then E contains $t = s$;
3. if E contains $r = s$ and $s = t$ then E contains $r = t$;
4. if E contains $s_1 = t_1, \ldots, s_n = t_n$ then E contains $f(s_1, \ldots, s_n) = f(t_1, \ldots, t_n)$
   for each n-ary function symbol $f \in \Sigma$.

The solvability problem for equational theories. We say that an equation $s = t$ has a *solution* σ in an equational theory E if σ is a substitution such that the equation $s\sigma = t\sigma$ belongs to E (where $t\sigma$ and $s\sigma$ denote the results of application of σ to s and t respectively). A *system of equations* is a finite set of equations. A system of equations is said to be *solvable* in E if there exists a substitution σ such that σ is a solution for each equation in the system. The *solvability problem* for an equational theory E is the following decision problem: given a system of equations, is the system solvable in E? We call an equational theory *NP-solvable* if the solvability problem for E is in NP. Similar definition can be given for other complexity classes (for example, P-solvable equational theories), but in this paper we consider only NP-solvable theories.

Herbrand interpretations with respect to equational theories. By a *Herbrand interpretation with respect to an equational theory* E we mean a Herbrand interpretation $\mathcal{I} \subseteq B_\Sigma$ such that if (i) an atom $p(s_1, \ldots, s_n)$ belongs to $\mathcal{I}$ and (ii) the equational theory E contains equations $s_1 = t_1, \ldots, s_n = t_n$, then $p(t_1, \ldots, t_n)$ belongs to $\mathcal{I}$ too. This notion can be viewed as a formalization of our data model.

Examples of complex values. *Trees* are data represented by ground terms, and the equality relation on them is defined as identity. Thus, trees correspond to equational theories that consist of equations $t = t$.

Lists are a special kind of trees. In logic programming, lists are constructed from the empty list [] by using *the list constructor* [s | t] whose second argument t is a list. The list $[t_1 \mid \ldots [t_n \mid []] \ldots]$ is denoted by $[t_1, \ldots, t_n]$. The equational theory for lists contains no additional axioms in comparison with trees.

Bags (or *finite multisets* in another terminology) are lists in which the order of elements is immaterial. We denote bags similar to lists, but using the empty bag $\langle\rangle$ instead of the empty list $[]$, *the bag constructor* $\langle s \mid t \rangle$ instead of the list constructor $[s \mid t]$ and $\langle t_1, \ldots, t_n \rangle$ instead of $[t_1, \ldots, t_n]$. Bags can be axiomatized by ground equations $\langle r \mid \langle s \mid t \rangle \rangle = \langle s \mid \langle r \mid t \rangle \rangle$, or alternatively by ground equations $\langle t_1, \ldots, t_m, t_{m+1}, \ldots, t_n \rangle = \langle t_1, \ldots, t_{m+1}, t_m, \ldots, t_n \rangle$.

Sets can be viewed as bags in which the number of occurrences of an element in immaterial. Similar to bags, we shall use the *set constructor* $\{s \mid t\}$ denoting the set obtained from the set t by adding s as an element. Sets can be axiomatized as follows. In the axioms for bags we replace the bag constructor by the set constructor and add all ground equations $\{s_1, s_1, \ldots, s_n\} = \{s_1, \ldots, s_n\}$. Note that we consider *hereditarily finite sets,* i.e. finite sets whose elements can be other (hereditarily finite) sets.

There are other ways of integrating finite sets into logic programming. For example, Dovier et.al. [8] represent sets by terms $\{s_1, \ldots, s_n \mid t\}$ where t can be any term not representing a set (the authors call such objects *colored sets*). Thus, a set is characterized not only by its elements but also by some "type". Our approach is simpler and more traditional: every tail subterm t in $\{s_1, \ldots, s_n \mid t\}$ is required to represent some set.

2.2 Logic programs with complex values

We fix a signature Σ and assume that all definitions are given with respect to Σ. A *clause* is any formula of the form $A_1 \wedge \ldots \wedge A_n \supset A$, where $A_1, \ldots, A_n, A$ are atoms and $n \geq 0$. Such a clause is usually written as

$$A :- A_1, \ldots, A_n.$$

and simply as A when $n = 0$. The atom A is called the *head*, and the sequence $A_1, \ldots, A_n$ is called the *body* of this clause. By the *length* of the body we mean the number n.

A *logic program* is a finite set of clauses in the signature Σ. A *goal* is a set $\{A_1, \ldots, A_n\}$ of atoms, where $n \geq 0$. When $n = 0$, it is called the *empty goal* and denoted by $\square$.

By a *logic program with complex values* we mean a logic program $\mathcal{P}$ over an *equational theory* E, i.e. in fact a pair $(\mathcal{P}, E)$. The semantics of such programs can be given in many ways by adapting standard logic programming semantics [3, 16, 10] for the corresponding equational theory. For example, one can define the least fixpoint semantics or a generalization of SLD-resolution. We define the semantics by using the notion of the least Herbrand model.

Let $A :- A_1, \ldots, A_n$ be a clause without variables. This clause is said to be *true* in a Herbrand interpretation $\mathcal{I}$ if A belongs to $\mathcal{I}$ whenever $A_1, \ldots, A_n$ belong to $\mathcal{I}$. We say that $\mathcal{I}$ is a *Herbrand model* for a logic program $\mathcal{P}$ if all ground instances of all clauses of $\mathcal{P}$ are true in $\mathcal{I}$. If, in addition, $\mathcal{I}$ is a Herbrand interpretation with respect to an equational theory E then we call $\mathcal{I}$ a *Herbrand model for* $\mathcal{P}$ *over* E.

Like the case of ordinary logic programs, Herbrand models for logic programs over equational theories have the following property: the intersection of Herbrand models for $\mathcal{P}$ over E is again a Herbrand model for $\mathcal{P}$ over E. This property allows us to define the *least Herbrand model* for a logic program with complex values: it is the smallest Herbrand model for $\mathcal{P}$ over E (with respect to the subset relation $\subseteq$).

We say that a logic program $\mathcal{P}$ over an equational theory E is a *logic program with sets* if E is the equational theory of sets. Similarly, we use the terms *logic programs with bags, trees* etc. To illustrate this notions, we shall define a simple logic program over sets and bags (at the same time), which defines the predicate set_to_bag. This predicate can be used to transform a set into a corresponding bag and conversely, bag into the corresponding set.

```
set_to_bag({},⟨⟩).
set_to_bag({x | y},⟨x | z⟩) :-
  set_to_bag(y,z).
```

We give no other examples of such programs here, there is a vast literature on this subject (e.g. [12, 14, 1, 4, 18, 20, 21, 2, 23]).

We introduce the following requirement concerning signatures and equational theories of bags and sets. Let E be an equational theories over a signature Σ. If Σ contains the bag constructor $⟨\ldots \mid \ldots⟩$, then Σ contains the constant $⟨⟩$ and E is required to include the axioms for bags. Similarly, if E contains the set constructor $\{\ldots \mid \ldots\}$, then Σ contains the constant $\{\}$ and E is required to include the axioms for sets.

Special kinds of logic programs. For every logic program $\mathcal{P}$, we define a directed graph $G_{\mathcal{P}}$ as follows. The set of vertices of $G_{\mathcal{P}}$ is the set of predicates symbols of $\mathcal{P}$; there is an edge from a vertex p_2 to a vertex p_1 if and only if there is a clause $A :- A_1, \ldots, A_n$ in $\mathcal{P}$ such that p_1 is the predicate symbol of A and p_2 occurs in $A_1, \ldots, A_n$. A logic program $\mathcal{P}$ is said to be *non-recursive* if the graph $G_{\mathcal{P}}$ contains no cycles.

A logic program $\mathcal{P}$ is called *range restricted* if every clause $A :- A_1, \ldots, A_n$ in $\mathcal{P}$ has the property: all variables of A appear in $A_1, \ldots, A_n$.

Range restricted logic programs have many desirable features, mainly related to *domain independence*. Range restricted non-recursive logic programs containing only constants and predicate symbols are equivalent to positive relational algebra. These classes of logic programs have been actively studied in database theory.

2.3 Query answering complexity

The SUCCESS problem. Let $\mathcal{P}$ be a logic program over an equational theory E and $\mathcal{G}$ be a goal. We say that $\mathcal{G}$ *succeeds with respect to* $\mathcal{P}$ *over* E if there exists a ground instance $\{A_1, \ldots, A_n\}$ of $\mathcal{G}$ such that $A_1, \ldots, A_n$ belong to the least Herbrand model of $\mathcal{P}$ over E. Let an equational theory E be fixed and $\mathfrak{C}$ be a set of logic programs (over E). By the *SUCCESS problem for* $\mathfrak{C}$ we mean the following decision problem: given a program $\mathcal{P} \in \mathfrak{C}$ and a goal $\mathcal{G}$, does $\mathcal{G}$ succeed with respect to $\mathcal{P}$ over E?

The complexity with respect to the combined size of the program and the goal is equivalent to the complexity with respect to the size of the program (as for polynomial-time reduction). Let $\mathcal{P}$ be a program, $\{A_1, \ldots, A_n\}$ be a goal and *yes* be a nullary predicate symbol foreign to Σ. Consider the program $\mathcal{P}' = \mathcal{P} \cup \{yes :- A_1, \ldots, A_n\}$. Evidently, $\mathcal{G}$ succeeds with respect to $\mathcal{P}$ if and only if the goal consisting of *yes* succeeds with respect to $\mathcal{P}'$. This transformation of $\mathcal{P}$ into $\mathcal{P}'$ preserves the classes $\mathfrak{C}$ of logic programs considered in this paper (non-recursive and/or range restricted programs).

3 NEXP-hardness for non-recursive range restricted logic programs

Theorem 1 *Let $\mathfrak{C}$ be the set of non-recursive range restricted logic programs with lists. Then the SUCCESS problem for $\mathfrak{C}$ is NEXP-hard.*

Proof. We reduce the TILING problem, which is known to be NEXP-complete (see e.g. [19], page 501), to the SUCCESS problem for $\mathfrak{C}$.

Informally, TILING is the problem of tiling the square of size $2^n \times 2^n$ by tiles (squares of size 1×1). More precisely, there is a finite set $\{t_0, \ldots, t_k\}$ of *tiles* and there are two binary relations on and to defined on the tiles. Tiles t_i and t_j are said to be *vertically compatible* if $on(t_i, t_j)$ holds and, similarly, *horizontally compatible* if $to(t_i, t_j)$ holds. A *tiling of the square of size* $2^n \times 2^n$ is a function f from $\{1, \ldots, 2^n\} \times \{1, \ldots, 2^n\}$ into $\{t_0, \ldots, t_k\}$ such that

1. $f(i,j)$ and $f(i+1,j)$ are vertically compatible, for all $1 \leq i < 2^n$ and $1 \leq j \leq 2^n$;
2. $f(i,j)$ and $f(i,j+1)$ are horizontally compatible, for all $1 \leq i \leq 2^n$ and $1 \leq j < 2^n$.

We also say that such f is a tiling *with t_i at the top left corner* if $f(1,1)$ is t_i. The TILING problem is defined as follows. Suppose that we are given a set $\{t_0, \ldots, t_k\}$ of tiles, compatibility relations on and to, and a number n (written in unary notation). The question we ask is whether there is a tiling of the square of size $2^n \times 2^n$ with t_0 at the top left corner.

The reduction we describe is a polynomial-time algorithm that transforms every instance I of the TILING problem into a logic program $\mathcal{P}$ and a goal $\mathcal{G}$ such that I has a tiling if and only if $\mathcal{G}$ succeeds with respect to $\mathcal{P}$. We think of tiles $t_0, \ldots, t_k$ in I as constants of the program $\mathcal{P}$. The compatibility relations on and to are represented in $\mathcal{P}$ by the corresponding predicates: namely, $\mathcal{P}$ contains clauses

$on(t_i, t_j)$.
$to(t_l, t_m)$.

for all pairs of compatible tiles.

To proceed the construction of $\mathcal{P}$ and $\mathcal{G}$, we generalize tiles in a natural way. A *hypertile of rank i* is defined by induction as a square of size $2^i \times 2^i$ consisting of tiles. A hypertile of rank 0 is any tile of the set $\{t_0, \ldots, t_k\}$. Let H_1, H_2, H_3 and H_4 be hypertiles of rank i, then the quadruple (H_1, H_2, H_3, H_4) is a hypertile of rank $i + 1$. We think of this hypertile as the square

H_1	H_2
H_3	H_4

and we represent it in $\mathcal{P}$ by the list $[H_1, H_2, H_3, H_4]$ where H_1, H_2, H_3 and H_4 are lists that represent the corresponding hypertiles of rank i. Here is an example of a hypertile of rank 2 (formed from tiles a, b and c) and its representation by a list:

b	a	c	a
c	a	b	b
c	a	a	a
b	b	a	c

$[[b,a,c,a],[c,a,b,b],[c,a,b,b],[a,a,a,c]]$

Clearly, we can identify hypertiles of rank i with functions from $\{1, \ldots, 2^i\} \times \{1, \ldots, 2^i\}$ into $\{t_0, \ldots, t_k\}$. We call a hypertile a *tiling* if the corresponding function is a tiling.

In addition to the predicates on and to, the program $\mathcal{P}$ defines n binary predicates $tiling_1, \ldots tiling_n$.

A statement $tiling_i(H, T_0)$ is intended to hold if and only if (i) H is a hypertile of rank i; (ii) H is a tiling with the tile T_0 at the top left corner. Thus, the goal $\mathcal{G}$ consisting of one atom $tiling_n(H, t_0)$ will succeed with respect to $\mathcal{P}$ if and only if there is a tiling for given $\{t_0, \ldots, t_k\}$, on, to and n.

The predicate $tiling_1$ is defined in the following obvious way:

X_1	X_2
X_3	X_4

$tiling_1([X_1, X_2, X_3, X_4], X_1)$:-
   $on(X_1, X_3)$,
   $on(X_2, X_4)$,
   $to(X_1, X_2)$,
   $to(X_3, X_4)$.

To define $tiling_{i+1}$, we use the following observation. Consider a hypertile H

$$[[X_1, X_2, X_3, X_4], [Y_1, Y_2, Y_3, Y_4], [Z_1, Z_2, Z_3, Z_4], [U_1, U_2, U_3, U_4]]$$

of rank $i + 1$ and 9 its subsquares, hypertiles of rank i, shown below

X_1	X_2		X_2	Y_1		Y_1	Y_2
X_3	X_4		X_4	Y_3		Y_3	Y_4

X_1	X_2	Y_1	Y_2
X_3	X_4	Y_3	Y_4
Z_1	Z_2	U_1	U_2
Z_3	Z_4	U_3	U_4

X_3	X_4		X_4	Y_3		Y_3	Y_4
Z_1	Z_2		Z_2	U_1		U_1	U_2

Z_1	Z_2		Z_2	U_1		U_1	U_2
Z_3	Z_4		Z_4	U_3		U_3	U_4

It is easy to see that H is a tiling if and only if these 9 hypertiles are tilings too. Thus, $\texttt{tiling}_{i+1}$ can be defined as follows:

$\texttt{tiling}_{i+1}([[X_1, X_2, X_3, X_4], [Y_1, Y_2, Y_3, Y_4],$
$\qquad\qquad [Z_1, Z_2, Z_3, Z_4], [U_1, U_2, U_3, U_4]], T_0)$:-
$\texttt{tiling}_i([X_1, X_2, X_3, X_4], T_0),$
$\texttt{tiling}_i([X_2, Y_1, X_4, Y_3], \_),$
$\texttt{tiling}_i([Y_1, Y_2, Y_3, Y_4], \_),$
$\texttt{tiling}_i([X_3, X_4, Z_1, Z_2], \_),$
$\texttt{tiling}_i([X_4, Y_3, Z_2, U_1], \_),$
$\texttt{tiling}_i([Y_3, Y_4, U_1, U_2], \_),$
$\texttt{tiling}_i([Z_1, Z_2, Z_3, Z_4], \_),$
$\texttt{tiling}_i([Z_2, U_1, Z_4, U_3], \_),$
$\texttt{tiling}_i([U_1, U_2, U_3, U_4], \_).$

Note that the program $\mathcal{P}$ described above is range restricted. Clearly, $\mathcal{P}$ and the goal $\mathcal{G}$ are constructed from the instance I of TILING in polynomial time.

Theorem 2 *The SUCCESS problem for the following classes of non-recursive range restricted logic programs is NEXP-hard:*

1. *logic programs with trees in any signature with at least one function symbol of arity ≥ 2;*
2. *logic programs with bags;*
3. *logic programs with sets;*
4. *logic programs with any combination of the above data (for example, logic programs with trees and bags).*

Proof. The program $\mathcal{P}$ from the proof of Theorem 1 contains the list constructor, but instead we could have used any function symbol of arity ≥ 2 if this symbol is different from the bag and set constructors. Thus, NEXP-hardness holds for non-recursive logic programs with trees.

How to construct $\mathcal{P}$ in the case when Σ contains only the bag constructor and/or the set constructor? We can simulate lists $[s_1, s_2, s_3, s_4]$ by means of bags or sets, i.e. we can use bags and sets to define predicates equivalent to the predicates of $\mathcal{P}$. For example, a list $[s_1, s_2, s_3, s_4]$ can be represented by the set of pairs $(s_1, 1), \ldots, (s_4, 4)$ where $1, \ldots, 4$ are, in turn, represented by sets of cardinalities $1, \ldots, 4$.

4 Inclusion in NEXP

Theorem 3 *Let E be an NP-solvable equational theory. The SUCCESS problem for the set of non-recursive logic programs over E is in NEXP.*

Proof. Let $\mathcal{P}$ be a logic program over E. We define a binary relation $\rightarrow$ on pairs $(\mathcal{G}, \mathcal{E})$, where $\mathcal{G}$ is a goal and $\mathcal{E}$ is a system of equations. We have

$$(\mathcal{G}, \mathcal{E}) \rightarrow (\mathcal{G}', \mathcal{E}')$$

if the following conditions hold:

1. $\mathcal{G}$ contains an atom $p(t_1, \ldots, t_n)$;
2. some clause in $\mathcal{P}$ has a variant $p(s_1, \ldots, s_n)$:- $B_1, \ldots, B_m$ whose variables are disjoint from variables of $(\mathcal{G}, \mathcal{E})$;
3. $\mathcal{G}'$ is $(\mathcal{G} \setminus \{p(t_1, \ldots, t_n)\}) \cup \{B_1, \ldots, B_m\}$;
4. $\mathcal{E}'$ is $\mathcal{E} \cup \{s_1 = t_1, \ldots, s_n = t_n\}$.

By an *SLD-derivation with constraints* we mean a sequence of pairs

$$(\mathcal{G}_1, \mathcal{E}_1), \ldots, (\mathcal{G}_N, \mathcal{E}_N)$$

such that $(\mathcal{G}_i, \mathcal{E}_i) \rightarrow (\mathcal{G}_{i+1}, \mathcal{E}_{i+1})$ holds for all $1 \leq i \leq N - 1$. By the standard technique (see e.g. [17, 9]) one can prove that a goal $\mathcal{G}$ succeeds with respect $\mathcal{P}$ if and only if there is an SLD-derivation with constraints such that (i) $\mathcal{G}_1 = \mathcal{G}$; (ii) $\mathcal{E}_1 = \emptyset$; (iii) $\mathcal{G}_N = \square$; and (iv) $\mathcal{E}_N$ is solvable in E.

It is easy to define a nondeterministic Turing machine M that decides the SUCCESS problem by generating SLD-derivations with constraints for every input $\mathcal{P}$ and $\mathcal{G}$. Namely, this machine nondeterministically makes transitions from $(\mathcal{G}_i, \mathcal{E}_i)$ to $(\mathcal{G}_{i+1}, \mathcal{E}_{i+1})$ (connected by the relation $\rightarrow$). When $\mathcal{G}_i$ is not empty and it is impossible to apply this step, M rejects the input. When $\mathcal{G}_i$ is empty, M nondeterministically checks whether $\mathcal{E}_i$ is solvable in E. If it is solvable then M accepts the input, otherwise the input is rejected. We show that M works within time $2^{n^{o(1)}}$ where n is the combined size of $\mathcal{P}$ and $\mathcal{G}$.

Consider the graph $G_\mathcal{P}$ defined in Subsection 2.2 (recall that we identify its vertices with predicate symbols of $\mathcal{P}$). For every predicate symbol p in $\mathcal{P}$, we define its *height* as the number of edges in the longest path in $G_\mathcal{P}$ coming from p. Since $\mathcal{P}$ is non-recursive, the height is bounded by the number of predicate symbols in $\mathcal{P}$. Let m be the maximum length of clause bodies in $\mathcal{P}$. We define the *weight* of an atom $p(t_1, \ldots, t_k)$ as the number $(m+1)^{h+1}$, where h is the height of p. The weight of a goal is the sum of weights of its atoms.

Consider any SLD-derivation generated by M. Evidently, the weight of the initial goal $\mathcal{G}$ is exponential in n. For each transition $(\mathcal{G}_i, \mathcal{E}_i) \to (\mathcal{G}_{i+1}, \mathcal{E}_{i+1})$, the weight of $\mathcal{G}_{i+1}$ is strictly less than the weight of $\mathcal{G}_i$. Therefore, the total size of all goals and the total size of all systems of equations is exponential in n.

The running time of M consists of the time of generating an SLD-derivation and the time of checking solvability of the final system of equation. Since the size of the SLD-derivation is exponential in n, the time of generating is exponential too. The time of checking solvability is exponential because the final system of equations has the size of the same order as the SLD-derivation, i.e. $2^{n^{O(1)}}$, and the theory E is NP-solvable by the condition of the theorem.

5 NEXP-completeness

Let Σ be a signature containing the bag and set constructors. Let E_Σ be the equational theory over Σ that consists of only the axioms for bags and sets and their logical consequences. We call E_Σ a *combined theory of trees, bags and sets*.

Lemma 1 *Any combined theory of trees, sets and bags is NP-solvable.*

The detailed proof may be found in our technical report [7].

Theorem 4 *The SUCCESS problem for the following classes $\mathfrak{C}$ of non-recursive logic programs is NEXP-complete:*

1. *logic programs with trees in any signature with at least one function symbol of arity ≥ 2;*
2. *logic programs with bags;*
3. *logic programs with sets;*
4. *logic programs with any combination of the above data (for example, logic programs with trees and bags).*

The same holds for the subclass of range restricted programs in $\mathfrak{C}$.

Proof. Immediate from Theorems 2 and 3 and Lemma 1.

References

1. S. Abiteboul and S. Grumbach. A rule-based language with functions and sets. *ACM Transactions on Database Systems*, 16(1):1–30, 1991.
2. S. Abiteboul and C. Beeri. The power of languages for the manipulation of complex values. *VLDB Journal*, 4:727–794, 1995.

3. K.R. Apt. Logic programming. In J. Van Leeuwen, editor, *Handbook of Theoretical Computer Science*, volume B: Formal Methods and Semantics, chapter 10, pages 493–574. Elsevier Science, Amsterdam, 1990.

4. C. Beeri, S. Naqvi, O. Schmueli, and S. Tsur. Set constructors in a logic database language. *Journal of Logic Programming*, 10:181–232, 1991.

5. E. Börger, E. Grädel, and Y. Gurevich. *The Classical Decision Problem*. Springer Verlag, 1997.

6. E. Dantsin, T. Eiter, G. Gottlob, and A. Voronkov. Complexity and expressive power of logic programming. In *CCC'97*, 1997. To appear.

7. E. Dantsin and A. Voronkov. Complexity of query answering in logic databases with complex values. UPMAIL technical report, Uppsala University, Computing Science Department, 1997. http://www.csd.uu.se/papers/reports.html

8. A. Dovier, E.G. Omodeo, E. Pontelli, and G. Rossi. {log}: A language for programming in logic with finite sets. *Journal of Logic Programming*, 28(1):1–44, 1996.

9. M. Gabbrielli, G.M. Dore, and G. Levi. Observable semantics for constraint logic programs. *Journal of Logic and Computation*, 5(2):133–171, 1995.

10. J.H. Gallier and S. Raatz. Extending SLD-resolution to equational Horn clauses using E-unification. *Journal of Logic Programming*, 6(3):3–44, 1989.

11. J. Jaffar and M. Maher. Constraint logic programming: a survey. *Journal of Logic Programming*, 19,20:503–581, 1994.

12. B. Jayaraman and D.A. Plaisted. Programming with equations, subsets and relations. In *Proc. NACLP'89*, Cleveland, 1989. MIT Press.

13. D. Kapur and P. Narendran. NP-completeness of the set unification and matching problems. In J. Siekmann, editor, *Proc. 8th CADE*, volume 230 of *Lecture Notes in Computer Science*, 1986.

14. G.M. Kuper. Logic programming with sets. *Journal of Computer and System Sciences*, 41:44–64, 1990.

15. G.M. Kuper and M.Y. Vardi. The logical data model. *ACM Transactions on Database Systems*, 18(3):379–413, 1993.

16. J.W. Lloyd. *Foundations of Logic Programming (2nd edition)*. Springer Verlag, 1987.

17. M.J. Maher. A CLP view of logic programming. In *Proc. Conf. on Algebraic and Logic Programming*, volume 632 of *Lecture Notes in Computer Science*, pages 364–383, October 1992.

18. T. Munakata. Notes on implementing sets in Prolog. *Communications of the ACM*, 35(3):112–120, 1992.

19. C.H. Papadimitriou. *Computational Complexity*. Addison-Wesley, 1994.

20. O. Shmueli, S. Tsur, and C. Zaniolo. Compilation of set terms in the logic data language (LDL). *Journal of Logic Programming*, 12(1):89–119, 1992.

21. G. Smolka and R. Treinen. Records for logic programming. *Journal of Logic Programming*, 18:229–258, 1994.

22. M. Vardi. The complexity of relational query languages. In *Proc. 14th ACM STOC*, 1982.

23. A. Voronkov. Logic programming with bounded quantifiers revisited. UPMAIL Technical Report, Uppsala University, Computing Science Department, 1997, to appear.

Recognition of Deductive Data Base Stability *

Michael I. Dekhtyar[1] and Alexander Ja. Dikovsky[2]

[1] Dept. of CS, Tver St. Univ. 33 Zheljabova str. Tver, Russia, 170013
dekhtyar@tversu.ac.ru
[2] Keldysh Inst. for Appl. Math. 4 Miusskaya sq. Moscow, Russia, 125047
dikovsky@spp.keldysh.ru

Abstract. A new concept is introduced of globally stable behavior of a deductive data base reacting to stimuli of its active medium. Various problems of integrity constraints restoration after updates fit this general frame. We explore the computational complexity of the problem of stability of a deductive data base in a given DB state with respect to its medium.

1 Introduction

In this work the deductive data bases (*DDBs*) serve as a model of interactive discrete dynamic systems with complex states described by relations over values of multiple parameters. At any given moment the system state is represented by a data base state (*DB state*), i.e. a finite set of extensional facts. Possible actions (moves) of the system in given states can be represented by some binary relation $\vdash$. External stimuli of the active medium of the system (*disturbances*) also change states and can be represented by another binary relation $\xrightarrow{d}$. In our approach the fundamental difference between these two relations is determined by the level of our knowledge about them. The theory of behavior of the system is assumed to be known. This being so, we describe it in the form of a logic program with updates $\mathcal{P}$ over DB states. More specifically, a move from state $\mathcal{E}_1$ to state $\mathcal{E}_2$, i.e. the action $\mathcal{E}_1 \vdash \mathcal{E}_2$ is described as a computation of some predefined goal $:- a$ of $\mathcal{P}$ transforming the first DB state to the second: $\mathcal{E}_1 \vdash_{\mathcal{P}}^{a} \mathcal{E}_2$. As for the disturbances, one can always estimate the set of possible disturbances and evaluate the effect of a committed disturbance on the current state. However *one cannot predict in the given state which disturbance is to be committed*. E.g., in a bank there exists the list of feasible services required by its clients (disturbances) and the corresponding transactions (actions), but one cannot predict exactly which orders for the services emerge at the given moment.

Local behavior of the system in the current state $\mathcal{E}_0$ is described as one interaction with its medium in this state. There are two types of interactions depending on what comes first, a medium disturbance or a system action: $\mathcal{E}_0 \xrightarrow{d}$

---

* This work was sponsored by INTAS (Grant 94-2412) and the Russian Fundamental Studies Foundation (Grants 96-01-00086 and 96-01-00395).

$\mathcal{E}_1^* \vdash_p^a \mathcal{E}_1$ and $\mathcal{E}_0 \vdash_p^a \mathcal{E}_1^* \xrightarrow{d} \mathcal{E}_1$. Normally, one can distinguish between feasible and not feasible interactions depending on a criterion of admissibility of system states. Each sensible interaction applies to an admissible state $\mathcal{E}_0$ and yields an admissible state $\mathcal{E}_1$. However, the intermediate state $\mathcal{E}_1^*$ may in general be inadmissible, in which case the reaction compensates for the ruinous stimuli. We represent the admissibility criterion by some *integrity constraints (IC)* expressed by a formula or a (logic) program Φ over DB states. In terms of the IC the feasibility of the interaction is expressed as follows: the interaction of the form $\mathcal{E}_0 \xrightarrow{d} \mathcal{E}_1^* \vdash_p^a \mathcal{E}_1$ or $\mathcal{E}_0 \vdash_p^a \mathcal{E}_1^* \xrightarrow{d} \mathcal{E}_1$ is *feasible* if $\mathcal{E}_0 \models \Phi$ and $\mathcal{E}_1 \models \Phi$. Thus, the discrete dynamic system is represented by the *deductive data base (DDB)* $\mathcal{B} =< \mathcal{P} \cup GOALS, \Phi >$, its medium is described by the relation $\xrightarrow{d}$, and their interaction is expressed in terms of feasible interactions.

Global behavior of the system in the current state $\mathcal{E}_0$ is represented by sequences of (feasible) interactions starting in $\mathcal{E}_0$: the *(feasible) trajectories*. Basically, we have two types of trajectories corresponding to the two types of interactions above: *disturbance-action* trajectories

$$\mathcal{E}_0 \xrightarrow{d} \mathcal{E}_1^* \overset{a_1}{\vdash} \mathcal{E}_1 \xrightarrow{d} \mathcal{E}_2^* \overset{a_2}{\vdash} \mathcal{E}_2 \ldots$$

and *action-disturbance* trajectories

$$\mathcal{E}_0 \overset{a_1}{\vdash} \mathcal{E}_1^* \xrightarrow{d} \mathcal{E}_1 \overset{a_2}{\vdash} \mathcal{E}_2^* \xrightarrow{d} \mathcal{E}_2 \ldots$$

Infinite feasible disturbance-action trajectories represent the homeostatic behavior of the DDB in spite of disturbances of its medium: the DDB always manages to restore the IC along this trajectory if and when they are violated by the medium disturbances. Such trajectories are called *homeostatic*. Dually, the infinite feasible action-disturbance trajectories represent the stable behavior of the DDB owing to disturbances of its medium: the medium always compensates along the trajectory for possible ruinous actions of the DDB. They are called *stable*.

Trajectories of each type form a tree with the root $\mathcal{E}_0$: $T_{da}(\mathcal{E}_0)$ and $T_{ad}(\mathcal{E}_0)$ respectively. A number of natural properties of interactive behavior of $\mathcal{B}$ in a given DB state can be formalized in terms of these trees.

Definition 1 *Let* $\mathcal{B}$ *be a DDB and* $\xrightarrow{d}$ *be a disturbance relation. Let* $Q_1, Q_2 \in \{ \forall, \exists \}$. *Then* $\mathcal{B}$ *is* $d - Q_1Q_2$-*stable in DB state* $\mathcal{E}_0$ *if in the tree* $T_{ad}(\mathcal{E}_0)$ *there is a* Q_1Q_2-*subtree in which all branches are infinite stable trajectories.* $\mathcal{B}$ *is* $d - Q_1Q_2$-*homeostatic in DB state* $\mathcal{E}_0$ *if in the tree* $T_{da}(\mathcal{E}_0)$ *there is a* Q_1Q_2-*subtree in which all branches are infinite homeostatic trajectories.*

In [4] we introduce $\exists\exists$-stability and explore its computational complexity. In [5] the $\forall\exists$-homeostaticity is introduced and investigated.

Another interesting class of properties of steady interactive behavior is the class of *total* properties of feasible interactions, i.e. properties which hold *in any* DB state. One such property has attracted widespread attention because it corresponds to an important and long known scenario in DDBs applications. This is the property of *total homeostaticity:*

$$\forall \mathcal{E}, \mathcal{E}_1^* \ (\mathcal{E} \models \Phi \ \& \ \mathcal{E} \xrightarrow{d} \mathcal{E}_1^* \Rightarrow \exists a, \mathcal{E}_1 \ (\mathcal{E}_1^* \overset{a}{\vdash} \mathcal{E}_1 \ \& \ \mathcal{E}_1 \models \Phi))$$

equivalent to $\forall \exists$-homeostaticity in any admissible DB state. The so called revision algorithms [10] are applied to DB states of a propositional DDB, violating the IC in order to transform them to DB states satisfying the IC. In fact the same problem is investigated in [11, 12]. Here the IC is expressed by a much more general revision program $\mathcal{P}$ and partial algorithms are proposed transforming a given inadmissible DB state $\mathcal{E}$ into a stable model of $\mathcal{P}$ (in the sense of [7]), "induced" by $\mathcal{E}$. In both cases updates are treated implicitly: the initial inadmissible DB state may be viewed as the result of a ruinous update. In [9] elementary updates (insertion or deletion of a ground literal) are explicit, the IC Φ is expressed by Datalog programs extended by simple productions (rules), and algorithms are described "constructing" a model of Φ consistent with an input update. So in these papers DB updates correspond to destructive disturbances and the algorithms enforcing the IC correspond to restoring DDB actions in our terms. This is in fact the basic scenario in the active data bases (cf. [3, 8]), where the disturbances are either the elementary updates or some external events recognizable by action rules. Here again the total homeostaticity is guaranteed by some strategy over actions.

It should be stressed that while in the cited literature one is always interested in models which **guarantee** the total homeostaticity, we are investigating the inverse problem: *given a DB state $\mathcal{E}$, a DDB $\mathcal{B}$ in certain class of DDBs $\mathcal{C}$, an IC Φ in a class of ICs $\mathcal{I}$, and a disturbance relation $\xrightarrow{d}$ restricted in some sense, one should find out whether $\mathcal{B}$ is stable (homeostatic) in $\mathcal{E}$ with respect to $\xrightarrow{d}$.*

As far as we know, the property of stable behavior went unnoticed in the literature. Meanwhile, this property is peculiar to many interactive systems whose operation involves consumption, compensation, and restoration of some resources. The typical examples are plants, trade enterprises, warehouses, etc. Their medium is formed of consumers and some financial or material sources necessary to support their successful and unlimited operation. In this paper we explore the computational complexity of the property of $\forall \exists$-stability and $\forall \forall$-stability of DDBs. The $\forall \exists$-stability means that for any DDB action violating the IC there exists some disturbance of the medium, properly restricted, which restores the IC. In the case of the $\forall \forall$-stability each restricted disturbance should restore the IC. Not surprisingly, these properties are undecidable in the class of all DDBs. However, they are solvable in various classes of DDBs via reasonable classes of IC of interest for real applications. E.g., we prove that in the nonrecursive $DATALOG^\neg$ case where DDBs are structureless production systems the $\forall \forall$-stability is exponential-space complete, while the $\forall \exists$-stability is deterministic-double-exponential-time complete. If DDBs are ground (i.e. variableless) production systems, then the first property is $PSPACE$-complete, while the other is deterministic-exponential-time complete. In some more narrow classes of DDBs the complexity of both properties is quite tractable. For example, in the class of ground production systems which never delete facts from states, and under monotonic integrity constraints they are solved in polynomial time. Although in many interesting situations the complexity of stability recognition appears to be too high, there are natural methods of factorization of the DB states space (described elsewhere), which can substantially decrease the size of the problems in practical DDB applications.

2 Basic Notation and Definitions

We consider a 1st order language $\mathbf{L}$ in a signature containing: pairwise disjoint sets $\mathbf{P}^e$ of *extensional predicates*, $\mathbf{P}^q$ of *intensional query predicates*, and $\mathbf{P}^u$ of *intensional update predicates*, and a set of function (and constant) symbols $\mathbf{F}$. $\mathbf{H}$ denotes the Herbrand universe of $\mathbf{L}$, and $\mathbf{B}^e$, $\mathbf{B}^i$ are for the extensional and intensional Herbrand bases. *DB states* are finite subsets of $\mathbf{B}^e$. We consider *logic programs with updates* in $\mathbf{L}$ with clauses of the form *Head :— Body*, where *Head* is an intensional atom $p(\bar{t})$ with $p \in \mathbf{P}^q \cup \mathbf{P}^u$, and *Body* is a (possibly empty) sequence of:

- literals of the form $q(\bar{u})$ or $\neg r(\bar{v})$ with $r \notin \mathbf{P}^u$, (i.e. only atoms with extensional and intensional query predicates, can be negated),

- elementary DB updates *insert(Fact)*, *delete(Fact)*, where *Fact* is an extensional atom,

- assignments of the form $X := e$ and arithmetical constraints of the form $e_1 < e_2$ with e, e_1, e_2 being arithmetical expressions.

So we subsume that numerals and arithmetic operators are included into $\mathbf{F}$. The semantic scheme below will show that they are interpreted in the standard way.

Definition 2 *Let $\mathcal{P}$ be a logic program with updates. We say that a predicate p refers to a predicate q if there is a clause $p(\bar{t}) :- \bar{\alpha}, q(\bar{s}), \bar{\beta}$ in $\mathcal{P}$. We consider the relation "depend on" which is the reflexive and transitive closure of the relation "refer to". Maximal strongly connected components of the graph $G(\mathcal{P})$ of the relation "depend on" are called* cliques. *A predicate (a goal, a subgoal) is* stationary *if it does not depend on elementary updates, assignments, and update predicates.*

We distinguish the following important class of logic programs with updates.

Definition 3 *A logic program with updates $\mathcal{P}$ is* stratified *if:*

1) all query predicates are stationary;

2) let $\mathcal{P}^q$ be the set of definitions of all query predicates in $\mathcal{P}$: $\mathcal{P}^q = \mathcal{P}(\mathbf{P}^q)$. Then for any DB state $\mathcal{E}$ the logic program $\mathcal{P}^s = \mathcal{P}^q \cup \mathcal{E}$ is stratified in the sense of [1].

A logic program with updates $\mathcal{P}$ is update (u-) stratified *if it is stratified and in every clause $p(\bar{u}) :- \alpha_1, ..., \alpha_i, q(\bar{v}), \alpha_{i+1}, ..., \alpha_r$ in which p, q are update predicates belonging to the same clique of $G(\mathcal{P})$, all predicates in α_i are stationary.*

The essence of the u-stratifiability is in point 3 which says that DB updates are available only at the steps where a clique is changed. This constraint provides tight bounds to the number of elementary DB updates fired in the course of a transaction caused by an update predicate call.

Example 1 *Let* $\mathbf{P}^e = \{e/1\}$ *and* $\mathbf{P}^u = \{a/0, b/0, c/0\}$. *Then the following program is u-stratified:*

$a :- e(0), b.$

$a :- insert(e(0)).$

$b :- \neg e(0), a.$

$b :- e(I), I_1 := I + 1, \neg e(I_1), delete(e(I)), insert(e(I_1)), c.$

$c :- e(I), I > 0, I_1 := 2 * I, \neg e(I_1), insert(e(I_1)).$

Here the dependency subgraph on $G(\mathbf{P}^u)$ consists of two cliques: $\{a, b\} \rightarrow \{c\}$.

Throughout this paper we consider only stratified logic programs with updates and call them simply logic programs with updates.

The operational semantics of a logic program with updates $\mathcal{P}$ is represented by a relation of the form $\mathcal{E} \cup \{ :- u\} \overset{\theta}{\vdash}_{\mathcal{P}} \mathcal{E}' \cup \{ :- v\}$. Intuitively it says that $\mathcal{P}$ reduces via the answer substitution θ the goal $:- u$ when in the input DB state $\mathcal{E}$, to the goal $:- v$ and changes $\mathcal{E}$ to the output DB state $\mathcal{E}'$. The following rule schemes 1 - 7 define this relation formally (index $\mathcal{P}$ is dropped). The letters u, v, ϕ in these schemes stand for sequences of subgoals, $\square$ is for the empty goal, θ, θ', and σ are for substitutions and for an MGU, ε denotes the identity substitution, e, e' denote arithmetical expressions, A, H stand for intensional atoms, $Fact$ is for an extensional atom, and $stable\_mod(\mathcal{E}' \cup \mathcal{P}^q)$ denotes the (unique) stable model (in the sense of [7]) of the stratified program $\mathcal{E}' \cup \mathcal{P}^q$.

1. $$\frac{}{\mathcal{E} \cup \{ :- u\} \overset{\varepsilon}{\vdash} \mathcal{E} \cup \{ :- u\}}$$

2. $$\frac{\mathcal{E} \cup \{ :- u\} \overset{\theta}{\vdash} \mathcal{E}' \cup \{ :- A, v\}, \ (H :- \phi) \in \mathcal{P} \ and \ A\theta\sigma = H\sigma, \ or \ A\theta\sigma \in \mathcal{E}' \ and \ \phi = \emptyset}{\mathcal{E} \cup \{ :- u\} \overset{\theta\sigma}{\vdash} \mathcal{E}' \cup \{ :- \phi, v\}}$$

3. $$\frac{\mathcal{E} \cup \{ :- u\} \overset{\theta}{\vdash} \mathcal{E}' \cup \{ :- X := e, v\}, \ e\theta \ is \ ground, \ \theta' = \theta[X \backslash val(e, \theta)]}{\mathcal{E} \cup \{ :- u\} \overset{\theta'}{\vdash} \mathcal{E}' \cup \{ :- v\}}$$

4. $$\frac{\mathcal{E} \cup \{ :- u\} \overset{\theta}{\vdash} \mathcal{E}' \cup \{ :- e < e', v\}, \ e\theta, \ e'\theta \ are \ ground \ and \ val(e, \theta) < val(e', \theta)}{\mathcal{E} \cup \{ :- u\} \overset{\theta}{\vdash} \mathcal{E}' \cup \{ :- v\}}$$

5. $$\frac{\mathcal{E} \cup \{ :- u\} \overset{\theta}{\vdash} \mathcal{E}' \cup \{ :- insert(Fact), v\}, \ Fact\theta \ is \ ground}{\mathcal{E} \cup \{ :- u\} \overset{\theta}{\vdash} (\mathcal{E}' \cup \{Fact\theta\}) \cup \{ :- v\}}$$

6. $$\frac{\mathcal{E} \cup \{ :- u\} \overset{\theta}{\vdash} \mathcal{E}' \cup \{ :- delete(Fact), v\}, \ Fact\theta \ is \ ground}{\mathcal{E} \cup \{ :- u\} \overset{\theta}{\vdash} (\mathcal{E}' \backslash \{Fact\theta\}) \cup \{ :- v\}}$$

7. $$\frac{\mathcal{E} \cup \{ :- u \} \stackrel{\theta}{\vdash} \mathcal{E}' \cup \{ :- \neg A, v \}, \ stable\_mod(\mathcal{E}' \cup \mathcal{P}^q) \models \neg A \circ \theta}{\mathcal{E} \cup \{ :- u \} \stackrel{\theta}{\vdash} \mathcal{E}' \cup \{ :- v \}}$$

These rules support leftmost strategy of subgoals evaluation. Rules scheme 1 introduces the identity answer substitution and serves as an axioms scheme in the derivations. The premises of rules schemes 2-7 consist of two parts: (i) a sequent of the form $\mathcal{E} \cup \{ :- u \} \stackrel{\theta}{\vdash}_{\mathcal{P}} \mathcal{E}' \cup \{ :- g, v \}$, where g is the subgoal to be resolved by the rule, and (ii) a condition of applicability of the rule. The conclusion of each scheme consists of the derived sequent of the same form. Rule 2 is the standard resolution step which resolves the leftmost subgoal A by a new variant of the clause $(H :- \phi) \in \mathcal{P}$, or by the fact $A \in \mathcal{E}'$. In this rule σ denotes an MGU. Rules 3 and 4 deal with arithmetic. All variables in expressions e, e' should be instantiated (by numbers). Rules 3, 4 do not change DB states. $val(e, \theta)$ denotes the value of the arithmetical expression e in the environment θ. The assignment $X := e$ changes the answer substitution θ binding X by $val(e, \theta)$. Rules 5 and 6 are the only rules changing DB states. They describe the effect of elementary updates. Specifically, rule 5 changes the DB state $\mathcal{E}'$ to the DB state $\mathcal{E}' \cup \{Fact \circ \theta)\}$. Rule 7 indicates that the negation is resolved in the (unique) stable model $stable\_mod(\mathcal{E}' \cup \mathcal{P}^q)$. It should be noted that our choice of negation semantics is quite arbitrary. Any negation semantics "effectively computable" on finite models will do here.

Rules 1 - 7 associate with each update predicate $a/0$ the following nondeterministic transaction operator $\stackrel{a}{\vdash}_{\mathcal{P}}$ on DB states: $\mathcal{E} \stackrel{a}{\vdash}_{\mathcal{P}} \mathcal{E}' \Longleftrightarrow \mathcal{E} \cup \{ :- a \} \stackrel{\theta}{\vdash}_{\mathcal{P}} \mathcal{E}' \cup \{\square\}$ for some θ.

A DDB $\mathcal{B} = < \mathcal{P} \cup \{ :- a_1, \ldots, :- a_n \}, \Phi >$ includes an intensional logic program with updates $\mathcal{P}$, a predefined set of 0-ary goals $\{a_1, \ldots, a_n\}$ implementing DB state transactions, and integrity constraints (IC) embodied by a relation Φ on DB states. The behavior of DDB $\mathcal{B}$ is defined by the transaction relation $\vdash_{\mathcal{B}}$ which is the union of relations $\stackrel{a_i}{\vdash}_{\mathcal{P}}$, $i = 1, \ldots, n$.

3 Problem Classes

We explore the computational complexity of the following problems. Given a DDB $\mathcal{B} = < \mathcal{P} \cup \{ :- a_1, \ldots, :- a_n \}, \Phi >$, a disturbance relation $\stackrel{d}{\longrightarrow}$ (in some representation), and a DB state $\mathcal{E}$, one should check whether $\mathcal{B}$ is $\forall \exists$-(respectively $\forall \forall$-) stable in $\mathcal{E}$ with respect to $\stackrel{d}{\longrightarrow}$.

It is readily seen that both problems are undecidable if no restrictions are imposed on their main parameters: logic programs with updates, ICs and disturbance relations. In this section we introduce various restrictions which come about naturally and guarantee the decidability of these problems.

Logic programs. Logic programs we consider are always u-stratified. However, this condition does not assure the solvability of the stability recognition problems by itself. We classify the restrictions to logic programs by the form of their clauses.

Definition 4 *A logic program* $\mathcal{P}$ *is* positive *if it does not use negation. It is called* ground *if all its clauses are ground. It is* flat *if all terms in its clauses are either variables or constants.* $\mathcal{P}$ *is* expanding *if the update delete/1 is not used in its clauses. We call* $\mathcal{P}$ branching *if it is not recursive, i.e. its dependency graph* $G(\mathcal{P})$ *has no cycles. And we call* $\mathcal{P}$ productional *if it defines the unique intensional predicate q/0 and all its clauses are productions, i.e. have the form* $q :- Con_1, \ldots, Con_k, Act_1, \ldots, Act_m$ *where each* Con_i *is an extensional literal and each* Act_j *is an elementary update. These are exactly the productions used in AI, so we keep their usual syntax:*

$$Con_1 \& \ldots \& Con_m \Longrightarrow Act_1, \ldots, Act_n.$$

We consider the following classes of u-stratified logic programs:
 - USF is the class of u-stratified flat programs;
 - USG is the class of u-stratified ground programs;
 - $BRAF$ is the class of branching flat programs;
 - $BRAG$ is the class of branching ground programs;
 - $BRAG^x$ is the class of expanding programs in $BRAG$;
 - $BRAG^+ = BRAG^p \cap BRAG^x$.

For each class of branching programs in this list we consider the corresponding class of productional logic programs: $PROF$, $PROG$, $PROG^x$, and $PROG^+$. The last class is the smallest one. Programs in $PROG^+$ have only productional rules which do not use negation and the update $delete/1$.

Integrity constraints. For a DB state $\mathcal{E}$ and an IC Φ we denote by $\mathcal{E} \models \Phi$ the relation "Φ is true in $\mathcal{E}$". The nature of IC Φ is immaterial here. What is essential is that Φ has a constructive representation in some formal language (a logical formula, a set of productions, or a polynomial time Turing machine, etc.) for which there is a universal algorithm checking the property $\mathcal{E} \models \Phi$.

Definition 5 *An IC* Φ *is preserved upwards if whenever* $\mathcal{E} \subseteq \mathcal{E}'$ *and* $\mathcal{E} \models \Phi$, *then* $\mathcal{E}' \models \Phi$.

Below we analyze the complexity of stability problems using ICs in the following three classes.

IC_0 : Φ is preserved upwards and the problem $\mathcal{E} \models \Phi$ is in **P**.

IC_1 : the problem $\mathcal{E} \models \Phi$ is in **P**.

IC_2 : the problem $\mathcal{E} \models \Phi$ is in **PSPACE**.

E.g., the class IC_0 contains ICs expressed by positive closed quantifier-free formulas, and some monotone graph properties (e.g., connectivity). The well-known functional dependencies and a number of properties of model size (e.g., parity) belong to the class IC_1. The class IC_2 contains the ICs expressed by the 1st order formulas, etc.

Disturbance relations. The constraints we impose on disturbance relations $\xrightarrow{d}$ are expressed in terms of the following *change set*:

$$c(d, \mathcal{E}) = \{(D^+, D^-) | \text{ there is } \mathcal{E}' : \mathcal{E} \xrightarrow{d} \mathcal{E}', D^+ = \mathcal{E}' \setminus \mathcal{E}, \text{ and } D^- = \mathcal{E} \setminus \mathcal{E}'\}.$$

Let $\delta = (\Delta^+, \Delta^-)$ be a pair of two finite subsets of $\mathbf{B}^e$. We define a δ-disturbance as a binary relation on DB states such that for every $\mathcal{E}$ $c(\delta, \mathcal{E}) = \{(D^+, D^-)|D^+ \subseteq \Delta^+, D^- \subseteq \Delta^-\}$.

We finish this section by a definition of stability problem. For a class of logic programs with updates $\mathbf{P}$, a class of ICs $\mathbf{I}$, and a pair of quantifiers $\mathbf{Q_1, Q_2}$
$STABLE^{\mathbf{Q_1 Q_2}}(\mathbf{P + I}) = $
$\{(<\mathcal{P}, \Phi>, \delta, \mathcal{E}) \mid \mathcal{P} \in \mathbf{P},\ \Phi \in \mathbf{I},\ <\mathcal{P}, \Phi> \ is\ \delta - Q_1 Q_2 - stable\ in\ \mathcal{E}\}.$

Example 2 *Imagine a population of organisms in which the breeding rules are phrased in terms of values of some characteristic feature as follows: a new individual emerges only if its parents have some different values of this feature, and exactly one value is inherited. For the case of three possible values a, b, c this rule is expressed very simply in our terms.*
Productional logic program $\mathcal{P}$:
$$\alpha \& \beta \Longrightarrow delete(\beta),\quad \alpha, \beta \in \{a, b, c\}.$$
IC $\Phi = (a\&b \lor a\&c \lor b\&c)\&\neg(a\&b\&c).\quad \delta = (\{a, b, c\}, \emptyset).$
It is clear that this biological system is δ-$\forall \exists$-stable in all admissible states. However, it is not δ_1-$\forall \exists$-stable in any DB state when $\delta_1 = \{a\}$.

4 Complexity of $\forall Q$-stability

In this section we establish complexity bounds to the problems of $\forall \forall-$ and $\forall \exists$-stability in the classes of DDBs defined above. Our results show that such features of DDBs as nonmonotone ICs, negations in conditions, deletions of facts, and the use of variables in logic programs can increase the complexity of these problems dramatically, whereas the restricted form (u-stratifiability) of recursion does not change the complexity bounds. Because of space limitations some proofs are omitted here, the others are only sketched.

In the theorems to follow the finite sets of facts $\mathcal{E}$ (Δ^+, Δ^-, etc.) are supposed to be represented in some standard coding. The size of the code $|\mathcal{E}|$ is $O(C_{\mathcal{E}} * FN_{\mathcal{E}})$, where $C_{\mathcal{E}}$ depends on maximal arity of extensional predicates and on constants used in $\mathcal{E}$, and $FN_{\mathcal{E}}$ denotes the number of facts in $\mathcal{E}$. This differs from the encoding used in the finite model complexity where predicates are represented by the sequences of their values on *all* possible data. Below for each Turing machine complexity class $CLASS(F)$ $ACLASS(F)$ denotes the corresponding complexity class for the alternating Turing machines [2].

Theorem 1
(1) The problem $STABLE^{\forall \forall}(PROG^+ + IC_0)$ is solved in polynomial time.
(2) The problem $STABLE^{\forall \forall}(BRAG^+ + IC_1)$ is co-NP-complete.
(3) The problem $STABLE^{\forall \forall}(PROG^x + IC_1)$ is PSPACE-complete.
(4) The problem $STABLE^{\forall \forall}(BRAG + IC_2)$ is PSPACE-complete.
(5) The problem $STABLE^{\forall \forall}(BRAF + IC_2)$ is $SPACE(2^{poly})$-complete.
(6) The problem $STABLE^{\forall \forall}(USG + IC_2)$ is PSPACE-complete.
(7) The problem $STABLE^{\forall \forall}(USF + IC_2)$ is $SPACE(2^{poly})$-complete.
(8) The problem $STABLE^{\forall \forall}(PROD + IC_0)$ is undecidable.

Sketch of the proof. (1) Let $\mathcal{P} \in PROG^+$, $\Phi \in IC_0$ and $\delta = (D^+, D^-)$. Then it follows from monotonicity of $\mathcal{P}$ and Φ that $(< \mathcal{P} \cup \{ :- g\}, \Phi >, \delta, \mathcal{E}) \in STABLE^{\forall\forall}(PROG^+ + IC_0)$ iff there is at least one production of $\mathcal{P}$ which applies to $\mathcal{E}$, and for all DB states $\mathcal{E}^*$ such that $\mathcal{E} \vdash_{\mathcal{P}} \mathcal{E}^*$ DB state $\mathcal{E}_1 = \mathcal{E}^* \backslash D^-$ satisfies the IC Φ. Since model checking $\mathcal{E}_1 \models \Phi$ can be done in polynomial time we get the polynomial time algorithm to solve the problem (1).

Due to the following reduction, in the proofs of all the other points of theorem 1 we can take advantage of the methods used in [4] for the problem of $\exists\exists$-stability and the following problem called in [4] a promise problem :
$PROMISE(\mathbf{P} + \mathbf{I}) = \{(< \mathcal{P}, \Phi >, \delta, \mathcal{E}) \mid \mathcal{P} \in \mathbf{P}, \Phi \in \mathbf{I}, \text{ and there is a finite}$
action-disturbance δ-trajectory starting in $\mathcal{E}$ and resulting in some DB state $\mathcal{E}'$ satisfying the IC $\Phi\}$.

Lemma 1 *Let* $\mathbf{P}$ *be any class of programs in the assertions (2)-(8) of theorem 1, and* $\mathbf{I} \in \{IC_1, IC_2\}$. *Then*

$$STABLE^{\forall\forall}(\mathbf{P} + \mathbf{I}) \equiv_{poly} \neg PROMISE(\mathbf{P} + \mathbf{I})$$

□

The problem of $\forall\exists$-stability turns to be more complex than that of $\forall\forall$-stability.

Theorem 2
   (1) The problem $STABLE^{\forall\exists}(PROG^+ + IC_0)$ is solved in polynomial time.
   (2) The problem $STABLE^{\forall\exists}(PROG^+ + IC_1)$ is PSPACE-complete.
   (3) The problem $STABLE^{\forall\exists}(BRAG^x + IC_2)$ is PSPACE-complete.
   (4) The problem $STABLE^{\forall\exists}(PROG + IC_1)$ is APSPACE- (i.e. DEXPTIME-) complete.
   (5) The problem $STABLE^{\forall\exists}(PROF + IC_2)$ is $ASPACE(2^{poly})$-complete.
   (6) The problem $STABLE^{\forall\exists}(USG + IC_1)$ is APSPACE- (i.e. DEXPTIME-) complete.
   (7) The problem $STABLE^{\forall\exists}(USF + IC_2)$ is $ASPACE(2^{poly})$-complete.
   (8) The problem $STABLE^{\forall\exists}(PROD + IC_0)$ is undecidable.

Sketch of the proof. (1) To get a polynomial time algorithm in this case it is enough to notice that $(< \mathcal{P} \cup \{ :- g\}, \Phi >, \delta, \mathcal{E}) \in STABLE^{\forall\exists}(PROG^+ + IC_0)$ iff there is at least one production of $\mathcal{P}$ which applies to $\mathcal{E}$ and for all DB states $\mathcal{E}^*$ such that $\mathcal{E} \vdash_{\mathcal{P}} \mathcal{E}^*$ DB state $\mathcal{E}_1 = \mathcal{E}^* \cup D^+$ satisfies the IC Φ.

The lower bounds (2)-(7) in this theorem are proven by simulation of successful computations of alternating Turing machines in terms of $\forall\exists$-stable trajectories of DDBs in the classes above. We sketch here the proof of only one lower bound, that of the point (2). To prove this lower bound we construct for any alternating Turing machine $\mathcal{M}$ a DDB $\mathcal{B}_{\mathcal{M}} = < \mathcal{P} \cup \{ :- g\}, \Phi >$ and some δ, such that $\mathcal{M}$ accepts an input word $x = a_{i_1}, ..., a_{i_n}$ in time bounded by a polynomial p, iff the DDB $\mathcal{B}_{\mathcal{M}}$ is $\delta - \forall\exists$-stable in a DB state $\mathcal{E}_x$ (constructed from x. Productions of $\mathcal{P}$ simulate $\forall$-instructions of $\mathcal{M}$ as follows: the instruction $q_l^{\forall} a_k \to q_r^{\exists} a_p S$ is simulated by the set of productions $q(i, l, t) \& a(i, k, t) \Longrightarrow$

$$insert(s(t + 1)),\ insert(q(i + S, r, t + 1)),\ insert(a(i, p, t + 1))$$
($N = p(n)$, $1 \leq i \leq N$, $1 \leq t \leq N - 1$).
Here $q(i, j, t)$ means that "at the step t the head of $\mathcal{M}$ visits the cell i in state q_j", $a(i, k, t)$ means that "at the step t the cell i contains the symbol a_k", and $s(t)$ remembers the current step number t. For the accepting final state q_a of $\mathcal{M}$ $\mathcal{P}$ includes the final productions:
$q(i, a, t) \implies$ **none** $(1 \leq i, t \leq N)$.
δ-disturbances simulate $\exists$-steps of $\mathcal{M}$ as follows: $\delta = (\mathcal{D}^+, \mathcal{D}^-)$, and
$\mathcal{D}^+ = \{q(i, l, 2t), a(i, k, t), s(2t)\ |q_l \in Q^{\vee}, 1 \leq i \leq N,\ 1 \leq t \leq N/2\}$,
$\mathcal{D}^- = \{q(i, l, 2t + 1), a(i, k, 2t + 1)\ |q_l \in Q^{\exists}, 1 \leq i \leq N,\ 0 \leq t \leq N/2\}$.

DB states represent prefixes of computations of $\mathcal{M}$, and the initial DB state $\mathcal{E}_x = \{s(0), q(1, 0, 0), a(1, i_1, 0), \ldots, a(n, i_n, 0), a(n+1, 0, 0), \ldots, a(N, 0, 0)\}$ represents the starting instantaneous description.
The following IC Φ filters out DB states representing feasible computations:
$$\Phi = A\ \&\ B\ \&\ C\ \&\ D\ \&\ E\ \&\ F\ \&\ G,$$
where: A expresses that a sequence of steps of $\mathcal{M}$ represented by DB state is continuous and finishes at an even ($\forall$-) step. B expresses that each cell contains a single symbol and the head visits a unique cell. C expresses that if $s(t)$ lacks in the DB-state then the step t is not represented by this state. D expresses that the symbols in the cells not visited by the head are not changed. E expresses that the states of $\mathcal{M}$ and the symbols in visited cells are changed by the corresponding instructions of $\mathcal{M}$. F expresses that $\mathcal{E}_x$ is included in the current DB state. G expresses that the computation represented by the DB state does not reject x.

Therefore, a DB state $\mathcal{E}$ satisfies the IC Φ iff it represents an initial segment of a computation of $\mathcal{M}$ starting in $\mathcal{E}_x$ and finishing in some $\forall$-state. Moreover, any stable trajectory of $\mathcal{B}_{\mathcal{M}}$ has to reach a DB state which represents a successful computation of $\mathcal{M}$ on x. Then it continues by a final production of $\mathcal{P}$ and by the empty disturbance. Now the theorem follows from the assertion that the $\forall\exists$-subtrees of stable trajectories in $T_{ad}(\mathcal{E}_x)$ one-to-one correspond to the $\forall\exists$-subtrees of successful computations of $\mathcal{M}$ on x. $\square$

5 Conclusion

The concept of $\forall\exists$-stable behavior of DDBs proposed in this paper substantially generalizes the following property: "for any action of the DDB violating the IC there exists a bounded external update which restores the IC". This concept applies naturally to the analysis of the behavior of interactive discrete dynamic systems with complex states in active medium, whose operation involves consumption, compensation, and restoration of resources, e.g. plants, trade enterprises, warehouses, etc. Among various possible types of steady behavior of such systems, we consider here only two kinds of stable behavior: $\forall\forall$- and $\forall\exists$-stability. Meanwhile, in some other application domains we encounter other types of steady behavior as well, e.g. $\exists\exists$-stability or homeostaticity of various types, which have been formalized and explored in our previous papers [4, 5].

2 Basic Concepts and Definitions

Let S be a finite set of symbols (an alphabet). A *conjunction* is a multiset over S. E.g., if $S = \{a, b, c\}$ then $X = \{a, a, a, b, b, c\}, Y = \{a, a, b, b, c, c, c,\}$ are conjunctions. We represent conjunctions by commutative words using exponential notations: $X = a^3 b^2 c$, $Y = a^2 b^2 c^3$. A *Horn implication* is a formula of the form $(X \to Y)$ where X and Y are nonempty conjunctions. A *Horn sequent* is an expression of the form $w\Gamma \vdash w'$ where w and w' are conjunctions, Γ is a sequence (multiset) of Horn implications. A Horn implication $(X \to Y)$ is *applicable* to a conjunction Z iff $Z = XU$ for some conjunction U (which may be empty). Then the result of application $(X \to Y)$ to Z is UY. A sequence of Horn implication $\Gamma = \gamma_1, \ldots, \gamma_n$, $\gamma_i = (\alpha_i \to \beta_i)$, $i = 1, \ldots, n$ is *applicable* to Z iff there exist a permutation (an application order) $\sigma = (j_1, \ldots, j_n)$ of $(1, \ldots, n)$ and a sequence of conjunctions $Z_0 = Z, Z_1, \ldots, Z_n$ such that for each $i = 1, \ldots, n$ Z_i is a result of application of γ_{j_i} to Z_{i-1}. It is easy to see that the result of application Γ to Z does not depend on the choice of σ (but some orders may be "unsuccessful")
. E.g. let $\Gamma = (ab \to b^2)(ab \to a^2 b)$ and $Z = ab$. Then an attempt to use the order $(1, 2)$ failes while the order $(2, 1)$ leads to the result ab^2.

It is easy to check that a sequent $w\Gamma \vdash w'$ is provable in Horn fragment of Multiplicative Linear Logic iff Γ is applicable to w and w' is the result of this application [1, 2].

The provability problem for Horn sequents (or its equivalent problem of applicability of a sequence of Horn implications to a conjunction) is *NP*-complete even for 2-letters alphabet S [1, 4]. It is proposed in [2] to consider some special subclasses of Γ for which only "small" part of the space of all application orders should be considered to decide if Γ is applicable to w. The approach is based on the idea of concurrency by simultaneous application of several implications of Γ on every computation step.

The following strongest property – superconcurrency – provides that implications can be applied in any order.

Definition 1 *[2] A sequence of Horn implications Γ is* superconcurrent *for a conjunction w if it is applicable to w and for any $(\alpha \to \beta) \in \Gamma$ if $w = \alpha\gamma$ then $\Gamma \setminus \{(\alpha \to \beta)\}$ is superconcurrent for $w' = \beta\gamma$.*

A sequence of Horn implications Γ is superconcurrent *if for each conjunction w either Γ is superconcurrent for w or Γ is not applicable to w. Let* **SUC** *denote the set of all superconcurrent sequences.*

The following definition explains which maximal subsets of Γ can be applied simultaneously to w.

Definition 2 *[2] A subsequence $\Delta = (\alpha_{i_1} \to \beta_{i_1}) \ldots (\alpha_{i_m} \to \beta_{i_m})$ of a sequence $\Gamma = (\alpha_1 \to \beta_1) \ldots (\alpha_n \to \beta_n)$ is called* maximal *for a conjunction w if two following conditions hold:*
(1) $w = \alpha_{i_1} \ldots \alpha_{i_m} U$ for some U (U may be empty);
(2) for any $(\alpha \to \beta) \in \Gamma \setminus \Delta$ $w \neq \alpha_{i_1} \ldots \alpha_{i_m} \alpha U$ for any conjunction U.

The following definition presents nondeterministic maximal concurrency.

Definition 3 *[2] A sequence of Horn implications Γ is concurrent for a conjunction w if one of following conditions holds:*
(1) Γ is maximal for w;
(2) there is a maximal subsequence Δ such that $\Gamma \setminus \Delta$ is concurrent for the result of applying Δ to w.
The sequence of Horn implications Γ is concurrent if it is concurrent for any conjunction w such that Γ is applicable to w. Let **CC** *denote the class of all concurrent sequences.*

The concurrency of Γ for w means that there is a sequence of applications of implications of Γ to w such that on each step some maximal subsequence is applied. We introduce also a notion of *strong concurrency* which represents the idea of deterministic maximal concurrency.

Definition 4 *A sequence of Horn implications is strong concurrent for w if one of two following conditions holds:*
(1) Γ is maximal for w;
(2) for any maximal subsequence Δ the sequence $(\Gamma \setminus \Delta)$ is strong concurrent for the result of applying Δ to w.
Γ is strong concurrent if it is strong concurrent for any conjunction w such that Γ is applicable to w. Let **STC** *denote the class of all strong concurrent sequences.*

The strong concurrency allows one to use any maximal subsequence on each step of computation.

It follows immediatly from the definitions that **SUC** $\subseteq$ **STC** $\subseteq$ **CC**. Both inclusions are proper even if 1-letter alphabet S is considered. It is easy to check that $\Gamma_1 = (a \to a^2)(a^2 \to a^3)(a^5 \to a) \in$ **STC** $\setminus$ **SUC**, and $\Gamma_2 = (a^2 \to a^2)(a^2 \to a) \in$ **CC** $\setminus$ **STC**.

3 Complexity of Provability and Concurrency Recognition

In this section we investigate in the classes **SUC, CC** and **STC** the complexity of two following algorithmic problems:
Provability. Given a class C of sequences of Horn implications, a sequence $\Gamma \in C$, and a conjunction w, one should check if Γ is applicapable to w.
Recognition. Given a class C of sequences of Horn implications and a sequence Γ one should check if $\Gamma \in C$.

Theorem 1 *(1) The provability problem for classes* **SUC** *and* **STC** *is decidable in polynomial time.*
(2) The provability problem for **CC** *is NP-complete.*

Sketch of the proof. (1) Follows immediately from the definitions.
(2) We reduce the 3-PARTITION [3] problem to the provability problem for
CC. Let $A = \{a_1, \ldots, a_{3m}\}$ be an instance of 3-PARTITION and $\sum_{i=1}^{3m} a_i = am$.
We define by A the following sequence of $(4m + 7)$ Horn implications:

$$\Gamma_A = (xy \to \varphi)(x^2 \to \varphi)(xu \to \varphi)(xz \to \varphi)(\varphi \to MAX)(x \to y^{9a})$$

$$(y^{3a} \to y^{a_1} z^{a-a_1}) \ldots (y^{3a} \to y^{a_{3m}} z^{a-a_{3m}})(y^a z^{2a} \to y^{9a} u)^m (y^{9a} u^m \to \varphi),$$

where MAX is a conjunction of left part of all implications without one symbol
x: $MAX = x^5 u^{m+1} z^{2am+1} y^{(9+10m)a+1}$. Γ_A is concurrent for any $a_1 \ldots, a_{3m}$.
However, it is applicable to x iff $A \in$ 3-PARTITION. $\square$

In [2] there was proposed a polynomial time verifiable sufficient condition
for the problem "$\Gamma \in$ SUC?" The following results shows that in general the
recognition problem is intractable for all three classes SUC, STC and CC.

Theorem 2 *The problems "$\Gamma \in$ SUC?" and "$\Gamma \in$ STC?" are co-NP-complete.*

Sketch of the proof. We reduce the complement of 3-PARTITION problem to the
problem "$\Gamma \in$ SUC?" ("$\Gamma \in$ STC?"). Let A be an instance of 3-PARTITION
as in the proof of Theorem 1.2. We define by A the following sequence of $4m + 2$
Horn implications $\Gamma_A = (y^{3a} \to y^{a_1} z^{a-a_1}) \ldots (y^{3a} \to y^{a_{3m}} z^{a-a_{3m}})(y^a z^{2a} \to$
$y^{9a})^m (y^{9a+1} \to y^{10am} z)(z \to y^{10am} z)$. Then $A \notin$ 3-PARTITION iff $\Gamma_A \in$ SUC
$(\Gamma_A \in$ STC). To prove this it is enough to check that Γ_A is superconcurrent (strong concurrent) for $w = y^{9a}$ iff $A \in$ 3-PARTITION. If $w = y^k$
and $k < 3a$ then Γ_A is not applicapable to w. In all other cases when
$w = y^k$, $k \geq 3a$, $k \neq 9a$, or $w = y^k z^m, m \geq 1$, Γ_A is superconcurrent
(strong concurrent) for w.

The upper bounds for both problems can be obtained by standard arguments.
$\square$

In [2] it was shown that the problem "$\Gamma \in$ CC?" is *co-NP*-hard. The following theorems refines this result.

Theorem 3 *The problem "$\Gamma \in$ CC?" is Π_2^P-complete.*

Sketch of the proof. Let Φ be some propositional formula depending of variables $x_1 \ldots, x_N, y_1 \ldots, y_M$. The validity problem for formulas of the form $\Psi =$
$\forall y_1 \ldots \forall y_M \exists x_1 \ldots \exists x_N \Phi$ is Π_2^P-complete [3, 6]. Given formula Ψ we construct
a sequence Γ which is in CC iff Ψ is true.

Let us fix a linear program $\mathcal{P}$ which computes the truth value of Φ from
values of $x_1 \ldots, x_N, y_1 \ldots, y_M$. We suppose that for every subformula φ of Φ
$\mathcal{P}$ includes only one operator to compute value of φ. It has one of the forms
$\varphi = \psi \circ \chi$, $\circ \in \{\&, \vee, \to\}$, or $\varphi = \neg \psi$. Let $\varphi_1 \ldots, \varphi_n$ be additional variables
used in $\mathcal{P}$. Let the result of $\mathcal{P}$ be in the output variable φ_n. For example, if
$\Phi = \neg(x_1 \to y_1) \vee x_2$ then $\mathcal{P}$ is

$\varphi_1 = x_1 \to y_1;$ $\varphi_2 = \neg \varphi_1;$ $\varphi_3 = \varphi_2 \vee x_2,$ and φ_3 is the output variable
of $\mathcal{P}$.

Let $S = \{x_1, \ldots, x_N, y_1, \ldots, y_M, \overline{x}_1, \ldots, \overline{x}_N, \overline{y}_1, \ldots, \overline{y}_M, y_1', \ldots, y_M', \overline{y}_1', \ldots, \overline{y}_M', \varphi_1, \ldots, \varphi_n, \overline{\varphi}_1, \ldots, \overline{\varphi}_n, b_1, \ldots, b_N, a_1, \ldots, a_M, a, v\}$. For every operator of $\mathcal{P}$ we include in the sequence Γ some implications defining by the following table.

Operator	Included implications
$\varphi = \psi \& \chi$	$(\psi\chi \to \varphi\psi\chi)(\overline{\psi}\chi \to \overline{\varphi}\psi\chi)(\psi\overline{\chi} \to \overline{\varphi}\psi\overline{\chi})(\overline{\psi}\overline{\chi} \to \overline{\varphi}\psi\overline{\chi})$
$\varphi = \psi \vee \chi$	$(\psi\chi \to \varphi\psi\chi)(\overline{\psi}\chi \to \varphi\overline{\psi}\chi)(\psi\overline{\chi} \to \varphi\psi\overline{\chi})(\overline{\psi}\overline{\chi} \to \overline{\varphi}\overline{\psi}\overline{\chi})$
$\varphi = \psi \to \chi$	$(\psi\chi \to \varphi\psi\chi)(\overline{\psi}\chi \to \varphi\overline{\psi}\chi)(\psi\overline{\chi} \to \overline{\varphi}\psi\overline{\chi})(\overline{\psi}\overline{\chi} \to \varphi\overline{\psi}\overline{\chi})$
$\varphi = \neg\psi$	$(\psi \to \overline{\varphi}\psi)(\overline{\psi} \to \varphi\overline{\psi})$

By the implications corresponding to operators of $\mathcal{P}$ we simulate a computation of Φ on known values of x_i and y_j. We include in Γ also the following implications:

1) $(v{y_i'}^2 \to \varphi_n)(v{\overline{y}_i'}^2 \to \varphi_n)(vy_i'\overline{y}_i' \to \varphi_n)$ $\quad i = \overline{1, M}$ $\quad$ 5) $(a_1 \ldots a_M \to b_1 \ldots b_N)$

2) $(vZ \to \varphi_n)$ $\quad Z \notin \{y_i'\} \cup \{\overline{y}_i'\}$ $\qquad\qquad$ 6) $(b_i \to x_i)(b_i \to \overline{x}_i)$ $\quad i = \overline{1, N}$

3) $(\ v \to a^M\ )$ $\qquad\qquad\qquad\qquad\qquad\qquad$ 7) $(\varphi_n \to MAX)$,

4) $(y_i'a \to y_i a_i)(\overline{y}_i' \to \overline{y}_i a_i)$ $\quad i = \overline{1, M}$ $\qquad$ 8) $(x_1 b_2 \to \varphi_n)$

where MAX is a conjunction of left parts of all implications without one symbol v.

It is easy to see that the size of Γ is polynomially bounded by the size of Ψ and that Γ can be constructed by Ψ in polynomial time. To finish the proof we show that Γ is concurrent iff the formula Ψ is true.

The crucial points for the assertion are those conjunctions which encode values of variables $y_1, \ldots, y_M$. They have form $v\tilde{y}_1' \ldots \tilde{y}_M'$, where $\tilde{y}_i' \in \{y_i', \overline{y}_i'\}$. y_i' corresponds to $y_i = true$, $\overline{y}_i'$ corresponds to $y_i = false$. Γ is applicapable to such conjunctions and it is concurrent iff Ψ is true. The concurrency of Γ for all other conjunctions does not depend on validity of Ψ. For every such w either Γ is concurrent or it is not applicapable to w.

To prove that the problem of concurrency recognition is in Π_2^P we reduce it to the validity problem for quantifying Boolean formulas of the form $\forall y_1, \ldots y_M \exists x_1, \ldots x_N \Phi$ which belongs to Π_2^P.

Let $\Gamma = (\alpha_1 \to \beta_1) \ldots (\alpha_n \to \beta_n)$ be a sequence of Horn implications. It is easy to see that Γ is concurrent if and only if it is concurrent for every sub-conjunction w of $\alpha_1 \ldots \alpha_n$ such that Γ is applicable to w. Now let $\mathcal{M}$ be a non-deterministic polynomial time Turing machine which given input w checks if Γ is applicable to w, and $\mathcal{N}$ be a non-deterministic polynomial time Turing machine which given input w checks if Γ is concurrent for w. Both machines can be constructed in polynomial time by Γ. As in the well-known Cook's proof of NP-completeness of SAT we construct Boolean formulas $\Phi(\overline{x}_1, \overline{x}_2)$ and $\Psi(\overline{y}_1, \overline{y}_2)$ such that $\mathcal{M}$ accepts w iff $\Phi(\overline{x}_1, \overline{x}_2) \in$ SAT and $\mathcal{N}$ accepts w iff $\Psi(\overline{x}_1, \overline{x}_2) \in$ SAT. We suppose that variables $\overline{x}_1$ and $\overline{y}_1$ in these formulas describe the start instantaneous descriptions of $\mathcal{M}$ and $\mathcal{N}$. Let formula $\Delta(\overline{x}_1)$ says that variables $\overline{x}_1$ correspond to some correct start instantaneous description of $\mathcal{M}$ and formula $\Lambda(\overline{y}_1)$ says the same for $\mathcal{N}$. And let $\Theta(\overline{x}_1, \overline{y}_1)$ be a formula that says that variables $\overline{x}_1$ and $\overline{y}_1$ describe the same input conjunction w.

Then it is easy to examine that the sequence Γ is concurrent if and only if the following formula A_Γ is true.

$$A_\Gamma = \forall \overline{x}_1 \forall \overline{y}_1 \left(\Lambda \& \Delta \& \Theta \to (\exists \overline{x}_2 \Phi \to \exists \overline{y}_2 \Psi) \right).$$

By the standard Boolean transformations we get that A_Γ is equivalent to the following $\forall \exists$-formula:

$$\forall \overline{x}_1 \forall \overline{y}_1 \forall \overline{x}_2 \exists \overline{y}_2 \left(\Lambda \& \Delta \& \Theta \to (\neg \Phi \vee \Psi) \right)$$

Hence the problem is in Π_2^P. $\square$

We notice that alphabet S in the proof of theorem 2 contains a limited number of symbols but in the proof of theorem 3 the size of S is unbounded. The following theorem shows that for the case of fixed size of S the problem may be easier.

Theorem 4 *Let* $\mathbf{CC}(k)$, $k = 1, 2, 3, \ldots$ *be a subclass of* $\mathbf{CC}$ *with sequences* Γ *over a k-letters alphabet* S. *Then for every* $k = 1, 2, \ldots$ *the problem* "$\Gamma \in \mathbf{CC}(k)$?" *is in* Δ_2^P.

Sketch of the proof. Let the alphabet of Γ be $a_1, \ldots a_k$. As in the previous proof we notice that to recognize concurrency it is enough to examine all sub-conjunctions w of $\alpha_1 \ldots \alpha_n = a_1^{m_1} \ldots a_k^{m_k}$. Since for every i $m_i < |\Gamma|$ the number N of different subconjunctions w is less than $|\Gamma|^k$.

For every such subconjunction w_i $(1 \leq i \leq N)$ we define the formula $\Phi_i(\overline{x}_i)$ which says that Γ is applicable to w_i, and the formula $\Psi_i(\overline{y}_i)$ which says that Γ is concurrent for w_i. Then the concurrency of Γ is described by the formula

$$\bigwedge_{i=1}^{N} [(\exists \overline{x}_i \Phi_i) \to (\exists \overline{y}_i \Psi_i)],$$

The validity of this formula can be checked in polynomial time with respect to NP-oracle SAT. Hence our problem is in Δ_2^P. $\square$

4 Maximal Concurrency

In this section we introduce a hierarchy of classes of Horn implication sequences which extends class **STC** and have the provability problem solvable in polynomial time. The idea is to use for a given k on each step of computation not less than k implications simultaneously.

Definition 5 *A sequence of Horn implications* Γ *is k-maximal concurrent for a conjunction w if one of two following conditions holds:*
(1) Γ is maximal for w;
(2) (a) If there is a maximal subsequence Λ of Γ and $|\Lambda| \geq k$ then for any maximal subsequence Δ such that $|\Delta| \geq k$ $(\Gamma \setminus \Delta)$ is k-maximal concurrent for the result of application Δ to w.

(b) If such a subsequence doesn't exist then for any maximal subsequence Δ of maximal length $(\Gamma \setminus \Delta)$ is k-maximal concurrent for the result of application Δ to w.

A sequence Γ of Horn implications is k-maximal concurrent iff for any w either Γ is k-maximal concurrent for w or Γ is not applicapable to w. A set of all k-maximal concurrent sequences we denote by $\mathbf{MC}_k$.
A sequence is maximal concurrent if it is k-maximal concurrent for some k. Let $\mathbf{MC} = \cup \{ \mathbf{MC}_k : k \in \omega \}$.

k-maximal concurrency garantees that on each step of application one can use any maximal subsequence of length k or more. If such "long" subsequence doesn't exist then any maximal subsequence of maximal length can be applied.

Theorem 5 (1) For every k the provability problem for class $\mathbf{MC}_k$ is solvable in polynomial time.
(2) The provability problem for class $\mathbf{MC}$ is NP-complete.
(3) Let $\mathbf{MC}(m)$, $m = 1, 2, 3, \ldots$ be a subclass of $\mathbf{MC}$ with sequences Γ over m-letters alphabet S. Then the provability problem for $\mathbf{MC}(m)$ is solvable in polynomial time.

Sketch of the proof. (1) Follows immediately from the definition of $\mathbf{MC}_k$.
(2) Let A is an instance of 3-PARTITION problem (as in the proof of Theorem 1.2). Denote by A the following sequence Γ :
$(y_i y_j y_k \to z)$ for all i, j, k such that $a_i + a_j + a_k = a$ and $i \neq j \neq k \neq i$,
$(z^m \to y_1^{9m^2} \ldots y_{3m}^{9m^2})$.
It is clear that Γ is applicable to $w = y_1 \ldots y_{3m}$ iff $A \in$ 3-PARTITION. It can be checked easily that $\Gamma \in \mathbf{MC}$.
(3) Let $\Gamma \in \mathbf{MC}(m)$ and $S = \{a_1, a_2, \ldots, a_m\}$ be the alphabet of Γ.
The main difficulty is to find a maximal subsequence $\Delta \subseteq \Gamma$ of maximal length . We show that the problem is solvable in time $O(|\Gamma|^2 |w|^m)$. Let ith implication of Γ be $a_1^{b_{i1}} a_2^{b_{i2}} \ldots a_m^{b_{im}} \to \beta_i$ and $w = a_1^{c_1} a_2^{c_2} \ldots a_m^{c_m}$. We reduce the problem of searching Δ to a linear optimization problem T of a special form and then solve it. Let T be the following problem:

$$\sum_{j=1}^{n} x_j \to \max$$

$$\begin{cases} \sum_{j=1}^{n} b_{ij} x_j \leq c_i & i = \overline{1, m} \\ x_j \in \{0, 1\} & j = \overline{1, n} \end{cases}$$

If x^* is a solution of T then $\Delta = \{(\alpha_i \to \beta_i) \in \Gamma : x_i^* = 1\}$ is the longest maximal subsequence of Γ.

When Δ is found we apply it to w. Then we repeat the procedure to $\Gamma_1 = (\Gamma \setminus \Delta)$ and so on, until for some i either $\Gamma_i = \emptyset$ (in this case the answer is "yes") or $\Gamma_i \neq \emptyset$ and $\Delta_i = \emptyset$ (the answer is "no"). The number of steps is less or equal to $|\Gamma|$. So, to prove that the problem of applicability of Γ to w can be evaluated in polynomial time it is enough to establish the following lemma.

Lemma 1 *The problem T can be solved in time $O(|\Gamma|^2|w|^m)$.*

The proof of the lemma is based on the dynamic programming technique.
We will compute the predicate $t(p, q, d_1, d_2, \ldots d_m)$ which says that the following system

$$
\begin{cases}
\sum_{j=1}^{p} x_j \geq q \\
\sum_{j=1}^{p} b_{ij} x_j \leq d_i \quad i = \overline{1, m} \\
x_j \in \{0, 1\} \quad j = \overline{1, p}
\end{cases}
$$

has a solution and simulteneously construct the sets $L(p, q, d_1, d_2, \ldots d_m) \subseteq \{x_1, x_2, \ldots x_p\}$ of it's solutions. The arguments have the following ranges: $p \in [1, n]$, $q \in [1, n]$, $d_i \in [0, c_i]$.   Values of t and L are defined by induction on p.

If $p = 1$ we set
$t(1, 0, d_1, d_2, \ldots d_m) = 1$,
$L(1, 0, d_1, d_2, \ldots d_m) = \emptyset$,
$t(1, 1, d_1, d_2, \ldots d_m) = 1$ if and only if $b_{11} \leq d_1, b_{21} \leq d_2, \ldots b_{m1} \leq d_m$,
$L(1, 1, d_1, d_2, \ldots d_m) = \{x_1\}$, if $t(1, 1, d_1, d_2, \ldots d_m) = 1$, and
$t(1, q, d_1, d_2, \ldots d_m) = 0$ for $q \geq 2$.

Now suppose that $t(p-1, q, d_1, \ldots d_m)$ and $L(p-1, q, d_1, \ldots d_m)$ are defined for all possible values of $q, d_1, \ldots, d_m$. Then the values of $t(p, q, d_1, \ldots d_m)$ and $L(p, q, d_1, \ldots d_m)$ can be defined as follows.
$t(p, q, d_1, d_2, \ldots d_m) = 1$ if and only if one of the two following conditions holds.

1. $t(p-1, q, d_1, d_2, \ldots d_m) = 1$, the solution exists for less number of variables. In this case the solution is the same:

$$
L(p, q, d_1, d_2, \ldots d_m) = L(p-1, q, d_1, d_2, \ldots d_m);
$$

2. $d_1 \geq b_{p1}, d_2 \geq b_{p2}, \ldots, d_m \geq b_{pm}$ and $t(p-1, q-1, d_1-b_{p1}, d_2-b_{p2}, \ldots d_m - b_{pm}) = 1$. In this case x_p is included into solution:

$$
L(p, q, d_1, d_2, \ldots d_m) = L(p-1, q-1, d_1 - b_{p1}, d_2 - b_{p2}, \ldots d_m - b_{pm}) \cup \{x_p\}.
$$

The complexity of computation of t and L for any p does not exceed $O(n \times \prod_{i=1}^{m} c_i)$. Therefore we can compute the maximal q such that $t(n, q, c_1, \ldots, c_m) = 1$ and the corresponding solution $L(n, q, c_1, \ldots, c_m)$ in time $O(n^2 \times \prod_{i=1}^{m} c_i) = O(|\Gamma|^2|w|^m)$. $\square$

The recognition problem for classes $\mathbf{MC}_k$ is difficult enough.

Theorem 6 *For every k the problem "$\Gamma \in \mathbf{MC}_k$?" is co-NP-complete.*

Sketch of the proof. It is easy to check that the problem is in *co-NP*. To prove the lower bound it is enough to consider the sequence Γ_A from the proof of theorem 2. $\square$

The following result shows that classes $\mathbf{MC}_k$ form a strict hierarchy.

Theorem 7 *For all* $k = 1, 2, 3, \ldots \mathbf{MC}_k \subset \mathbf{MC}_{k+1}$.

Sketch of the proof. We show for every k and Γ if $\Gamma \in \mathbf{MC}_k$ then $\Gamma \in \mathbf{MC}_{k+1}$. Suppose that Γ is k-maximal concurrent for w. Then one of conditions of the definition 5 holds. If it is condition 1 or 2(b) then it is easy to see that the sequence Γ is $k+1$-maximal concurrent. If Γ satisfies condition 2(a) then one of two following cases holds .

(i) There is a maximal subsequence of the length $\geq k + 1$. In this case the definition holds for $k + 1$ as well.

(ii) There is no maximal subsequence of the length $\geq k + 1$. In this case the definition says that we can use any maximal subsequence of the length k. Such subsequences are the longest among the all maximal subsequences. Hence condition 2(b) holds for the definition of $(k + 1)$-maximal concurrency.

To show that $\mathbf{MC}_k \neq \mathbf{MC}_{k+1}$ let us consider the sequence $\Gamma = (a \to b)^{k+1}(a^2 \to b)(b^k \to c)(b \to a^2)$. Let $w = a^{k+1}c^l$ for any l. Then Γ is $(k + 1)$-maximal concurrent but not k-maximal concurrent for w. It is easy to check that for any other w the two properties of Γ are equivalent. Hence $\Gamma \in \mathbf{MC}_{k+1} \setminus \mathbf{MC}_k$. $\square$

It follows immediately from the definitions that $\mathbf{MC} \subseteq \mathbf{CC}$. It is easy to see that sequence $\Gamma = (a^2 \to a)(a^2 \to a^2) \in \mathbf{CC} \setminus \mathbf{MC}$ and therefore $\mathbf{MC} \neq \mathbf{CC}$. ¿From theorem 7 it follows that for every k $\mathbf{MC}_k \neq \mathbf{MC}$. So we get the following relationships between the classes we consider.

$$\mathbf{SUC} \subset \mathbf{STC} = \mathbf{MC}_1 \subset \mathbf{MC}_2 \subset \ldots \subset \mathbf{MC}_k \subset \mathbf{MC}_{k+1} \subset \ldots \subset \mathbf{MC} \subset \mathbf{CC}$$

5 Conclusions

In this work we have introduced several new natural subclasses of multisets of Horn implications with polynomial time provability problem. The proposed definitions explore the idea of parallel application of a maximal number of implications. The complexity of concurrency recognition for these classes as well as for classes defined in [2] was shown to be high enough. So it would be interesting to find out some conditions which guarantee polynomial time algorithms for the problem.

Acknowledgments

I would like to thank M.I. Dekhtyar for numerous propositions and comments, and to M.A. Taitslin for the fruitful discussion of this work.
This work was sponsored by the Russian Foundation for Basic Research (grant 96-01-00086).

References

1. Archangelsky D.A., Taitslin M.A., *Linear Logic With Fixed Resources*. Annals of Pure and Applied Logic, v. 67 (1994),pp. 3–28.

2. Archangelsky D.A., Dekhtyar M.I., Kruglov E., Musikaev I.Kh., and Taitslin M.A. *Concurrency problem for Horn fragment of Girard's Linear Logic.* Logical Foundation of Computer Science, St.Petersburg'94, Lecture Notes in Computer Science, N 813, 1994, pp. 18–22.

3. Garey M.R., Johnson D.S. *Computers and Intractability. A Guide to the Theory of NP-Completeness.* W.H.Freeman and Company, San Francisco, 1979.

4. Kanovich M.I. *Horn Programming in Linear Logic is NP-complete.* Proc.7-th Annual IEEE Symposium on Logic in Computer Science, 1992, pp. 200–210.

5. Lincoln P., Mitchell J., Scerdov A., and Shankar N. *Decision Problems for Propositional Linear Logic.* Proc.31-th IEEE Symposium on Foundation of Computer Science, 1990, pp. 662–671.

6. Meyer A.R., Stockmeyer L.J. *The equivalence problem for regular expressions with squaring requires exponential time.* Proc.13th Ann. Symp. on Switching and Automata Theory, 1972, pp. 125–129.

Studying Algorithmic Problems for Free Semi-groups and Groups

Valery Durnev*

Yaroslavl State Univesity, Yaroslavl, Russia

Abstract. In this paper we study some algorithmic problems for free semi-groups and groups.

1 Unsolvability of positive $\forall\exists^3$-theory of a free semi-group

Let Π_m denote the *free semi-group of rank m with free generators $a_1, \ldots, a_m$*. In 1946, W. Quine [25] proved the unsolvability of the elementary theory of the semi-group Π_m for $m \geq 2$. Since then, various parts of this theory have been investigated. In 1973, the author in [3] proved the unsolvability of the fragment of the elementary theory of Π_m consisting of negation-free formulae with quantor prefixes of type $\exists x \forall y \exists z_1 \exists z_2 \exists z_3$. In [13], S. S. Marchenkov proved the unsolvability of the positive $\forall\exists^4$-theory of a free semi-group, which improves the result of [3] from the standpoint of a number of the quantifier blocks in the considered formulae, but the total number of quantifiers used in [3] and [13] is the same.

In [8], we further improved the results of [3] and [13] and obtained the following theorem.

Theorem 1. *For $m \geq 2$ the positive $\forall\exists^3$-theory of the free semi-group Π_m is algorithmically unsolvable.*

Note that by G. S. Makanin's theorem [11], both the universal and existential theories of the semi-group Π_m are algorithmically solvable, i.e. there exists an algorithm which, given an arbitrary closed formula Φ in prenex form, in which the quantifier prefix consists of only one quantifier block, determines if Φ is true of the semi-group Π_m.

2 Solvability of equations with right-hand parts in free semi-groups and groups

We will consider the following problem in an arbitrary semi-group Π:
  "*Consider an equation of the form*

---

* Supported by the grant # 96-01-00525 of the Russian Foundation for Fundamental Research

$$w(x_1, \ldots, x_n) = g, \tag{1}$$

where $w(x_1, \ldots, x_n)$ is a word in the alphabet of unknowns

$$\{x_1, \ldots, x_t, \ldots\},$$

and g is an element of semi-group Π, and determine whether it has a solution".

Equations of form (1) have been studied in a number of papers and were called: *equations with right-hand parts* or *equations solved with respect to unknowns*.

If Π is the free semi-group Π_2 with free generators a_1, a_2, then we consider the number

$$L = |w(x_1, \ldots, x_n)| + |g(a_1, a_2)|$$

where $|A|$ is the length of word A, as *the dimension of the solvability problem for equations of form (1)* .

We have proved the following theorem.

Theorem 2. *The solvability problem for equations of form (1) in the free semi-group Π_2 is NP-complete.*

Equations of form (1) in the case where the semi-group Π is a group are called *equations with one coefficient.* The solvability problem for equations of form (1) in a free group was investigated by R. Lyndon [24] and P. Schupp [26] and was called *the substitution problem for free groups.*

It was generally considered that the solvability problem for equations of form (1) in a free group was a "simpler" problem than the solvability problem of arbitrary equations. Thus the following result is somewhat unexpected.

Theorem 3. *The solvability problem for equations in the free group F_m is linearly reducible to the solvability problem for equations with one coefficient in the group F_k, and the solvability problem for systems of equations in the group F_m is polynomially reducible to the solvability problem for equations with one coefficient in the group F_k where*

$$k = 2[\frac{m+1}{2}] \quad (m \geq 2).$$

We conclude this section with the following theorem.

Theorem 4. *The solvability problem for equations with one coefficient in the free group F_2 is NP-difficult.*

3 Equations in words and generalized lengths in free semi-groups

Since G. S. Makanin [11] gave an algorithm solving the problem of consistency for systems of equations in free groups, the question of the existence of an analogous algorithm for systems of equations in words and lengths has been of special interest. In this connection, the following two theorems are of interest for us.

Let $M(n)$ denote the set of all non-ordered pairs $\{i, j\}$ $(i, j = 1, \ldots, n)$ for an arbitrary natural number n.

If X is a word in the semi-group Π_m, then let $|X|$ denote the length of word X, and let $|X|_i$ denote the number of entries of the letter a_i in the word X.

Theorem 5. *We can produce a number n, a subset B of $M(n)$ and a one-parametric class of equations*

$$w(x, x_1, \ldots, x_n, a_1, a_2) = v(x, x_1, \ldots, x_n, a_1, a_2)$$

with unknowns $x_1, \ldots, x_n$, constants a_1, a_2 and parameter x, such that there does not exist an algorithm that decides of an arbitrary natural number k, whether the equation

$$w(a_1^k, x_1, \ldots, x_n, a_1, a_2) = v(a_1^k, x_1, \ldots, x_n, a_1, a_2)$$

has a solution $\langle X_1, \ldots, X_n \rangle$ in the free semi-group Π_m $(m \geq 2)$, satisfying $|X_i|_1 = |X_j|_1$ and $|X_i|_2 = |X_j|_2$, for all $\{i, j\} \in B$.

Theorem 6. *We can produce a number n, a subset B of $M(n)$ and a one-parametric class of equations*

$$w(x, x_1, \ldots, x_n, a_1, a_2) = v(x, x_1, \ldots, x_n, a_1, a_2)$$

with unknowns $x_1, \ldots, x_n$, constants a_1, a_2 and parameter x, such that there does not exist an algorithm that decides of an arbitrary natural number k, whether the equation

$$w(a_1^k, x_1, \ldots, x_n, a_1, a_2) = v(a_1^k, x_1, \ldots, x_n, a_1, a_2)$$

has a solution $\langle X_1, \ldots, X_n \rangle$ in the free semi-group Π_m $(m \geq 2)$, satisfying $|X_i| = |X_j|$ and $|X_i|_2 = |X_j|_2$, for all $\{i, j\} \in B$.

4 Some algorithmic problems for Diophantine sets in Π_2

Let a and b denote the free generators of the semi-group Π_2.

A subset S of the set Π_2^m is called *Diophantine* if there exist words u, v in the alphabet

$$\{a, b, x_1, \ldots, x_m, y_1, \ldots, y_n, \ldots\},$$

such that for all elements $g_1, \ldots, g_m \in \Pi_2$, the equivalence

$\langle g_1, \ldots, g_m \rangle \in S \iff$

$\Pi_2 \models (\exists y_1, \ldots, y_n) \; u(g_1, \ldots, g_m, y_1, \ldots, y_n, a, b) = v(g_1, \ldots, g_m, y_1, \ldots, y_n, a, b)$

is valid. We call such a pair of words $\langle u, v \rangle$ a *representaton* of the set S.

The intersection and union of two Diophantine sets is also a Diophantine set. Every one-element set and its complement are Diophantine, therefore any finite set and its complement are Diophantine.

We have constructed a Diophantine set in Π_2^2 whose complement is *not Diophantine*.

We have considered algorithmic questions concerning the number of elements belonging to a Diophantine set or to its complement, and proved the following theorems.

Theorem 7. *There exists an algorithm that decides, given any representation of a Diophantine set S, i.e. the corresponding pair of words $\langle u, v \rangle$, and given any natural number k, whether the set S contains not less than k elements, or not more than k elements, or exactly k elements.*

The following problem remains open: *Does there exist an algorithm that decides, given an arbitrary representation $\langle u, v \rangle$ of a Diophantine set S, whether the set S is finite?*

If a given Diophantine set S is finite, we can establish this fact. Difficulties arise if S is infinite. The following theorem throws light on the nature of these difficulties.

Theorem 8. *For any fixed k, there does not exist an algorithm that decides, given an arbitrary representation $\langle u, v \rangle$ of a Diophantine set S, whether the complement of S consists of k elements (consists of not less than k elements if $k > 0$, or consists of not more than k elements).*

A further problem remains open: *Does there exist an algorithm that decides, given an arbitrary representation $\langle u, v \rangle$ of a Diophantine set S, whether the complement of S is finite?*

5 Equations in free groups with subgroup restrictions on the solutions

Let F_m be the free group with free generators $a_1, \ldots, a_m$, and let F be the free group of denumerable rank, the free generators of which are denoted $x_1, x_2, \ldots$.

Expressions of the form:

$$w(x_1, \ldots, x_n, a_1, \ldots, a_m) = 1,$$

where $w \in F_m * F$ is called *an equation in the free group F_m with unknowns* $x_1, \ldots, x_n$.

A set of elements $g_1, \ldots, g_n$ of the group F_m is called a *solution* of the indicated equation if

$$w(g_1, \ldots, g_n, a_1, \ldots, a_m) = 1$$

in F_m.

G. S. Makanin [12] has given an algorithm that decides whether an equation in a free group has a solution.

A number of questions about free groups can be reduced to questions concerning the existence of solutions of equations with various properties. The question of the solvability of the positive theory of a free group, positively solved by G. S. Makanin, was reduced using the known result of Yu. I. Merzlyakov about concurrence of positive theories of free non-Abelian groups to the following problem:

To construct an algorithm that decides of an arbitrary equation

$$w(x_1, \ldots, x_n, a_1, \ldots, a_m) = 1,$$

whether it has a solution $g_1, \ldots, g_n \in F_m$ *satisfying the condition*

$$g_1 \in F_{m_1}, \ldots, g_t \in F_{m_t}$$

where $m_1 \leq \ldots \leq m_t$ *and* F_{m_i} *is a free group with generators* $a_1, \ldots, a_{m_i}$.

In "Kourov's Notebook" G. S. Makanin formulated the following problem:

"9.25. *To give an algorithm that decides, given an equation*

$$w(x_1, \ldots, x_n, a_1, \ldots, a_m) = 1$$

in the free group F_m *and given a list of finitely generated subgroups* $H_1, \ldots, H_n$ *of the group* F_m, *whether there exists a solution of this equation satisfying the condition*

$$x_1 \in H_1, \ldots, x_n \in H_n".$$

Finitely generated subgroups appear in the statement of problem 9.25 because the occurrence problem is solvable for finitely generated subgroups of a free group. In fact, the occurrence problem is solvable for certain infinitely generated subgroups of the free group F_m, and, for example, the occurrence problem for the commutator $[F_m, F_m]$ of the free group F_m can be solved very simply, indeed, more simply than for certain finitely generated subgroups. Therefore, it is natural to give the following generalization of problem 9.25.

"9.25a. *To give an algorithm that decides, given an equation*

$$w(x_1, \ldots, x_n, a_1, \ldots, a_m) = 1$$

in the free group F_m *and given a list of subgroups* $H_1, \ldots, H_n$ *of the* F_m *with solvable occurrence problems, whether there exists a solution of this equation satisfying the condition*

$$x_1 \in H_1, \ldots, x_n \in H_n".$$

We have proved the algorithmic unsolvability of problem 9.25a.

Theorem 9. *For any free group F_m with free generators $a_1, \ldots, a_m$ $(m \geq 2)$ we can construct an equation*

$$w(x, x_1, \ldots, x_n, a_1, a_2) = 1$$

with unknowns $x_1, \ldots, x_n$, constants a_1, a_2 and parameter x, such that there does not exist an algorithm that decides of an arbitrary natural number k, whether there exists a solution of the equation

$$w(a_1^k, x_1, \ldots, x_n, a_1, a_2) = 1$$

satisfying the condition

$$x_1 \in [F_m, F_m], \ldots, x_t \in [F_m, F_m],$$

where $[F_m, F_m]$ is the commutator of the group F_m, and t is some fixed number between 1 and n.

Let $G^{(s)}$ denote the s commutator of a group G, and let G_s denote the s term of the lower central series of a group G.

Recall that

$$G^{(0)} \rightleftharpoons G, \quad G^{(n+1)} \rightleftharpoons [G^{(n)}, G^{(n)}],$$

$$G_0 \rightleftharpoons G, \quad G_{n+1} \rightleftharpoons [G_n, G].$$

Corollary 10. *For any free group F_m with free generators $a_1, \ldots, a_m$ $(m \geq 2)$ and for any $s \geq 1$, we can construct an equation*

$$w_s(x, x_1, \ldots, x_n, a_1, a_2) = 1$$

with unknowns $x_1, \ldots, x_n$, constants a_1, a_2 and parameter x, such that there does not exist any algorithm that decides, for an arbitrary natural number k, whether there exists a solution $g_1, \ldots, g_n$ of the equation

$$w_s(a_1^k, x_1, \ldots, x_n, a_1, a_2) = 1$$

such that $g_1 \in F_m^{(s)}, \ldots, g_t \in F_m^{(s)}$.

Corollary 10 remains valid if we replace the s commutant $F_m^{(s)}$ of group F_m by $(F_m)_s$, the s term of the lower central series of F_m.

The following corollary throws light on the causes of the difficulties that arise even in the case of equations without coefficients.

Corollary 11. *There does not exist an algorithm that decides of an arbitrary equation without coefficients $w(x_1, \ldots, x_n) = 1$ in the group F_2, whether it has a solution $g_1, \ldots, g_n$ satisfying*

$$g_1 \in F_2^{(1)}, \ldots, g_t \in F_2^{(1)} \text{ and } [g_{n-1}, g_n] = [a_1, a_2].$$

Notes and Comments. The conditions imposed on the solution of the equation without coefficients

$$w(x_1, \ldots, x_n) = 1$$

considered in the corollary, can be naturally divided into two classes:

Class One: $x_1 \in F_2^{(1)}, \ldots, x_t \in F_2^{(1)}$;

Class Two: the single condition: $[x_{n-1}, x_n] = [a_1, a_2]$.

It is interesting to note that if we omit the conditions of one class or the other, we obtain an algorithmically solvable problem.

In conclusion, we note that the situation is different for the equations with one unknown.

Let N be either the s commutator $F_m^{(s)}$ of the free group F_m or the s term $(F_m)_s$ of its lower central series. We have proved the following theorem.

Theorem 12. *There exists an algorithm that decides of any equation with one unknown*

$$w(x_1, a_1, \ldots, a_m) = 1$$

in the group F_m, whether it has such solution x_1 such that $x_1 \in N$.

Corollary 13. *There exists an algorithm that decides of any equation without coefficients $w(x_1, x_2, x_3) = 1$ in three unknowns, whether it has a solution g_1, g_2, g_3 satisfying $g_1 \in N$ and $[g_2, g_3] = [a_1, a_2]$, where N is a subgroup of group F_2, the same as in the theorem.*

6 Equations with automorphisms and endomorphisms in free groups

In "Kourov's Notebook" G. S. Makanin raised the following question:

"*10.26... Docs there exists an algorithm recognizing the solvability in the free group of equations of the following form:*

$$w(\varphi_1(x_1), \ldots, \varphi_n(x_n)) = 1$$

where $\varphi_1, \ldots, \varphi_n$ are automorphisms of this group?"

The purpose of this section is to show that if we consider endomorphisms along with automorphisms then the answer to the reformulated question will be negative.

Let a, b denote the free generators of the group F_2, as before.

Theorem 14. *We can give an automorphism ψ and an endomorphism φ of the free group F_2 as well as a set of the equations*

$$w(x, x_1, \ldots, x_n, \psi(x_n), \varphi(x_1), \ldots, \varphi(x_t), a, b) = 1$$

with unknowns $x_1, \ldots, x_n$, constants a, b and parameter x such that there does not exist an algorithm that decides of an arbitrary natural number k whether the equation

$$w(a^k, x_1, \ldots, x_n, \psi(x_n), \varphi(x_1), \ldots, \varphi(x_t), a, b) = 1$$

has a solution in the free group F_2.

Endomorphism φ in the previous theorem is applied to many variables, but, moving up from group F_2 to group F_3, we can ensure that the endomorphisms are applied only once, and that the total number of the used endomorphisms remains the same.

Let F_3 be the free group with generators a, b, c, and φ_1, φ_2 be endomorphisms of F_3 determined by the following equalities:

$$\varphi_2(b) \rightleftharpoons \varphi_1(a) \rightleftharpoons 1, \ \varphi_2(a) \rightleftharpoons a, \ \varphi_1(b) \rightleftharpoons b, \ \varphi_1(c) \rightleftharpoons \varphi_2(c) \rightleftharpoons c.$$

Then we have the following theorem.

Theorem 15. *There does not exist an algorithm that decides of an arbitrary equation*

$$w(x_1, \ldots, x_n, a, b, c) = 1$$

in the group F_3, whether it has a solution $x_1, \ldots, x_n$ such that $\varphi_1(x_1) = 1$ and $\varphi_2(x_1) = 1$.

Let H denote the intersection of the kernels of the endomorphisms φ_1 and φ_2 defined above. Then the following is a direct corollary of the theorem.

Corollary 16. *There does not exist an algorithm that decides of an arbitrary equation*

$$w(x_1, \ldots, x_n, a, b, c) = 1$$

in the group F_3, whether it has a solution $x_1, \ldots, x_n$ such that $x_1 \in H$.

The theorem formulated above is interesting in connection with the second part of question 10.26 from "Kourov's Notebook":
"... does there exist an algorithm recognizing the solvability in the free group of equations of the form

$$w(\varphi_1(x_{i_1}), \ldots, \varphi_n(x_{i_n})) = 1$$

where $\varphi_1, \ldots, \varphi_n$ are automorphisms of this group?"

Corollary 16, in its turn, is interesting in connection with question 9.25 from "Kourov's Notebook", discussed in the previous section. Although the subgroup H is not finitely generated, the occurrence problem for H can be solved for linear time.

Let ψ denote the following automorphism of the group F_3:

$$\psi(a) \rightleftharpoons b, \ \psi(b) \rightleftharpoons a, \ \psi(c) \rightleftharpoons c.$$

Corollary 17. *There does not exist an algorithm that decides of an arbitrary equation*

$$w(x_1, \ldots, x_{n+2}, a, b, c) = 1$$

whether it has a solution $x_1, \ldots, x_{n+2}$ such that

$$\psi(x_1) = x_{n+1} \qquad \varphi_1(x_{n+2}) = 1.$$

Corollary 18. *There does not exist an algorithm that decides of an arbitrary equation*

$$v(x_1, \ldots, x_{n+2}, \psi(x_1), \varphi(x_{n+2}), a, b, c) = 1$$

whether it has a solution in F_3.

There is a certain "non-symmetry" of the generators a, b, c in the definition of the endomorphisms φ_1 and φ_2: endomorphisms φ_1, φ_2 "delete" a or b, but do not change c.

As it turns out, this "non-symmetry" can be removed in the following way.

As before, let F_m denote the free group of rank m with free generators $a_1, \ldots, a_m$, and let φ_i denote the following endomorphisms:

$$\varphi_i(a_j) \rightleftharpoons a_j \text{ for } j \neq i, \quad \varphi_i(a_i) \rightleftharpoons 1. \tag{2}$$

By analogy with braid groups, the endomorphism φ_i can be called *the endomorphism plucking the i-th generator*, then, naturally, we will call $Ker\,\varphi_i$ the *subgroup of i-pure elements*, and

$$P_m \rightleftharpoons \bigcap_{i=1}^{m} Ker\,\varphi_i$$

as the *subgroup of pure or smooth elements*.

It is clear that $P_m \subseteq F_m^{(1)}$ and $P_2 = F_2^{(1)}$, but for $m \geq 3$ $P_m \neq F_m^{(1)}$.

Subgroups P_m are dual to the commutators $F_n^{(1)}$ of the groups F_m in the following sense. Define endomorphisms ψ_i by the following equalities (3) dual to equalities (2):

$$\psi_i(a_j) \rightleftharpoons 1 \text{ for } j \neq i, \quad \psi_i(a_i) \rightleftharpoons a_i; \tag{3}$$

then

$$F_m^{(1)} = \bigcap_{i=1}^{m} Ker\,\psi_i.$$

Therefore, the following theorem is interesting for us.

Theorem 19. *For $m \geq 3$, there does not exist an algorithm that decides of an arbitrary equation in the group F_m*

$$w(x_1, \ldots, x_n, a_1, \ldots, a_m) = 1$$

whether it has a solution $x_1, \ldots, x_n$ such that $x_1 \in P_m$.

Theorem 20. *There exists an algorithm that decides of any equation with one unknown*

$$w(x_1, a_1, \ldots, a_m) = 1$$

in the group F_m, whether it has a solution x_1 such that $x_1 \in P_m$.

As above, let $F_m^{(s)}$ denote the s commutant of free group F_m of rank m, and let $(F_m)_s$ denote the s term of the lower central series of F_m.

It was proved that for $m \geq 2$ and $s \geq 1$, we could construct an equation

$$w_s(x, x_1, \ldots, x_n, a_1, a_2) = 1$$

with unknowns $x_1, \ldots, x_n$, constants a_1, a_2 and parameter x, such that there does not exist an algorithm that decides of an arbitrary natural number k, whether the equation

$$w_s(a_1^k, x_1, \ldots, x_n, a_1, a_2) = 1$$

has a solution $g_1, \ldots, g_n$ such that

$$g_1 \in F_m^{(s)}, \ldots, g_t \in F_m^{(s)}$$

or correspondingly

$$g_1 \in (F_m)_s, \ldots, g_t \in (F_m)_s$$

where t is a proper fixed number.

The following two theorems considerably strengthen the indicated result: they show that for $s \geq 2$ we can take $t = 1$.

Theorem 21. *For any $s \geq 2$, there does not exist an algorithm that decides of an arbitrary equation*

$$w(x_1, \ldots, x_n, a_1, a_2) = 1$$

in the group F_2, whether it has a solution $g_1, \ldots, g_n$ such that $g_1 \in F_2^{(s)}$.

Notes and Comments. It is easy to understand why the theorem holds for any group F_m for $m \geq 2$. If we consider the terms of lower central series $(F_m)_s$ of group F_m instead of the commutators $F_m^{(s)}$, then for $s \geq 5$ the validity of the theorem immediately follows from the result of N. N. Repin [16], replacing the condition $x_1 \in F_2^{(s)}$ by the condition $x_1 \in (F_2)_s$. However, there arise principal difficulties for "small" s, particularly, if $s = 2$. Thus the following theorem is of interest.

Theorem 22. *We can give an r such that there does not exist an algorithm that decides of an arbitrary equation*

$$w(x_1, \ldots, x_n, a_1, \ldots, a_r) = 1$$

in the free group F_r, whether it has a solution $g_1, \ldots, g_n$ such that $g_1 \in (F_r)_2$.

Notes and Comments. It is easy to show that the theorem remains valid if we replace the condition $g_1 \in (F_r)_2$ by any of the conditions:

$$g_1 \in (F_r)_s$$

for $s \geq 2$.

7 Equations in words, lengths and with endomorphisms in free semi-groups

Note that the free semi-group Π_2 has only two automorphisms: the identity and automorphisms ψ such that

$$\psi(a_1) = a_2, \ \psi(a_2) = a_1.$$

Therefore, studying systems of equations in words, lengths and with injective endomorphisms in the semi-group Π_2 is of interest. In this direction we have obtained the following results.

Theorem 23. *We can construct an injective endomorphism φ of the semi-group Π_2 and a system of equations*

$$\Psi(x, x_1, \ldots, x_n)$$

of the form

$$w = u \ \ and \ \ \bigwedge_{\{i,j\} \in B} |x_i| = |x_j| \ \ and \ \ \varphi(x_1) = x_2,$$

such that there does not exist an algorithm that decides of an arbitrary natural number p whether the system

$$\Psi(a_1^p, x_1, \ldots, x_n)$$

has a solution in Π_2.

Corollary 24. *For the injective endomorphism φ of the semi-group Π_2 indicated in the theorem, there does not exist an algorithm that decides of an arbitrary system of equations in words, lengths and with endomorphism φ of the form*

$$w(x_1, \ldots, x_n, \varphi(x_1), a_1, a_2) = u(x_1, \ldots, x_n, a_1, a_2)$$

$$and \ \ \bigwedge_{\{i,j\} \in B} |x_i| = |x_j|$$

whether it has a solution in Π_2.

Theorem 25. *We can construct an automorphism ψ of the semi-group Π_2 and an endomorphism φ such that there does not exist an algorithm that decides of an arbitrary equation of the form*

$$w(x_1, \ldots, x_n, \psi(x_n), \varphi(x_1), \ldots, \varphi(x_t), a_1, a_2) = u(x_1, \ldots, x_n, a_1, a_2)$$

whether it has a solution in Π_2.

The equations considered above contain automorphism ψ and endomorphism φ only in their left-hand parts, and the automorphism ψ is applied only to one variable.

If we go up from the semi-group Π_2 to the semi-group Π_3, we can obtain the following result in which the endomorphism φ is applied only once.

Theorem 26. *We can construct an automorphism ψ of the semi-group Π_3 and an endomorphism φ such that there does not exist an algorithm that decides of an arbitrary equation of the form*

$$w(x_1, \ldots, x_n, \psi(x_1), \varphi(x_2), a_1, a_2, a_3) = u(x_1, \ldots, x_n, a_1, a_2, a_3)$$

whether it has a solution in Π_3.

8 Equations in words and lengths with endomorphisms in free groups

The naturalness of including the length function in the signature when studying elementary theories of free products of groups is substantiated in [14] and [15]. In particular, the elementary theory of the free group in a signature that includes the predicate for equality of lengths is studied in these papers. In addition, elementary theories of groups and semi-groups in signatures that include fixed endomorphisms and automorphisms are studied in a number of papers listed in the References. Thus the following results are of interest.

Let $|g|$ denote the length of an element g of the group F_2 with respect to its fixed free generators a_1, a_2, i.e. the length of the corresponding irreducible word.

Theorem 27. *We can construct an automorphisms φ of the group F_2 and a system of equations $\Psi(x, x_1, \ldots, x_n)$ of the form*

$$w(x, x_1, \ldots, x_n, a_1, a_2) = 1 \quad and \quad \bigwedge_{\{i,j\} \in B} |x_i| = |x_j| \quad and \quad \varphi(x_1) = x_2,$$

such that there does not exist an algorithm that decides of an arbitrary natural number p, whether the system of equations

$$\Psi(a_1^p, x_1, \ldots, x_n)$$

has a solution in F_2.

Corollary 28. *Does there exist an automorphism φ of the group F_2 such that there does not exist an algorithm that decides of an arbitrary system of equations in words, lengths and with automorphism φ of the form*

$$u(x_1, \ldots, x_n, \varphi(x_1), a_1, a_2) = 1 \quad and \quad \bigwedge_{\{i,j\} \in B} |x_i| = |x_j|$$

whether it has a solution in F_2.

References

1. Adian S. I., Makanin G. S., Investigations on algorithmic questions on algebra, Trudy Matematicheskogo Instituta Akademii Nauk SSSR. V. 168. Algebra, mathematical logics, theory of numbers, topology. Sbornik obzornyh statey. 1. K 50-letiyu Instituta. Moscow: Nauka, 1984. P.197 - 217.

2. Beltyukov A. P., Solvability of universal theory of natural numbers with addition and divisibility, Zapiski hauchnyh seminarov LOMI Akademii Nauk SSSR. 1976. V. 60, N 7. P.15 - 28.

3. Durnev V. G., Positive theory of free semi-group, Doklady Akademii Nauk SSSR. 1973. V. 211, N 4. P.772 - 774.

4. Durnev V. G., About equtions on free semi-groups and groups, Mathematicheskie zametki. 1974. V. 16, N 5. P.717 - 724.

5. Durnev V. G., About equations with endomorphisms in free semi-groups, Diskretnaya mathematika. 1992. V. 4, N 2. P.136 - 141.

6. Durnev V. G., About equations with restrictions on solutions in free groups, Mathematicheskie zametki. 1993. V. 53, N 1. P.36 - 40.

7. Durnev V. G., About equations with semi-group restrictions on solutions in free groups, Diskretnaya mathematika. 1995. V.7, N 4. P.60-67.

8. Durnev V. G., Unsolvability of positive $\forall\exists^3$-theory of free semi-group, Sibirsky mathematichesky jurnal. 1995. V.36, N 5. P.1067 - 1080.

9. Durnev V. G., To problem of solvability of equations with one coefficient, Mathematicheskie zametki. 1996. V. 59, N 6. P.832 - 846.

10. Kosovsky N. K., Elements of mathematical logic and its application to theory of sub-recursive algorithms. Leningrad: Izdatelstvo Leningradskogo Gosudarstvennogo Universiteta, 1981. 192 p.

11. Makanin G. S., Problem of solvability of equations in free semi-group, Mathematichesky sbornik. 1977. V. 103, N 2. P.147 - 236.

12. Makanin G. S., Equations in free groups, Izvestiya Akademii Nauk SSSR. Seriya mathematiki. 1982. V. 46, N 6. P.1199 - 1274.

13. Marchenkov S. S., Unsolvability of positive $\forall\exists$-theory of free semi-group, Sibirsky mathematichesky jurnal. 1982. V.23, N 1. P.196 - 198.

14. Myacnikov A. G., Remeslennikov V. N., Elementary equivalence of free products of groups with length function in signature, X Vsesoyuzniy simpozium po teorii grupp. Tezisy dokladov. Minsk, 1986. P.160.

15. Myacnikov A. G., Remeslennikov V. N., Elementary equivalence of free products. Sibirskoe otdelenie Akademii Nauk SSSR, VC, Preprint N 718. Novosibirsk, 1987. 20 p.

16. Repin N. N., Certain simply defined groups for which an algorithm recognizing solvability of equations is impossible, Boprocy kibernetiki. Slojnoct vychisleniy i prikladnaya mathematicheskaya logika. Moscow, 1988. P.167 - 174.

17. Romankov V. A., About unsolvability of problem of endomorphic reducibility in free nilpotent groups and in free rings, Algebra i logika. 1977. V. 16, N 4. P.457 - 471.

18. Taytslin M. A., Some examples of unsolvable theories, Algebra i logika. 1967. V.6, N 3. P.105 - 111.

19. Taytslin M. A., About algorithmic problem for coomutative semi-groups, Doklady Akademii Nauk SSSR. 1968. V. 178, N 4. P.786 - 789.

20. Hmelevsky Yu. I., Systems of equations in free group, Izvestiya Akademii Nauk. Seriya mathematiki. 1971. V. 35, N 6. P.1237 - 1268.

21. Büchi J. R., Senger S., Definability in the existential theory of concatenation and undecidable extensions of this theory, Z. Math. Log. und Grundl. Math. 1988. Bd 34, N 4. P. 337 - 342.
22. Büchi J. R., Senger S., Coding in the existential theory of concatenation, Arch. Math. Logik. 1986/87. Bd 26. P. 101 - 106.
23. Lipshitz L., The Diophantine problem for addition and divisibility, Trans. Amer. Math. Soc. 1978. V. 235. P. 271 - 283.
24. Lyndon R. C., Dependence in groups, Colloq. Math. 1966. N 4. P. 275 - 283.
25. Quine W., Concatenation as a basis for arithmetic, J. Symbolic Logic. 1946. V.11. P. 105 - 114.
26. Schupp P.E., On the substitution problem for free groups, Proc. Amer. Math. Soc. 1969. V. 23, N 2. P. 421 - 423.

Learning Small Programs
with Additional Information*

Rūsiņš Freivalds**[1], Gints Tervits***[1], Rolf Wiehagen[2] and Carl Smith†[3]

[1] Institute of Mathematics and Computer Science,
University of Latvia
Raiņa bulvāris 29, LV-1459, Riga, Latvia
[2] Department of Computer Science
University of Kaiserslautern
D-67653 Kaiserslautern, Germany
[3] Department of Computer Science, University of Maryland,
College Park, MD 20742, USA

Abstract. This paper was inspired by [FBW 94]. An arbitrary upper bound on the size of some program for the target function suffices for the learning of some program for this function. In [FBW 94] it was discovered that if "learning" is understood as "identification in the limit," then in some programming languages it is possible to learn a program of size not exceeding the bound, while in some other programming languages this is not possible.

We have studied three other learning types, namely, "finite identification," "co-learning" and "confidence-learning." These three types are very different. Co-learning with the considered additional information in the form "an arbitrary upper bound for the size of the minimal program" allows the learning of the class of all recursive functions. "Finite identification" does not allow this. "Confidence-learning" is strong enough to learn the class of all recursive functions even without the additional information. However, the results of our paper show exactly the opposite rating for the capabilities of learning programs not exceeding the size given by the bound.

For finite identification it is still possible to identify small programs with additional information in some programming languages but not in all of them. For co-learning it is not possible in any programming language. These results contrast to the result in [FKS 94] showing that an arbitrary class of recursive functions is co-learnable if and only if it is identifiable in the limit. Finally, for confidence-learning it is in general not possible to identify small programs with additional information.

* This work was facilitated by an international agreement under NSF Grants 9119540 and 9421640
** Supported by Latvian Science Council Grant No.96.0282
*** Supported by Latvian Science Council Grant No.96.0282
† Supported in part by NSF Grants 9020079 and 9301339

1 Introduction

Inductive Inference is the term used for reconstruction of programs from sample computations. It is the part of Computational Learning Theory heavily based on the Recursive Functions Theory. Started by the well-known paper [Go 67] nowadays Inductive Inference is the most developed part of Computational Learning Theory (see the survey [AS 83] and the monograph [OSW 86]). Far from development of effective machine learning algorithms, inductive inference has actively supplied the practicioners with ideas on how to learn, what to learn, what difficulties to avoid.

The main object of the research in Inductive Inference is consideration, comparison of the identification types and finding deeper relations among the types and among the identifiable classes of functions and languages. Since Gold [Go67] many different types of identification have been defined by various authors, cf. [AS83], [OSW86], for some examples. We follow the idea first used in Freivalds and Wiehagen [FW79], namely we consider the inference process where the input data are not only the graph of the target function but also an arbitrary upper bound of the minimal number of the target function. The latter can be considered as additional information in terms of [FW79]. Intuitively, this kind of additional information can be interpreted as an upper bound on the length of a program of the target function. In a sense, the availability of such an information seems to be realistic. The usefulness of this kind of additional information was made explicit already in [Fr78].

We use the following restriction, however. The result of the inference process has to be smaller or equal to the value of a fixed recursive function from the additional information. It is obvious that if there is a strategy identifying in the limit the *minimal* indices for a class U of recursive functions, then U is identifiable with additional information as well (where the fixed function is the identity function). The results below show that in general this is not so for classes U for which *arbitrary* indices can be identified in the limit.

To describe the identification types widely used in Inductive Inference we need some notions and notation.

Let N denote the set of natural numbers. For any sets L and M, let $L \subset M$ denote the proper inclusion of L in M. Let P, R denote the sets of all partial recursive and recursive functions of one argument. For a Gödel numbering $\varphi \in P^2$ of P, cf. [Ro67], and $f \in P$, let $\min_\varphi f$ denote the minimal number of f in φ. If $f(x)$ is defined for all $x \leq n$ then $f[n]$ denotes an encoding of $(f(0), f(1), \ldots, f(n))$. We say that the sequence of natural numbers $(x_n)_{n \in N}$ converges to x ($x = \lim_n x_n$) iff there exists $n_0 \in N$ such that $x_n = x$ for all $n > n_0$. Finally, let $id(x) = x$ for any $x \in N$.

The following identification types from [Go67, OSW86, AS83] will serve as a basis for investigation.

Definition 1. Let $U \subseteq R$ and let φ be any Gödel numbering. U is called identifiable in the limit (written: $U \in EX$) iff there is a strategy $S \in P$ such that for any $f \in U$,

1) $S(f[n])$ is defined for all n;
2) $a = lim_n S(f[n])$ exists;
3) $\varphi_a = f$.

Definition 2. Let $U \subseteq R$ and let φ be any Gödel numbering. U is called identifiable in the limit by a consistent strategy (written: $U \in CONS$) iff there is a strategy $S \in P$ such that $U \in EX$ by S and

$\varphi_{S(f[n])}(x) = f(x)$ for every n and all $x \leq n$.

Note that Definitions 1, 2 do not depend on the choice of the Gödel numbering φ.

In [FW79] every upper bound of $min_\varphi f$ was considered as additional information. It was proved that the presence of this additional information can considerably influence the identifiability of function classes. For this purpose the following identification type was introduced.

Definition 3. Let $U \subseteq R$ and let φ be any Gödel numbering. $U \in EX^+$ iff there is a strategy $S \in P^2$ such that for any function $f \in U$ and any $b \geq min_\varphi f$,
1) $S(f[n], b)$ is defined for all n;
2) $a = S(f, b) = lim_n S(f[n], b)$ exists;
3) $\varphi_a = f$.

In a similar way one can define $CONS^+$. It was proved that $EX \subset pR$ where pR denotes the power set of R, cf. [Go67], and that $CONS \subset EX$, cf. [WZ95]. The following theorem proved in [FW79] leads to a characterization of both EX^+ and $CONS^+$.

Theorem 4. $R \in CONS^+$

Consequently, $R \in EX^+$ and $CONS^+ = EX^+ = pR$. Note that these results hold for arbitrary Gödel numberings. However, as it will be clear from the proof of Theorem 1, the Gödel numbers synthesized are in general greater than the given bounds for the corresponding minimal numbers. Therefore the question arises whether the additional information is suffcient for constructing Gödel numbers which are "small" (smaller than the additional information, possibly modulo a given recursive function). This leads to the definition of the identification type studied in the following.

Definition 5. Let $U \subseteq R$, φ a Gödel numbering and $h \in R$. $U \in EX_\varphi^{+,h}$ iff there is a strategy $S \in P^2$ such that for all $f \in U$ and for all $b \geq min_\varphi f$,
1) $S(f[n], b) \leq h(b)$ for all $n \in N$;
2) $a = S(f, b) = lim_n S(f[n], b)$ exists;
3) $\varphi_a = f$.

The special case when the function h in this definition is the identity function is denoted by $EX_\varphi^{+,id}$. There were several results in [FBW94] but the main ones are captured by the following two theorems.

Theorem 6. *There is a Gödel numbering φ such that $R \in EX_\varphi^{+,id}$.*

Theorem 7. *For every $h \in R$, there is a Gödel numbering φ such that $R \notin EX_\varphi^{+,h}$.*

Below we consider analogous types $I_\varphi^{+,h}$ that arise from various functions h and three different identification types I, namely finite identification [Go67], co-learning [FKS94] and confidence-learning [BFS96]. All these types I are defined for Gödel numberings. However, they do not depend on the specific Gödel numbering used. On the other hand, the types $I_\varphi^{+,h}$ *may depend* on the numbering φ, as we shall see below. Actually, we show that for finite identification it is still possible to identify small programs with additional information in some programming languages but not in all of them. For co-learning it is not possible in any programming language. These results contrast to the result in [FKS 94] showing that an arbitrary class of recursive functions is co-learnable iff it is identifiable in the limit. Finally, for confidence-learning it is in general not possible to identify small programs with additional information.

2 Finite Identification

In this section we introduce the notions of finite identification that we consider in this paper. Intuitively, finite learners do not alter their original hypothesis. Sometimes finite learning is called "one shot learning."

Definition 8. Let $U \subseteq R$ and let φ be any Gödel numbering. U is called finitely identifiable (written: $U \in FIN$) iff there is a strategy $S \in P$ such that for any $f \in U$, there is an $n \in N$ such that
1) for any $x < n$, $S(f[x]) = ?$,
2) $S(f[n]) \in N$ and $\varphi_{S(f[n])} = f$.

As above, one can define FIN^+. Notice that in the definition of FIN^+-identification it is not demanded that the result of the identification (the output of the strategy) does not exceed the given bound of the minimal φ-index. Indeed, in most FIN^+-strategies we have $S(f, b) > b$, cf. [FW79] where $FIN \subset FIN^+ \subset pR$ was proved. We wish to find out whether or not we can restrict ourselves to strategies S for which always $S(f, b) \leq b$. The main result of this section is that this is possible for some Gödel numberings, but not for all.

Definition 9. Let $U \subseteq R$ and let φ be a Gödel numbering. $U \in FIN_\varphi^{+,id}$ iff there is a FIN-strategy $S \in P^2$ such that for all $f \in U$ and for all $b \geq \min_\varphi f$:
1) $S(f, b) \leq b$,
2) $\varphi_{S(f,b)} = f$.

Definition 10. Let $U \subseteq R$ and let φ be a Gödel numbering. $U \in FIN_\varphi^{+,lin}$ iff there is a FIN-strategy $S \in P^2$ and a positive constant c such that for all $f \in U$ and for all $b \geq \min_\varphi f$:
1) $S(f, b) \leq c \cdot b$,
2) $\varphi_{S(f,b)} = f$.

Theorem 11. *There is a Gödel numbering φ such that $FIN_\varphi^+ = FIN_\varphi^{+,lin}$.*

Comment. Since $R \notin FIN^+$, there can be infinitely many $U \in FIN^+$ with very much different strategies involved. The proof of our Theorem 11 provides no uniform procedure how to get a $FIN_\varphi^{+,lin}$-strategy from the given FIN_φ^+-strategy. However, the numbering φ does not depend on the class U.

Proof. Let ψ be an arbitrary Gödel numbering of P. However, we demand that ψ_0 be the empty function.

Let $S_0, S_1, S_2, \ldots$ denote the FIN^+-strategies in the numbering ψ. (If a class of functions is FIN^+-identifiable in one Gödel numbering, then this class is FIN^+-identifiable in any other Gödel numbering as well).

We define a new Gödel numbering φ such that:

1. for all $n > 0$ and all x, $\varphi_{n^3}(x) = \psi_n(x)$;
2. if the FIN_ψ^+-strategy $S_0(\psi_1, 1)$ outputs a result j, then $\varphi_0(x) = \psi_j(x)$, for all x;
3. if the values $\psi_1(0), \psi_1(1), \ldots, \psi_1(t)$ are defined such that the FIN_ψ^+-strategy $S_0(\psi_1[t], 1)$ outputs a result j, then

$$\varphi_0(x) = \begin{cases} \psi_j(x), & \text{if the values} \\ & \psi_1(0) = \psi_j(0), \psi_1(1) = \psi_j(1), \ldots, \psi_1(x) = \psi_j(x), \\ \text{undefined,} & \text{otherwise;} \end{cases}$$

4. if the values $\psi_i(0), \psi_i(1), \ldots, \psi_i(t)$ are defined such that the FIN_ψ^+-strategy $S_k(\psi_i[t], n)$ outputs a result j, and $1 \le i \le n$, and $0 \le k \le n$, then

$$\varphi_{(n-1)^3 + (i-1)\cdot n + k}(x) = \begin{cases} \psi_j(x), & \text{if there is a function } \psi_m \\ & \text{such that } m \le n \text{ and, for all } y \le x, \\ & \psi_m(y) = \psi_j(y) \\ \text{undefined,} & \text{otherwise.} \end{cases}$$

Notice that rather many functions φ_z are not defined by 1)–4). These functions are empty.

We observe that, if a recursive function f has the minimal ψ-index n, then f has a φ-index less than $(n-1)^3 + 1$.

Now, let the class U of recursive functions be FIN_ψ^+-identifiable by strategy S_k and let the minimal ψ-index n of a function $f \in U$ exceed k. Let y be the minimal φ-index of f. In this case $(n-1)^3 < y < (n-1)^3 + n^2$. However, if $n < k$, then y can be larger. However, U contains only a *finite* number of such functions.

Now let the minimal φ-index of a function $f \in U$ be y. We wish to estimate the minimal ψ-index n of the same function f. Either y is a complete cube (and then $n^3 \le y$) or y is not a complete cube, and, for some m, $m < y < (m-1)^3 + m^2$. Anyway, let m be the last positive integer such that $y < m^3$. In this case f is among the functions $\psi_1, \psi_2, \ldots, \psi_m$. If U is FIN_ψ^+-identified by strategy S_k and the minimal ψ-index n of f exceeds k, then f must have at least one φ-index among $(m-1)^3 + 1, (m-1)^3 + 2, \ldots, (m-1)^3 + m^2$.

We consider the following strategy S for $FIN_\varphi^{+,lin}$- identification of the class U. Since $U \in FIN_\psi^+$, there is a strategy S_k for FIN_ψ^+- identification of U. Our strategy $S(f, b)$ works as follows. At first, it finds the last m such that $b < m^3$. Then S computes $S_k(f, m) = j$. After that S considers the functions $\varphi_{(m_1)^3+k}, \varphi_{(m_1)^3+m+k}, \ldots, \varphi_{(m_1)^3+(m-1)\cdot m+k}$ which one of these first shows up to be non-empty. The φ- index of this function is the output of our strategy S.

First, S always outputs a result not exceeding $b + O(b)$.

Second, S is correct for nearly all functions in U (for all but those with ψ-indices less than k). $\qquad\qquad\qquad\qquad\qquad\qquad\qquad\qquad\qquad\qquad\qquad\qquad$ □

Theorem 12. *For every $h \in R$, there is a Gödel numbering φ and a class U of recursive functions such that $U \in FIN$ but $U \notin EX_\varphi^{+,h}$.*

Proof. Without loss of generality, we consider a strictly increasing h. Let φ' be a Gödel numbering of P. We construct a sequence of natural numbers $n_1, n_2, \ldots$ as follows:

$$n_k = \begin{cases} h(2) + 1 & \text{if } k = 1 \\ h(n_{k-1} + 2) + 1 & \text{if } k > 1 \end{cases}$$

A new Gödel numbering φ can be defined in the following way:

1. $\varphi_0 = \varphi'_0$;
2. $\varphi_{n_k} = \varphi_{k-1}$ for all k;
3. (a) φ_{n_k+1} and φ_{n_k+2} are defined as the empty function if $\varphi'_n(0)$ is not defined;
   (b) φ_{n_k+1} and φ_{n_k+2} are defined from $\varphi'_k(0)$ and $S_{\varphi'_k(0)}$ by means of Construction M;
4. φ_n is the empty function for all other n.

Construction M (M for mistake).

Let $S_0, S_1, S_2, \ldots$ be an effective enumeration of all strategies. We shall make every strategy make a mistake on a particular *constant* function. For this purpose let us define the special pair φ_{n_k+1} and φ_{n_k+2}.

If $\varphi'_n(0) = \varphi_{n_k+1}(0)$ is not defined, then φ_{n_k+1} and φ_{n_k+2} are defined nowhere.

If $\varphi'_n(0) = a$ then in order to compute the values of these functions, we consider the work of the strategy $S_a(n_k + 2, a[x])$ with $a[x]$ being the initial fragment of the constant function a and x assuming all the values of the sequence $0, 1, 2, \ldots$.

We define $\varphi_{n_k+1}(0) = a$ in the first step, $\varphi_{n_k+1}(1) = a$ in the second step, $\varphi_{n_k+1}(2) = a$ in the third step, ...

We continue in the same way until for some x it turns out that $S_a(n_k + 2, a[x]) = n_k + 1$. Then we stop defining the function φ_{n_k+1} (assume that it was defined up to x_1) and begin to define $\varphi_{n_k+1}(0) = a$ in the x_1-th step, $\varphi_{n_k+1}(1) = a$ in the $x_1 + 1$-th step, $\varphi_{n_k+1}(2) = a$ in the $x_1 + 2$-th step, ...

And so on until $S_a(n_k + 2, a[x]) = n_k + 2$. At this point let φ_{n_k+2} be defined up to x_2. Then $\varphi_{n_k+1}(x_1 + 1) = a$ in the $x_1 + x_2$-th step, $\varphi_{n_k+1}(x_1 + 2) = a$ in the $x_1 + x_2 + 1$-th step, etc. until $S_a(n_k + 2, a[x]) = n_k + 1$.

Then we again stop defining φ_{n_k+1} and continue with φ_{n_k+2}.... At least one of these functions receives infinitely many values, becomes the constant function a and, consequently, no stabilization of the strategy S_a on this number can take place. That means that no strategy can identify even the class of all constant functions in the $EX_\varphi^{+,id}$-sense. Actually, assume that S_a is a candidate for such a strategy. Then let n be the smallest number of a function with $\varphi_n'(0) = a$. Let us see what the strategy $S_a(n_k + 2, a[x])$ does. First, the upper bound $n_k + 2$ in the new Gödel numbering is correct for the constant function a. Second, a stabilization on a number that is smaller or equal to $n_k + 2$ is impossible (the right number cannot be smaller than $n_k + 1$, since there, even at zero, the value is different from a).

In the Gödel numbering φ the class of all constant functions cannot be identified in $EX_\varphi^{+,h}$-sense. Hence the class of all the constants can be taken as U. $\square$

Corollary 13. *For every $h \in R$, there is a Gödel numbering φ and a class U of recursive functions such that $U \in FIN$ but $U \notin FIN_\varphi^{+,h}$.*

3 Co-Learning

There was a recent paper [FKS 94] where an identification type *CoLearn* was introduced.

Definition 14. We say that S co-converges on f to p iff $[N - \{S(f[m])|m \in N\} = \{p\}]$. If there exists a p such that S co-converges on f to p, then we say that S co-converges on f (to p). Otherwise, we say that S co-diverges on f.

Definition 15. [FKS 94]

(a) A learning strategy $S \in P$ is said to *CoLearn* f (written: $f \in CoLearn(S)$) just in case $(\exists p | \varphi_p = f)$ $[S$ co-converges on f to $p]$.
(b) $CoLearn = \{U \mid (\exists S)\ [U \subseteq CoLearn(S)]\}$

It was proved in [FKS 94] that, for any Gödel numbering φ, $CoLearn_\varphi = EX_\varphi$. The next result follows immediately.

Theorem 16. $EX^+ = CoLearn^+$.

However, Theorem 6 does not hold for *CoLearn*-identification.

Definition 17. Let $U \subseteq R$, and φ be a Gödel numbering. $U \in CoLearn_\varphi^{+,h}$ iff there is a *CoLearn*-strategy $S \in P^2$ such that for all $f \in U$ and for all $b \geq \min_\varphi f$:

1. $S(f,b) \leq h(b)$,
2. $\varphi_{S(f,b)} = f$.

Theorem 18. *There is a Gödel numbering φ such that for an arbitrary recursive function h, $R \notin CoLearn_\varphi^{+,h}$.*

Proof. Let ψ be an arbitrary Gödel numbering of P. We construct our Gödel numbering φ by defining $\varphi_{g(n)}$ as equal to ψ_n plus making the new numbering uniformly computable.

We diagonalize against all the possible pairs (strategy of co-learning, recursive function h bounding the increase of the Gödel number). Hence, we establish a fixed order of pairs (i, j), determined by standard pairing functions $c(x, y) = z, l(z) = x, r(z) = y$ establishing one-to-one relation between the set of all natural numbers and the set of all pairs of natural numbers.

We organize the functions in blocks. The functions $\varphi_{g(n)+1}, \varphi_{g(n)+2}, \dots,$ $\varphi_{g(n+1)-1}$ constitute one block. Each block is (temporarily) associated with a movable marker (i, j). During the process of computation the markers can be moved. However, the movement of markers is always monotonic with respect to $c(i, j)$:

(i) Every marker can be moved to later (larger) blocks only.
(ii) Once a block has got a marker, the marker can be removed or replaced by an earlier marker but never the marker can be replaced by a later marker.

Every block $(\varphi_{g(n)+1}, \varphi_{g(n)+2}, \dots, \varphi_{g(n+1)-1})$ is defined to contain $g(n) + 2$ functions which we try to make to be constants

$$\varphi_{g(n)+1}(x) = g(n + 1)$$
$$\varphi_{g(n)+2}(x) = g(n + 2)$$
$$\dots$$
$$\varphi_{g(n)+g(n)+2}(x) = g(n) + g(n) + 2$$

and one more function

$$\varphi_{g(n)+g(n)+3} = \varphi_{g(n+1)-1}$$

which may be defined in the process of computation, or maybe not.

The process of computation is organized in stages.

Stage 0. Go to stage 1.

Stage t $(t > 0)$. We consider the block associated with the marker $(l(l(t)),$ $r(l(t)))$. If this block has already got the special marker "DONE," then do nothing and go to the stage $t + 1$.

Otherwise, follow the instructions below.

If the computation of $\varphi_{r(l(t))}(g(n + 1) - 1)$ does not terminate within t steps, then we define

$$\varphi_{g(n)+1}(x) = g(n + 1)$$
$$\varphi_{g(n)+2}(x) = g(n + 2)$$
$$\dots$$
$$\varphi_{g(n+1)-2}(x) = g(n + 1) - 2$$

for $x = 0, 1, 2, \dots, t$, and go to the stage $t + 1$.

If the computation of $\varphi_{r(l(t))}(g(n + 1) - 1)$ terminates in no more than t steps of computation, but it does not terminate in $c(l(t), r(t) - 1)$ steps, then we move all the markers (i, j) such that $c(i, j) > c(l(l(t)), r(l(t))) = l(t)$ (i. e. all the markers associated to blocks coming after the currently considered block) to later blocks starting with the first block $\varphi_{g(n')+1}, \varphi_{g(n')+2}, \ldots, \varphi_{g(n'+1)-1}$ such that:

(i) $g(n') > \varphi_{r(l(t))}(g(n+1)-1)$
(ii) no function in this block and in later blocks has got any value in stages $1, 2, \ldots, t-1$

If the computation of $\varphi_{r(l(t))}(g(n+1)-1)$ terminates in no more than t steps, and even in no more than $c(l(t), r(t)-1)$ steps, then we simulate the work of the co-learning strategy $S_{l(l(t))}$ on the first t finite strings of integers (in some standard computable numbering) using t steps of computation on each of the strings. If the strategy working on some finite string asks for a new input value, we input the next element of the string (using the elements of the string as if they were $f(0), f(1), f(2), \ldots$). If there are no more elements in the string but the strategy asks for another input, we stop this simulation.

We look for the first string $\alpha_0 \alpha_1 \alpha_2 \ldots \alpha_n$ such that:

1. the co-learning strategy with the additional information $\varphi_{r(l(t))}(g(n+1)-1)$ crosses out all the possible integers $z = 0, 1, 2, 3, \ldots, \varphi_{r(l(t))}(g(n+1)-1)$ but one (we denote this integer as w),
2. the function φ_w takes values $\varphi_w(0) = \alpha_0, \varphi_w(1) = \alpha_1, \varphi_w(2) = \alpha_2, \ldots,$ $\varphi_w(n) = \alpha_n$,
3. $\varphi_w(n+1)$ is defined.

If there is no such string $\alpha_0 \alpha_1 \alpha_2 \ldots \alpha_n$, then we define

$$\varphi_{g(n)+1}(x) = g(n+1)$$
$$\varphi_{g(n)+2}(x) = g(n+2)$$
$$\ldots$$
$$\varphi_{g(n)+g(n)+2}(x) = g(n) + g(n) + 2$$

for $x = 0, 1, 2, \ldots, t$, and go to stage $t+1$.

If $\alpha_0 \alpha_1 \alpha_2 \ldots \alpha_n$ is the first such string found, then we define

$$\varphi_{g(n)+g(n)+3}(x) = \varphi_{g(n+1)-1}(x) = \alpha_x$$

for $x = 0, 1, 2, \ldots, n$, and

$$\varphi_{g(n)+g(n)+3}(x) = \varphi_{g(n+1)-1}(x) = \varphi_w(n+1) + 1$$

for $x \geq n+1$.

Finally, we associate the marker "DONE" with the block under consideration. No markers are moved. This completes the instructions for stage t.

Assume from the contrary that there exists a co-learning strategy S_i such that $R \in CoLearn_\varphi^{+,h}$. Let j be an arbitrary φ-index for the function h. Consider the marker (i, j) in the construction of the Gödel numbering φ. Since every marker can be moved only finitely many times, there is a final position of the marker (i, j). Consider the block $\varphi_{g(n)+1}, \varphi_{g(n)+2}, \ldots, \varphi_{g(n+1)-1}$ corresponding to this final position of the marker (i, j).

If this block finally gets the special marker "DONE," then the strategy S_i on the function $\varphi_{g(n+1)-1}$ is to co-learn it with the additional information $h(g(n+1)-1) = \varphi_{r(l(t))}(g(n+1)-1)$ (if $t = c(c(i,j),y)$ and $y \geq 0$). On the other hand, the strategy S_i on $\varphi_{g(n+1)-1}$ crosses out all integers $0, 1, 2, \ldots, h(g(n+1)-1)$ but one w. However, φ_w differs from $\varphi_{g(n+1)-1}$ on $x = n+1$ where n is the length of the string of the values of the function $\varphi_{g(n+1)-1}$ such that these all (but one) were crossed out based on the finite string of the values of $\varphi_{g(n+1)-1}$.

If the block never gets the special marker "DONE," the functions $\varphi_{g(n)+1}$, $\varphi_{g(n)+2}, \ldots, \varphi_{g(n+1)-2}$ become pairwise distinct functions. The number of these functions exceeds the total number of functions in the preceeding blocks. Hence, there is at least one of these functions $\varphi_{g(n)+1}, \varphi_{g(n)+2}, \ldots, \varphi_{g(n+1)-2}$ having exactly one φ-index not exceeding $h(g(n+1)-1)$. However, the strategy S_i fails to cross out all but one of the integers $0, 1, 2, \ldots, h(g(n+1)-1)$.

Since the function h is total, there is no such a possibility when $\varphi_{r(l(t))}(g(n+1)-1)$ does not terminate. $\qquad\square$

Corollary 19. *For any Gödel numbering φ and for any recursive function h, $R \notin CoLearn_\varphi^{+,h}$.*

Proof. Assume the contrary. Let ψ be any Gödel numbering for which the contrary holds. Let φ be the Gödel numbering from Theorem 18. Since ψ is reducible to φ and φ is reducible to ψ with a recursive increase of the index, it follows that $R \in CoLearn_\varphi^{+,h}$ for some recursive h. This contradicts Theorem 18. $\qquad\square$

Corollary 20. *For any Gödel numbering φ and for any recursive function h,*

$$CoLearn_\varphi^{+,h} \subset CoLearn_\varphi^+.$$

Proof. Immediately from Corollary 19. $\qquad\square$

4 Confidence-Learning

A new and very different identification type was introduced in [BFS96].

A *Confidence inductive inference machine* (CIIM) works like a usual inductive inference machine except that it outputs a sequence of hypotheses with confidence levels: $s = p_0/c_0$, p_1/c_1, p_2/c_2, ... We say that a CIIM M learns a recursive function f iff on input from the graph of f it outputs a sequence of hypotheses paired with confidence levels such that for any i, the confidence sequence $C(s, p_i)$ is monotone and there exists a unique p with $C(s, p)$ converging to 1 and $\varphi_p = f$. In this way, each CIIM learns a set of recursive functions. The collection of all such sets is denoted by CEX.

Theorem 21. [BFS96] $R \in CEX$.

Hence this identification differs from all the other identification types considered above. The additional information is not at all needed. The Confidence inductive inference machine can do without the additional information. Nonetheless some Gödel numberings are very bad for learning small programs.

Theorem 22. *For every $h \in R$, there is a Gödel numbering φ and a class U of recursive functions such that $U \in FIN$ but $U \notin CEX_\varphi^{+,h}$.*

Proof. Immediately from Theorem 12. □

References

[AS 83] Dana Angluin, and Carl H. Smith. Inductive inference: Theory and methods, Computing Surveys, v. 15, 1983, pp. 237–269.

[BFS 96] Jānis Bārzdiņš, Rūsiņš Freivalds, and Carl H. Smith. Learning with confidence, Lecture Notes in Computer Science, v.1046, 1996, pp. 207-218.

[Fr 75] Rūsiņš Freivalds. Minimal Gödel numbers and their identification in the limit. Lecture Notes in Computer Science, v. 32, 1975, pp. 219–225.

[Fr 78] Rūsiņš Freivalds. Effective operations and functionals computable in the limit. Zeitschrift für Mathematische Logik und Grundlagen der Mathematik, v. 24, 1978, pp. 193–206 (in Russian)

[FBW 94] Rūsiņš Freivalds, Ognian Botuscharov, and Rolf Wiehagen. Identifying nearly minimal Gödel numbers from additional information. Lecture Notes in Computer Science, v. 872, 1994, pp. 91–99.

[FKS 94] Rūsiņš Freivalds, Marek Karpinski, and Carl H. Smith. Co-learning of total recursive functions. In Proceedings of the Seventh Annual Conference on Computational Learning Theory, New Brunswick, New Jersey, pp. 190–197. ACM Press, July 1994.

[FS 93] Rūsiņš Freivalds, and Carl H. Smith. The role of procrastination in machine learning. Information and Computation, vol. 107, 1993, pp. 237–271.

[FW 79] Rūsiņš Freivalds, and Rolf Wiehagen. Inductive inference with additional information. Journal of Information Processing and Cybernetics, v. 15, 1979, pp. 179–185.

[Go 67] E. M. Gold. Language identification in the limit. Information and Control, v. 10, 1967, pp. 447–474.

[OSW 86] Daniel N. Osherson, Michael Stob, and Scott Weinstein. Systems that Learn. MIT Press, 1986.

[Ro 67] Hartley Rogers Jr. Theory of Recursive Functions and Effective Computability. McGraw-Hill, 1967.

[WZ 95] Rolf Wiehagen, and Thomas Zeugmann. Learning and consistency. Lecture Notes in Artificial Intelligence, v. 961, 1995, pp. 1–24.

Cut Elimination for the Second Order Propositional Logic with Hilbert's ε-symbol, Extensionality, and Full Comprehension

Michael Gavrilovich[1]

St-Petersburg State University, Russia

Abstract. The cut-elimination technique is well developed for classical higher order systems, but, since the normal form (not normalization) theorem was established by Mints[1, 2, 3, 4, 5], not much further progress has been achieved for the logical systems with an ε-symbol.

We give a cut elimination procedure for the second order propositional logic with Hilbert's ε-symbol, extensionality and full comprehension.

We prove that selective elimination of the ε-symbol and quantifiers is possible, so that a cut formula becomes essentially quantifier- and epsilon-free, after which cuts can be eliminated by the standard Gentzen procedure.

1 Definitions

The language of our logic has the connectives $\{\neg, \&, \forall, \varepsilon\}$ and constants $1(true)$, $0(false)$, (propositional) variables $X, Y, Z, \dots$ that are allowed to be both *free* and *bound*. We define $\sim$ (equivalence), $\supset$ (implication), $\vee$ (disjunction) in the usual way.

We call the formulas of the form $\varepsilon Y F$ ε-*terms*. Note that there is no difference between terms and formulas in our logic. Intuitively, the ε-term $\varepsilon Y F$ is a value for which $F[Y \backslash \varepsilon Y F]$ is true, if it exists. We define four logics S, S_-^T, S^T, S^+.

By steps, we reduce each logic to the preceding one. In the three sections $S^+ \sim S^T$, $S^T \sim S_-^T$, $S_-^T \sim S$ we prove the corresponding statements.

Now we formulate the exact rules of these logics.

Definition 1. The axioms of these logics are Gentzen-type logical axioms, i.e. sequents of the form $\Gamma, A \mapsto \Delta, A$, $\Gamma, 0 \mapsto \Delta$, $\Gamma \mapsto \Delta, 1$. For all four logics, rules for introduction of connectives $\&, \neg$ are defined as usual. Theory S has the following rules:

1.
$$\frac{\Gamma \mapsto A[Z], \Delta}{\Gamma \mapsto \forall X A[X], \Delta},$$

Z is not contained in $\Gamma, \forall X A[X] \mapsto \Delta$.

2.
$$\frac{\Gamma, \forall X A[X], A[G] \mapsto \Delta}{\Gamma, \forall X A[X] \mapsto \Delta},$$

where G is an arbitrary formula.

3.
$$\frac{\Gamma \mapsto F[G], \Delta \quad \Gamma, F[\varepsilon X F[X]] \mapsto \Delta}{\Gamma \mapsto \Delta} \varepsilon \mapsto$$

and $\varepsilon X F[X]$ is contained in $\Gamma \mapsto \Delta$.

This restriction is necessary, otherwise we could just "simulate" the cut rule.

4.
$$\frac{\Gamma \mapsto F[X] \sim G[X], \Delta \quad \Gamma, \varepsilon X_1 F[X_1] \sim \varepsilon X_2 G[X_2] \mapsto \Delta}{\Gamma \mapsto \Delta} ext$$

and $\varepsilon X_1 F[X_1]$, $\varepsilon X_2 G[X_2]$ are contained in $\Gamma \mapsto \Delta$.

The restriction has the same meaning as in the previous item. This concludes the list of rules of the logic S. The logic S^+ has all the rules of S and the cut rule:

5.
$$\frac{\Gamma \mapsto C, \Delta \quad \Gamma, C \mapsto \Delta}{\Gamma \mapsto \Delta}$$

In addition to the rules of S, the logic S_-^T has no cut rule but a supplement to *ext* is allowed in the following two cases :

6.
$$\frac{\Gamma \mapsto 0 \sim G[X], \Delta \quad \Gamma, \varepsilon Z 0 \sim \varepsilon X G[X] \mapsto \Delta}{\Gamma \mapsto \Delta}$$

and $\varepsilon X G[X]$ is contained in $\Gamma \mapsto \Delta$ and

7.
$$\frac{\Gamma \mapsto 1 \sim G[X], \Delta \quad \Gamma, \varepsilon Z 1 \sim \varepsilon X G[X] \mapsto \Delta}{\Gamma \mapsto \Delta}$$

and $\varepsilon X G[X]$ is contained in $\Gamma \mapsto \Delta$.

Formulas containing no quantifiers and no ε-terms except for $\varepsilon Z 0$ and $\varepsilon Z 1$ are called *simple*. The logic S^T is the same as S^+, except that cut is allowed for simple formulas only.

8.
$$\frac{\Gamma \mapsto C, \Delta \quad \Gamma, C \mapsto \Delta}{\Gamma \mapsto \Delta}$$

and C is simple.

2 The proof of $S^+ \sim S^T$

We define an operation $*$ on formulas. This operation is the key of this section. The operation $*$ transforms formulas into "equivalent" simple ones. The exact sense of "equivalence" will be given in the following lemmas. Let Z denote a fixed variable from the language of the logic.

Definition 2. Define $*$ by induction on height of the formula.

1. $X^* = X$
2. $(\neg A)^* = \neg A^*$
3. $(A \& B)^* = A^* \& B^*$
4. $(A \vee B)^* = A^* \vee B^*$

5. $(\forall X A[X])^* = A[0]^* \& A[1]^*$
6. $(\varepsilon X A)^* = A[0]^* \& A[1]^* \& \varepsilon Z1 \vee \neg A[0]^* \& A[1]^* \vee \neg A[0]^* \& \neg A[1]^* \& \varepsilon Z0$

Note that $(F[G])^* = F^*[G^*]$.

(Note that a simpler definition $\varepsilon X A[X] = A[1]$ does not work because of the extensionality rule.)

Lemma 3. *In any of the systems* S^T, S^T_-, S^+ *the following implications hold:*

1. $\vdash \Gamma \mapsto \Delta, F[0]$ *implies* $\vdash \Gamma \mapsto \Delta, F[G], G$
2. $\vdash \Gamma \mapsto \Delta, F[1]$ *implies* $\vdash G, \Gamma \mapsto \Delta, F[G]$
3. $\vdash F[1], \Gamma \mapsto \Delta$ *implies* $\vdash G, F[G], \Gamma \mapsto \Delta$
4. $\vdash F[0], \Gamma \mapsto \Delta$ *implies* $\vdash F[G], \Gamma \mapsto \Delta, G$
5. $\vdash F[0], F[1] \mapsto F[Y]$
6. $\vdash F[Y] \mapsto F[0], F[1]$

Proof. (1)-(4): Indeed, take a proof (in any of the systems) and just replace all the occurrences of the predecessors of 1 or 0 by G and add G to the left or right side of each sequent, renaming variables if needed. The figure obtained is obviously a valid proof.

(5) and (6): Easy to prove by induction on the height of F.

Lemma 4. *(Main properties of operation* *.*)*

1. *For any formulas* F, G, *in* S^T *it is provable that*

$$(F[0]^* \sim G[0]^*) \& (F[1]^* \sim G[1]^*) \mapsto (\varepsilon X F[X])^* \sim (\varepsilon X G[X])^* \ .$$

2. *For each formula* F, *the following sequents are provable in* S^T: $F \mapsto F^*$ *and* $F^* \mapsto F$.
3. *In* S^T *it is provable that* $F^*[Y] \mapsto F^*[(\varepsilon F)^*]$.

Proof. (1) and (2): Making cuts on $F[0]^*, G[0]^*, F[1]^*, G[1]^*, Y$ and $(\varepsilon X F[X])^*$ and further simplifications gives the proof.

(3) Making cuts by Y and $(\varepsilon F)^*$ and further simplifications gives the proof. The cuts are allowed because Y and $(\varepsilon F)^*$ do not contain occurrences of ε-terms other than $\varepsilon Z1$ and $\varepsilon Z0$.

Definition 5. A *variant* of a formula A is A^*. A *variant* of a sequent $\Gamma \mapsto \Delta$ is a sequent obtained by replacing some of the occurrences of formulas in $\Gamma \mapsto \Delta$ by their variants.

Theorem 6. *For any proof* P *in* S^+ $P : \Gamma \mapsto \Delta$ *there exists a proof* P' *in* S^T *of the same sequent.*

Proof. By induction on the height of S^+-proof P we prove a slightly stronger statement:

For any proof P in S^+ $P : \Gamma \mapsto \Delta$ and any variant $\Sigma \mapsto \Omega$ of $\Gamma \mapsto \Delta$ there exists a proof P' in S^T of the sequent $\Sigma \mapsto \Omega$.

Basis follows from Lemma 4.

Induction step.

1. If the last rule is in P is a logical one, the statement is rather easy.
2. If the last rule in P is *ext*, let P end with

$$\frac{\Gamma' \mapsto \Delta', F[Y] \sim G[Y] \quad \varepsilon X F[X] \sim \varepsilon X G[X], \Gamma' \mapsto \Delta'}{\Gamma' \mapsto \Delta'}.$$

Now consider the next figure:

$$\frac{\Gamma' \mapsto \Delta', (F[0] \sim G[0])^* \& (F[1] \sim G[1])^* \quad (\varepsilon X F[X] \sim \varepsilon X G[X])^*, \Gamma' \mapsto \Delta'}{\Gamma' \mapsto \Delta'}.$$

The upper sequents provability follows from the induction hypothesis. Since by Lemma 4 we know

$$\vdash_{\mathrm{S^T}} (F[0] \sim G[0])^* \& (F[1] \sim G[1])^* \mapsto (\varepsilon X F[X] \sim \varepsilon X G[X])^*$$

the lower-sequent of this figure can be derived from the upper ones with the help of a cut on

$$(F[0] \sim G[0])^* \& (F[1] \sim G[1])^* \supset (\varepsilon X F[X] \sim \varepsilon X G[X])^* .$$

3. If the last inference of a rule is cut, we just replace the cut formula by its variant:

transforms to

$$\frac{\Gamma \mapsto \Delta, C \quad C, \Gamma \mapsto \Delta}{\Gamma \mapsto \Delta}$$

$$\frac{\Gamma \mapsto \Delta, C^* \quad C^*, \Gamma \mapsto \Delta}{\Gamma \mapsto \Delta} .$$

4. The $\varepsilon \mapsto$ rule can be handled similarly.

3 The Proof of $\mathrm{S^T} \sim \mathrm{S^T_-}$

Lemma 7. *Let $P : \Gamma \mapsto \Delta$ be a proof in $\mathrm{S^T}$. Then there exists a $\mathrm{S^T}$-proof P' of the same sequent that does not have applications of ext of the form*

$$\frac{\Gamma \mapsto F \sim G, \Delta \quad \Gamma, \varepsilon Z F \sim \varepsilon Z G \mapsto \Delta}{\Gamma \mapsto \Delta}$$

where both formulas F, G are constants $0, 1$.

Proof. Indeed, each inference of this form could be replaced by a cut on $\varepsilon Z 0$ or $\varepsilon Z 1$.

Theorem 8. $\mathrm{S^T} \sim \mathrm{S^T_-}$

Proof. Apply usual cut elimination procedure to the proof P. Since the cut formula contain no quantifiers and no ε-terms other than $\varepsilon Z 0$ and $\varepsilon Z 1$, there will be none of the difficulties usually arising in a second order theory. So, the procedure will work successfully. Of course, the procedure does not take into account the restrictions that some particular ε-terms should be contained in the lower sequent of a rule. But $\mathrm{S^T_-}$ has no such restriction on the occurrences of $\varepsilon Z 0$ and $\varepsilon Z 1$ and the cut formulas in the proof of $\mathrm{S^T}$ do not contain any terms except for these. Thus this does not invalidate the procedure.

However, this does invalidate the procedure to prove that $\mathrm{S} \sim \mathrm{S^T}$.

4 The Proof of $S^T_- \sim S$

Definition 9. The sequent $\Gamma^F_G \mapsto \Delta^F_G$ is the result of replacing all the occurrences of F by G in the sequent $\Gamma \mapsto \Delta$.

Lemma 10. *1. If $\vdash_S G, \Sigma \mapsto \Omega$ and $\vdash_{S^T_-} G \mapsto \Delta$ and $\Sigma \mapsto \Omega$ contains $\varepsilon X G$ then $\vdash_S \Sigma, \Gamma^{\varepsilon Z0}_{\varepsilon X G} \mapsto \Delta^{\varepsilon Z0}_{\varepsilon X G}, \Delta$.*
   2. If $\vdash_S \Sigma \mapsto \Omega, G$ and $\vdash_{S^T_-} G \mapsto \Delta$ and $\Sigma \mapsto \Omega$ contains $\varepsilon X G$ then $\vdash_S \Sigma, \Gamma^{\varepsilon Z1}_{\varepsilon X G} \mapsto \Delta^{\varepsilon Z1}_{\varepsilon X G}, \Delta$.

Proof. Just replace all the predecessors of the occurrences of $\varepsilon Z0$ by $\varepsilon X G$ renaming variables if necessary. It is not hard to check that such a procedure leaves all the interferences of rules valid except for *ext* and $\varepsilon \mapsto$. But because $\varepsilon X G$ is contained in $\Sigma \mapsto \Delta$, the ε-term in interference of *ext* and $\varepsilon \mapsto$ belongs to the end-sequent of the interference, thus the interference stays valid.

Provability of $\vdash_S \Sigma \mapsto \Delta, G$ assures that the axioms are provable in S.

Theorem 11. $S \sim S^T_-$

Proof. Let us prove it by induction on the height of the S^T_--proof $P : \Gamma \mapsto \Delta$. If the last inference in P is an inference of a rule of S, there is nothing to prove. If it is not, then it is an application of the *ext*:

$$\frac{\Gamma \mapsto F[X] \sim G[X], \Delta \quad \Gamma, \varepsilon X_1 F[X_1] \sim \varepsilon X_2 G[X_2] \mapsto \Delta}{\Gamma \mapsto \Delta},$$

and one of formulas F, G is a constant. With the help of the previous lemma we could prove that the end-sequent is provable in S:

$$\frac{\Gamma \mapsto F[X] \sim G[X], \Delta \quad \Gamma, \varepsilon X_1 F[X_1] \sim \varepsilon X_2 G[X_2] \mapsto \Delta}{\Gamma \mapsto \Delta}.$$

Because it does not belong to S, one of the formulas G and F is 0 or 1, For example, assume that $F = 0$ and that $\varepsilon Z0$ do not belong to the sequent $\Gamma \mapsto \Delta$. Then the interference is of the following form:

$$\frac{\Gamma \mapsto 0 \sim G[X], \Delta \quad \Gamma, \varepsilon Z0 \sim \varepsilon X G[X] \mapsto \Delta}{\Gamma \mapsto \Delta}.$$

The sequent $\Gamma \mapsto 0 \sim G[X], \Delta$ is provable in S. Then it follows that $\Gamma, G[X] \mapsto \Delta$ is also provable in S. The sequent $\Gamma \mapsto \Delta$ is provable in S^T_-. Now applying the previous lemma to this two formulas we have that

$$\vdash_S \Gamma, (\Gamma, \varepsilon Z0 \sim \varepsilon X G[X])^{\varepsilon Z0}_{\varepsilon X G[X]} \mapsto \Delta, \Delta^{\varepsilon Z0}_{\varepsilon X G[X]}.$$

Since $\Gamma \mapsto \Delta$ does not contain $\varepsilon Z0$, we have

$$\vdash_S \Gamma, (\varepsilon Z0 \sim \varepsilon X G[X])^{\varepsilon Z0}_{\varepsilon X G[X]} \mapsto \Delta.$$

Now applying *ext* yields the desired result:

$$\frac{\Gamma \mapsto G[X] \sim G[X], \Delta \quad \Gamma, \varepsilon X G[X] \sim \varepsilon G[X] \mapsto \Delta}{\Gamma \mapsto \Delta}.$$

The four other cases can be done analogously.

Acknowledgements I thank very much Prof. Mints who have given me this problem for many helpful discussions and for his continuous help. I thank Prof. Hatcher and Prof. Orevkov for their help in editing the paper.

References

1. Mints, G.: Normalization theorems for the intuitionistic systems with choice principles. Math Logic, 59-66 Plenum, New York 1990
2. Mints, G.: Hilbert's Substitution Method and Gentzen-type systems. Proceedings of 9-th International Congress on Logic, Method. and Philos. of Sci, (1991)
3. Mints, G.: Epsilon Substitution Method for Elementary Analysis. Report No. CSLI-93-175, Stanford University, February 1993
4. Mints, G.: Strong Termination for the Epsilon Substitution Method. (to appear)
5. Mints, G.: Simplified Consistency Proof for Arithmetic. Proc. Eston. Acad. of Sci Fiz.-Math. **31** No. 4 (1982) 376-164

Finite Bases of Admissible Rules for the Logic $S5_2C$

Mikhail Golovanov*

Mathematics Department, Krasnoyarsk University,
av. Svobodnyi 79, 660 041 Krasnoyarsk, Russia
e-mail: glvnv@math.kgu.krasnoyarsk.su

Abstract. We study bimodal logic system $S5_2C$ having two modal operators $\Box_0$ and $\Box_1$, each of which satisfies the axioms of $S5$ and in addition, an axiom for commutability of modal operators: $\Box_0\Box_1 p \equiv \Box_1\Box_0 p$. The main result of this paper establishes that the bimodal logic $S5_2C$ and all its extensions have finite bases for admissible inference rules. We also show that even though the logic $S5_2C$ is not locally finite, any proper extension of $S5_2C$ is already locally finite. Moreover, the universal theory of the free algebra of any $S5_2C$-logic is decidable. It is shown also that any $S5_2C$-logic λ with the adjoined inference rule $\Diamond_0\Diamond_1 x \wedge \Diamond_0\Diamond_1 \neg x/y$ is structurally complete and that logic has the same set of theorems as the logic λ.

1 Introduction

Modal logics related to strong systems placed near the top in the lattice of modal logics have been under active consideration for a long time. For instance, K. Fine [2] proved that all extensions of the mono-modal logic $S4.3$ are finitely axiomatizable and have the finite model property. Continuing this affirmative tradition V. V. Rybakov [1] has shown that the set of all admissible inference rules for any logic containing $S4.3$ has a finite bases. We cintinue in this tradition, transferring methods developed by these authors for monomodal logics to the bimodal logic $S5_2C$. The study of the modal system $S5_2C$ looks as an natural and necessary step towards the study of poly-modal logics since this logic, being a fusion of two monomodal simple Lewis's systems $S5$ with the weakest condition on interaction between distinct modal operators, already reveals many of the difficultiesof polymodalities and certain essential ways to handle them.

Our primary focus is on questions concerning bases for admissible inference rules of the bimodal logic $S5_2C$ with the law of commutating modalities. Recall that $S5_2C$ is a logic having two modal logical connectives $\Box_0$ and $\Box_1$, each of which satisfies the axioms of $S5$ and in addition, the following axiom of commutability for modal operators: $\Box_0\Box_1 p \equiv \Box_1\Box_0 p$. We will prove that the logic $S5_2C$, as well as any of its extensions, have a finite basis of admissible inference

---

* The research was supported by Grant Center of Novosibirsk State University and RFFI grant 96-01-00228

rules. We also show that, even though $S5_2C$ is not locally finite, any proper extension of $S5_2C$ is locally finite.

2 Main results

Let K be a cone of some frame which is adequate for the logic $S5_2C$ and is generated by one element, i. e. for some w, $K = \{x|wR_0R_1x\}$. An R_i-cluster is a set $\{x|wR_ix\}$ for some element w of a frame. An $R_0 \cap R_1$-cluster is an intersection of an R_0-cluster and an R_1-cluster. We define the $\square_1$-length of the cone K to be the number of $R_0 \cap R_1$-clusters contained in some R_1-cluster of the cone K. Similarly, we define $\square_0$-length of the cone K. Let $\mathcal{M}$ be a finite model adequate to the logic $S5_2C$ and consisting of only one cone. Denote by $C_0^i (i = 1, \ldots, n)$ and by $C_1^j (j = 1, \ldots, m)$ the equivalence classes of the universe of the model $\mathcal{M}$ with respect to R_0 and R_1 respectively. And denote by X_{ij} the set $C_0^i \cap C_1^j$. Let r, s be natural numbers, where $1 \leq r \leq n$ and $1 \leq s \leq m$. It is convenient to represent the frame of the model $\mathcal{M}$ as a matrix

$$\begin{pmatrix} X_{11} & X_{12} & \ldots & X_{1s} & \ldots & X_{1m} \\ \ldots & \ldots & \ldots & \ldots & & \ldots \\ X_{r1} & X_{r2} & \ldots & X_{rs} & \ldots & X_{rm} \\ \ldots & \ldots & \ldots & \ldots & & \ldots \\ X_{n1} & X_{n2} & \ldots & X_{ns} & \ldots & X_{nm} \end{pmatrix} .$$

Consider a new model $\mathcal{M}'$ which is obtained from $\mathcal{M}$ by its transformation as follows. A set $|\mathcal{M}'|$ ($|\mathcal{M}'|$ denotes the universe of the model $\mathcal{M}'$) consists of $n+1$ classes of equivalences with respect to R_0, and this set also consists of $m + 1$ classes of equivalences with respect to R_1. We denote by

$$D_0^1, \ldots, D_0^{r-1}, D_0^{1r}, D_0^{2r}, D_0^{r+1}, \ldots, D_0^n$$

the classes of equivalences of $|\mathcal{M}'|$ with respect to R_0 The classes of equivalences of $|\mathcal{M}'|$ with respect to R_1 are denoted by

$$D_1^1, \ldots, D_1^{s-1}, D_1^{1s}, D_1^{2s}, D_1^{s+1}, \ldots, D_1^m.$$

As a set $X'_{ij} = D_0^i \cap D_1^j$, we take X_{ij} if $i \neq r$ and $j \neq s$. Further, if only one from the equalities $i = r$, $j = s$ holds, then we set $D_0^{tr} \cap D_1^j = X_{rj}^t = \{w^t|w \in X_{rj}\}$ for $t = 1, 2$ and $D_0^i \cap D_1^{ts} = X_{is}^t = \{w^t|w \in X_{is}\}$ for $t = 1, 2$. And further, let $Y_{rs} = X_{rs} \setminus \{x_0\}$, where $\{x_0\}$ is a fixed element of the set X_{rs}. We pose by definition $D_0^{tr} \cap D_1^{ts} = Y_{rs}^t = \{w^t|w \in Y_{rs}\}$ for $t = 1, 2$, $D_0^{1r} \cap D_1^{2s} = \{x_0^1\}$, $D_0^{2r} \cap D_1^{1s} = \{x_0^2\}$ which concludes our definition of the frame for the model $\mathcal{M}'$. For simplicity we exhibit the constructed frame as a matrix:

$$\begin{pmatrix} X_{11} & X_{12} & \ldots & X_{1s}^1 & X_{1s}^2 & \ldots & X_{1m} \\ \ldots & \ldots & & \ldots & \ldots & & \ldots \\ X_{r1}^1 & X_{r2}^1 & \ldots & Y_{rs}^1 & \{x_0^1\} & \ldots & X_{rm}^1 \\ X_{r1}^2 & X_{r2}^2 & \ldots & \{x_0^2\} & Y_{rs}^2 & \ldots & X_{rm}^2 \\ \ldots & \ldots & & \ldots & \ldots & & \ldots \\ X_{n1} & X_{n2} & \ldots & X_{ns}^1 & X_{ns}^2 & \ldots & X_{nm} \end{pmatrix} .$$

The valuation V' on a frame $\mathcal{M}'$ is defined as follows: the domain of V' coincides with the domain of V, $V'(p)$ consists of all elements of $V(p)$ and theirs duplicates. It is not hard to prove the following lemma by induction on the length of formulas.

Lemma 1. *If* $w \in C_0^i (i \neq r)$ *and* $w \notin C_1^s$, *then*

$$\forall \varphi (\mathcal{M}, w \Vdash \varphi \iff \mathcal{M}', w \Vdash \varphi).$$

If $w \in C_0^r$ *or* $w \in C_1^s$, *then*

$$\forall \varphi (\mathcal{M}, w \Vdash \varphi \iff \mathcal{M}', w^1 \Vdash \varphi \iff \mathcal{M}', w^2 \Vdash \varphi).$$

Lemma 2. *If* λ *is a proper extension of the logic* $S5_2C$ *then there exists a number* n *depending only on the logic* λ *such that, for any cone* K *of its canonical Kripke model,* $\square_0$*-length or* $\square_1$*-length of* K *does not exceed* n.

Proof. Prove by contradiction. Suppose that in the canonical Kripke model $\mathcal{C}_\lambda$, for any natural number n, there exists a cone K for which its $\square_0$-length $d_0(K) > n$ and its $\square_1$-length $d_1(K) > n$. Note that the frame of the model $\mathcal{C}_\lambda$ is adequate to the logic $S5_2C$. Let $\alpha \in \lambda \setminus S5_2C$. Since $S5_2C$ has fmp and $\alpha \notin S5_2C$, there exists some finite model $\mathcal{M}$ adequate to $S5_2C$ such that $\mathcal{M} \not\Vdash \alpha$. According to Lemma 1 there exists a model $\mathcal{A}$ adequate to the $S5_2C$ with the single-element $\square_0 \cap \square_1$-clusters for which $\mathcal{A} \not\Vdash \alpha$. We enumerate its $\square_0$-clusters: $D_1^0, \ldots, D_{n_0}^0$ and then its $\square_1$-clusters: $D_1^1, \ldots, D_{n_1}^1$. Let $D_i^0 \cap D_j^0 = \{a_{ij}\}$. According to the our assumption there exists a cone K in canonical model $\mathcal{C}_\lambda$ with $d_0(K) > n_0$ and $d_1(K) > n_1$. But then there exists in K pairwise different $\square_0$-clusters $C_1^0, \ldots, C_{n_0}^0$ and $\square_1$-clusters $C_1^1, \ldots, C_{n_1}^1$. Note that $C_i^0 \cap C_j^1 \neq \emptyset$ since $\forall x, y \in K (x R_0 R_1 y)$. For each C_i^t, where $t = 0, 1$, exists a formula φ_i^t of the type $\square_t \delta$ or $\lozenge_t \delta$ which is true on every element of C_i^t and is false on every element of C_j^t for $j \neq i$. Note that $\{x | x \Vdash \varphi_i^0 \wedge \varphi_j^1\} \neq \emptyset$ for $i = 1, \ldots, n_0, j = 1, \ldots, n_1$ since $C_i^0 \cap C_j^1 \neq \emptyset$. Let $\psi_i^t = \varphi_i^t \wedge \bigwedge_{j \neq i} \neg \varphi_j^t$, $\psi^t = \bigvee_i \psi_i^t$,

$$\Psi_k = (\bigwedge_i \psi_i^0 \wedge \bigwedge_j \psi_j^1) \wedge (\bigvee_{i,j,a_{ij} \Vdash p_k} (\psi_i^0 \wedge \psi_j^1) \vee$$

$$\bigvee_{i,a_{i1}p_k} (\psi_i^0 \wedge \neg\psi^1) \vee \bigvee_{j,a_{1j} \Vdash p_k} (\psi_j^1 \wedge \neg\psi^0)).$$

Now let

$$\Phi_k = \begin{cases} \Psi_k & \text{if } a_{11} \not\Vdash p_k \\ \Psi_k \vee (\neg\psi^0 \wedge \neg\psi^1) & \text{if } a_{11} \Vdash p_k \end{cases}$$

We define a definable valuation $V(p_k) = \{x | x \Vdash \Phi_k\}$ on canonical model $\mathcal{C}_\lambda$. It can be proved by induction on the length of a formula that $\forall \delta (\mathcal{A} \Vdash \delta \iff \mathcal{C}_\lambda \Vdash_V \delta)$. Hence $\mathcal{C}_\lambda \not\Vdash_V \alpha$. Since the valuation V is definable $\alpha \notin \lambda$. This contradicts the choice of α. $\square$

Theorem 3. *Any* $S5_2C$*-logic* λ *has fmp and hence* λ *is Kripke complete.*

Proof. For the logic $S5_2C$ the assertion of this lemma is true. Let λ be a proper extension of $S5_2C$. According to the Lemma 2, for λ there exists such natural n that, for any cone K of canonical Kripke model C_λ, $d_0(K) \leq n$ or $d_1(K) \leq n$. Examine the case $d_1(K) \leq n$, the case $d_0(K) \leq n$ is similar. If $d_1(K) = d$ then there exists a family of R_0-clusters $C_1, \ldots, C_d$ for which $K = \bigcup_i C_i$. Then there exist the formulas $\square_0\psi_1, \ldots, \square_0\psi_d$ of such type that, for any $x \in K$ $x \in C_i \iff x \Vdash \square_0\psi_i$. Let a set Y consists of all subformulas of formulas $\square_0\psi_1, \ldots, \square_0\psi_d$ and their negations and let X consists of all formulas of types φ and $\Diamond_i\varphi$, $i = 0, 1$, where φ is a conjunction of the formulas from the set Y, and of all their subformulas. Now we filtrate the submodel K of the model C_λ by the set X. Let M_i be the set of all formulas of the type $\square_i\varphi$ or $\Diamond_i\varphi$, $i = 0, 1$. Pose by definition $[x]R_i[y]$ if $x \cap X \cap M_i = y \cap X \cap M_i$. Denote by $\bar{K}$ the filtrated model. Since $\square_0\psi_i \in X$ for $i = 1, \ldots, d$, we have $[x]R_0[y] \iff xR_0y$, therefore the map $x \to [x]$ is a p-morphism of K on $\bar{K}$ relatively the relation R_0. Let $[x]R_1[y]$ and $y \Vdash \varphi \wedge \bigwedge_j \Diamond_1\varphi_j$, where φ is a conjunction of all formulas from the set Y which are valid on y. Since $y \Vdash \Diamond_1\varphi$ and $\Diamond_1\varphi \in X$, we have $x \Vdash \Diamond_1\varphi$ according to the definition of relation R_1 on the set $\bar{K}$. Therefore there exists y' for which $y' \Vdash \varphi$ and xR_1y'. But then $y' \Vdash \varphi \wedge \bigwedge_j \Diamond_1\varphi_j$ and $[y'] = [y]$. Hence the map $x \to [x]$ is a p-morphism of K on $\bar{K}$. For the case $d_0(K) \leq n$, we can construct the similar p-morphism of K on some finite model. Since the universe of the canonical Kripke model C_λ is a disjoint union of cones of types examined above, we can put in correspondence a p-morphism $f_K : K \to \bar{K}$ for each K. Let $\bar{C}_\lambda$ be the disjoint union of all $\bar{K}$ and let $f : C_\lambda \to \bar{C}_\lambda$ be a map defined by condition $f|_K = f_K$. Then f is a p-morphism. Hence $\forall\alpha(C_\lambda \Vdash \alpha \iff \bar{C}_\lambda \Vdash \alpha)$. All cones in universe of the model $\bar{C}_\lambda$ are finite hence the logic λ has finite model property and is Kripke complete. $\square$

If $A = (a_{ij})$ and $B = (b_{ij})$ are rectangular matrix of a dimension $n \times m$ then we set by definition $A \leq B$ iff $\forall i, j a_{ij} \leq b_{ij}$. For certain matrixes A and B of distinct dimensions we pose $A \leq B$ if there exists in B a submatrix A' of dimension equal to dimension of A such that $A \leq A'$. We set $A \preceq B$ if $A \leq B$ and, for any column b of the matrix B, there exists a column a of matrix A such that $a \leq b$. By induction on k we can prove the following two lemmas.

Lemma 4. *If $B_1, B_2, \ldots$ is infinite sequence of matrix with k rows then exists an infinite subsequence of matrix $B'_1, B'_2, \ldots$, which satisfies to the condition $B'_1 \leq B'_2 \leq \ldots$.*

In the case $k = 1$ Lemma 4 is proved by K.Fine [2]. Our proof of this Lemma is a generalization on case $k > 1$ of the proof of V.V.Rybakov for the case $k = 1$.

Lemma 5. *If $B_1 \leq B_2 \leq \ldots$ is infinite chain of matrices with k rows then there exists an infinite subsequence of matrices $B'_1, B'_2, \ldots$, which satisfies to the condition $B'_1 \preceq B'_2 \preceq \ldots$.*

Proof. Induction on k. Let $B_1 \leq B_2 \leq \ldots$ is an infinite increasing sequence of matrix. We may assume $\forall\beta^l_{ij} \in B_l (\beta^l_{ij} \leq \beta^{l+1}_{ij})$.

Investigate the case $k = 1$. Let $\mu_t = \min_j(\{\beta_j^t | \beta_j^t \in B_t\})$. Two cases are possible : a) the sequence μ_t does not have supremum, b) the sequence μ_t has a supremum.

If case a) holds then in sequence $B_1, B_2, \ldots$ there exists infinite subsequence $B_{j_1}, B_{j_2}, \ldots$, for which $\mu_{j_s} < \mu_{j_{s+1}}$. Proceeding to examination of this subsequence we may believe that $B_1, B_2, \ldots$ coincide with this subsequence, that is $\mu_s < \mu_{s+1}$ and the conclusion of the Lemma holds.

In case b), there is a number m for which the inclusion $\mu_t \leq m$ holds. Then there exists some infinite subsequence $B_{j_1}, B_{j_2}, \ldots$ in sequence $B_1, B_2, \ldots$ for which $\mu_t = const$. We may proceed to examination of this subsequence and also set that this assumption holds for the sequence $B_1, B_2, \ldots \mu_t = const$ already, that is the conclusion of lemma holds.

Now proceed to the case $k \geq 2$. Suppose that the lemma is proved for the matrices having less than k rows. Denote by C_t^i a submatrix of matrix B_l which is obtained from the B_l by deleting the row with number i. By inductive hypotheses we can assume that the number of columns of the matrix B_l is an increasing function of l and, for any i, $C_1^i \preceq C_2^i \preceq \ldots$. Let $\mu_l^i = \min(\{\beta_{it}^l | \beta_{it}^l \in B_l\})$. Examine two possible cases: a) one of the sequences μ_t^i, $i = 1, \ldots, k$, does not have supremum, b) each from the sequences μ_t^i, $i = 1, \ldots, k$ has a supremum.

Case a), we can assume that μ_t^1 has no supremum. Let

$$\delta_t = \max(\{\alpha_{1j}^t | \alpha_{1j}^t \in B_t\}).$$

Under these assumptions, in our sequence of matrix there is a subsequence of matrices with the numbers $t_1, t_2, \ldots$ such that $\delta_{t_j} < \mu_{t_{j+1}}$. Proceeding to the chosen subsequence and changing the numeration we may assume that for the sequence $B_1, B_2, \ldots$, $\delta_t < \mu_{t+1}^1$ is true already, that is, for all i, j if $i < j$ then $\alpha_{1i}^l < \alpha_{1j}^{l+1}$. Applying the inductive hypotheses to the sequence of submatrices $C_1^1, C_2^1, \ldots$ and choosing the corresponding subsequence from the sequence $B_1, B_2, \ldots$ we obtain a sequence satisfying to the conclusion of our Lemma.

Case b). The sequence μ_t^1 has an infinite constant subsequence. Going to the corresponding subsequence in starting sequence of matrices we can assume that, for the sequence $B_1, B_2, \ldots$, we have $\mu_t^1 = const = \mu^1$ already. For constructed sequence of matrices the sequence μ_t^2 has a supremum. Repeating this procedure we get that, for the sequence $B_1, B_2, \ldots$, $\mu_t^2 = const = \mu^2$. After repeating this procedure k times we obtain that, for all $i = 1, \ldots, k$, $\mu_t^i = const = \mu^i$. Let $B_l^{\mu^i}$ be a maximal submatrix of matrix B_l which has in i-th row the number μ^i. If the matrices $B_l^{\mu^i}$ $(i = 1, \ldots, k)$ have a common column $\bar{b}$ then the sequence of matrices $B_1, B_2, \ldots$ satisfies the conclusion of Lemma since $\forall l (\bar{b} \in B_l)$ and $\forall l (\bar{b} \preceq B_l)$.

Now assume that the matrices $B_l^{\mu^i} (i = 1, \ldots, k)$ have not the common columns. By the inductive hypotheses the sequence of matrices $B_1^{\mu^1}, B_2^{\mu^1} \ldots$ has a subsequence with the numbers $j_1, j_2, \ldots$ which satisfies the conclusion of our Lemma. We may go to the subsequence with the numbers $j_1, j_2, \ldots$ instead of sequence $B_1, B_2 \ldots$. After that we may assume that $B_1, B_2 \ldots$ is such that

$B_1^{\mu^1}, B_2^{\mu^1} \ldots$ satisfies the conclusion of our Lemma. Repeating successively the same reasoning for the submatrices $B_l^{\mu^i}$ for $i = 2, \ldots, k$ we get that $B_1^{\mu^i}, B_2^{\mu^i} \ldots$ satisfies the conclusion of our Lemma for all $i = 1, \ldots, k$.

If $\forall l \exists t > l (B_l \preceq B_t)$ then this sequence has an infinite subsequence satisfying the conclusion of lemma and lemma is proved. In the converse case there exists a matrix B_{l_0}, for which there are no matrices $B_l (l > l_0)$ such that $B_{l_0} \preceq B_l$. We may pose $l_0 = 1$. Let $\gamma_i = \min(\{\beta_{it}^1 | \beta_{it}^1 \in B_1^{\mu^i}\})$. Then the other matrices must contain the columns of type $\bar{b} = (\beta_{1j}, \ldots, \beta_{kj})^{tr}$, where $\beta_{ij} < \gamma_i$ for $i = 1, \ldots, k$. Since the number of columns of this type is finite it is possible to select from our sequence of matrices some infinite subset W, consisting of matrices such that, for any column $\bar{b}$ of described type, either all matrices from the set W contain this column or all matrices from the set W do not contain this column. Since we may assume that the matrices B_l do not contain equal columns there exists the maximal submatrix C, consisting of columns included in all matrices of the set W. By supposition, for each matrix $B \in W$, $B_1 \not\preceq B$ hence the matrix C is nonempty. Denote $m_i = \max_j(\beta_{ij} \in C)$. Let $A \in W$ and let $\bar{b} = (u_1, \ldots, u_k)^{tr}$ be an arbitrary column of the matrix A. If $u_i \leq m_i$, for $i = 1, \ldots, k$ and the column $\bar{b}$ is contained only in finite number of matrices from W, then we remove these matrices from W. Otherwise, when infinite number of matrices from W contain the column $\bar{b}$, we remove from W the matrices which do not contain this column. Then we repeat this procedure for every column $\bar{b}$ satisfying the condition $u_i \leq m_i$ for $i = 1, \ldots, k$ and then redefine the matrix C as the submatrix consisting of columns of the type $u_i \leq m_i$, for $i = 1, \ldots, k$ contained in all matrices from W. If $\bar{b}$ is an arbitrary column of the matrix $A \in W$ then the following cases: 1) $\bar{b} \in C$, 2) $u_i < m_i$ and $u_j \geq m_j$ for some $i, j (i \neq j)$, 3) $\forall i u_i \geq m_i$ are possible.

For matrix $A \in W$ by A^{ij} we denote the submatrix of A consisting of columns satisfying the condition 2) for fixed $i, j (i \neq j)$. Fixe some pair of numbers $1 \leq i, j \leq k$. If the set $V^{ij} = \{A^{ij} | A \in W)\}$ is finite the following cases: a) the set $W^{ij} = \{A \in W | A^{ij} \in V^{ij}\}$ is finite, b) the set $W^{ij} = \{A \in W | A^{ij} \in V^{ij}\}$ is finite are possible. Case a), we remove from W all the matrices contained in W^{ij} and denote the obtained set by W_{ij}. Case b), we take in W the infinite subset W_{ij} satisfying the condition $\forall A, A' \in W'(A^{ij} = (A')^{ij})$. If the set $V^{ij} = \{A^{ij} | A \in W\}$ is infinite then by inductive hypotheses in V^{ij} there exists an infinite chain $A_1^{ij} \preceq A_2^{ij} \preceq \ldots$. In this case we choose from W the subset W_{ij} consisting of the matrices for which the submatrices satisfying the condition 2) for fixed $i, j (i \neq j)$ belong to the chosen chain. So we constructed an infinite set W_{ij} in which for any two matrices $A, A' \in W_2$, either $A^{ij} \preceq (A')^{ij}$ or $(A')^{ij} \preceq A^{ij}$. Further, for each matrix $A \in W_{ij}$, by $A^{i_1 j_1}$ we denote a submatrix of matrix A consisting of columns satisfying the condition 2) for other pair of numbers $i_1, j_1 (i_1 \neq j_1)$. By repeating the same reasoning as for the columns of type 2) for the numbers $i, j (i \neq j)$ we construct the set $W_{ij i_1 j_1}$ for which the sets $\{A^{ij} | A \in W_{ij i_1 j_1}\}$ and $\{A^{i_1 j_1} | A \in W_{ij i_1 j_1}\}$ are linearly ordered by relation $\preceq$. Using reasoning of this type for all possible pairs $1 \leq i, j \leq k$, satisfying the conditions of type 2), we construct an infinite set W_0 in which, for any pair of numbers $1 \leq i, j \leq k (i \neq j)$, the subset $\{A^{ij} | A \in W_0\}$ is linearly ordered by relation $\preceq$. We enumerate the elements of the set W_0 in such a way that the numeration is co-ordinated with the order $\preceq$, i.e. $A_i \preceq A_j \iff i < j$. The matrix $B_l \in W_0$ has the following structure $B_l = (C, B_l^2, B_l^3)$, where the matrix B_l^2 consists of the columns of matrix B_l satisfying the condition 2) for some i, j and hence $\forall l (B_l^2 \preceq B_{l+1}^2)$ and B_l^3 consists of the columns of matrix B_l satisfying the condition 3) and hence $\forall l (C \preceq B_l^3)$. Therefore for any l $B_l \preceq B_{l+1}$. $\square$

If the frames F_1, F_2 are exhibited by the corresponding matrices B_1, B_2 and $B_1 \preceq B_2$ then exists a p-morphism F_2 onto F_1. Therefore using Lemma 5, we can show that the following holds.

Lemma 6. *Every increasing chain of $S5_2C$-logics is finite.*

Corollary 7. *All $S5_2C$-logics are finitely axiomatizable and decidable.*

Lemma 8. *Let B be a certain finitely generated modal algebra from the variety $Var(S5_2C)$, $\|B\| > 1$ and let $a_1, \ldots, a_n$ be a minimal set of generators of algebra B. And let in B^+ the valuation be defined as $V(p_i) = \{\nabla | a_i \in \nabla\}$. Then, for any proper cone from B^+, the valuation V distinguish certain two distinct elements of that cone.*

Proof. Let C be a non-single-element cone and the valuation V does not distinguish the elements of C. Since any generator a_i can be replaced by $\neg a_i$ and by assumption the valuation V does not distinguish the elements of C we get $\forall a_i \forall \nabla (a_i \in \nabla)$. Let $\nabla_1, \nabla_2 \in C$. Further by induction on the length of term it follows that $t(a_1, \ldots, a_n) \in \nabla_1 \iff t(a_1, \ldots, a_n) \in \nabla_2$. Let $v_1 \wedge v_2 \in \nabla_1$ then $v_1 \in \nabla_1$ and $v_2 \in \nabla_1$. By inductive hypothesis $v_1 \in \nabla_2$ and $v_2 \in \nabla_2$ hence $v_1 \wedge v_2 \in \nabla_2$. Let $\neg v \in \nabla_1$ then $v \notin \nabla_1$ and by inductive hypotheses $v \notin \nabla_2$ that is $\neg v \in \nabla_2$.

Let $\square_0 v \in \nabla_1$ and $\square_0 v \notin \nabla_2$. Denote $\square_0[\nabla_2] = \{c | \square_0 c \in \nabla_2\}$. Let $\Delta = \{z | \exists c \in \square_0[\nabla_2] \neg v \wedge c \leq z\}$. If $\square_0 c_1 \in \nabla_2$ and $\square_0 c_2 \in \nabla_2$ then $\square_0(c_1 \wedge c_2) = \square_0 c_1 \wedge \square_0 c_2 \in \nabla_2$. Hence Δ is a filter. Let $\neg v \wedge c_0 = \bot$ then $(\neg v \wedge c_0) \vee v = v = (\neg v \vee v) \wedge (c_0 \vee v) = c_0 \vee v$, i.e. $c_0 \leq v$ and $c_0 \to v = \top$ hence $\square_0 c_0 \to \square_0 v = \top$ and $\square_0 c_0 \leq \square_0 v$. Since $\square_0 c_0 \in \nabla_2$, we have $\square_0 v \in \nabla_2$ but this contradicts to supposition. So Δ is non-trivial filter. Let Δ^* be an ultrafilter containing Δ. Then Δ^* belongs to the cone C and by construction $\neg v \in \Delta^*$ but this contradicts inductive hypotheses. Hence $\square_0 v \in \nabla_2$. The case of $\square_1$ can be shown similarly. Hence all elements of the cone coincide. $\square$

We will denote the composition $\square_0 \square_1$ by $\square$. Then, in logic $S5_2C$, operator $\square$ satisfies the axioms of logic $S5$. If $\langle U, R_0, R_1 \rangle$ is a frame adequate to $S5_2C$ then $\langle U, R_0 R_1 \rangle$ is a frame adequate to the logic $S5$ with the operator $\square = \square_0 \square_1$. Due to this connection we can use the following lemma.

Lemma 9. *Let $F = \langle U, R \rangle$ is a frame adequate to the logic $S5$. If the valuation V of variables $p_1, \ldots, p_n$ on the set U distinguish is not constant for every cluster of frame F then exists a formula α so that on the model $M = \langle U, R, V \rangle$ the formula $\Diamond \alpha \wedge \Diamond \neg \alpha$ is true.*

Lemma 10. *Let B be a finitely generated modal algebra belonging to the variety $Var(S5_2C)$ and $\|B\| > 1$. If $B \models \Diamond_0\Diamond_1 x \wedge \Diamond_0\Diamond_1 \neg x = \top \Rightarrow y = \bot$ then the frame B^+ has a single-element cone.*

Lemma 2 and lemma 3 allow us to construct a finite n-characterizing model for any proper extension of $S5_2C$. Note that the frame $\mathcal{E} = \langle U, R_0, R_1\rangle$ with $|U| = 1$ and reflexive relations R_0, R_1 is adequate to the logic $S5_2C$ and hence n-characterizing model $Char_\lambda(n)$ contains the single-element open submodels.

Lemma 11. *Let $F_i\ 1 \leq i \leq n$ are the different cones of the model $Char_\lambda(n)$. The algebra $(\bigsqcup_{i=1}^{n} F_i \circ 1)^+$ is isomorphic to a subalgebra of free algebra $Char_\lambda(n + m)^+(V(p_1), \ldots, V(p_{n+m}))$ of the variety $Var(\lambda)$ for $m \geq 1$.*

Proof. Evidently that $(\bigsqcup_{i=1}^{n} F_i \circ 1)$ is open subframe of the frame $Char_\lambda(n+1)$ and we can consider its elements as the elements of the model $Char_\lambda(n+1)$. Let e be an element of the model $Char_\lambda(n + 1)$ corresponding to the single-element cone 1 of the frame $(\bigsqcup_{i=1}^{n} F_i \circ 1)$. All elements of the model $Char_\lambda(n + 1)$ are definable. For every element $x \in \bigsqcup_{i=1}^{n} F_i$ we assign a formula α_x which is true only on the element x of the model $Char_\lambda(n+1)$. Let B be a subalgebra of the algebra $Char_\lambda(n + 1)^+$ generated by the set of elements $\{\alpha_x | x \in \bigsqcup_{i=1}^{n} F_i\}$. We will prove that the subalgebra B is isomorphic to the algebra $(\bigsqcup_{i=1}^{n} F_i \circ 1)^+$. We introduce the mapping $f : Char_\lambda(n + 1)^+ \to (\bigsqcup_{i=1}^{n} F_i \circ 1)^+$ as:

$$\forall X \subset |Char_\lambda(n + 1)|\ f(X) = X \cap |\bigsqcup_{i=1}^{n} F_i \circ 1|.$$

Since $\bigsqcup_{i=1}^{n} F_i \circ 1$ is an open subframe of the frame of the model $Char_\lambda(n+1)$ f is a homomorphism on algebra $(\bigsqcup_{i=1}^{n} F_i \circ 1)^+$. Prove that the restriction of reflexion f to the subalgebra B is an isimorphism of the algebras B and $(\bigsqcup_{i=1}^{n} F_i \circ 1)^+$. It is evident that this restriction is a homomorphism of B in $(\bigsqcup_{i=1}^{n} F_i \circ 1)^+$. Let $x \in \bigsqcup_{i=1}^{n} F_i$. Since $x \Vdash \vee \alpha_x$ and the formula α_x is false on the others elements of the characteristic model we have $V(\alpha_x) = \{x\}$ and $f(V(\alpha_x)) = \{x\}$. According to the choice of the formulas α_x $e \not\Vdash \vee \alpha_x$ for each formula α_x. Since the inclusion $x \in V(\alpha_x)$ holds for each $x \in \bigsqcup_{i=1}^{n} F_i$ we have

$$\{e\} = V(\bigwedge_{\alpha_x} \neg\alpha_x) \cap |\bigsqcup_{i=1}^{n} F_i \circ 1|.$$

Hence $f(V(\bigwedge_{\alpha_x} \neg\alpha_x)) = \{e\}$. Thus we proved that f maps B on $(\bigsqcup_{i=1}^{n} F_i \circ 1)^+$. Since f is a homomorphism, for proof of the assertion that the restriction f on B is one-to-one correspondence it is sufficient to prove that f-image of any nonempty set Y belonging to B is nonempty. Let Y be an element of B and $Y \neq \emptyset$. Then $Y = V(t(\alpha_{x_1}, \ldots, \alpha_{x_k}))$, where $t(v_1, \ldots, v_k)$ is a term on the variables $v_1, \ldots, v_k$. Hence there exists an element $y \in Y$ for which $y \Vdash \vee t(\alpha_{x_1}, \ldots, \alpha_{x_k})$. If from y by one of the relations R_0 or R_1 is accessible some element from $|\bigsqcup_{i=1}^{n} F_i \circ 1|$ then $y \in |\bigsqcup_{i=1}^{n} F_i \circ 1|$ and $y \in f(Y)$ that is $f(Y) \neq \emptyset$. Let for any

$z \in |\bigsqcup_{i=1}^{n} F_i \circ 1|$ $(y,z) \notin R_0 R_1$. For any formula α_x, $y \not\Vdash_V \alpha_x$ and $e \not\Vdash_V \alpha_x$. Since each formula α_x is true only on a single corresponding element x it is possible to prove by induction on the length of formula, that for any bimodal formula $\delta(v_1, \ldots, v_k)$ the assertion holds:

$$e \Vdash_V \delta(\alpha_{x_1}, \ldots, \alpha_{x_k}) \iff y \Vdash_V \delta(\alpha_{x_1}, \ldots, \alpha_{x_k}).$$

Hence $e \Vdash_V t(\alpha_{x_1}, \ldots, \alpha_{x_k})$ and $e \in V(t(\alpha_{x_1}, \ldots, \alpha_{x_k}))$ i.e. $e \in f(Y)$, $f(Y) \neq \emptyset$. $\qquad \square$

Lemma 12. *The universal theory of free algebra $F_\lambda(\omega)$ of countable rank from $Var(\lambda)$ and universal theory of all modal algebras of type $(\bigsqcup_{i=1}^{n} F_i \circ 1)^+$ coincide.*

Proof. It is proved in lemma 11 that all algebras of type $(\bigsqcup_{i=1}^{n} F_i \circ 1)^+$ are isomorphic to subalgebras of free algebra $F_\lambda(g)$ of finite rank g. Hence the universal theory of free algebra $F_\lambda(\omega)$ is a subtheory of the class of algebras of type $(\bigsqcup_{i=1}^{n} F_i \circ 1)^+$.

Conversely, let ϕ be an universal formula which is false on $F_\lambda(\omega)$. If ψ is a prenex normal form of ϕ then

$$\psi = \forall x_1 \ldots \forall x_l \bigwedge_k ((f^k = \top) \to \bigvee_{j=1}^{m_k} (\Box_0 \Box_1 g_j^k = \top)).$$

Since ϕ false on some algebra $F_\lambda(d)$ of finite rank d and the algebra

$$Char_\lambda(d)^+ (V(p_1), \ldots, V(p_d w))$$

is isomorphic to the free algebra $F_\lambda(d)$, we have, for some valuation S of variables of formula ψ on frame of characteristic model $Char_\lambda(d)$,:

$$\exists k [(S(f^k) = |Char_\lambda(d)|) \,\&\, \bigwedge_{j=1}^{m_k} (\Box_0 \Box_1 g_j^k \neq |Char_\lambda(d)|].$$

This implies that in $Char_\lambda(d)$ there are certain cones C_j of such type that $\forall a_j \in C_j (a_j \not\Vdash_S \Box_0 \Box_1 g_j^k$. ¿From this we infer that the formula ψ is false on the algebra $(\bigsqcup_{j=1}^{m_k} C_j \circ 1)^+$. But ϕ is equivalent to ψ, hence ϕ false on that algebra too. $\qquad \square$

Theorem 13. *For any $S5_2C$-logic λ the universal theory of free algebra $F_\lambda(\omega)$ is decidable.*

Proof. Let Th_λ be the universal theory of the free algebra $F_\lambda(\omega)$ and $\overline{Th_\lambda}$ is a complement of Th_λ to the set of all universal formulas. In accordance with Lemma 12 the set $\overline{Th_\lambda}$ is recursively enumerable. The set Th_λ is also recursively enumerable according to Lemma 7. Hence the set Th_λ is recursive and the theorem holds. $\qquad \square$

Lemmas 12 and 2 allow as to describe some algorithm of recognizing for belonging of universal formula to Th_λ and summarizing that lemmas we get the following theorem.

Theorem 14. *Every proper extension of the logic $S5_2C$ is locally finite.*

By virtue of connection between the inference rules and quasi-identities can conclude

Corollary 15. *The problem of admissibility of inference rules is decidable for every $S5_2C$-logic.*

Theorem 16. *Free algebra $F_\lambda(\omega)$ of any $S5_2C$-logic λ has a finite basis of quasi-identities consisting from finite basis of identities for $F_\lambda(\omega)$ and quasi-identity $\Diamond_0\Diamond_1 x \wedge \Diamond_0\Diamond_1\neg x = \top \Rightarrow y = \top$*

Proof. According to Lemma 7 a basis of identities for the logic λ and the free algebra $F_\lambda(\omega)$ is finite. Since the quasi-identity $\Diamond_0\Diamond_1 x \wedge \Diamond_0\Diamond_1\neg x = \top \Rightarrow y = \top$ holds on each algebra of type $(\bigsqcup_{i=1}^n F_i \circ 1)^+$ we have according to Lemma 12 $F_\lambda(\omega) \models \Diamond_0\Diamond_1 x \wedge \Diamond_0\Diamond_1\neg x = \top \Rightarrow y = \top$. Let B be finitely generated algebra from $Var(\lambda)$ and $B \models \Diamond_0\Diamond_1 x \wedge \Diamond_0\Diamond_1\neg x = \top \Rightarrow y = \top$. Suppose also $B \not\models \bigwedge_{i=1}^m f_i(x_1,\dots,x_n) = \top \Rightarrow g(x_1,\dots,x_n) = \top$. It is evident that each quasi-identity in $Var(S5_2C)$ is equivalent to quasi-identity of previous type. We can assume that the algebra B is generated by the elements $a_1,\dots,a_n$ and $\bigwedge_{i=1}^m f_i(a_1,\dots,a_n) = \top$, $g(a_1,\dots,a_n) \neq \top$. Then $\Box_0\Box_1 \bigwedge_{i=1}^m f_i(x_1,\dots,x_n) \rightarrow g(x_1,\dots,x_n) \notin \lambda$. Since λ has fmp (Lemma 3) there is an adequate to λ frame F_1 for which $F_1 \not\Vdash \Box_0\Box_1 \bigwedge_{i=1}^m f_i(x_1,\dots,x_n) \rightarrow g(x_1,\dots,x_n)$. But in this case there exists an open subframe F in F_1 such that

$$F \not\Vdash (\bigwedge_{i=1}^m f_i(x_1,\dots,x_n)) = \top \Rightarrow g(x_1,\dots,x_n) = \top. \tag{1}$$

According to Lemma 10 the frame B^+ has a single-element open subframe. Hence the formula $\bigwedge_{i=1}^m f_i(p_1,\dots,p_n)$ is true under a certain valuation of variables on a singleton reflexive frame. This and (1) yield $(F \circ 1)^+ \not\models \bigwedge_{i=1}^m f_i(x_1,\dots,x_n) = \top \Rightarrow g(x_1,\dots,x_n) = \top$. According to Lemma 12 the last assertion means $F_\lambda(\omega) \not\models \bigwedge_{i=1}^m f_i(x_1,\dots,x_n) = \top \Rightarrow g(x_1,\dots,x_n) = \top$. Hence a finite basis of identities of variety $Var(\lambda)$ and the quasi-identity $\Diamond_0\Diamond_1 x \wedge \Diamond_0\Diamond_1\neg x = \top \Rightarrow y = \top$ form a finite basis of quasi-identities of algebra $F_\lambda(\omega)$. □

Corollary 17. *Any $S5_2C$-logic λ has the finite basis of inference rules consisting of single inference rule $\Diamond_0\Diamond_1 x \wedge \Diamond_0\Diamond_1\neg x/y$.*

Proved properties of $S5_2C$-logics, criterions for admissibility of inference rules and an easy modification of a proof from the paper [1] allow to prove for the $S5_2C$-logics the V.V.Rybakov's results for $S4.3$-logics.

Lemma 18. *Let λ be a $S5_2C$-logic and $A(p_1,\ldots,p_n)/B(p_1,\ldots,p_n)$ be an admissible but nonderivable rule in logic λ. Then, for any family of formulas $\gamma_1,\ldots,\gamma_n$, $\neg A(\gamma_1,\ldots,\gamma_n) \in \lambda(\mathcal{F})$, where $\mathcal{F}$ is a single-element frame with two reflexive relations. In particular, $A(\gamma_1,\ldots,\gamma_n) \notin \lambda$ if the logic λ is consistent.*

Theorem 19. *Any $S5_2C$-logic λ with adjoined inference rule $\Diamond_0\Diamond_1 x \wedge \Diamond_0\Diamond_1\neg x/y$ is structurally complete and has the same set of theorems as the logic λ.*

Proof. Suppose A/B is some admissible inference rule of the logic λ with adjoined inference rule $\Diamond_0\Diamond_1 x \wedge \Diamond_0\Diamond_1\neg x/y$. According to Corollary 17 the rule A/B is admissible in logic λ. By Lemma 18 either the rule A/B is derivable in logic λ or $\neg A \in \lambda(F_1)$, where F_1 is a single-element reflexive frame. If the rule A/B is derivable in λ then A/B is derivable in logic λ with adjoined inference rule $\Diamond_0\Diamond_1 x \wedge \Diamond_0\Diamond_1\neg x/y$. We will prove that otherwise

$$\square_0\square_1 A \to \square_0\square_1 \bigvee_{i=1}^{n}(\Diamond_0\Diamond_1 p_i \wedge \Diamond_0\Diamond_1\neg p_i) \in \lambda, \qquad (2)$$

where $p_1\ldots,p_n$ are propositional variables of the formula A. Suppose otherwise, then by fmp there exists a finite frame F with permutable relations, on which our formula is false for some valuation. Then in F there is a cone F_c, on which this formula is false. Then the valuation V refuting the formula is such that $V(p_i) = F_c$ or $V(p_i) = \emptyset$. Hence there exists a p-morphism of the model $\langle F_c, V\rangle$ on single-element reflexive model. But this yields $\neg A \notin \lambda(F_1)$. That contradiction shows the correctness of considering inclusion.

Let C_h be a submodel of the model $Char_\lambda(n)(V(p_1),\ldots,V(p_n))$ consisting of all non-single-element cones. According to our construction of n-characteristic model the valuation distinguish the elements of its non-single-element cones. By Lemma 9 there exists a formula α such that the formula is true on the model C_h $\Diamond_0\Diamond_1\alpha \wedge \Diamond_0\Diamond_1\neg\alpha$. Then $Char_\lambda(n) \Vdash \square_0\square_1 \bigvee_{i=1}^{n}(\Diamond_0\Diamond_1 p_i \wedge \Diamond_0\Diamond_1\neg p_i) \to \Diamond_0\Diamond_1\alpha \wedge \Diamond_0\Diamond_1\neg\alpha$. Since the model $Char_\lambda(n)$ is n-characterizing for λ, a formula true on it belongs to the logic λ. ¿From the formula 2 we obtain $\square_0\square_1 A \to \Diamond_0\Diamond_1\alpha \wedge \Diamond_0\Diamond_1\neg\alpha \in \lambda$. Hence using postulated inference rules of the logic λ we can derive from A the formula $\Diamond_0\Diamond_1\alpha \wedge \Diamond_0\Diamond_1\neg\alpha$. After this by means of the rule $\Diamond_0\Diamond_1 x \wedge \Diamond_0\Diamond_1\neg x/y$ we derive B. Hence the rule A/B is derivable in $\lambda + \Diamond_0\Diamond_1 x \wedge \Diamond_0\Diamond_1\neg x/y$. $\square$

References

1. V.V.Rybakov, Admissible rules for logics, which are extension of $S4.3$, Siberian math. j., 25, N 5, (1984), 141–145.
2. FINE K. Logics Containing S4.3.Z. für Math. Logic and Grundl. Math., V.17, 1971, 371 - 376.

An Algebraic Correctness Criterion for Intuitionistic Proof-Nets

Philippe de Groote

Projet Calligramme
INRIA-Lorraine & CRIN-C.N.R.S.
615, rue du Jardin Botanique - B.P. 101
54602 Villers-lès-Nancy Cedex – FRANCE
e-mail: degroote@loria.fr

1 Introduction

We consider intuitionistic fragments of multiplicative linear logic for which we define appropriate notions of proof-nets.

Intuitionistic proof-nets may be easily defined by first introducing intuitionistic (or polarised) proof-structures [1, 5] and then by using any usual correctness criterion [2, 3]. Nevertheless, when using a criterion such as Girard's or Danos-Regnier's, one does not take any advantage of the intuitionistic nature of the polarised proof-nets. Indeed, the aforementioned criteria have been formulated in the classical framework.

In this paper, we formulate a new criterion, which is intrinsically intuitionistic. This criterion consists in decorating the proof-structures with algebraic terms that must obey some constraints reminiscent of phase semantics. These constraints are defined according to the polarities of the proof-structure, which explains the intuitionistic nature of our criterion.

We first state our criterion for intuitionistic implicative multiplicative linear logic (that is the fragment of linear logic whose only connective is "⊸"). Then we explain how to accommodate the multiplicative conjunction "⊗". Finally, we adapt our criterion to the non-commutative case, i.e., the Lambek calculus [8]. In this last case, the criterion is particularly interesting, as we explain at the end of the paper.

2 Implicative linear logic

We first consider the intuitionistic implicative multiplicative fragment of linear logic (which we call implicative linear logic, for short). This fragment, which concerns the only connective "⊸" (linear implication), obeys the following grammar:

$$\mathcal{F} ::= \mathcal{A} \mid \mathcal{F} \multimap \mathcal{F}$$

where $\mathcal{A}$ is the alphabet of atomic formulas.

The deduction rules are specified by the sequent calculus that follows.

Identity rules

$$A \vdash A \quad \text{(ident)} \qquad \frac{\Gamma \vdash A \quad A, \Delta \vdash B}{\Gamma, \Delta \vdash B} \quad \text{(cut)}$$

Logical rules

$$\frac{\Gamma \vdash A \quad B, \Delta \vdash C}{A \multimap B, \Gamma, \Delta \vdash C} \quad (\multimap \text{ left}) \qquad \frac{A, \Gamma \vdash B}{\Gamma \vdash A \multimap B} \quad (\multimap \text{ right})$$

Structural rule

$$\frac{\Gamma, A, B, \Delta \vdash C}{\Gamma, B, A, \Delta \vdash C} \quad \text{(Exchange)}$$

3   Intuitionistic proof-structures

In order to define a notion of proof-structure for implicative linear logic, we first introduce the notion of polarised multiplicative formula. Let $\mathcal{A}^+$ and $\mathcal{A}^-$ stand respectively for $A \times \{+\}$ and $A \times \{-\}$. For any $a \in \mathcal{A}$, we write a^+ (respectively, a^-) for $\langle a, + \rangle$ (respectively, $\langle a, - \rangle$). Polarised formulas ($\mathcal{PN}$) are defined as follows:

$$\begin{aligned} \mathcal{PN} &::= \mathcal{P} \mid \mathcal{N} \\ \mathcal{P} &::= \mathcal{A}^+ \mid \mathcal{N} \,\wp\, \mathcal{P} \\ \mathcal{N} &::= \mathcal{A}^- \mid \mathcal{P} \otimes \mathcal{N} \end{aligned}$$

where $\mathcal{P}$ and $\mathcal{N}$ are respectively called *positive* and *negative* formulas.

In fact, by interpreting a^- as $a^\perp$ (and a^+ as a itself), the polarised formulas form a proper subset of the formulas of classical multiplicative linear logic, and the notion of *positive* and *negative* polarities correspond to Danos' notion of *output* and *input* formulas [1]. Hence, by translating the formulas of implicative linear logic into polarised formulas, we will get a notion of proof-structure adapted to implicative linear logic.

Consider the following positive and negative translations:

$$\begin{aligned} (a)^+ &= a^+ &&(\text{when } a \text{ is atomic}) \\ (A \multimap B)^+ &= A^- \,\wp\, B^+ \end{aligned}$$

$$\begin{aligned} (a)^- &= a^- &&(\text{when } a \text{ is atomic}) \\ (A \multimap B)^- &= A^+ \otimes B^- \end{aligned}$$

These translations allow each intuitionistic sequent "$\Gamma \vdash A$" to be transformed into the sequence of polarised formulas "$(\Gamma)^-, (A)^+$". Then, by combining Girard's notion of link with the above translations, one obtains the polarised links given in Figure 1, where negative and positive polarities are emphasised by black and white circles, respectively.

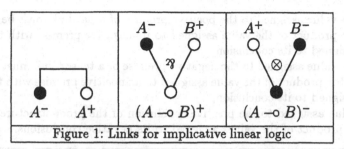

Figure 1: Links for implicative linear logic

The above links are respectively called *axiom-link, heterogeneous par-link* and *heterogeneous tensor-link*. The formulas A^- and A^+ are defined to be the conclusions of the axiom-link; the formula $(A \multimap B)^+$ is defined to be the conclusion of the par-link while the formulas A^- and B^+ are defined to be its premises; one defines the conclusion and the premises of the tensor-link similarly.

Finally, an *intuitionistic proof-structure* is defined to be a set of (occurrences of) polarised formulas connected by polarised links, such that:

1. every (occurrence of a) formula is a conclusion of exactly one link and is a premise of at most one link;
2. the resulting graph is connected;
3. the resulting graph as exactly one positive conclusion (i.e., exactly one occurrence of a positive formula that is not the premise of any link).

Remark that condition 3, in the above definition, corresponds to the fact that the succedent of any positive intuitionistic sequent is made of exactly one formula. Proof-structures corresponding to graphs whose vertices are (occurrences of) formulas, we will freely use the terminology of graph theory in the sequel. In particular, we will write $P = \langle V, E \rangle$ for a proof-structure P whose set of vertices is V, and set of edges is E.

Given an intuitionistic proof-structure, we define its *principal inputs* to be its negative conclusions (i.e., the negative vertices that are not the premises of any link) together with those vertices that appear as the negative premises of its heterogeneous par-links. This notion of principal input correspond to the notion of (free or bound) variable in the λ-calculus.

4 An algebraic correctness criterion

Let $\mathbf{M} = \langle M, \cdot, 1 \rangle$ be some freely generated commutative monoid *with sufficiently many generators* (in a technical sense that will be made precise in the sequel).

We define a proof-net to be an intuitionistic proof-structure $\langle V, E \rangle$ together with an application $\rho : V \to M$ such that:

1. the values assigned by ρ to the principal inputs are pairwise coprime (i.e., do not have any common factor);
2. the values assigned by ρ obey the constraints given in Figure 2, i.e.:
   (a) the values assigned to the two conclusions of an axiom-link must be equal,

(b) the value assigned to the positive premise of a par-link must be equal to the product of the value assigned to its negative premise with the value assigned to its conclusion

(c) the value assigned to the negative premise of a tensor-link must be equal to the product of the value assigned to its positive premise with the value assigned to its conclusion;

3. the value assigned to the positive conclusion of the proof-structure is equal to the product of the values assigned to its negative conclusions.

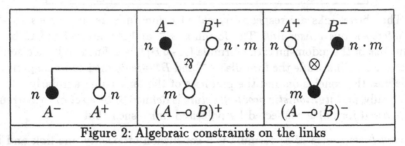

Figure 2: Algebraic constraints on the links

Condition 1, in the above definition of a proof-net, cannot be satisfied if the considered monoid does not have, at least, as many generators as there are principal inputs in the proof-structure. This explains what we meant by *sufficiently many generators*. Practically we will work with the strictly positive integers and the usual multiplication.

As an example, consider the proof-structure given in Figure 3:

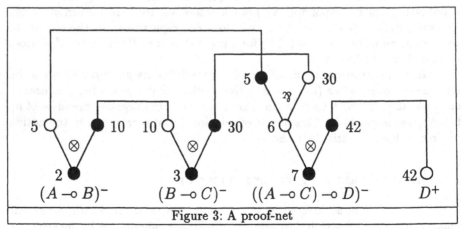

Figure 3: A proof-net

This proof-structure is a proof-net: the values assigned to the principal inputs $(2, 3, 5, 7)$ are pairwise coprime; the algebraic constraints of Figure 2 are satisfied for each link; it is the case that $2 \cdot 3 \cdot 7 = 42$.

In order to show that our definition of an intuitionistic proof-net makes sense, we must prove that:

1. any formal derivation of a sequent $\Gamma \vdash A$ may be transformed into a proof-net whose conclusions are $(\Gamma)^-, (A)^+$;

2. any proof-net whose conclusions are $(\Gamma)^-, (A)^+$ may be *sequentialised* into a formal derivation of the sequent $\Gamma \vdash A$.

Establishing Property 1 consists of a routine induction whose details are left to the reader. Property 2, which amounts to Girard's sequentialisation theorem, will be proven in Section 6.

5 A dynamic view of the criterion

Given some proof-structure how can we check whether it is (or is not) a proof-net? In other words, how can we prove that there exist, for that proof-structure, a valuation ρ satisfying the constraints in which our criterion consists?

Consider again Figure 3 and try to figure out how the given valuation could have been found. Here is a possible solution:

- assign pairwise coprime numbers $(2, 3, 5, 7)$ to the principal inputs of the proof-structure;
- propagate 5 along the axiom link;
- knowing the values assigned to the positive premise (5) and to the conclusion (2) of the left-most tensor-link, assign $10 = 5 \cdot 2$ to its negative premise;
- by steps similar to the previous ones, assign $30 = 10 \cdot 3$ to the negative premise of the second tensor-link, and propagate this value along the axiom;
- check that 30 is divisible by 5 and, consequently, assign 6 to the conclusion of the par-link;
- this allows the value assigned to the premise of the last tensor-link to be computed as $42 = 6 \cdot 7$;
- propagate 42 along the axiom-link and check that $42 = 2 \cdot 3 \cdot 7$.

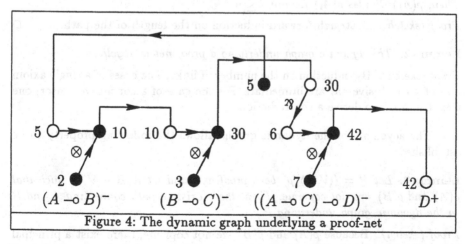

Figure 4: The dynamic graph underlying a proof-net

It can be proven that the above procedure obeys a general algorithm. Any proof-net may be assigned a valuation ρ by propagating the values assigned to its principal inputs. This propagation follows the paths of a directed graph that we call the *dynamic graph underlying the proof-net*. Figure 4 exemplifies this concept.

The notion of dynamic graph may be easily defined by introducing a notion of switch:

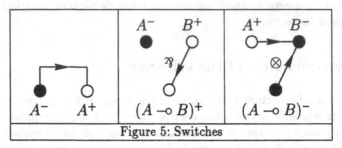

Figure 5: Switches

The dynamic graph underlying a proof-net (or a proof-structure) is defined to be the directed graph obtained by replacing each link of the proof-net by the corresponding switch.

Our dynamic graphs correspond (up to their orientation) to the paths of Lamarche in [5], which he derives from his game semantics [6].

Important properties of the dynamic graphs are given by the following lemmas.

Lemma 1. *Let $P = \langle\langle V, E\rangle, \rho\rangle$ be a proof-net, and let $\langle A_1, \ldots A_n\rangle \in V^n$ be a path in the dynamic graph underlying P such that:*

1. *A_1 is a principal input of P;*
2. *in the case that A_1 is the input premise of a heterogeneous par-link, the path does not go through the corresponding output premise.*

Then, $\rho(A_1)$ divides $\rho(A_i)$ for any $i \leq n$.

Proof (sketch). A straightforward induction on the length of the path. □

Lemma 2. *The dynamic graph underlying a proof-net is acyclic.*

Proof (sketch). By induction on the number of links. The cases of a single axiom and of a conclusive par are immediate. For the case of a conclusive tensor, one uses Lemma 1 to derive a contradiction. □

The acyclicity of the dynamic graphs allows the following property to be established.

Lemma 3. *Let $P = \langle\langle V, E\rangle, \rho\rangle$ be a proof-net, and let $A, B \in V$ be such that $\rho(A)$ and $\rho(B)$ are not coprime. Then, there exists a path connecting A and B in the dynamic graph underlying P.*

Proof (sketch). Because $\rho(A)$ and $\rho(B)$ are not coprime, there exist a principal input I such that $\rho(I)$ divides both $\rho(A)$ and $\rho(B)$. It is easy to prove that there exist two path connecting I to $\rho(A)$ and $\rho(B)$ respectively. Moreover one of these paths must be a sub-path of the other. □

Using the two lemmas above, we may prove that the checking algorithm sketched at the beginning of this section is general. Another consequence of Lemma 3 is the following.

Lemma 4. *The premises of any heterogeneous par-link occurring in a proof-net are connected by a path of the underlying dynamic graphs. This path goes from the negative premise to the positive one.* □

This last lemma will be useful when establishing the sequentialisation property.

6 Sequentialisation

Our sequentialisation proof follows the method of the *splitting tensor* [3, 4].

Given a proof-net, we define a *splitting tensor* to be a tensor-link such that:

1. its conclusion is not a premise of any other link (in other words, its conclusion is one of the conclusions of the proof-net);
2. the value assigned to its positive premise is equal to the product of the values assigned to some of the negative conclusions of the proof-net

The next lemma justifies the above definition.

Lemma 5. *Let P be a proof-net that contains a splitting tensor. Then, removing this tensor-link splits P into two disconnected proof-nets.*

Proof (sketch). It is immediate that the splitting tensor splits the graph underlying P into two disconnected subgraphs G_1 and G_2. Therefore, if the splitting tensor does not split the proof-net, there must exist a par-link one premise of which belongs to G_1 and the other premise of which belongs to G_2. But then, by Lemma 4, there would exist a path going from one of the premises of this par to the other one. Because G_1 and G_2 are connected only by the switch corresponding to the splitting tensor, this path would go through this switch. But this, by Lemma 1, conflicts with Condition 2 in the definition of a splitting tensor. □

The key lemma of the sequentialisation proof is the following.

Lemma 6. *Let P be a proof-net whose no conclusion is the conclusion of a par-link. If P contains at least one tensor-link then it contains a splitting tensor.*

Proof (sketch). Since P does not contain any conclusive par, its output conclusion must be the output conclusion of an axiom link. Consider the input conclusion (say A) of this axiom link. This input conclusion A must be the premise of some link (say l) otherwise P would only consists of one axiom link, which would contradict the fact that it contains at least one tensor-link. Because of Lemma 4, l cannot be a par-link, therefore, it is a tensor-link. Consider the conclusion of this tensor-link (which is an input conclusion) and iterate the same kind of argument. One eventually finds a conclusive tensor-link. It is easy to show that this tensor-link must be a splitting tensor. □

Proposition 7. *Any proof-net is sequentialisable.* □

7 Adding multiplicative conjunction

Intuitionistic multiplicative linear logic is obtained from implicative linear logic by adding the following formation rule:

$$\mathcal{F} ::= \mathcal{F} \otimes \mathcal{F},$$

together with the two inference rules that follows:

$$\frac{A, B, \Gamma \vdash C}{A \otimes B, \Gamma \vdash C} \quad (\otimes \text{ left}) \qquad \frac{\Gamma \vdash A \quad \Delta \vdash B}{\Gamma, \Delta \vdash A \otimes B} \quad (\otimes \text{ right})$$

Our correctness criterion may be easily adapted to intuitionistic multiplicative linear logic by enriching the free commutative monoid $\mathbf{M}$ with two operations, $\frac{1}{2}(\cdot)$ and $(\cdot)^{\frac{1}{2}}$, that obey the following law:

$$\tfrac{1}{2}(n) \cdot (n)^{\frac{1}{2}} = n$$

Then, the notion of polarised formula of Section 3 is extended by the following rules:

$$\begin{aligned} \mathcal{N} &::= \mathcal{N} \,\mathfrak{N}\, \mathcal{N} \\ \mathcal{P} &::= \mathcal{P} \otimes \mathcal{P}, \end{aligned}$$

which allows one to add the following clauses to the positive and negative translations of Section 3:

$$\begin{aligned} (A \otimes B)^- &= A^- \,\mathfrak{N}\, B^- \\ (A \otimes B)^+ &= A^+ \otimes B^+. \end{aligned}$$

This gives rise to two additional kinds of links, which are respectively called *homogeneous par-link* and *homogeneous tensor-link*. These links together with the corresponding algebraic constraints and switches are given by Figure 6.

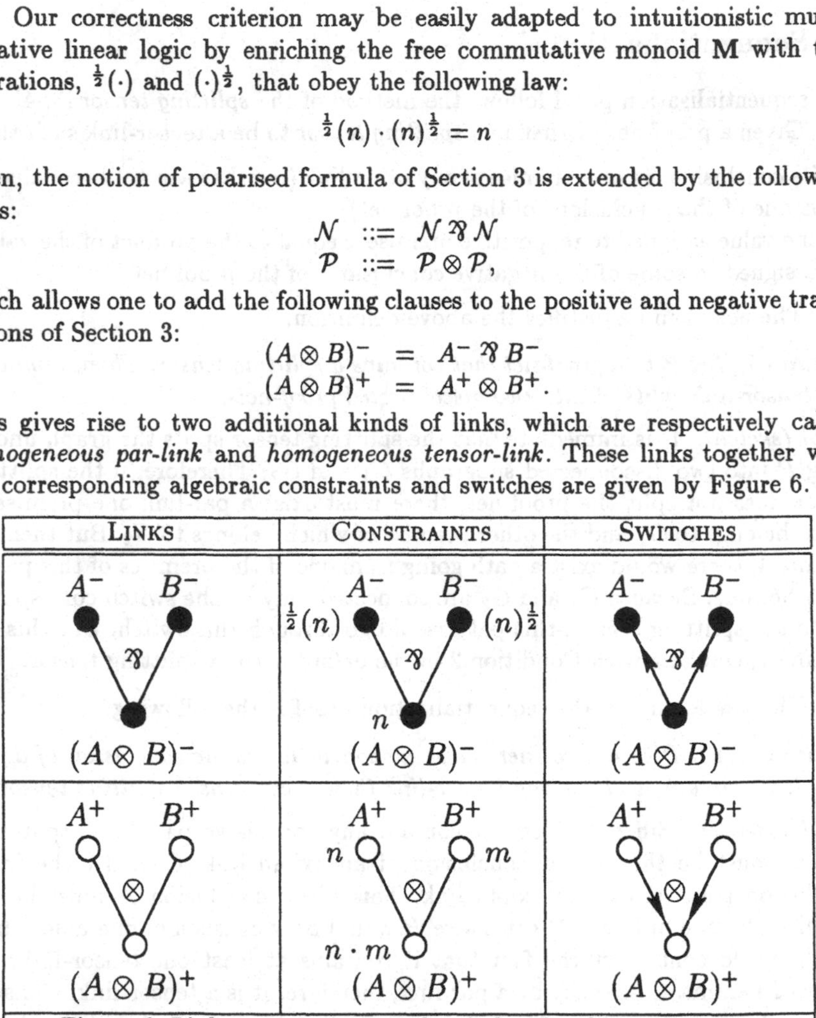

Figure 6: Links, constraints, and switches for the conjunction

The idea behind the adaptation of our criterion to the case of the multiplicative conjunction is straightforward. It is to be noted, however, that to adapt our sequentialisation proof to this new setting requires some work.

8 The non-commutative case: the Lambek calculus

By rejecting the exchange rule, which is is the only structural rule of intuitionistic multiplicative logic, one obtains a non-commutative logic known as the Lambek calculus [8].

The formulas of the Lambek calculus are built according to the following grammar:

$$\mathcal{F} ::= A \mid \mathcal{F} \bullet \mathcal{F} \mid \mathcal{F} \backslash \mathcal{F} \mid \mathcal{F}/\mathcal{F}$$

where formulas of the form $A \bullet B$ correspond to conjunctions (or products), formulas of the form $A \backslash B$ correspond to direct implications (i.e., A *implies* B), and formulas of the form A/B to retro-implications (i.e., A *is implied by* B).

The deduction relation of the calculus is defined by means of the following system:

Identity rules

$$A \vdash A \quad \text{(ident)} \qquad \frac{\Gamma \vdash A \quad \Delta_1, A, \Delta_2 \vdash B}{\Delta_1, \Gamma, \Delta_2 \vdash B} \quad \text{(cut)}$$

Logical rules

$$\frac{\Gamma, A, B, \Delta \vdash C}{\Gamma, A \bullet B, \Delta \vdash C} \quad (\bullet \text{ left}) \qquad \frac{\Gamma \vdash A \quad \Delta \vdash B}{\Gamma, \Delta \vdash A \bullet B} \quad (\bullet \text{ right})$$

$$\frac{\Gamma \vdash A \quad \Delta_1, B, \Delta_2 \vdash C}{\Delta_1, \Gamma, A \backslash B, \Delta_2 \vdash C} \quad (\backslash \text{ left}) \qquad \frac{A, \Gamma \vdash B}{\Gamma \vdash A \backslash B} \quad (\backslash \text{ right})$$

$$\frac{\Gamma \vdash A \quad \Delta_1, B, \Delta_2 \vdash C}{\Delta_1, B/A, \Gamma, \Delta_2 \vdash C} \quad (/ \text{ left}) \qquad \frac{\Gamma, A \vdash B}{\Gamma \vdash B/A} \quad (/ \text{ right})$$

In order to adapt our criterion to the Lambek calculus, it suffices to work in a freely generated monoid Σ^* (enriched with the left and right square roots, when the product is present) *that is not commutative*. Then, because the calculus is not commutative, one must carefully distinguish between the direct and the retro implication, between the left and the right premises of the corresponding links, and between left and right cancellation in the monoid.

The translation of the Lambek formulas into polarised formulas is the following:

$$
\begin{aligned}
(a)^- &= a^- & (a)^+ &= a^+ \\
(A \backslash B)^- &= A^+ \otimes B^- & (A \backslash B)^+ &= B^+ \,\invamp\, A^- \\
(A/B)^- &= A^- \otimes B^+ & (A/B)^+ &= B^- \,\invamp\, A^+ \\
(A \bullet B)^- &= A^- \,\invamp\, B^- & (A \bullet B)^+ &= B^+ \otimes A^+
\end{aligned}
$$

This gives rise to the links, the constraints, and the switches of Figure 7 and 8.

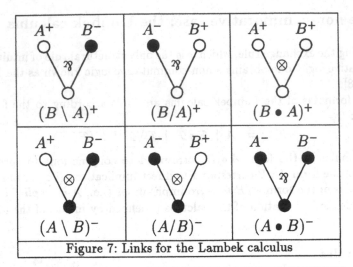

Figure 7: Links for the Lambek calculus

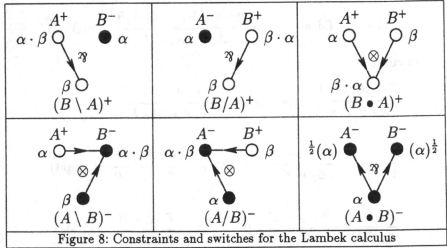

Figure 8: Constraints and switches for the Lambek calculus

9   Concluding remarks

As we said in the introduction, our criterion is intrinsically intuitionistic, which is also the case of Lamarche's [5]. Similarly, we could say that the non commutative version of our criterion is intrinsic to the Lambek calculus, which solves an open question raised by Retoré [7]. Indeed, in the literature, proof-nets for the Lambek calculus are defined in terms of conditions that ensure commutative correctness, together with an additional condition that ensures non-commutativity. The latter is, most often, a planarity condition [7, 9]. In contrast, when using our criterion, commutative correctness and non-commutativity are not checked independently.

In [9, CHAP. III, §6, pp. 38–40], Roorda defines a way of decorating proof-nets that is almost identical to ours. He then observes that the existence of such a decoration is necessary, and raises the question whether it is sufficient (in fact,

140

he conjectures it is not). Consequently, our paper solves Roorda's open question (in the unexpected sense).

Another difference between Roorda's work and ours lies in the dynamic interpretation of our criterion. Indeed, Roorda's decorating algorithm involves associative (commutative) unification. In this paper, we have avoided this unnecessary complexity by introducing the notion of *underlying dynamic graph* and the two *square root* operators.

References

[1] V. Danos. *Une application de la logique linéaire à l'étude des processus de normalisation et principalement du lambda calcul.* Thèse de doctorat, Université de Paris VII, 1990.

[2] V. Danos and L. Regnier. The structure of multiplicatives. *Archive for Mathematical Logic*, 28:181–203, 1989.

[3] J.-Y. Girard. Linear logic. *Theoretical Computer Science*, 50:1–102, 1987.

[4] J.-Y. Girard. Quantifiers in linear logic II. Technical Report 19, Equipe de Logique Mathématique, Université de Paris VII, 1991.

[5] F. Lamarche. Proof nets for intuitionistic linear logic 1: Essential nets. Technical report, Imperial College, April 1994.

[6] F. Lamarche. Games semantics for full propositional linear logic. In *Ninth Annual IEEE Symposium on Logic in Computer Science*. IEEE Press, 1995.

[7] F. Lamarche and C. Retoré. Proof nets for the lambek calculus. In M. Abrusci, C. Casadio, and G. Sandri, editors, *Third Roma Workshop: Proofs and Linguistic Categories*, Rapporto di Ricerca del Dipartimento de Filosofia. Università di Bologna, 1996.

[8] J. Lambek. The mathematics of sentence structure. *Amer. Math. Monthly*, 65:154–170, 1958.

[9] D. Roorda. *Resource Logics: proof-theoretical investigations.* PhD thesis, University of Amsterdam, 1991.

Towards a Theory of Recursive Structures*

(Extended Abstract)

David Harel

Dept. of Applied Mathematics and Computer Science
The Weizmann Institute of Science, Rehovot, Israel
harel@wisdom.weizmann.ac.il

Abstract. In computer science, one is interested mainly in finite objects. Insofar as infinite objects are of interest, they must be computable, i.e., recursive, thus admitting an effective finite representation. This leads to the notion of a recursive graph, or, more generally, a recursive structure, model or data base. We summarize our recent work on recursive structures and data bases, including (i) high undecidability of specific problems, (ii) connections between the descriptive complexity of finitary problems and the computational complexity of their infinitary analogues, (iii) completeness for query languages, (iv) descriptive and computational complexity, and (v) zero-one laws.

This paper provides a summary of work — all but the first part of which is joint with Tirza Hirst — on infinite recursive (i.e., computable) structures and data bases, and attempts to put it in perspective. The work itself is contained in four papers [H, HH1, HH2, HH3].

When computer scientists become interested in an infinite object, they require it to be computable, i.e., recursive, so that it possesses an effective finite representation. Given the prominence of finite graphs in computer science, and the many results and open questions surrounding them, it is very natural to investigate recursive graphs too. Moreover, insight into finite objects can often be gleaned from results about infinite recursive variants thereof. Without loss of generality, an infinite recursive graph can be thought of simply as a recursive binary relation over the natural numbers. Recursive graphs can then be represented by the (finite) algorithms, or Turing machines, that recognize their edge sets, so that it makes sense to investigate the complexity of problems concerning them.

Indeed, a significant amount of work has been carried out in recent years regarding the complexity of problems on recursive graphs. Some of the first papers were written in the 1970s by Manaster and Rosenstein [MR] and Bean

---

* A full, but older, version of this paper appeared as an invited paper in STACS '94, *Proc. 11th Ann. Symp. on Theoretical Aspects of Computer Science*, Lecture Notes in Computer Science, Vol. 775, Springer-Verlag, Berlin, 1994, pp. 633–645.

[B1, B2]. Following that, a variety of problems were considered, including ones that are NP-complete for finite graphs, such as k-colorability and Hamiltonicity [B1, B2, BG2, Bu, GL, MR] and ones that are in P in the finite case, such as Eulerian paths [B2, BG1] In most cases (including the above examples) the problems turned out to be undecidable. This is true even for highly recursive graphs [B1], i.e., ones for which node degree is finite and the set of neighbors of a node is computable. Beigel and Gasarch [BG1] and Gasarch and Lockwood [GL] investigated the precise level of undecidability of many such problems, and showed that they reside on low levels of the arithmetical hierarchy. For example, detecting the existence of an Eulerian path is Π_3^0-complete for recursive graphs and Π_2^0-complete for highly recursive graphs [BG1].

The case of Hamiltonian paths seemed to be more elusive. In 1976, Bean [B2] had shown that the problem is undecidable (even for planar graphs), but the precise characterization was not known. In response to this question, posed by R. Beigel and B. Gasarch, the author was able to show that Hamiltonicity is in fact *highly* undecidable, *viz*, Σ_1^1-complete. The result, proved in [H], is shown to hold even for highly recursive graphs with degree bounded by 3. (It actually holds for planar graphs too.) Hamiltonicity is thus an example of an interesting graph problem that becomes highly undecidable in the infinite case.[2]

The question then arises as to what makes some NP-complete problems highly undecidable in the infinite case, while others (e.g., k-colorability) remain on low levels of the arithmetical hierarchy. This was the starting point of the joint work with T. Hirst. In [HH1], we provide a general definition of infinite recursive versions of NP optimization problems, in such a way that MAX CLIQUE, for example, becomes the question of whether a recursive graph contains an infinite clique. Two main results are proved in [HH1], one enables using knowledge about the infinite case to yield implications to the finite case, and the other enables implications in the other direction. The results establish a connection between the descriptive complexity of (finitary) NP optimization problems, particularly the syntactic class MAX NP, and the computational complexity of their infinite versions, particularly the class Σ_1^1. Taken together, the two results yield many new problems whose infinite versions are highly undecidable and whose finite versions are outside MAX NP. Examples include MAX CLIQUE, MAX INDEPENDENT SET, MAX SUBGRAPH, and MAX TILING.

The next paper, [HH2], puts forward the idea of infinite recursive relational data bases. Such a data base can be defined simply as a finite tuple of recursive relations (not necessarily binary) over some countable domain. We thus obtain a natural generalization of the notion of a finite relational data base. This is not an entirely wild idea: tables of trigonometric functions, for example, can be viewed as a recursive data base, since we might be interested in the sines or cosines of infinitely many angles. Instead of keeping them all in a table, which is impossible, we keep rules for computing the values from the angles, and vice versa, which is really just to say that we have an effective way of telling whether an edge is present between nodes i and j in an infinite graph, and this is precisely the notion of a recursive graph.

---

[2] Independent work in [AMS] showed that perfect matching is another such problem.

In [HH2], we investigate the class of computable queries over recursive data bases, the motivation being borrowed from [CH1]. Since the set of computable queries on such data bases is not closed under even simple relational operations, one must either make do with a very humble class of queries or considerably restrict the class of allowed data bases. The main parts of [HH2] are concerned with the completeness of two query languages, one for each of these possibilities. The first is quantifier-free first-order logic, which is shown to be complete for the non-restricted case. The second is an appropriately modified version of the complete language QL of [CH1], which is proved complete for the case of "highly symmetric" data bases. These have the property that their set of automorphisms is of finite index for each tuple-width.

While the previous topic involves languages for *computable* queries, our final paper, [HH3], deals with languages that express *non*-computable queries. In the spirit of results for finite structures by Fagin, Immerman and others, we sought to connect the computational complexity of properties of recursive structures with their descriptive complexity, i.e, to capture levels of undecidability syntactically as the properties expressible in various logical formalisms. We consider several formalisms, such as first-order logic, second-order logic and fixpoint logic. One of our results is analogous to that of Fagin [F1]; it states that, for any $k \geq 2$, the properties of recursive structures expressible by Σ_k^1 formulas are exactly the generic properties in the complexity class Σ_k^1 of the analytical hierarchy.

[HH3] also deals with zero-one laws. It is not too difficult to see that many of the classical theorems of logic that hold for general structures (e.g., compactness and completeness) fail not only for finite models but for recursive ones too. Others, such as Ehrenfeucht–Fraisse games, hold also for finite and recursive structures. Zero-one laws, to the effect that certain properties (such as those expressible in first-order logic) are either almost surely true or almost surely false (see, e.g., [F2]), are considered unique to finite model theory, since they require counting the number of structures of a given finite size. We introduce a way of extending the definition of these laws to recursive structures, and prove that they hold for first-order logic, strict Σ_1^1 and strict Π_1^1. We then use this fact to show non-expressibility of certain properties of recursive structures in these logics.

While recursive structures and models have been investigated quite widely by logicians (see, e.g., [NR]), the kind of issues that computer scientists are interested in have not been addressed prior to the work mentioned above. We feel that that this is a fertile area for research, and raises theoretical and practical questions concerning the computability and complexity of properties of recursive structures, and the theory of queries and update operations over recursive data bases. We hope that the work summarized here will stimulate more research on these topics.

References

[AMS]   R. Aharoni, M. Magidor and R. A. Shore, "On the Strength of König's Duality Theorem", *J. of Combinatorial Theory (Series B)* **54**:2 (1992), 257–290.

[B1]   D.R. Bean, "Effective Coloration", *J. Sym. Logic* **41** (1976), 469–480.

[B2]   D.R. Bean, "Recursive Euler and Hamiltonian Paths", *Proc. Amer. Math. Soc.* **55** (1976), 385–394.

[BG1]   R. Beigel and W. I. Gasarch, unpublished results, 1986-1990.

[BG2]   R. Beigel and W. I. Gasarch, "On the Complexity of Finding the Chromatic Number of a Recursive Graph", Parts I & II, *Ann. Pure and Appl. Logic* **45** (1989), 1–38, 227–247.

[Bu]   S. A. Burr, "Some Undecidable Problems Involving the Edge-Coloring and Vertex Coloring of Graphs", *Disc. Math.* **50** (1984), 171–177.

[CH1]   A. K. Chandra and D. Harel, "Computable Queries for Relational Data Bases", *J. Comp. Syst. Sci.* **21**, (1980), 156–178.

[F1]   R. Fagin, "Generalized First-Order Spectra and Polynomial-Time Recognizable Sets", In *Complexity of Computations* (R. Karp, ed.), SIAM-AMS Proceedings, Vol. 7, 1974, pp. 43–73.

[F2]   R. Fagin, "Probabilities on Finite Models", *J. of Symbolic Logic*, **41**, (1976), 50 – 58.

[GL]   W. I. Gasarch and M. Lockwood, "The Existence of Matchings for Recursive and Highly Recursive Bipartite Graphs", Technical Report 2029, Univ. of Maryland, May 1988.

[H]   D. Harel, "Hamiltonian Paths in Infinite Graphs", *Israel J. Math.* **76**:3 (1991), 317–336. (Also, *Proc. 23rd Ann. ACM Symp. on Theory of Computing*, New Orleans, pp. 220–229, 1991.)

[HH1]   T. Hirst and D. Harel, "Taking it to the Limit: On Infinite Variants of NP-Complete Problems", *J. Comput. Syst. Sci.*, to appear. (Also, *Proc. 8th IEEE Conf. on Structure in Complexity Theory*, IEEE Press, New York, 1993, pp. 292–304.)

[HH2]   T. Hirst and D. Harel, "Completeness Results for Recursive Data Bases", *J. Comput. Syst. Sci.*, to appear. (Also, *12th ACM Ann. Symp. on Principles of Database Systems*, ACM Press, New York, 1993, 244–252.)

[HH3]   T. Hirst and D. Harel, "More about Recursive Structures: Descriptive Complexity and Zero-One Laws", *Proc. 11th Symp. on Logic in Computer Science*, New Brunswick, NJ, July, 1996.

[MR]   A. Manaster and J. Rosenstein, "Effective Matchmaking (Recursion Theoretic Aspects of a Theorem of Philip Hall)", *Proc. London Math. Soc.* **3** (1972), 615–654.

[NR]   A. Nerode and J. Remmel, "A Survey of Lattices of R. E. Substructures", In *Recursion Theory*, Proc. Symp. in Pure Math. Vol. 42 (A. Nerode and R. A. Shore, eds.), Amer. Math. Soc., Providence, R. I., 1985, pp. 323–375.

On the Complexity of Prefix Formulas in Modal Logic of Subset Spaces

Bernhard Heinemann

FernUniversität, 58084 Hagen, Germany

Abstract. Many logics used in computer science have an intractable satisfiability (model–checking, derivability) problem. This restricts their general applicability severely. May be restricting to smaller classes of admissible formulas can bring down the complexity bounds. In the present paper we follow this strategy for the *subset space logic* proposed by MOSS and PARIKH recently [Moss and Parikh 1992], [Dabrowski et al. 1996]. Forming *nSAT* as in sentential logic, but with *prefix formulas* instead of literals, we obtain nice generalizations of the following results well–known from the propositional case: $nSAT \in P$ if $n \leq 2$ and $nSAT$ is *NP*–complete if $n \geq 3$. Thus $nSAT$ is "feasible", iff $n \leq 2$. Moreover, full *SAT* turns out to be *PSPACE*–hard (as usual).

1 Introduction

Modal logics of time, computation, or knowledge, have proved their suitability for specifying properties in several branches of computer science and Artificial Intelligence. E.g., *dynamic logic* is an appropriate tool in case of sequential programs [Harel 1979], *temporal logic* serves corresponding purposes for concurrent programs [Manna and Pnueli 1992], and *logic of knowledge* for protocols in multi–agent systems [Fagin et al. 1995].

Unfortunately, any of these logics has the disadvantage of an unfeasibly high complexity. Thus, for example, testing the consistency of a specification is not possible in reasonable time generally. This need no longer be true if the admissible formula classes are reduced: Interesting feasible subclasses have been singled out even for *propositional logic* recently [Heusch 1995].

In the present paper we are concerned with a general logical framework proposed by MOSS and PARIKH [Moss and Parikh 1992], which in particular can be applied for *reasoning about knowledge*, but is also applicable to other topics such as specifying the behaviour of programs which compute infinite objects approximatively (see [Heinemann 1997] for a more detailed discussion.) Let us briefly revisit the system of MOSS and PARIKH. The single–agent case of classical logic of knowledge is formalized by the modal system **S5** mostly [Fagin et al. 1995]. Corresponding frames are equivalence relations, and the points equivalent to the actual state represent the knower's present *view* of the world, i.e. the states she considers possible at the moment. Spending *effort* – e.g. computational –, this view may change. In many situations the change is monotone in the sense that no new alternative states occur, but some of the states may be considered impossible at a later time point. Thus a successive shrinking of the set of knowledge states results if one describes the development of knowledge in the course of time.

MOSS and PARIKH defined a *bimodal language* comprising the operators K and $\Box$ for *knowledge* and for *effort (in time)*, respectively. Their interpretations are, however, quite different from ordinary multimodal logic. Now underlying frames consist of a non–empty set X and a distinguished set $\mathcal{O}$ of subsets of X, called *opens* (although they need not be open sets in the sense of topology). K then varies over the elements of an open set, whereas $\Box$ captures the shrinking of an open. While K retains its S5–like character the modality of $\Box$ is S4–like, as it expresses descent in a system of sets (w.r.t. inclusion). Moreover, the operators interact suitably.

In the meantime various systems of this *topological modal logic* have been studied. Sound and complete axiomatizations were proposed e.g. for the basic *subset space logic* (where $\mathcal{O}$ may be an arbitrary set of subsets of X) [Dabrowski et al. 1996], [Moss and Parikh 1992], and for *topologic* (where $\mathcal{O}$ is actually a topology on X) [Dabrowski et al. 1996], [Georgatos 1994a]. Moreover, *decidability* of the respective logics could be shown. As a further topic the topological modal theory of *treelike spaces*, in which no two opens on the same "level" overlap, was determined by GEORGATOS [Georgatos 1994b]; as to trees of finite height and a generalization of the well–known modal system **G** see [Heinemann 1996b].

The complexity results of the present paper could be obtained for the *subset space logic (SSL)*. They are inspired by recent progress for $\mathbf{S5}_m$, the usual multi–agent logic of knowledge [Scherer 1995]. We consider *prefix formulas* which are formed from a block of modal operators followed by a literal. Our main results can be summarized as follows: *w.r.t. satisfiability, prefix formulas behave like literals in propositional logic*; i.e. if one let $nSAT$ be the set of all satisfiable finite clause sets such that each clause contains at most n prefix formulas, then $nSAT \in P$ if $n \leq 2$, and $nSAT$ is *NP*–complete if $n \geq 3$. (But, contrary to propositional logic, the set SAT of all satisfiable SSL–formulas is *PSPACE*–hard.) In particular, $1SAT$ and $2SAT$ are "feasible" sets of SSL–formulas. — For ordinary logic of knowledge, the fact "$1SAT \in P$" was established by SCHERER [Scherer 1995]; to a large extent, the methods applied to obtain the corresponding case of the present paper are due to him.

To get the results, we proceed as follows: In Section 2 we deal with SSL in a more precise way. We define the logical language, introduce a corresponding axiomatization, and state some facts about the logic which are needed later on. In Section 3 we define prefix formulas and present a useful characterization of the satisfiable sets X of prefix formulas: satisfiability already holds if each two–element subset of X is satisfiable (Theorem 15). In Section 4 we determine the complexity of deciding the word problem for a special Semi–Thue system, for which the rules originate from the axioms of the subset space logic. The main result is the polynomial time bound here (Theorem 21). Combining these auxiliary results, we obtain the complexity assertions stated above in Section 5. The paper is finished with some concluding remarks. — Almost all proofs are omitted in this draft. They are carried out in [Heinemann 1996a].

2 Prerequisites

First we introduce a language, called *subset space logic (SSL)*, in the following way. The *syntax* of *SSL* is based upon a recursively enumerable set of *propositional variables, PV* (denoted by upper case Roman letters). Then the set $\mathcal{F}$ of *SSL*–formulas is defined by the following clauses:

- $PV \cup \{\top\} \subseteq \mathcal{F}$;
- $\alpha, \beta \in \mathcal{F} \Longrightarrow \neg \alpha, K\alpha, \Box\alpha, (\alpha \wedge \beta) \in \mathcal{F}$;
- no other strings belong to $\mathcal{F}$.

We omit brackets whenever possible, and use the following abbreviations (besides the usual ones from sentential logic): $L\alpha$ for $\neg K \neg \alpha$, $\Diamond \alpha$ for $\neg \Box \neg \alpha$. — The *semantical domains* of *SSL* are generally triples $(X, \mathcal{O}, \sigma)$, wherein X is a non–empty set, $\mathcal{O}$ is a set of subsets of X, and $\sigma : PV \times X \longrightarrow \{0, 1\}$ is a mapping (called X-*valuation*). The pair $\mathcal{S} = (X, \mathcal{O})$ is referred to as a *subset frame* subsequently, whereas $\mathcal{M} = (X, \mathcal{O}, \sigma)$ is called a *subset space* or a *model* (*based on* $(X, \mathcal{O})$). The elements of $\mathcal{O}$ are often called *opens* (*of* $\mathcal{S}$) although they need not be open sets in the sense of topology. — We now define *validity* of *SSL*–formulas in models based on subset frames.

Definition 1 Semantics of *SSL*. Let $\mathcal{M} = (X, \mathcal{O}, \sigma)$ be a subset space.

(1) $X \otimes \mathcal{O} := \{(x, U) \mid U \in \mathcal{O}, x \in U\}$ is the set of *neighbourhood situations* (of the underlying subset frame).

(2) *Validity* of an *SSL*–formula *in model* $\mathcal{M}$ *at a neighbourhood situation* x, U (brackets are omitted) is defined by recursion on the structure of formulas; we only state the crucial cases:

$$x, U \models_{\mathcal{M}} A \quad :\Longleftrightarrow \quad \sigma(A, x) = 1$$
$$x, U \models_{\mathcal{M}} K\alpha : \Longleftrightarrow \quad (\forall y \in U)\ y, U \models_{\mathcal{M}} \alpha$$
$$x, U \models_{\mathcal{M}} \Box\alpha : \Longleftrightarrow \quad (\forall V \in \mathcal{O})(x \in V \subseteq U \Longrightarrow x, V \models_{\mathcal{M}} \alpha)$$

for all $A \in PV$ and all formulas $\alpha, \beta \in \mathcal{F}$.

(3) Formula $\alpha \in \mathcal{F}$ *holds in* $\mathcal{M}$ (denoted by $\models_{\mathcal{M}} \alpha$), iff it holds in $\mathcal{M}$ at every neighbourhood situation.

If there is no ambiguity, we omit the index $\mathcal{M}$ subsequently. — Next we present by a list of axioms and rules, respectively, a logical system **MP**.

Axioms

(1) All $\mathcal{F}$ – instances of propositional tautologies
(2) $(A \to \Box A) \wedge (\neg A \to \Box \neg A)$
(3) $K(\alpha \to \beta) \to (K\alpha \to K\beta)$

(4) $K\alpha \rightarrow \alpha$

(5) $K\alpha \rightarrow KK\alpha$

(6) $L\alpha \rightarrow KL\alpha$

(7) $\Box(\alpha \rightarrow \beta) \rightarrow (\Box\alpha \rightarrow \Box\beta)$

(8) $\Box\alpha \rightarrow \alpha$

(9) $\Box\alpha \rightarrow \Box\Box\alpha$

(10) $K\Box\alpha \rightarrow \Box K\alpha$

for all $A \in PV$ and all $\alpha, \beta \in \mathcal{F}$.

The **rules** present in the system are *modus ponens*, K*-necessitation*, and $\Box-$
necessitation; e.g., $\Box$–necessitation is given by

$$\frac{\alpha}{\Box\alpha}$$

for all $\alpha \in \mathcal{F}$. — In [Dabrowski et al. 1996], the following *completeness theorem*
is proved (Section 2, Theorem 2.4).

Theorem 2. *A formula $\alpha \in \mathcal{F}$ is derivable in the system* **MP**, *iff α holds in all
subset spaces.*

The logical system **MP** also determines a class of certain bimodal structures
in the usual sense called *SSL–frames* and *SSL–models*, respectively.

Definition 3. (1) Let $\mathcal{F} := (W, \{R, S\})$ be a bimodal frame (i.e. W is a non–
empty set and R, S are binary relations on W). Then $\mathcal{F}$ is called an *SSL–
frame*, iff
  - R is an equivalence relation on W;
  - S is reflexive and transitive;
  - $(\forall s, t, u \in W)((s, t) \in S \wedge (t, u) \in R \Longrightarrow (\exists v \in W)[(s, v) \in R \wedge (v, u) \in S])$ (this is called the *cross property*).
(2) A model $\mathcal{M} := (W, \{R, S\}, \sigma)$ based on an *SSL*–frame $(W, \{R, S\})$ is called
  an *SSL–model*, iff for all $s, t \in W$ and all $A \in PV$ it holds that

$$(s, t) \in S \text{ implies } [\sigma(A, s) = 1 \iff \sigma(A, t) = 1].$$

In fact, it is not hard to see that **MP** is sound and complete w.r.t. the just
defined structures as well.

Theorem 4. *Every* **MP**–*derivable formula holds in every SSL–model. On the
other hand, every $\alpha \in \mathcal{F}$ which is not* **MP**–*derivable is falsified in some SSL–
model at some point.*

Further notions are used as in ordinary modal logic subsequently (for which
we use [Chellas 1980] and [Goldblatt 1987] as references): We let $\vdash$ designate
MP–derivability, call non–derivability of the formula $\bot$ from a set $X \subseteq \mathcal{F}$ *consistency* of X, and validity of all formulas of X at some neighbourhood situation of
some model *satisfiability* of X. Note that it suffices to solve the *SSL*–satisfiability
problem for *SSL*–models in view of Theorems 2 and 4. Furthermore, satisfiability
corresponds with consistency.

3   Satisfiability of prefix formulas

In this section we define the set of *prefixes* first. Then we introduce several (deleting, producing, commuting) relations on prefixes, which are induced by the **MP**–axioms and their duals. Our first result (Proposition 7) connects these relations with logic, showing that the reflexive and transitive closure $\xrightarrow{p^\bullet}$ of their union can be interpreted as a strong implication. Next we define *prefix formulas*. We use them to prove that $\xrightarrow{p^\bullet}$ *exactly* corresponds with implication (Corollary 16). As a by–product we get a satisfiability criterion for sets of prefix formulas: such a set X is satisfiable, iff every two–element subset of X is satisfiable (Theorem 15). This result will be used later on. — Let Σ be a finite set of symbols. Denote by Σ^* the set of words over Σ, and let Λ designate the empty word.

Definition 5. The set $\mathcal{P} \subseteq \{K, \Box, \neg\}^*$ of *prefixes* (*of SSL*) is defined by the following clauses:

- $\Lambda \in \mathcal{P}$;
- if $P \in \mathcal{P}$, then also $KP, LP, \Box P, \Diamond P \in \mathcal{P}$;
- no other strings belong to $\mathcal{P}$.

We now define the relations on prefixes announced above.

Definition 6. Let $P, P' \in \mathcal{P}$.

(1) $P \xrightarrow{-} P' :\iff$ there are $Q, R \in \mathcal{P}$ such that

- $P = QOR$ and $P' = QR$ $(O \in \{K, \Box\})$ or
- $P = QLOR$ and $P' = QOR$ $(O \in \{K, L\})$ or
- $P = Q\Diamond\Diamond R$ and $P' = Q\Diamond R$ or
- $P = Q\Diamond$ and $P' = Q$.

(2) $P \xrightarrow{+} P' :\iff$ there are $Q, R \in \mathcal{P}$ such that

- $P = QR$ and $P' = QOR$ $(O \in \{L, \Diamond\})$ or
- $P = QOR$ and $P' = QKOR$ $(O \in \{K, L\})$ or
- $P = Q\Box R$ and $P' = Q\Box\Box R$ or
- $P = Q$ and $P' = Q\Box$.

(3) $P \xrightarrow{c} P' :\iff$ there are $Q, R \in \mathcal{P}$ such that

- $P = QK\Box R$ and $P' = Q\Box KR$ or
- $P = Q\Diamond LR$ and $P' = QL\Diamond R$.

(4) $\xrightarrow{p} := \xrightarrow{-} \cup \xrightarrow{+} \cup \xrightarrow{c}$.

Let $\xrightarrow{=}$ denote the respective reflexive closure, and $\xrightarrow{\bullet}$ the respective reflexive and transitive closure. We write $\xrightarrow{(\ldots)_i}$ to indicate that only rule i of the relation $\xrightarrow{}$ has been applied.

The relationship between these notions will be studied in the next section. Here we point out that the relation $\xrightarrow{p^\bullet}$ is connected with consistency of sets of special formulas to be introduced soon. — With the aid of some modal proof theory the following proposition can be shown.

Proposition 7. *Let $P, P' \in \mathcal{P}$ be prefixes such that $P \overset{p^\bullet}{\longrightarrow} P'$. Then, for all $\alpha \in \mathcal{F}, \vdash P\alpha \rightarrow P'\alpha$.*

The aim of this section is to show that the converse of this proposition also holds. For this purpose we consider *prefix formulas*.

Definition 8. *Let $\mathcal{L}$ be the set of* literals, *i.e. the set PV joined with the set $\{\neg A \mid A \in PV\}$. A formula $\alpha \in \mathcal{F}$ is called a* prefix formula, *iff $\alpha = P\lambda$ for some prefix $P \in \mathcal{P}$ and some $\lambda \in \mathcal{L}$.*

Let $\mathcal{PF}$ denote the set of prefix formulas. $\mathcal{PF}$ is closed under negation in the following sense: if $\alpha = P\lambda \in \mathcal{PF}$, then $\neg \alpha$ is equivalent to $\bar{P}\bar{\lambda}$, where $\bar{P}$ is obtained from P by substituting each modal operator $O \in \{\Box, \Diamond, K, L\}$ occurring in P by its dual, and

$$\bar{\lambda} = \begin{cases} \neg\lambda & \text{if } \lambda \in PV \\ A & \text{if } \lambda = \neg A \text{ for some } A \in PV; \end{cases}$$

now, $\bar{P}\bar{\lambda}$ is a prefix formula (which, by abuse of notation, will be written as $\neg\alpha$). — Next we connect the relation $\overset{p^\bullet}{\longrightarrow}$ with prefix formulas.

Definition 9. *Define a relation $\overset{s}{\longrightarrow}$ on $\mathcal{PF}$ in the following way: Let $P\lambda, P'\lambda' \in \mathcal{PF}$. Then*

$$P\lambda \overset{s}{\longrightarrow} P'\lambda' :\Longleftrightarrow P \overset{p^\bullet}{\longrightarrow} P' \text{ and } \lambda = \lambda'.$$

Note that $\overset{s}{\longrightarrow}$ is a "strong" implication between prefix formulas in view of Proposition 7. Furthermore, it can easily be seen that $\overset{s}{\longrightarrow}$ is transitive and respects duals. The latter means that $\alpha \overset{s}{\longrightarrow} \beta$ implies $\neg\beta \overset{s}{\longrightarrow} \neg\alpha$ $(\alpha, \beta \in \mathcal{PF})$. — Following [Scherer 1995], we now view the relation $\overset{s}{\longrightarrow}$ as a kind of consistency.

Definition 10. *Let $X \subseteq \mathcal{PF}$ be a set of prefix formulas.*

(1) *If $X = \{\alpha, \beta\}$, then X is called* pseudo consistent, *iff not $\alpha \overset{s}{\longrightarrow} \neg\beta$.*
(2) *X is called* pseudo consistent, *iff every subset of X consisting of at most two elements is pseudo consistent.*

Pseudo consistent sets satisfy formally stronger properties than usually consistent sets:

Lemma 11. *Let $X \subseteq \mathcal{PF}$ be pseudo consistent and $\alpha \in \mathcal{PF}$ an arbitrary prefix formula. Then the following conditions are satisfied:*

(a) *if $X \cup \{\neg\alpha\}$ is not pseudo consistent, then there is some $\beta \in X$ such that $\beta \overset{s}{\longrightarrow} \alpha$;*
(b) *if $\beta \overset{s}{\longrightarrow} \alpha$ for some $\beta \in X$, then $X \cup \{\alpha\}$ is pseudo consistent.*

As in ordinary modal proof theory we can introduce the concept of *maximal pseudo consistent* sets of formulas.

Definition 12. Let $X \subseteq \mathcal{PF}$ be a set of prefix formulas. Then X is called *maximal pseudo consistent*, iff

- X is pseudo consistent, and
- $X \cup \{\alpha\}$ is not pseudo consistent for all $\alpha \in \mathcal{PF} \setminus X$.

Using Lemma 11, we get the following property of maximal pseudo consistent sets:

Lemma 13. *Let $X \subseteq \mathcal{PF}$ be maximal pseudo consistent. Then, for all $\alpha \in \mathcal{PF}$, either $\alpha \in X$ or $\neg \alpha \in X$ holds.*

Note that every pseudo consistent set of prefix formulas is contained in a maximal pseudo consistent set by Zorn's Lemma. Thus, we are in a position to define a "canonical" model for prefix formulas now. The following notational convention will be used subsequently: For $O \in \{K, L\}$ let $O\mathcal{PF} := \{O\alpha \mid \alpha \in \mathcal{PF}\}$. Define a structure $\mathcal{M}_{pr} = (C_{pr}, \{R_{pr}, S_{pr}\}, \sigma_{pr})$ through

- $C_{pr} := \{t \subseteq \mathcal{PF} \mid t \text{ maximal pseudo consistent}\}$,

and, for all $t, u \in C_{pr}$ and $A \in PV$,

- $(t, u) \in R_{pr} : \Longleftrightarrow t \cap K\mathcal{PF} = u \cap K\mathcal{PF}$,
- $(t, u) \in S_{pr} : \Longleftrightarrow \{\alpha \in \mathcal{PF} \mid \Box\alpha \in t\} \subseteq u$,
- $\sigma_{pr}(A, t) = 1 : \Longleftrightarrow A \in t$.

Then the following crucial *truth lemma* is valid.

Lemma 14. *The just defined structure $\mathcal{M}_{pr}$ is an SSL–model. Moreover, for all prefix formulas $\alpha \in \mathcal{PF}$ and every $t \in C_{pr}$ we have that $\mathcal{M}_{pr} \models \alpha[t]$, iff $\alpha \in t$.*

For convenience of the reader, we prove the first statement of the lemma.

Proof. We first prove that $\mathcal{M}_{pr}$ is an *SSL*–model. Clearly, R_{pr} is an equivalence relation on the set C_{pr}. We next show that S_{pr} is reflexive. So let $t \in C_{pr}$ and $\alpha \in \mathcal{PF}$ be given such that $\Box\alpha \in t$. From Definition 6(1) we get that $\Box\alpha \xrightarrow{s} \alpha$. Lemma 11(b) and the maximality of t yield $\alpha \in t$. Consequently, $(t, t) \in S_{pr}$, i.e. S_{pr} is reflexive. In a similar way, using 6(2), one obtains transitivity of S_{pr}. As to the *cross property*, take $t, u, v \in C_{pr}$ such that $(t, u) \in S_{pr}$ and $(u, v) \in R_{pr}$. Define

$$X := [t \cap (K\mathcal{PF} \cup L\mathcal{PF})] \cup \{\Diamond\beta \mid \beta \in v\}.$$

We claim that X is pseudo consistent. Assume towards a contradiction that this is not the case. Then there is a two–element subset $Y \subseteq X$ which is not pseudo consistent. We first convince ourselves that $Y \not\subseteq \{\Diamond\beta \mid \beta \in v\}$. For otherwise there would exist $\beta_1, \beta_2 \in v$ such that $\Diamond\beta_1 \xrightarrow{s} \neg\Diamond\beta_2$. Since $\beta_1 \xrightarrow{s} \Diamond\beta_1$ by Definition 6(2), we get $\beta_1 \xrightarrow{s} \Box\neg\beta_2$ because of the transitivity of the relation

$\overset{s}{\longrightarrow}$. Moreover, $\Box\neg\beta_2 \overset{s}{\longrightarrow} \neg\beta_2$ by Definition 6(1). Thus $\beta_1 \overset{s}{\longrightarrow} \neg\beta_2$ follows. Maximality of v together with Lemma 11(b) implies $\neg\beta_2 \in v$. This contradicts the pseudo consistency of v. — As t is likewise pseudo consistent, Y can be written as $Y = \{\alpha, \Diamond\beta\}$, where $\alpha \in t \cap (KP\mathcal{F} \cup LP\mathcal{F})$ and $\beta \in v$.

Now, because of the pseudo inconsistency of Y, $O\gamma \overset{s}{\longrightarrow} \Box\neg\beta$ holds, where $O \in \{K, L\}$ and $\gamma \in \mathcal{PF}$ are such that $\alpha = O\gamma$. We get $KO\gamma \overset{s}{\longrightarrow} K\Box\neg\beta$ as well, and, with the aid of 6(2), 6(3), and the transitivity of $\overset{s}{\longrightarrow}$,

$$\alpha = O\gamma \overset{s}{\longrightarrow} \Box K\neg\beta$$

follows. Using the same arguments as above we obtain $\Box K\neg\beta \in t$. Thus $K\neg\beta \in u$. We conclude that $K\neg\beta \in v$ is valid, and, by known arguments, we get that $\neg\beta \in v$; contradiction! Consequently, X is pseudo consistent.

Let w be a maximal pseudo consistent set of prefix formulas containing X. We want to show that $(t, w) \in R_{pr}$ and $(w, v) \in S_{pr}$. The first conjunct follows easily from the definition of the first component of X (applying Lemma 13). As to the second, assume that there exists a prefix formula γ such that $\Box\gamma \in w$ and $\gamma \notin v$. Then $\neg\gamma \in v$ by Lemma 13, whence $\Diamond\neg\gamma \in w$ by the definition of X. This contradicts the pseudo consistency of w. It follows that $(w, v) \in S_{pr}$. Altogether, the *cross property* is proved.

Finally, let $A \in PV$ and $(t, u) \in S_{pr}$. Assume first that $\sigma_{pr}(A, t) = 1$ holds. By the definition of σ_{pr}, $A \in t$. According to 6(2) and the maximality of t, $\Box A \in t$ holds as well. Thus $A \in u$, i.e. $\sigma_{pr}(A, u) = 1$. Now let on the other hand $\sigma_{pr}(A, t) = 0$. Then $A \notin t$. Therefore $\neg A \in t$, which implies $\Box\neg A \in t$. Thus $\neg A \in u$, i.e. $\sigma_{pr}(A, u) = 0$. This shows all *SSL*-model properties for $\mathcal{M}_{pr}$.

The main result of this section can be proved with the aid of the previous lemma and reads as follows:

Theorem 15. *Let $X \subseteq \mathcal{PF}$ be a set of prefix formulas. Then X is consistent, iff it is pseudo consistent.*

As a corollary, the converse of Proposition 7 in fact follows.

Corollary 16. *Let $P, P' \in \mathcal{P}$ be prefixes. Then $P \overset{p*}{\longrightarrow} P'$, iff $\vdash P\alpha \to P'\alpha$ for all $\alpha \in \mathcal{F}$.*

4 Analyzing a Special Semi–Thue System

We now investigate the relation $\overset{p*}{\longrightarrow}$ in more detail. In particular, we determine the computational complexity of deciding the *word problem* for the special Semi–Thue system given by the rules in Definition 6. To this end we introduce further designations based on 6, (1) – (3).

Definition 17. We also consider the following unions of relations:

- $\xrightarrow{q_1} := \xrightarrow{-} \cup \xrightarrow{+} \cup \xrightarrow{(c)_1}$;
- $\xrightarrow{q_2} := \xrightarrow{-} \cup \xrightarrow{+} \cup \xrightarrow{(c)_2}$;
- $\xrightarrow{q} := \xrightarrow{-} \cup \xrightarrow{+}$.

We are going to state a couple of lemmas, which enable us to represent $\xrightarrow{p^\bullet}$ in a computably feasible manner eventually.

Lemma 18. (a) $\xrightarrow{p^\bullet} = \xrightarrow{(c)_2^\bullet} \circ \xrightarrow{q_1^\bullet}$.

(b) $\xrightarrow{p^\bullet} = \xrightarrow{q_2^\bullet} \circ \xrightarrow{(c)_1^\bullet}$.

The next lemma concerns the relation $\xrightarrow{q^\bullet}$. It can be composed by the relations $\xrightarrow{-^\bullet}$ and $\xrightarrow{+^\bullet}$ (in this order). The result is due to [Scherer 1995].

Lemma 19. $\xrightarrow{q^\bullet} = \xrightarrow{-^\bullet} \circ \xrightarrow{+^\bullet}$.

As a consequence of these two lemmas we get:

Corollary 20. $\xrightarrow{p^\bullet} = \xrightarrow{(c)_2^\bullet} \circ \xrightarrow{-^\bullet} \circ \xrightarrow{+^\bullet} \circ \xrightarrow{(c)_1^\bullet}$.

We next have to study the computational properties of the relation $\xrightarrow{q^\bullet}$. It turns out that this relation behaves very similar to the corresponding one in usual (multi–agent) logic of knowledge, where no interchange between knowledge operators of different agents occurs. Thus the results here are as in [Scherer 1995], Section V 2. As a consequence, we get:

Theorem 21. *The word problem for the relation $\xrightarrow{p^\bullet}$ is decidable in polynomial time.*

5   Main Results

In the present section we state our main results on the complexity of the satisfiability problem for various classes of (prefix) *SSL*–formulas. In particular, we obtain nice generalizations of well–known facts from sentential logic.

Definition 22. Let $n \in \mathbb{N}$, $n \geq 1$. Define

- $nSAT := \{\alpha \in \mathcal{F}\} \mid \alpha$ is a conjunction of disjunctions of prefix formulas such that each conjunct contains at most n prefix formulas, and α is satisfiable}.

Moreover, let $SAT := \{\alpha \in \mathcal{F} \mid \alpha$ is satisfiable}.

Note that these sets are built as in propositional logic, substituting literals by prefix formulas. — The key result for the determination of the complexity of $nSAT$ is contained in our first theorem. Its proof results from Theorems 4, 15, and 21. Note that its propositional counterpart holds trivially.

154

Theorem 23. *1SAT is decidable in polynomial time.*

As to $n \geq 2$, the corresponding results for $nSAT$ generalize from propositional logic to SSL. We start with the case $n = 2$.

Theorem 24. *2SAT is polynomial–time decidable.*

In the proof of this theorem a generalization of propositional resolution ([Ben-Ari 1993], Section 2.10) is used, which is possible after some preparations. — Turning from $n = 2$ to $n = 3$, the complexity increases in the same manner as in propositional logic.

Theorem 25. *For $n \geq 3$, the set $nSAT$ is NP–complete.*

In fact, a non–deterministic Turing machine *guesses* a prefix formula in each clause and checks satisfiability of the resulting conjunction according to Theorem 23. In this way the asserted upper bound is established. — Using *PSPACE*-completeness of the **S4**–satisfiability problem ([Ladner 1977], 5.3) and ideas which are related to those occurring in [Halpern and Moses 1992], Theorem 6.6 (final part of the proof), we can prove the following additional result:

Theorem 26. *SAT is PSPACE–hard.*

6 Concluding Remarks

Using methods due to SCHERER [Scherer 1995] among other things, we examined the complexity of the satisfiability problem for several subclasses of formulas of the modal logic of subset spaces (SSL) developed by MOSS and PARIKH. These formula classes are given as certain clause sets of prefix formulas. We showed that $nSAT \in P$ if $n \leq 2$, and that $nSAT$ is NP–complete if $n \geq 3$ (wherein n is the upper bound for the cardinality of the clauses). The results generalize well–known facts from propositional logic. Satisfiability with no restriction on the formulas was proved to be *PSPACE*–hard additionally. — It should be remarked that Theorems 24 and 25 hold for $S5_m$ analogously, as it was conjectured in [Scherer 1995].

References

[Ben-Ari 1993] Ben-Ari, M. 1993. *Mathematical Logic for Computer Science.* Prentice Hall.
[Chellas 1980] Chellas, B. F. 1980. *Modal Logic: An Introduction.* Cambridge: Cambridge University Press.
[Dabrowski et al. 1996] Dabrowski, A., L. S. Moss, and R. Parikh. 1996. Topological Reasoning and The Logic of Knowledge. *Ann. Pure Appl. Logic* 78:73–110.
[Fagin et al. 1995] Fagin, R., J. Y. Halpern, Y. Moses, and M. Y. Vardi. 1995. *Reasoning about Knowledge.* Cambridge(Mass.): MIT Press.

[Georgatos 1994a] Georgatos, K. 1994a. Knowledge Theoretic Properties of Topological Spaces. In *Knowledge Representation and Uncertainty*, eds. M. Masuch and L. Polos, 147–159. Springer. LNCS 808.

[Georgatos 1994b] Georgatos, K. 1994b. Reasoning about Knowledge on Computation Trees. In *Proc. Logics in Artificial Intelligence (JELIA '94)*, eds. C. MacNish, D. Pearce, and L. M. Pereira, 300–315. Springer. LNCS 838.

[Goldblatt 1987] Goldblatt, R. 1987. *Logics of Time and Computation*. CSLI Lecture Notes Number 7. Stanford: Center for the Study of Language and Information.

[Halpern and Moses 1992] Halpern, J. Y., and Y. Moses. 1992. A Guide to Completeness and Complexity for Modal Logics of Knowledge and Belief. *Artificial Intelligence* 54:319–379.

[Harel 1979] Harel, D. 1979. *First-Order Dynamic Logic*. Lecture Notes in Computer Science 68. Springer.

[Heinemann 1996a] Heinemann, B. 1996a. On the Complexity of Prefix Formulas in Modal Logic of Subset Spaces. Informatik Berichte 202. Hagen: Fernuniversität.

[Heinemann 1996b] Heinemann, B. 1996b. 'Topological' Modal Logic of Subset Frames with Finite Descent. In *Proc. 4th Intern. Symp. on Artificial and Mathematics, AI/MATH-96*, 83–86. Fort Lauderdale.

[Heinemann 1997] Heinemann, B. 1997. On Binary Computation Structures. *MLQ* 43. to appear.

[Heusch 1995] Heusch, P. 1995. The Complexity of the Falsifiability Problem for Pure Implicatinal Formulas. In *Mathematical Foundations of Computer Science*, eds. J. Wiedermann and P. Hájek, 221–226. Springer. Lecture Notes in Computer Science 969.

[Ladner 1977] Ladner, R. E. 1977. The Computational Complexity of Provability in Systems of Modal Propositional Logic. *SIAM J. Comput.* 6:467–480.

[Manna and Pnueli 1992] Manna, Z., and A. Pnueli. 1992. *The Temporal Logic of Reactive and Concurrent Systems*. New York: Springer.

[Moss and Parikh 1992] Moss, L. S., and R. Parikh. 1992. Topological Reasoning and The Logic of Knowledge. In *Proc. 4th Conf. on Theoretical Aspects of Reasoning about Knowledge (TARK 1992)*, ed. Y. Moses, 95–105. Morgan Kaufmann.

[Scherer 1995] Scherer, B. G. 1995. *Atome und Präfixe in der Wissenslogik*. Doctoral dissertation, ETH Zürich.

The Undecidability of Second Order Linear Affine Logic

Alexei P. Kopylov*

Department of Mathematics and Mechanics
Moscow State University
119899, Moscow, Russia
e-mail: alexei@lpcs.math.msu.ru

Abstract. Quantifier-free propositional linear affine logic (i.e. linear logic with weakening) is decidable [Kop, Laf2]. Lafont and Scedrov proved that the multiplicative fragment of *second-order* linear logic is undecidable [LS]. In this paper we show that second order linear affine logic is undecidable as well. At the same time it turns out that even its multiplicative fragment is undecidable. Moreover, we obtain a whide class of undecidabile second order logics which lie between the Lambek calculus (LC) and linear affine logic. The proof is based on an encoding of two-counter Minsky machines in second order linear affine logic. The faithfulness of the encoding is proved by means of the phase semantics.

1 Introduction and summary

Our notation. Linear logic has been introduced by Girard [Gir]. The inference rules of second order linear logic are represented in Table 1. Linear affine logic is linear logic with the weakening rule (see Table 2) [T]. Non-commutative linear logic is linear logic without the permutation rule [Abr91]. Note than non-commutative linear logic has two implications: the right one and the left one (−∘ and ∘−). We shall abbreviate second order linear logic and linear affine logic as LL2 and LLW2 correspondingly, and the non-commutative versions of these logics as N-LL2 and N-LLW2. We shall use the abbreviations LL, LLW, N-LL, N-LLW for the quantifier-free fragments of the corresponding logics. There are also intuitionistic versions of all the logics mentioned above. As usual an intuitionistic derivation is a derivation containing only sequents which have no more than one formula in the consequent. The letter **I** stands for intuitionistic logics (e.g. ILL, ILLW and so on). Connectives and constants of LL are divided into three groups: the multiplicatives ($\otimes$, $\wp$, −∘, 1 and $\bot$), the additives ($\oplus$, &, 0 and $\top$) and the exponentials (! and ?).

In referring to linear logic fragments,

---

* The research described in this publication was made possible in part by Grant No. NFQ300 from the International Science Foundation and by the Russian Foundation for Basic Researh (grant 96–01–01395).

$$(I)\ \frac{}{A \Rightarrow A} \qquad\qquad (CUT)\ \frac{\Gamma_1 \vdash \Delta_1, A \quad A, \Gamma_2 \vdash \Delta_2}{\Gamma_1, \Gamma_2 \vdash \Delta_1, \Delta_2}$$

$$(L\perp)\ \frac{\Gamma \vdash A, \Delta}{A^\perp, \Gamma \vdash \Delta} \qquad\qquad (R\perp)\ \frac{A, \Gamma \vdash \Delta}{\Gamma \vdash A^\perp, \Delta}$$

$$(L\otimes)\ \frac{A, B, \Gamma \vdash \Delta}{A \otimes B, \Gamma \vdash \Delta} \qquad (R\otimes)\ \frac{\Gamma_1 \vdash A, \Delta_1 \quad \Gamma_2 \vdash B, \Delta_2}{\Gamma_1, \Gamma_2 \vdash A \otimes B, \Delta_1, \Delta_2}$$

$$(L\wp)\ \frac{A, \Gamma_1 \vdash \Delta_1 \quad B, \Gamma_2 \vdash \Delta_2}{A \wp B, \Gamma_1, \Gamma_2 \vdash \Delta_1, \Delta_2} \qquad (R\wp)\ \frac{\Gamma \vdash A, B, \Delta}{\Gamma \vdash A \wp B, \Delta}$$

$$(L\multimap)\ \frac{\Gamma_1 \vdash A, \Delta_1 \quad B, \Gamma_2 \vdash \Delta_2}{A \multimap B, \Gamma_1, \Gamma_2 \vdash \Delta_1, \Delta_2} \qquad (R\multimap)\ \frac{A, \Gamma \vdash B, \Delta}{\Gamma \vdash A \multimap B, \Delta}$$

$$(L1)\ \frac{\Gamma \vdash \Delta}{1, \Gamma \vdash \Delta} \qquad\qquad (R1)\ \frac{}{\vdash 1}$$

$$(L\perp)\ \frac{}{\perp \vdash} \qquad\qquad (R\perp)\ \frac{\Gamma \vdash \Delta}{\Gamma \vdash \perp, \Delta}$$

$$(L\oplus)\ \frac{A, \Gamma \vdash \Delta \quad B, \Gamma \vdash \Delta}{A \oplus B, \Gamma \vdash \Delta} \qquad (R\oplus)\ \frac{\Gamma \vdash A, \Delta}{\Gamma \vdash A \oplus B, \Delta} \quad \frac{\Gamma \vdash B, \Delta}{\Gamma \vdash A \oplus B, \Delta}$$

$$(L\&)\ \frac{A, \Gamma \vdash \Delta}{A \& B, \Gamma \vdash \Delta} \quad \frac{B, \Gamma \vdash \Delta}{A \& B, \Gamma \vdash \Delta} \qquad (R\&)\ \frac{\Gamma \vdash A, \Delta \quad \Gamma \vdash B, \Delta}{\Gamma \vdash A \& B, \Delta}$$

$$(L0)\ \frac{}{0, \Gamma \vdash \Delta} \qquad\qquad (R\top)\ \frac{}{\Gamma \vdash \top, \Delta}$$

$$(L!)\ \frac{A, \Gamma \vdash \Delta}{!A, \Gamma \vdash \Delta} \qquad\qquad (R!)\ \frac{!\Gamma \vdash A, ?\Delta}{!\Gamma \vdash !A, ?\Delta}$$

$$(W!)\ \frac{\Gamma \vdash \Delta}{!A, \Gamma \vdash \Delta} \qquad\qquad (C!)\ \frac{!A, !A, \Gamma \vdash \Delta}{!A, \Gamma \vdash \Delta}$$

$$(L?)\ \frac{A, !\Gamma \vdash ?\Delta}{?A, !\Gamma \vdash ?\Delta} \qquad\qquad (R?)\ \frac{\Gamma \vdash A, \Delta}{\Gamma \vdash ?A, \Delta}$$

$$(W?)\ \frac{\Gamma \vdash \Delta}{\Gamma \vdash ?A, \Delta} \qquad\qquad (C?)\ \frac{\Gamma \vdash ?A, ?A, \Delta}{\Gamma \vdash ?A, \Delta}$$

$$(L\forall)\ \frac{\Gamma, A[B/\alpha] \vdash \Delta}{\Gamma, \forall \alpha A \vdash \Delta} \qquad (R\forall)^1\ \frac{\Gamma \vdash A[\beta/\alpha], \Delta}{\Gamma \vdash \forall \alpha A, \Delta}$$

$$(L\exists)^1\ \frac{\Gamma, A[\beta/\alpha] \vdash \Delta}{\Gamma, \exists \alpha A \vdash \Delta} \qquad (R\exists)\ \frac{\Gamma \vdash A[B/\alpha], \Delta}{\Gamma \vdash \exists \alpha A, \Delta}$$

[1] In the rules $(R\forall)$ and $(L\exists)$ β must be a fresh variable.

Table 1: The inference rules of LL2.

$$(W)\ \frac{W, \Gamma_0 \vdash \Delta_0}{W, \Gamma \vdash \Delta}, \quad \text{where } \Gamma_0 \subseteq \Gamma, \Delta_0 \subseteq \Delta$$

Table 2: The weakening rule for LLW2.

$$(I)\ \frac{}{A \vdash A} \qquad\qquad (CUT)\ \frac{\Gamma, A, \Delta \vdash Z \quad \Pi \vdash A}{\Gamma, \Pi, \Delta \vdash Z}$$

$$(L\otimes)\ \frac{\Gamma, A, B, \Delta \vdash Z}{\Gamma, A \otimes B, \Delta \vdash Z} \qquad (R\otimes)\ \frac{\Gamma \vdash A \quad \Delta \vdash B}{\Gamma, \Delta \vdash A \otimes B}$$

$$(L\multimap)\ \frac{\Gamma, B, \Delta \vdash Z \quad \Pi \vdash A}{\Gamma, \Pi, A \multimap B, \Delta \vdash Z} \qquad (R\multimap)^2\ \frac{A, \Gamma \vdash B}{\Gamma \vdash A \multimap B}$$

$$(L\circ\!\!-)\ \frac{\Gamma, B, \Delta \vdash Z \quad \Pi \vdash A}{\Gamma, B \circ\!\!- A, \Pi, \Delta \vdash Z} \qquad (R\circ\!\!-)^2\ \frac{\Gamma, A \vdash B}{\Gamma \vdash B \circ\!\!- A}$$

$$(L\forall)\ \frac{\Gamma, A[B/\alpha], \Delta \vdash Z}{\Gamma, \forall \alpha A, \Delta \vdash Z} \qquad (R\forall)^1\ \frac{\Gamma \vdash A[\beta/\alpha]}{\Gamma \vdash \forall \alpha A}$$

$$(L\exists)^1\ \frac{\Gamma, A[\beta/\alpha], \Delta \vdash Z}{\Gamma, \exists \alpha A, \Delta \vdash Z} \qquad (R\exists)\ \frac{\Gamma \vdash A[B/\alpha]}{\Gamma \vdash \exists \alpha A}$$

[1] In the rules $(R\forall)$ and $(L\exists)$ β must be a fresh variable.

[2] In the rules $(R\multimap)$ and $(R\circ\!\!-)$ Γ is not empty.

Table 3: The inference rules of LC2.

M stands for the multiplicative fragment (i.e. the fragment containing only multiplicatives),

A stands for the additive fragment (i.e. the fragment containing only additives),

E stands for the exponential fragment, (i.e. the fragment containing only exponentials).

For example, MLL abbreviates the multiplicative fragment of LL, MALLW denotes multiplicative-additive fragment of LLW, and so on.

Also we shall consider the Lambek calculus (LC). In contrast to the traditional notations for connectives of LC: $\backslash$, $/$, $\cdot$, we shall use the following notations: $\multimap$ and $\multimapinv$ stands for the left and right implications, and $\otimes$ for the tensor product, (see Table 3 for the inference rules). It is clear that LC $\subseteq$ N-ILL $\subseteq$ ILL $\subseteq$ LL $\subseteq$ LLW.

Main results. Lincoln, Scedrov, and Shankar showed the undecidability of IMLL2 and IMALL2 by an embedding of LJ2 [LSS]. Lafont proved the undecidability of MALL2 [Laf1]. Then Lafont and Scedrov proved that MLL2 is undecidable too [LS]. Emms demonstrated an embedding of LJ2 into N-IMLL2. Kanovich demonstrated in [Kan2] the undecidability of N-MLL2, cyclic LL and second order Lambek Calculus (LC2). On the other hand, quantifier-free linear affine logic is decidable [Kop, Laf2]. The decidability problem for second order linear affine logic remained open.

In the current paper we prove the undecidability of LLW2. Also we prove that for any logic L if LC2 $\subseteq$ L $\subseteq$ LLW2, then L is undecidable. In particular, all second-order logics mentioned above are undecidable as well as MLLW2, MALLW2, LLW2, IMLLW2, N-MLLW2, etc. The main ideas of the proof are similar to the ideas of [LS]. Namely, we encode two-counter machines (Minsky machines) in LC2 and LLW2. This encoding is similar to the encodings from [Kan1, Laf1, LS]. In order to obtain the faithfulness of the encoding we use (as in [Laf1, LS]) the phase semantics, but here we need the phase semantics for linear affine logic.

2 Phase semantics

Let us recall some definitions concerning phase semantics [Gir, Laf2]. Phase space is the triple $(M, \perp, K)$, where M is a commutative monoid, $\perp \subseteq M$ and K is a submonoid of the submonoid $J(M) = \{x \in \perp^{\perp} \mid x \in \{x^2\}^{\perp\perp}\}$. For instance, K may be $\{1\}$.

Let $X, Y \subseteq M$, then, by definition,

$$XY = \{xy \mid x \in X, y \in Y\},$$
$$X \multimap Y = \{z \in M \mid \forall x \in X \quad xz \in Y\},$$
$$X^{\perp} = X \multimap \perp.$$

We say that X is a fact, when there is a set Y such that $X = Y^{\perp}$. It is easy to see that X is a fact if and only if $X = X^{\perp\perp}$. If X is any subset of M, then $X^{\perp\perp}$

is the smallest fact containing X. By definition,

$$X \otimes Y = (XY)^{\perp\perp}, \quad X \wp Y = (X^\perp \otimes Y^\perp)^\perp,$$
$$X \& Y = X \cap Y, \quad X \oplus Y = (X^\perp \& Y^\perp)^\perp,$$
$$1 = \perp^\perp, \quad \top = M, \quad 0 = \top^\perp.$$

If all atoms p are interpreted by facts $p^\bullet$, then for any formula A we can naturally define a fact $A^\bullet$. Namely, $^\bullet$ commutes with all connectives, and $(\forall \alpha \, A[\alpha])^\bullet$ is defined as

$$\bigcap_{X \text{ is a fact}} A[X]^\bullet,$$

where $A[X]^\bullet$ is an interpretation of $A[\alpha]$, where $\alpha^\bullet = X$. By definition, a formula A is satisfied, if $1 \in A^\bullet$. A sequent $A_1, \ldots, A_n \vdash B_1, \ldots, B_k$ is satisfied, if the formula $(A_1 \otimes \ldots \otimes A_n) \multimap (B_1 \wp \ldots \wp B_k)$ is satisfied. This is equivalent to $(A_1 \otimes \ldots \otimes A_n)^\bullet \subseteq (B_1 \wp \ldots \wp B_k)^\bullet$.

Theorem 1 [Gir]. *If a sequent Φ is derivable in MALL, then any phase space $(M, \perp)$ satisfies Φ.*

For linear affine logic there is an analogous theorem.

Definition 2 [Laf2]. We say that the phase space is an affine phase space if $\perp$ is an ideal, i.e. $M \cdot \perp \subseteq \perp$.

Theorem 3 [Laf2]. *If a sequent Φ is derivable in LLW2, then any affine phase space $(M, \perp, K)$ satisfies Φ.*

Remark For quantifier-free linear affine logic without modalities the converse theorems hold as well.

If $(M, \perp)$ is an affine phase space, then $M \cdot \perp = \perp$, and $\{1\}^{\perp\perp} = \perp^\perp = M$. Hence, for any fact X, if $1 \in X$ then $X = M$. Moreover, $\perp = M^\perp$ is the least fact. Now let us prove a useful lemma. Assume we have a homomorphism $\gamma : M \to \mathbf{N}$ ($\mathbf{N}$ is the monoid of natural numbers with the operation $+$). Define the sets $S_n = \gamma^{-1}(\{n\})$, and $M_n = \bigcup_{k \geq n} S_k$. In particular, $M_n = M$, whenever $n \leq 0$.

Lemma 4. *Let $X = X' \cup X''$ and $Y = Y' \cup M_{k+1}$ be two sets, where $\emptyset \neq X' \subseteq S_n$; $X'' \subseteq M_{n+1}$ and $Y' \subseteq S_k$. Then $X' \multimap Y' \subseteq S_{k-n}$, and*

$$X \multimap Y = (X' \multimap Y') \cup M_{k-n+1}.$$

Proof. If $z \in X' \multimap Y'$, then $zx \in Y'$, for any $x \in X'$. Therefore $\gamma(zx) = k$, because of $Y' \subseteq S_k$. But $\gamma(x) = n$, so $\gamma(z) = k - n$, and $z \in S_{k-n}$.

Now let us prove that $(X' \multimap Y') \cup M_{k-n+1} \subseteq X \multimap Y$. Indeed, from

$$M_{k-n+1}X \subseteq M_{k-n+1}M_n \subseteq M_{k+1} \subseteq Y$$

we conclude $M_{k-n+1} \subseteq X \multimap Y$. Moreover,

$$(X' \multimap Y')X = (X' \multimap Y')X' \cup (X' \multimap Y')X'' \subseteq Y' \cup S_{k-n}M_{n+1} \subseteq Y' \cup M_{k+1} = Y.$$

So, $X' \multimap Y' \subseteq X \multimap Y$.

Conversely, let $z \in X \multimap Y$. Then $zx \in Y$, for any $x \in X$. Therefore, $\gamma(zx) \geq k$, because of $Y \subseteq M_k$. But $\gamma(x) = n$, for $x \in X'$. Hence, $\gamma(z) \geq k - n$. Consider two cases. If $\gamma(z) > k - n$, then $z \in M_{k-n+1}$. Otherwise, $\gamma(z) = k - n$. In the last case one obtain $zX' \subseteq zX \subseteq Y$ and $zX' \subseteq S_{k-n}S_n \subseteq S_k$. Therefore, $zX' \subseteq Y \cap S_k = Y'$, and $z \in X' \multimap Y'$. In both cases $z \in (X' \multimap Y') \cup M_{k-n+1}$. ∎

Corollary 5. *If $\bot = \bot' \cup M_{k+1}$, where $\bot' \subseteq S_k$, then any fact is a set of the kind: $X' \cup M_{n+1}$, where $X' \subseteq S_n$.*

3 Two-counter Minsky machines

A configuration of a two-counter Minsky machine is determined by three numbers (i, p, q). Here $p, q \in \mathbf{N}$ are states of the machine's counters, and $i \in [0, m]$ is a label. A program of the machine is given by a map $\tau : [1, m] \to \{+\} \times \{1, 2\} \times [0, m] \cup \{-\} \times \{1, 2\} \times [0, m] \times [0, m]$. If the current configuration of the machine is (i, p, q), then the next configuration is:

- $(j, p+1, q)$ if $\tau(i) = (+, 1, j)$ (increment the first counter),
- $(j, p-1, q)$ if $\tau(i) = (-, 1, j, k)$ and $p > 0$ (decrement the first counter),
- (k, p, q) if $\tau(i) = (-, 1, j, k)$ and $p = 0$ (test for zero the first counter),
- $(j, p, q+1)$ if $\tau(i) = (+, 2, j)$ (increment the second counter),
- $(j, p, q-1)$ if $\tau(i) = (-, 2, j, k)$ and $q > 0$ (decrement the second counter),
- (k, p, q) if $\tau(i) = (-, 2, j, k)$ and $q = 0$ (test for zero the second counter).

In other words a Minsky machine has transitions of the following types:

- $(i, p, q) \to (j, p+1, q)$,
- $(i, p, q) \to (j, p-1, q)$ *if* $p > 0$,
- $(i, 0, q) \to (k, 0, q)$,
- $(i, p, q) \to (j, p, q+1)$,
- $(i, p, q) \to (j, p, q-1)$ *if* $q > 0$,
- $(i, p, 0) \to (k, p, 0)$.

The machine stops when $i = 0$. A configuration (i, p, q) is accepted by the machine if, starting from (i, p, q), it eventually stops on $(0, 0, 0)$.

Theorem 6 [M, Lk]. *There is a Minsky machine for wich the set of accepted configurations is not recursive.*

4 Encoding two-counter Minsky machines

We can encode Minsky machines in the following way. Let us consider two formulas:

$$\varphi[\alpha] = (\alpha \multimap f) \multimap h,$$
$$\psi[\alpha] = (\alpha \multimap g) \multimap e.$$

We construct the following infinite sets of formulas:

$$\varphi_{-1} = a, \quad \varphi_n = \varphi[\varphi_{n-1}], n \in \mathbf{N},$$
$$\psi_{-1} = b, \quad \psi_n = \psi[\psi_{n-1}], n \in \mathbf{N}.$$

Any machine configuration (i, p, q) is encoded by the following formula:

$$c_i \otimes \varphi_p \otimes \psi_q.$$

Here a, b, c_i, e, f, g, h are literals. An increment transition $(i, p, q) \to (j, p+1, q)$ is encoded by the formula

$$\forall \alpha, \beta(c_i \otimes \varphi[\alpha] \otimes \beta \multimap c_j \otimes \varphi[\varphi[\alpha]] \otimes \beta).$$

A decrement transition $(i, p, q) \to (j, p-1, q)$ if $p > 0$ is encoded by the formula

$$\forall \alpha, \beta(c_i \otimes \varphi[\varphi[\alpha]] \otimes \beta \multimap c_j \otimes \varphi[\alpha] \otimes \beta).$$

And a test-for-zero transition $(i, 0, q) \to (k, 0, q)$ is encoded by the formula

$$\forall \beta(c_i \otimes \varphi[a] \otimes \beta \multimap c_k \otimes \varphi[a] \otimes \beta).$$

By analogy, the transitions for the second counter are encoded by the following formulas

$$\forall \alpha, \beta(c_i \otimes \alpha \otimes \psi[\beta] \multimap c_j \otimes \alpha \otimes \psi[\psi[\beta]]),$$
$$\forall \alpha, \beta(c_i \otimes \alpha \otimes \psi[\psi[\beta]] \multimap c_j \otimes \alpha \otimes \psi[\beta]),$$
$$\forall \alpha(c_i \otimes \alpha \otimes \psi[b] \multimap c_k \otimes \alpha \otimes \psi[b]).$$

Let $T_1, \ldots, T_n$ be formulas of the above kind which encode all transitions of our machine. Let $T = T_1 \otimes \ldots \otimes T_n$. We may assume without lost of generality, that $n \geq 1$. Consider the following formulas:

$$W = \forall x, y, z(x \otimes y \otimes z \multimap y),$$
$$W' = \forall y, z(y \otimes z \multimap y),$$
$$C = \forall x(x^4 \multimap x^5),$$
$$U = (x \multimap x).$$

The first two formulas encode weakening, the third formula encodes a restricted kind of contraction, and the last formula encodes an "unit" in LC. Here α^n denotes $\alpha \otimes \alpha \otimes \ldots \otimes \alpha$ (n times).

Theorem 7. *The following four assertions are equivalent:*

(i) *The configuration (i, p, q) is accepted by the machine.*

(ii) *The following sequent is derivable in LC2:*

$$c_i \otimes \varphi_p \otimes \psi_q, (U \otimes T \otimes U \otimes W)^4, C^9, W' \vdash c_0 \otimes \varphi_0 \otimes \psi_0$$

(iii) *The above sequent is derivable in any logic L, such that $LC2 \subseteq L \subseteq LLW2$.*
(iv) *This sequent is derivable in LLW2.*

This theorem and theorem 6 provide

Corollary 8. *Any logic L such that $LC2 \subseteq L \subseteq LLW2$ is undecidable.*

Proof of theorem 7. The implications (ii)→(iii) and (iii)→(iv) are trivial. Let us prove the implication (i)→(ii). We begin with

Lemma 9. *The following rules are derivable in LC2:*

$$\frac{\Pi, A^5, C^9, \Gamma \vdash Z}{\Pi, A^4, C^9, \Gamma \vdash Z}(\tilde{C}) \qquad \frac{\Pi, A, \Gamma \vdash Z}{\Pi, S_1 \otimes A \otimes S_2, W, \Gamma \vdash Z}(\tilde{W})$$

$$\frac{\Pi, A, \Gamma \vdash Z}{\Pi, A, \Sigma, W', \Gamma \vdash Z}(\tilde{W}')$$

In the last rule Σ is not empty.

Proof. Note that the following sequents are easily derivable: $A^4, C \vdash A^5$; $S_1 \otimes A \otimes S_2, W \vdash A$ and $A, \Sigma, W' \vdash A$. The derivation of the rule $(\tilde{C})$ is the following:

$$\frac{(C^2)^4, C \vdash (C^2)^5 \quad \dfrac{\dfrac{\dfrac{A^4, C \vdash A^5 \quad \Pi, A^5, C^9, \Gamma \vdash Z}{\Pi, A^4, C, C^9, \Gamma \vdash Z}(CUT)}{\Pi, A^4, C^{10}, \Gamma \vdash Z}(L\otimes)}{\Pi, A^4, C^8, C, \Gamma \vdash Z}(CUT)}{\Pi, A^4, C^9, \Gamma \vdash Z}(L\otimes)$$

And here is the derivations of the rules $(\tilde{W})$ and $(\tilde{W}')$:

$$\frac{S_1 \otimes A \otimes S_2, W \vdash A \quad \Pi, A, \Gamma \vdash Z}{\Pi, S_1 \otimes A \otimes S_2, W, \Gamma \vdash Z}(CUT)$$

$$\frac{A, \Sigma, W' \vdash A \quad \Pi, A, \Gamma \vdash Z}{\Pi, A, \Sigma, W', \Gamma \vdash Z}(CUT) \qquad \blacksquare$$

Lemma 10. *For any transition $(i, p, q) \to (i', p', q')$, if T_k is a formula encoding this transition then the sequent $c_i \otimes \varphi_p \otimes \psi_q, T_k \vdash c_{i'} \otimes \varphi_{p'} \otimes \psi_{q'}$ is derivable in LC2.*

Proof. It is easy to see that the following sequents are derivable in LC2:

$$c_i \otimes \varphi_p \otimes \varphi_q, \forall \alpha, \beta(c_i \otimes \varphi[\alpha] \otimes \beta \multimap c_j \otimes \varphi[\varphi[\alpha]] \otimes \beta) \vdash c_j \otimes \varphi_{p+1} \otimes \varphi_q,$$

$$c_i \otimes \varphi_p \otimes \varphi_q, \forall \alpha, \beta(c_i \otimes \varphi[\varphi[\alpha]] \otimes \beta \multimap c_j \otimes \varphi[\alpha] \otimes \beta) \vdash c_j \otimes \varphi_{p-1} \otimes \varphi_q,$$

$$c_i \otimes \varphi_0 \otimes \varphi_q, \forall \beta(c_i \otimes \varphi[a] \otimes \beta \multimap c_k \otimes \varphi[a] \otimes \beta) \vdash c_k \otimes \varphi_0 \otimes \varphi_q,$$

$$c_i \otimes \psi_p \otimes \psi_q, \forall \alpha, \beta(c_i \otimes \alpha \otimes \psi[\beta] \multimap c_j \otimes \alpha \otimes \psi[\psi[\beta]]) \vdash c_j \otimes \psi_p \otimes \psi_{q+1},$$

$$c_i \otimes \psi_p \otimes \psi_q, \forall \alpha, \beta(c_i \otimes \alpha \otimes \psi[\psi[\beta]] \multimap c_j \otimes \alpha \otimes \psi[\beta]) \vdash c_j \otimes \psi_p \otimes \psi_{q-1},$$

$$c_i \otimes \psi_p \otimes \psi_0, \forall \alpha(c_i \otimes \alpha \otimes \psi[b] \multimap c_k \otimes \alpha \otimes \psi[b]) \vdash c_k \otimes \psi_p \otimes \psi_0.$$

∎

We prove the implication (i)→(ii) by induction on the length of the computation. The base of induction holds because of the rule $(\tilde{W}')$:

$$\frac{c_0 \otimes \varphi_0 \otimes \psi_0 \vdash c_0 \otimes \varphi_0 \otimes \psi_0}{c_0 \otimes \varphi_0 \otimes \psi_0, (U \otimes T \otimes U \otimes W)^4, C^9, W' \vdash c_0 \otimes \varphi_0 \otimes \psi_0}(\tilde{W}')$$

Now let us verify the induction step. If the computation has the following form:

$$(i, p, q) \to (i', p', q') \to \ldots \to (0, 0, 0)$$

then by the induction hypothesis the following sequent is derivable:

$$c_{i'} \otimes \varphi_{p'} \otimes \psi_{q'}, (U \otimes T \otimes U \otimes W)^4, C^9, W' \vdash c_0 \otimes \varphi_0 \otimes \psi_0.$$

On the other hands, the by Lemma 10 the following sequent is also derivable:

$$c_i \otimes \varphi_p \otimes \psi_q, T_k \vdash c_{i'} \otimes \varphi_{p'} \otimes \psi_{q'}.$$

One can apply the rule (Cut) to these two sequents, and then obtain the derivation:

$$\frac{\dfrac{\dfrac{c_i \otimes \varphi_p \otimes \psi_q, T_k, (U \otimes T \otimes U \otimes W)^4, C^9, W' \vdash c_0 \otimes \varphi_0 \otimes \psi_0}{c_i \otimes \varphi_p \otimes \psi_q, U \otimes T \otimes U, W, (U \otimes T \otimes U \otimes W)^4, C^9, W' \vdash c_0 \otimes \varphi_0 \otimes \psi_0}(\tilde{W})}{c_i \otimes \varphi_p \otimes \psi_q, (U \otimes T \otimes U \otimes W)^5, C^9, W' \vdash c_0 \otimes \varphi_0 \otimes \psi_0}(L\otimes)}{c_i \otimes \varphi_p \otimes \psi_q, (U \otimes T \otimes U \otimes W)^4, C^9, W' \vdash c_0 \otimes \varphi_0 \otimes \psi_0}(\tilde{C})$$

∎

5 Faithfulness

Now let us check the implication (iv)→(i) using the phase space. Consider the free commutative monoid M generated by infinite set of atoms (c_i), $\overline{f}$, (f_n), $\overline{h}$, (h_n), $\overline{g}$, (g_n), $\overline{e}$, (e_n), where $i = 0, \ldots, m$; $n \in \mathbf{Z}$. Let $\gamma : M \to \mathbf{N}$ be the homomorphism defined by the following equations:

$$\gamma(c_i) = \gamma(\overline{f}) = \gamma(f_n) = \psi(\overline{h}) = \gamma(h_n) = \gamma(\overline{g}) = \gamma(g_n) = \gamma(\overline{e}) = \gamma(e_n) = 1,$$

l	$\alpha=\alpha'\cup M_{l+1}$	$\alpha-\!\circ f^\bullet$	$\varphi[\alpha]=$ $(\alpha-\!\circ f^\bullet)-\!\circ h^\bullet$	$\varphi[\varphi[\alpha]]$	$(c_i\varphi[\alpha])^\perp$	$(c_j\varphi[\varphi[\alpha]])^\perp$
0	M	$f^\bullet$	M_1	M_1	M_2	M_2
1	$\{f_{n-1}\}\cup M_2$	$\{h_n\}\cup M_2$	$\{f_n\}\cup M_2$	$\{\mathbf{f_{n+1}}\}\cup\mathbf{M_2}$	$\{g_q\|(i,n,q)$ is accepted$\}\cup M_2$	$\{g_q\|(j,n+1,q)$ is accepted$\}\cup M_2$
	$\{h_n\}\cup M_2$	$\{f_{n-1}\}\cup M_2$	$\{h_{n-1}\}\cup M_2$	$\{h_{n-2}\}\cup M_2$	M_2	M_2
	$otherwise$	M_2	M_1	M_1	M_2	M_2
2	$\alpha\subseteq f^\bullet$	M_0	$h^\bullet$	M_2	M_1	M_1
	$otherwise$	M_1	M_2	M_2	M_1	M_1
$OTHERWISE$		M_0	$h^\bullet$	M_2	M_1	M_1

Table 4. $\varphi[\alpha]$, $\varphi[\varphi[\alpha]]$, $(c_i\varphi[\alpha])^\perp$, $(c_j\varphi[\varphi[\alpha]])^\perp$ for $\alpha=\alpha'\cup M_{l+1}$.

and $\gamma(xy)=\gamma(x)+\gamma(y)$. As in lemma 4, let M_n be a prototype of the set $\{k|k\ge n\}$.

$\perp$ is defined in the following way:

$$\perp'=\{c_if_pg_q|(i,p,q)\text{ is accepted by the machine}\}\cup$$
$$\{\overline{f}f_{n-1}h_n|n\in\mathbf{Z}\}\cup\{\overline{h}h_nf_n|n\in\mathbf{Z}\}\cup$$
$$\{\overline{g}g_{n-1}e_n|n\in\mathbf{Z}\}\cup\{\overline{e}e_ng_n|n\in\mathbf{Z}\},$$
$$\perp=\perp'\cup M_4.$$

K is defined as $\{1\}$. It is clear that $\perp$ is an ideal, and so $(M,\perp,K)$ is an affine phase space. Note, that $\perp'\subseteq S_3$. So, we can use corollary 5, and define the interpretation of literals as follows:

$$f^\bullet=\{\overline{f}\}^\perp=\{f_{n-1}h_n|n\in N\}\cup M_3,\quad h^\bullet=\{\overline{h}\}^\perp=\{h_nf_n|n\in N\}\cup M_3,$$
$$g^\bullet=\{\overline{g}\}^\perp=\{g_{n-1}e_n|n\in N\}\cup M_3,\quad e^\bullet=\{\overline{e}\}^\perp=\{e_ng_n|n\in N\}\cup M_3,$$
$$a^\bullet=\{\overline{f}h_0\}^\perp=\{f_{-1}\}\cup M_2,\qquad\qquad b^\bullet=\{\overline{g}e_0\}^\perp=\{g_{-1}\}\cup M_2,$$
$$c_i^\bullet=\{c_i\}^{\perp\perp}.$$

Lemma 11. *For the interpretation defined above the following holds*

$$\varphi_n^\bullet=\{f_n\}\cup M_2,\qquad\psi_n^\bullet=\{g_n\}\cup M_2.$$

Lemma 12. *M satisfies all formulas T_k encoding our transitions.*

Proof of 11 and 12. Using lemma 4, we can calculate $\varphi[\alpha]$, $\varphi[\varphi[\alpha]]$, $(c_i\varphi[\alpha])^\perp$, $(c_j\varphi[\varphi[\alpha]])^\perp$ for any set $\alpha \subseteq M$, where $\alpha = \alpha' \cup M_{l+1}$ and $\alpha' \subseteq M_l$ (see Table 4).

We prove that $\varphi_n^\bullet = \{f_n\} \cup M_2$ by induction on n. For $n = -1$ we have that $\varphi_{-1}^\bullet = a^\bullet = \{f_{-1}\} \cup M_2$ by definition. The induction step holds because of $\varphi[\alpha] = \{f_n\} \cup M_2$, for $\alpha = \{f_{n-1}\} \cup M_2$. The second assertion of lemma 11 can be proved similarly.

One can see from Table 4 that $(c_i\varphi[\alpha])^\perp = (c_j\varphi[\varphi[\alpha]])^\perp$ for any $\alpha \neq \{f_{n-1}\} \cup M_2$. Moreover, if the increment transition $(i, p, q) \to (j, p+1, q)$ is accepted, then $(c_i\varphi[\alpha])^\perp \supseteq (c_j\varphi[\varphi[\alpha]])^\perp$ for $\alpha = \{f_{n-1}\} \cup M_2$. Hence, the formula

$$\forall \alpha, \beta(c_i \otimes \varphi[\alpha] \otimes \beta \multimap c_j \otimes \varphi[\varphi[\alpha]] \otimes \beta)$$

is satisfied. By analogy, if the decrement transition $(i, p, q) \to (j, p-1, q)$ is accepted, then $(c_j\varphi[\alpha])^\perp \subseteq (c_i\varphi[\varphi[\alpha]])^\perp$ for $\alpha = \{f_{n-1}\} \cup M_2$. So, the formula

$$\forall \alpha, \beta(c_i \otimes \varphi[\varphi[\alpha]] \otimes \beta \multimap c_j \otimes \varphi[\alpha] \otimes \beta)$$

is satisfied. Finally, if the test-for-zero transition $(i, 0, q) \to (k, 0, q)$ is accepted, then $(c_k\varphi[a^\bullet])^\perp \subseteq (c_i\varphi[a^\bullet])^\perp$. So the formula

$$\forall \beta(c_i \otimes \varphi[a] \otimes \beta \multimap c_k \otimes \varphi[a] \otimes \beta)$$

is satisfied. Clearly, all formulas encoding transitions for the second counter are satisfied as well. ∎

Lemma 13. *M satisfies the formula C.*

Proof. Let X be a fact. If $1 \in X$, then $X = M$ and $X^4 = X^5$. Otherwise, $XXXX \subseteq M_4 \subseteq \perp$, and $X^4 = (XXXX)^{\perp\perp} = \perp$, because $\perp$ is the least fact. In the same way $X^5 = \perp$. In both cases we have $X^4 = X^5$. Therefore, C is satisfied. ∎

Lemma 14. *M satisfies the formulas W, W' and U.*

Proof. The formulas W, W' and U are derivable in LLW2. Hence, they are satisfied. ∎

Now let us prove (iv)→(i). It follows from (iv) and lemmas 12, 13 and 14 that M satisfies the implication

$$c_i \otimes \varphi_p \otimes \psi_q \multimap c_0 \otimes \varphi_0 \otimes \psi_0.$$

By definition $c_0 f_0 g_0 \in \perp$. Hence, $(c_0 \otimes \varphi_0 \otimes \psi_0)^\bullet = \{c_0 f_0 g_0\}^{\perp\perp} = \perp$, because $\perp$ is the least fact. Therefore, we have

$$c_i f_p g_q \in (c_i \otimes \varphi_p \otimes \psi_q)^\bullet \subseteq (c_0 \otimes \varphi_0 \otimes \psi_0)^\bullet = \perp.$$

So, (i) holds. ∎

Acknowledgments

I am grateful to Sergei Artemov for his supervision. I would like to thank Yves Lafont for his helpful remarks on a draft of this paper. I am also indebted to Max Kanovich for his inspiring introduction to the problem and discussions on it.

References

[Abr91] V.M. Abrusci. Phase semantics and sequent calculus for pure noncommutative classical linear propositional logic. Jornal of Symbolic logic 56, 1403–1451. 1991.

[Gir] J.-Y. Girard. Linear Logic. *Theoretical Computer Science*, 50, 1–102, 1987.

[Kan1] M.I. Kanovich. The Direct Simulation of Minsky machine in Linear logic. To appear in *Advances in Linear Logic,* edited by J.-Y. Girard, Y. Lafont & L. Regnier, London Mathematical Society Lecture Note Series, Cambridge University Press. 1995.

[Kan2] M.I. Kanovich. Second order Lambek is undecidable. Message to LL list. 3 July 1995.

[Kop] A.P. Kopylov. Decidability of Linear Affine Logic. 10-th Annual IEEE Symposium on Logic in Computer Science, San Diego, California, IEEE Computer Society Press. 1995.

[Laf1] Y. Lafont. The Undecidability of Second Order Linear Logic without Exponentials. To apear in Jornal of Symbolic Logic. Available by anonymous ftp from lmd.univ-mrs.fr as pub/lafont/undecid.[dvi,ps].Z. 1995.

[Laf2] Y. Lafont. The Finite Model Property for Various Fragments of Linear Logic. Submited to publication. Available by anonymous ftp from lmd.univ-mrs.fr as pub/lafont/model.[dvi,ps].Z. 1995.

[LS] Y. Lafont, A. Scedrov. The Undecidability of Second Order Multiplicative Linear Logic. To apear in Information and Computation. Available by anonymous ftp from lmd.univ-mrs.fr as pub/lafont/mll2.[dvi,ps].Z. 1995.

[LSS] P. Lincoln, A. Scedrov, N. Shankar. Decision Problem for Second Order Linear Logic. 10-th Annual IEEE Symposium on Logic in Computer Science, San Diego, California, IEEE Computer Society Press. 1995.

[Lk] J. Lambek. How to program an infinite abacus. Canadian Math. Bulletin 4. 295–302. 1961.

[M] M. Minsky. Recursive unsolvability of Post's problem of 'tag' and other topics in the theory of Turing machines. *Annals of Mathematics*, 74:3:437-455, 1961.

[T] A.S. Troelstra. Lectures on Linear Logic. *CSLI lecture notes; no.29*, 1992

Operational Logic of Proofs with Functionality Condition on Proof Predicate

Vladimir N. Krupski *

Moscow State University,
Faculty of Mechanics and Mathematics,
Moscow 119899, Russia
e-mail: krupski@lpcs.math.msu.ru

Abstract. The extended operational (term-labeled modal) language is used to give the axiomatic description for functional proof predicate supplied with effective operations on proofs induced by modus ponens and necessitation rules. An additional operation is involved which restores a statement from its proof. The arithmetical completeness and decidability theorems are proved. The cut-elimination property for Gentzen style reformulation of corresponding logic is established.

1 Introduction

The *Operational Modal Logic of Proofs* $\mathcal{LP}$ was introduced by S. Artemov in [1]. It describes the nondeterministic (nonfunctional) version of proof predicate "x proofs F" together with operations on proofs induced by propositional PA-sound inference rules. $\mathcal{LP}$ provides a provability interpretation for the modal logic $S4$, which was intended informal semantics for $S4$ in Gödel's paper [2]:

$$S4 \hookrightarrow \mathcal{LP} \hookrightarrow PA.$$

In the $\mathcal{LP}$-language the $S4$ -modality $\Box(\cdot)$ is split into an infinite set of labeled modalities $[t](\cdot)$ where t is a term combined from proof variables and axiom constants (i.e. notations for the trivial proofs of axioms) using a finite set of functional symbols which denote some particular computable operations on proofs. Thus a term t is some kind of program which computes a proof given the proofs denoted by its atomic components. The constructor $[\cdot](\cdot)$ in $\mathcal{LP}$-language corresponds to the arithmetical proof predicate $Prf(\cdot, \cdot)$. $\mathcal{LP}$ is sound and complete with respect to arithmetical proof interpretations and is the exact realization of $S4$, i.e. $S4 \vdash F$ iff there is a way $(\cdot)^r$ to label all boxes in F by terms for which $\mathcal{LP} \vdash F^r$.

The operational language provides a PA-sound formalization for the *verification property* of proofs:

"if F has a proof then F is valid".

---

* Partially supported by Grant INTAS-RFBR No. 95-0095 and by Grant RFBR No.96-01-01470.

In [2] this principle was chosen as the general property of provability notion, but its formalization in the uni-modal language via the reflexivity axiom $\Box F \to F$ turned out to be incorrect with respect to the straightforward arithmetical interpretation of $\Box F$ as "F is provable in PA", so it is inconsistent with the formal provability logic (see [3]). In $\mathcal{LP}$-language the verification property of proofs is formalized by the q-reflexivity axiom

$$[t]F \to F,$$

which is correct with respect to arithmetical proof interpretations of operational language.

There is another common feature of any "real" proof system (Gilbert style, Gentzen style, natural derivations, ets.): the proofs are structured in such a way that it is possible to extract from a proof p the formula which is proved by p, so "real" proof predicates are functional. There is no way to express the functionality property of proof predicates in the propositional uni-modal language, but the operational labeled modal language (like $\mathcal{LP}$-language) is already sufficient. In [4], [5] this problem was considered for more primitive labeled-modal language without any operations on proofs. We extend this approach and define the *Operational Logic of Functional Proof Predicates* ($\mathcal{FLP}$) which combines the reflexivity axioms with the functionality condition on proof predicate in operational language. The involved operations on proofs correspond to *modus ponens* and *necessitation* rules. We prove that $\mathcal{FLP}$ is *decidable, sound and complete* with respect to arithmetical proof interpretation based on functional proof predicates.

Note that $\mathcal{LP}$ describes the *nondeterministic* versions of proof predicates, so $\mathcal{FLP}$ is incompatible with $\mathcal{LP}$. The logics like $\mathcal{FLP}$ also cannot be a realization of any normal uni-modal logic: a formula of the form

$$\neg[a]F \vee \neg[a](F \to F)$$

is valid for any functional interpretation of $[\cdot](\cdot)$, but its uni-modal variant

$$\neg\Box F \vee \neg\Box(F \to F)$$

is inconsistent with any normal logic when F is a tautology. So $\mathcal{FLP}$ provides *the most abstract* axiomatic description of functional proof predicates.

When a deductive system with functional proof predicate is considered the particular proofs may be used as special names for the formulas they prove. We extend the operational labeled modal language with ι-terms of the form $v_t = \iota F \cdot [t]F$ (denotes the formula which is proved by t) and study the logic of functional proof predicates in the extended language. The Hilbert style calculus $\mathcal{FLP}^{ext}$ is introduced as axiomatic description for this logic. We prove it to be *sound* and *complete* with respect to arithmetical interpretations and supply it with *adequate Gentzen style cut-free formalization* which gives the *decidability* of $\mathcal{FLP}^{ext}$.

2 Operational logic of functional proof predicate

The language of $\mathcal{FLP}$ contains boolean constants $\top, \bot$, sentence letters $S_1, S_2, \ldots$, proof letters $p_1, p_2, \ldots$, boolean connective $\to$ (other connectives denote the corresponding combinations of $\to, \bot$), functional symbols $!(\cdot)$ and $\times(\cdot, \cdot)$ (acting on proofs, the result is a proof too) and an operator symbol $[\cdot](\cdot)$. The following notation is used:

$$ts =_{def} \times(s,t), \quad !t =_{def} !(t), \quad [t]F =_{def} [t](F).$$

Let Tm be the set of all terms in the signature $\{!, \times\}$ with the variables from $\{p_1, p_2, \ldots\}$. The set Fm of formulas is defined as follows:

$$\{\top, \bot, S_1, S_2, \ldots\} \subseteq Fm,$$
$$F, G \in Fm \Rightarrow (F \to G) \in Fm,$$
$$t \in Tm, \ F \in Fm \Rightarrow [t]F \in Fm.$$

In order to give a formula $[t]F$ the meaning "t computes a proof for F" and understand $\times, !$ as operations on proofs induced by modus ponens and necessitation rules we borrow the following axioms from $\mathcal{LP}$ (see [1]) :

(A0) Tautologies in the language of $\mathcal{FLP}$
(A1) $[t]F \to F$
(A2) $[t](F \to G) \to ([s]F \to [ts]G)$
(A3) $[t]F \to [!t][t]F$

We wish to express in the $\mathcal{FLP}$ -axioms the functionality condition: *"each proof determines the formula which it proves uniquely"*. Our approach comes from [4], [5] where the similar problem is considered for more primitive language without any operations over proofs.

Any formula of the form $[t_1]F_1 \wedge \ldots \wedge [t_n]F_n$ has an associated "conditional" unification problem:

$$t_i = t_j \Rightarrow F_i = F_j, \quad i, j = 1, \ldots, n. \tag{1}$$

A solution, or a *unifier* of (1) is a substitution σ with the property $F_i\sigma = F_j\sigma$ when $t_i\sigma = t_j\sigma, \quad i, j = 1, \ldots, n$. We shall write

$$A = B \ mod \ S,$$

where $A, B \in Tm \cup Fm$ and S is a system (1) if $A\sigma = B\sigma$ for any unifier σ of S.

Lemma 1. *The unifiability problem for the systems of the form (1) is decidable. There is an algorithm which computes a unifier σ for any unifiable system S of the form (1). The resulting substitution σ has the following properties:*

- *σ is most general (m.g.u.): any unifier θ of (1) has the form $\theta = \sigma\lambda$ for some substitution λ.*
- *σ is idempotent ($\sigma^2 = \sigma$) and conservative ($Var(\sigma) \subseteq Var(S)$).*

Corollary 2. *The relation $A = B \bmod S$ is decidable.*

Definition 3. *The operational logic of functional (deterministic) proof predicate* is the calculus $\mathcal{FLP}$ with axioms (A0),(A1),(A2),(A3) together with *the unification axiom*

(A4) $[t_1]F_1 \wedge \ldots \wedge [t_n]F_n \rightarrow (A \leftrightarrow B)$ if $A = B \bmod (S)$.

The only rule is *modus ponens*.

We consider the Peano Arithmetic $\mathcal{PA}$ with ι-terms included into term system of the language (cf.[7],[8]). A ι-term is called *recursive* if it has the form $\mu z.\varphi(z)$ where $\varphi(z) \wedge (\forall x < z)\neg\varphi(x)$ is provably Σ_1-formula. A recursive term $\mu z.\varphi(z)$ is called *closed* if the formula $\exists z \varphi(z)$ is closed and provable. The recursive terms are used as notations for partial recursive functions: $\mu z.\varphi(x, z)$ denotes the function f where

$$Dom(f) = \{x \mid \exists z \varphi(x, z)\},$$

$$x \in Dom(f) \Rightarrow (\ \varphi(x, f(x)) \wedge (\forall y < f(x))\neg\varphi(x, y) \text{ is valid}).$$

If $\varphi(x, z)$ is provably Σ_1-formula and $\mathcal{PA} \vdash \varphi(x, z_1) \wedge \varphi(x, z_2) \rightarrow z_1 = z_2$ then the term $\mu z.\varphi(x, z)$ is recursive, so Kleene's normal form gives in fact the recursive term representation for any partial recursive function.

For a recursive term $f = \mu z.\varphi(z)$ and arithmetical formula $F(x)$ we define

$$F(f) = \exists z(\text{``}z = f\text{''} \wedge F(z)) \quad \text{where} \quad \text{``}z = f\text{''} = (\varphi(z) \wedge (\forall y < z)\neg\varphi(y)).$$

Note that the formulas $z = f$ and "$z = f$" are equivalent and if $\mathcal{PA} \vdash \bar{k} = f$ for some $k \in \omega$ then $\mathcal{PA} \vdash F(\bar{k}) \leftrightarrow F(f)$.

Lemma 4. *1. If f is closed recursive term then $\mathcal{PA} \vdash \bar{k} = f$ for some $k \in \omega$.*
2. If $\mu z.\varphi(w, z)$, $f = \mu u.\psi(u)$ are recursive terms and z is not free in $\psi(u)$ then $\mu z.\varphi(f, z)$ is recursive term too.
3. If $F(x)$ is provably Σ_1-formula (Δ_1-formula) and f is closed recursive term then $F(f)$ is provably Σ_1-formula (Δ_1-formula) too.

A *functional proof predicate* is a Δ_1-formula $Prf(x, y)$ with the properties:

1. $\mathcal{PA} \vdash \varphi \Leftrightarrow$ for some $n \in \omega$ $\mathcal{PA} \vdash Prf(\bar{n}, \lceil \varphi \rceil)$,
2. Prf is functional, i.e. $Prf(x, y) \wedge Prf(x, z) \rightarrow y = z$,
3. the function

$$T(x) = \text{ the canonical index of the finite set } \{y \mid Prf(x, y)\}$$

is total recursive. (Note that the set $D_{T(x)} = \{y \mid Prf(x, y)\}$ is either empty or contains a single element and it is sufficient to know its canonical index $T(x)$ to provide an effective checking procedure for emptiness [9] .)

Any functional proof predicate already has the recursive terms which implement the operations $!$, $\times$. They represent the recursive functions $c(x)$, $m(x, y)$.

$$c(x) \simeq \begin{cases} \mu l.Prf(l, \lceil Prf(f, y) \rceil) \text{, if } x = \lceil f \rceil \text{ for some closed recursive term} \\ \qquad\qquad\qquad\quad f \text{ and } D_{T(f)} = \{y\}, \\ f \qquad\qquad\qquad\text{, if } x = \lceil f \rceil \text{ for some closed recursive term} \\ \qquad\qquad\qquad\quad f \text{ and } D_{T(f)} = \emptyset, \\ \text{undefined} \qquad\quad\text{, otherwise} \end{cases}$$

$$m(x, y) = \begin{cases} \mu l.Prf(l, \lceil \psi \rceil) \text{, if } D_{T(x)} = \{\lceil \varphi \to \psi \rceil\} \text{ and } D_{T(y)} = \{\lceil \varphi \rceil\} \\ \qquad\qquad\qquad \text{for some statements } \varphi, \psi, \\ 0 \qquad\qquad\quad\text{, otherwise.} \end{cases}$$

For closed recursive terms $f = \mu z.F(z)$, $g = \mu z.G(z)$ and arithmetical statements φ, ψ the following holds:

$$\begin{aligned} &c(\lceil f \rceil) \text{ is defined (and so it is a closed recursive term),} \\ &\mathcal{PA} \vdash Prf(f, \lceil \varphi \rceil) \to Prf(c(\lceil f \rceil), \lceil Prf(f, \lceil \varphi \rceil) \rceil), \end{aligned} \tag{2}$$

$$\begin{aligned} &m \text{ is total,} \\ &\mathcal{PA} \vdash Prf(f, \lceil \varphi \to \psi \rceil) \to (Prf(g, \lceil \varphi \rceil) \to Prf(m(f, g), \lceil \psi \rceil). \end{aligned} \tag{3}$$

So, c is a search procedure aimed on the search of a proof for statements of the form $Prf(f, \lceil \varphi \rceil)$ and m is the operation on proofs induced by the rule *modus ponens*. An example of functional proof predicate is the standard Gödel proof predicate.

Definition 5. A *functional proof interpretation* is $* = \langle Prf, c, m, (\cdot)^* \rangle$ where Prf is a functional proof predicate, $c = \mu z.C(x, z)$ is a recursive term for a partial recursive function which satisfies (2), $m = \mu z.M(x, y, z)$ is a recursive term for a total recursive function which satisfies (3) and $(\cdot)^*$ is an evaluation of sentence letters by arithmetical statements and proof letters by closed provably recursive terms. We extend this evaluations to all terms and formulas of $\mathcal{FLP}$-language: $\top^* = (0 = 0)$, $\perp^* = (0 = 1)$, $*$ commutes with boolean connectives, $(t \times s)^* = m(t^*, s^*)$, $(!t)^* = c(\lceil t^* \rceil)$ and $([t]F)^*$ is $Prf(t^*, \lceil F^* \rceil)$.

Remark. Any functional proof interpretation $* = \langle Prf, c, m, (\cdot)^* \rangle$ translates the axioms (A0) – (A4) into arithmetical formulas which are provable in $\mathcal{PA}$. Indeed, by Lemma 4 for any $F \in Fm$ its translation F^* is closed Δ_1-formula. But closed Δ_1-formulas are $\mathcal{PA}$-provable iff they are valid, so it is sufficient to prove that the axioms are translated into valid formulas. For (A0), (A2) and (A3) schemes this fact follows immediately from the Definition 5. For (A1) axiom its translation A^* is $Prf(t^*, \lceil F^* \rceil) \to F^*$.

$$\begin{aligned} Prf(t^*, \lceil F^* \rceil) \text{ is not valid} &\Longrightarrow A^* \text{ is valid;} \\ Prf(t^*, \lceil F^* \rceil) \text{ is valid} &\Longrightarrow Prf(\bar{k}, \lceil F^* \rceil) \text{ is valid for } k \text{ being} \\ &\qquad\quad \text{the value of } t^*, \\ &\Longrightarrow \mathcal{PA} \vdash F^*, \\ &\Longrightarrow F^* \text{ is valid,} \\ &\Longrightarrow A^* \text{ is valid.} \end{aligned}$$

Consider the unification axiom (A4). It has the form $G \to (A \leftrightarrow B)$. If the formula

$$G^* = ([t_1]F_1 \wedge \ldots \wedge [t_n]F_n)^* = Prf(t_1^*, \lceil F_1^* \rceil) \wedge \ldots \wedge Prf(t_n^*, \lceil F_n^* \rceil)$$

is not valid then $(G \to C)^*$ is valid for any $C \in Fm$. Let G^* be valid. Consider the equivalence relation $\sim$ on $Tm \cup Fm$:

$$t \sim s \Leftrightarrow \quad t^* \text{ and } s^* \text{ coincide up to renaming of bound variables.}$$

The relation $\sim$ has the following properties:

1. $\top \not\sim \bot$;
2. $f(u_1, \ldots, u_k) \sim f(v_1, \ldots, v_k) \Leftrightarrow \bigwedge_i (u_i \sim v_i)$ where $f \in \{ \to, !, \times, [\cdot](\cdot) \}$;
3. $f(u_1, \ldots, u_k) \not\sim g(v_1, \ldots, v_l)$ for $f \neq g$;
4. $x \not\sim t(x)$ where x is a variable and a term or a formula $t(x) \in Tm \cup Fm$ contains x but is different from it;
5. $t_i \sim t_j \Rightarrow F_i \sim F_j, \quad i, j = 1, \ldots, n$.

For any equivalence relation satisfying 1.– 5. there exists a substitution σ which unifies (1) and

$$s\sigma = t\sigma \Rightarrow s \sim t$$

holds (σ can be obtained by a suitable modification of Algorithm A from [6]). As $A = B \, mod \, (S)$ where S is the system (1) we have $A\sigma = B\sigma$ and $A \sim B$. Thus $A^* \leftrightarrow B^*$ is valid and $(G \to (A \leftrightarrow B))^*$ is valid too.

Theorem 6. *(Arithmetical completeness of $\mathcal{FLP}$)*

$$\mathcal{FLP} \vdash F \Leftrightarrow \mathcal{PA} \vdash F^* \text{ for every functional proof interpretation } *$$
$$\Leftrightarrow F^* \text{ is valid for every functional proof interpretation } * .$$

Theorem 7. *The logic $\mathcal{FLP}$ is decidable.*

Remark. We extract the completeness proof and the decision algorithm from the analysis of a special nondeterministic saturation procedure based on $\mathcal{FLP}$. A computation of the saturation procedure starts from a formula F and may terminate or not. If every computation terminates then $\mathcal{FLP} \vdash F$. An infinite computation defines an arithmetical functional proof interpretation $*$ for which F^* is not valid. This proves the completeness theorem. We also prove that it is sufficient to make a finite number of steps in order to understand that the computation is infinite. It gives the decision procedure because the branching is binary.

3   Language extension

Given a functional proof predicate the particular proofs may be used as special names for the formulas they proof. This is a way to enrich a propositional labeled modal language with some weak form of quantification (see [10], [11]). We add to the $\mathcal{FLP}$-language new term-indexed sentence variables

$$v_t, \quad t \in Tm$$

denoting the formulas which are proved by corresponding t's. These variables are dependent: *if t and s denote the same proof then v_t and v_s must denote the same formula.* (Note that the converse in general is not true, so in the presence of the unification axioms we cannot express the same notion via an ordinary functional symbol $v(t)$.) Syntactically the variables of the form v_t are treated in the same way as the sentence letters S_i. Let Fm^{ext} be the set of all formulas of the extended language.

Definition 8. A functional proof interpretation for the extended language is $* = \langle Prf, c, m, l, (\cdot)^* \rangle$ where $\langle Prf, c, m, (\cdot)^* \rangle$ is a functional proof interpretation of the basic $\mathcal{FLP}$ -language and the interpretation for v_t is given by a total recursive function $l(x)$ which maps natural numbers into arithmetical statements and satisfies the condition:

$$\text{if } Prf(n, \lceil \varphi \rceil) \text{ for some statement } \varphi \text{ then } l(n) = \varphi.$$

(such a function does exist for any functional proof predicate);

$$(v_t)^* = l(t^*).$$

Examples. Here are the examples of predicates which can be expressed in the extended operational language:

$$[t]v_t \qquad \Leftrightarrow \text{ "t is a (correct) proof"};$$
$$[t]v_s \wedge [s]v_t \qquad \Leftrightarrow \text{ "t and s prove the same formula"};$$
$$[t](v_s \rightarrow v_{ts}) \wedge [s]v_s \Leftrightarrow \text{ "t and s prove the premises for \textit{modus ponens}"}.$$

The variants of the verification axioms (A1), (A2), (A3) in the extended language are

(A1') $[t]v_t \rightarrow v_t,$

(A2') $[t](v_s \rightarrow Y) \wedge [s]v_s \leftrightarrow [ts]Y,$

(A3') $[t]v_t \rightarrow [!t][t]v_t.$

The schemes (A1), (A3) are rewritten in equivalent forms without metavariables over formulas. The scheme (A2') describes the operation $\times$ as a *strict* implementation of the *modus ponens* rule. The word "strict" means that $\times$ cannot give a correct proof if its operands are not the proofs of the premises for the *modus ponens* rule. This refinement cannot be expressed in the basic $\mathcal{FLP}$ -language, so (A2') is stronger then (A2). It corresponds to the notion of strict functional proof interpretation.

Definition 9. A functional proof interpretation $* = \langle Prf, c, m, l, (\cdot)^* \rangle$ is called strict if the following condition holds for every $x, y \in \omega$

$$Prf(m(x,y), k) \Rightarrow k = \lceil \psi \rceil, \; Prf(x, \lceil \varphi \to \psi \rceil) \text{ and } Prf(y, \lceil \varphi \rceil)$$
$$\text{for some statements } \varphi, \psi.$$

The unification axioms need the special treatment. With a formula $[t_1]F_1 \wedge \ldots \wedge [t_n]F_n$ from Fm^{ext} we associate an *infinite* unification problem:

$$
\begin{aligned}
v_{t_i} &= F_i, & i &= 1, \ldots, n, \\
t = s &\Rightarrow v_t = v_s, & t, s &\in Tm.
\end{aligned}
\tag{4}
$$

Now in the definition of the relation "$A = B \, mod \, (S)$" where S is the system (4) we have to consider *infinite* unifiers for (4), i.e. the infinite mappings of Tm into Tm and Fm^{ext} into Fm^{ext} which commute with $\to$, $\times$, $!$, $[\cdot](\cdot)$, do not change $\top, \bot$, and satisfy each (simple or conditional) equality from (4). (Note that there is no difference in the treatment of term-indexed variables v_t and sentence letters S_i.)

The unification axioms for the extended operational language are all the formulas of the form

(A4′) $[t_1]F_1 \wedge \ldots \wedge [t_n]F_n \to (A \leftrightarrow B)$ where $A = B \, (S)$ and S is the infinite unification problem (4).

The set of unification axioms for the extended language is decidable. Indeed, the infinite unification problem (4) can be reduced to a finite one (this reduction was proposed by I. Tandetnik, see his paper in this volume). Let $M \subseteq Tm \cup Fm^{ext}$ be a finite set with the following *closure properties*: (i) $F_i, v_{t_i} \in M$, $i = 1, \ldots, n$; (ii) if $A \in M$ then all subterms and subformulas of A belong to M; (iii) if $v_t \in M$ then $t \in M$. Consider a finite unification problem

$$
\begin{aligned}
v_{t_i} &= F_i, & i &= 1, \ldots, n, \\
t = s &\Rightarrow v_t = v_s, & v_t, v_s &\in M.
\end{aligned}
\tag{5}
$$

Lemma 10. *The problem (4) is unifiable iff (5) is. If a finite substitution σ is an idempotent conservative m.g.u. for (5) then the infinite substitution*

$$
x\bar{\sigma} := \begin{cases}
x\sigma, & \text{if } x \text{ is a sentence or proof letter or } x = v_t \in Dom(\sigma), \\
v_s\sigma, & \text{if } x = v_t \notin Dom(\sigma) \text{ and } t\sigma = s\sigma \text{ for some } v_s \in Dom(\sigma), \\
v_{t\sigma}, & \text{if } x = v_t \notin Dom(\sigma) \text{ and } t\sigma \neq s\sigma \text{ for every } v_s \in Dom(\sigma)
\end{cases}
$$

is an idempotent m.g.u. for (4).

Corollary 11. *The relation $A = B \, mod \, (S)$ where $A.B \in Tm \cup Fm^{ext}$ and S has the form (4) is decidable.*

Let (A0′) be the set of all tautologies in the extended language. We define $\mathcal{FLP}^{ext}$ – the operational logic of functional proof predicate in the extended language:

Axioms: (A0′), (A1′), (A2′), (A3′), (A4′).

Rule: *modus ponens.*

Theorem 12. *(Arithmetical completeness of $\mathcal{FLP}^{ext}$)*

$\mathcal{FLP}^{ext} \vdash F \Leftrightarrow \mathcal{PA} \vdash F^*$ for every strict functional proof interpretation $*$
$\qquad\qquad\qquad \Leftrightarrow F^*$ is valid for every strict functional proof interpretation $*$.

Remark. A similar completeness theorem can be proved for the calculus $\mathcal{FLP}_0^{ext}$ – the variant of $\mathcal{FLP}^{ext}$ with the axiom scheme (A2′) replaced by

$$[t](v_s \to Y) \wedge [s]v_s \to [ts]Y$$

which is equivalent form of (A2). $\mathcal{FLP}_0^{ext}$ is complete with respect to all functional proof interpretations of the extended language. It seems that $\mathcal{FLP}^{ext}$ describes the difference between the primitive "syntactical" operation $\times$ and more complex search procedure ! better than $\mathcal{FLP}_0^{ext}$.

$\mathcal{FLP}^{ext}$ has an adequate Gentzen style cut-free reformulation, which provides more standard decision algorithm for $\mathcal{FLP}^{ext}$. The corresponding saturation procedure always terminates, so the decision algorithm may simply construct the finite saturation tree and then check whether all the leafs are labeled with axioms.

$\mathcal{FLP}_G^{ext}$ is the following sequent calculus:

Axioms:

- $\Gamma \Longrightarrow \Delta$ such that $\Gamma \cap \Delta \neq \emptyset$ or $\bot \in \Gamma$ or $\top \in \Delta$.
- $\Gamma \Longrightarrow \Delta$ such that $\Xi \subseteq \Gamma$, where $\Xi = \{[t_i]F_i \mid i = 1, \ldots, n\}$ and the system (4) for Ξ is not unifiable.

Rules:

The classical rules for $\to$ and structural rules together with the cut-rule;

$$\frac{X, \Gamma \Longrightarrow \Delta}{[t]X, \Gamma \Longrightarrow \Delta} \qquad\qquad\qquad \frac{\Gamma \Longrightarrow [t]X, \Delta}{\Gamma \Longrightarrow [!t][t]X, \Delta}$$

$$\frac{[t](v_s \to Y), [s]v_s, \Gamma \Longrightarrow \Delta}{[ts]Y, \Gamma \Longrightarrow \Delta} \qquad \frac{\Gamma \Longrightarrow [t](v_s \to Y), \Delta \quad \Gamma \Longrightarrow [s]v_s, \Delta}{\Gamma \Longrightarrow [ts]Y, \Delta}$$

$$\frac{\Xi, Y(x\bar{\sigma}), \Gamma \Longrightarrow \Delta}{\Xi, Y(x), \Gamma \Longrightarrow \Delta} \qquad\qquad \frac{\Xi, \Gamma \Longrightarrow Y(x\bar{\sigma}), \Delta}{\Xi, \Gamma \Longrightarrow Y(x), \Delta}$$

where $\Xi = \{[t_i]F_i \mid i = 1, \ldots, n\}$, a substitution $\bar{\sigma}$ is an idempotent m.g.u. of (4) for Ξ and $Y(x\bar{\sigma})$ is the result of the replacement of a chosen occurrence of a variable x in $Y(x)$ (it may be in the index of some variable v_t too) by $x\bar{\sigma}$.

In the last pair of rules any particular choice of $\bar{\sigma}$ is valid. For some technical reasons we specify it in the following *standard* way. Let $M \subseteq Tm \cup Fm^{ext}$ be the least set with the closure properties which contains Y and all the formulas from Ξ, Γ, Δ. We pick a (finite) idempotent conservative m.g.u. σ for (5) and set $\bar{\sigma}$ to be its standard modification from lemma 10. This setting makes the set of rules decidable because the list of all idempotent conservative most general unifiers for (5) is finite and can be effectively computed from (5).

$\mathcal{FLP}^{ext}_{G-}$ is the calculus obtained from $\mathcal{FLP}^{ext}_G$ by the replacement of the cut rule with the following rules which are admissible for $\mathcal{FLP}^{ext}_G$:

$$\frac{Y(v_t), \Gamma \Longrightarrow \Delta}{[t]X, Y(X), \Gamma \Longrightarrow \Delta}, \qquad \frac{\Gamma \Longrightarrow Y(v_t), \Delta}{[t]X, \Gamma \Longrightarrow Y(X), \Delta}$$

where $Y(X)$ is the result of the replacement of some occurrence of v_t in $Y(v_t)$ by X;

$$\frac{[ts]Y, \Gamma \Longrightarrow \Delta}{[t](X \to Y), [s]X, \Gamma \Longrightarrow \Delta} \quad \text{if } ts \text{ occurs in } X, Y, \Gamma \text{ or } \Delta \text{ (may be in index);}$$

$$\frac{[!t][t]X, \Gamma \Longrightarrow \Delta}{[t]X, \Gamma \Longrightarrow \Delta} \quad \text{if } !t \text{ occurs in } X, \Gamma \text{ or } \Delta \text{ (may be in index).}$$

Theorem 13. $\mathcal{FLP}^{ext}_G \vdash \Gamma \Longrightarrow \Delta$ *iff* $\mathcal{FLP}^{ext}_{G-} \vdash \Gamma \Longrightarrow \Delta$
iff $\mathcal{FLP}^{ext} \vdash \bigwedge \Gamma \to \bigvee \Delta$.

Theorem 14. *The logic* $\mathcal{FLP}^{ext}$ *is decidable.*

References

1. S. Artëmov, Operational Modal Logic, Tech. Rep. 95-29, Mathematical Sciences Institute, Cornell University, December 1995.
2. K. Gödel, Eine Interpretation des intuitionistischen Aussagenkalküls , *Ergebnisse Math. Colloq.*, Bd. 4 (1933). S. 39-40.
3. R.M. Solovay, Provability interpretations of modal logic, *Israel J. Math.*, 25 (1976), pp. 287-304.
4. S. Artëmov and T. Straßen, Functionality in the basic logic of proofs, Tech. Rep. IAM 92-004, Department for Computer Science, University of Bern, Switzerland, Jan. 1993.
5. S. Artëmov, Logic of proofs, *Annals of Pure and Applied Logic*, v. 67 (1994), pp. 29-59.
6. M.S. Paterson and M.N. Wegman, Linear unification, *J. Comput. System Sci.*, v.16 (2) (1978), pp. 158-167.
7. D. Hilbert and P. Bernays, Grundlagen der Mathematik, Springer, 1934-1939.
8. D. van Dalen, Logic and Structure, Springer-Verlag, 1994.
9. H. Rogers, Theory of recursive functions and effective computability, McGraw-Hill Book Company, New York, 1967.

10. S. Artëmov and V. Krupski, Referential data structures and labeled modal logic, *Lecture Notes in Computer Science*, v.813 (1994), pp. 23-33.
11. S. Artëmov and V. Krupski, Data storage interpretation of labeled modal logic, *Annals of Pure and Applied Logic* , v.78 (1996), pp. 57-71.

On Linear Ordering of Strongly Extensional Finitely-Branching Graphs and Non-well-founded Sets

Alexei Lisitsa and Vladimir Sazonov *

Program Systems Institute of Russian Academy of Sciences,
Pereslavl-Zalessky, 152140, Russia
e-mail: {sazonov,lisitsa}@logic.botik.ru

1 Introduction

Definability of a linear order in finite structures of some given class by logical means is an important issue of finite model theory and descriptive complexity theory. For example, first-order logic extended by least fixed point operator (FO + LFP) describes exactly PTIME-computability over finite *linear ordered* structures [10, 12, 23, 11] (cf. also [15, 7] where recursive finite functions and the successor operation are considered instead of FO + LFP and the linear order, respectively). If the structures considered are not linearly ordered in advance, but some order may be uniformly defined in FO + LFP then the same result holds for this class of structures, too.

We show in this paper that this is the case for the class of *strongly extensional (SE) finite graphs* (i.e. arbitrary finite graphs considered up to bisimulation equivalence relation). The vertices of such graphs serve as a faithful representation of hereditarily-finite non-well-founded sets (with finite transitive closure) which constitute a universe HFA. Actually, we define in FO+IFP (which is a version of FO + LFP) a linear ordering on arbitrary SE *finitely branching* graphs. These graphs may be infinite and so represent a more general class of hereditarily-finite non-well-founded sets HFA^∞ (possibly with infinite transitive closure). Non-well-founded sets are also called *hypersets* (cf. [1, 3]).

Our interest to these questions arose from a work on describing the complexity classes of computable set-theoretic operations by some bounded set theory (BST) languages [16]–[22]. So, definability in the language Δ_D considered in [19, 20] over HFA-sets was characterized in terms of definability in FO+LFP over finite graphs of the abovementioned kind. Now, having definable linear ordering, we answer affirmatively the question in op. cit. on coincidence of Δ_D-definability with PTIME-computability over HFA.

We propose two different definitions of linear ordering, one of which is an application of the method of A. Dawar, S. Lindell and S. Weinstein [4] to the case of SE finite graphs, and give a comparison of both approaches in terms of coherence of the orders on different graphs.

---

* Both authors are supported by RFBR (grant 96-01-01717) and by INTAS (grant 93-0972).

Also note, that these definability results may be naturally considered in the framework of a set-theoretic approach to "nested" and "circular" semi-structured Data Bases or to *Web-like* Data Bases (WDB), by using the Δ_D-language regarded as a *query language*. Cf. [18] and also more abstract and mathematically oriented paper [19] for "non-circular" and "circular" cases, respectively. A paper on the connection to the World-Wide Web and WDB using the main result of this work is in preparation.

Agreement. We will use abbreviations slo and slp-o for *strict linear order* and, respectively, *strict linear pre-order*.

2 Graphs and Sets, or Sets as Graphs

Remember, that *hereditarily-finite sets* are defined as finite sets whose elements are finite sets, etc., until the empty set or an *urelement* (i.e. an atomic object which is not a set) will be obtained. These sets constitute the universe HF. Analogously, it is defined the universe HC of *hereditarily-countable sets*. For simplicity, we will assume that there are no urelements involved in HF- and HC-sets, i.e. that all considered sets are *pure*.[2] It is implicit also that the universes $HF \subset HC$ are *well-founded*, i.e. that there are no infinite chains of sets satisfying

$$\ldots \in v_{i+1} \in v_i \in \ldots \in v_1 \in v_0. \tag{1}$$

We will represent HF- and HC-sets (and also sets of more general nature) by vertices of graphs g. Formally, a *graph* is defined here as arbitrary map of the form $g : |g| \to \mathcal{P}(|g|)$ with $|g| = \mathbf{dom}(g)$ being a set of *vertices*, and $\mathcal{P}(|g|)$ denoting the powerset of $|g|$. For any $u, v \in |g|$ we write $u \, \epsilon^g \, v$ or $g \models u \in v$ instead of $u \in gv$ and say that "u is an element of v in the sense of g". This is just a binary relation between vertices u and v which says that there exists an edge between u and v in a fixed direction you choose, say $u \leftarrow v$.

Depending on the cardinality of sets gv or $|g|$, there are three natural classes of graphs: *countably branching*, $\mathcal{CBG}$, *finitely branching*, $\mathcal{FBG}$, and just *finite* graphs, $\mathcal{FG}$. (Finite) *well-founded* graphs, $(\mathcal{F})\mathcal{WG}$ have no infinite chains of vertices like (1). $\mathcal{FWG}$ are just *finite acyclic graphs*, $\mathcal{FAG}$.

Let us define *denotational semantics* $[\![v]\!]_g$ (called also *decoration* [1] or *collapsing*, in the case of $\mathcal{WG}$ [2, 16, 18]) for the vertices v of any graph g. Let $[\![v]\!]_g$ be a *set of sets of sets, etc.* defined recursively by the identity

$$[\![v]\!]_g = \{[\![u]\!]_g | u \, \epsilon^g \, v\} \quad \text{i.e.} \quad \forall z(z \in [\![v]\!]_g \Leftrightarrow \exists u \, \epsilon^g \, v(z = [\![u]\!]_g)) \tag{2}$$

For example, if g contains exactly five edges $u \, \epsilon^g \, v$, $u \, \epsilon^g \, v'$, $u \, \epsilon^g \, w$, $v \, \epsilon^g \, w$ and $v' \, \epsilon^g \, w$ then $[\![u]\!]_g = \emptyset$, $[\![v]\!]_g = [\![v']\!]_g = \{\emptyset\}$, and $[\![w]\!]_g = \{\{\emptyset\}, \emptyset\}$. If g is any well-founded graph, then it is easy to "calculate" $[\![v]\!]_g$ so that, for g countably/finitely branching, $[\![v]\!]_g$ is an HC-, respectively, HF-set.

---

[2] The case of urelements or, more generally, of sets of *labelled* elements is important for application to databases; cf. the corresponding approach in [18].

However, in non-well-founded case, such as in the graph ○, there is a problem. We need sets like $\Omega = \{\Omega\}$. This leads to rather unusual *non-well-founded set theory* whose universe may contain sets serving as their own members, etc. [1, 3]. We confine ourselves to its subuniverse HCA of hereditarily-countable (or even hereditarily-finite) sets. Its main property, besides (no more than) countability of each set $x \in$ HCA, consists in the following "external" *anti-foundation axiom* (which is just a *finality* property in an appropriate category $CB\mathcal{G}$):

AFA: *For any $CB\mathcal{G}$ g there exists a unique denotational semantics (or decoration) map $[\![\cdot]\!]_g : |g| \to$ HCA satisfying the above identity (2).*

If g ranges only over $\mathcal{FG}$, respectively, over $\mathcal{FBG}$ then $[\![v]\!]_g$ will range over the corresponding subuniverses HFA $\subset$ HFA$^\infty$ of HCA.[3] HFA consists of (possibly non-well-founded) hereditarily-finite sets $x \in$ HCA whose *transitive closure*

$$TC(x) \rightleftharpoons \{z | z \in z_1 \in \ldots \in z_n \in x, \quad n \geq 0, \quad \text{and} \quad z_1, \ldots, z_n \in \text{HCA}\}$$

is finite. HFA$^\infty$ consists also of hereditarily-finite sets. However, their transitive closure may be infinite (just countable). Also note, that HFA is a countable universe, whereas HFA$^\infty$ is of continual cardinality [6, 13]. Evidently, $x \in y \in$ HFA implies $x \in$ HFA for all $x, y \in$ HCA and the same for HFA$^\infty$ in place of HFA. Also sets from the subuniverses HF $\subseteq$ HFA and HC $\subseteq$ HCA correspond to vertices of finite and, respectively, countable well-founded graphs. It can be proved in the line of [1] the following theorem and proposition.

Theorem 1. *There exists the unique, up to isomorphism, universe HCA consisting only of finite or countable sets and satisfying anti-foundation axiom (and analogously also for HFA$^\infty$ and HFA).*

□

Consider a binary relation $\approx^{g,g'} \subseteq |g| \times |g'|$

$$u \approx^{g,g'} u' \rightleftharpoons [\![u]\!]_g = [\![u']\!]_{g'}.$$

It can be redefined in terms of the graphs g, g' only, without any mentioning (the semantics $[\![\cdot]\!]$ in) a universe of sets like HCA, HFA or HFA$^\infty$.

A set of pairs $R \subseteq |g| \times |g'|$ is called *bisimulation relation* between g and g' if it satisfies the implication

$$uRu' \Rightarrow \forall v\, \epsilon^g\, u \exists v'\, \epsilon^{g'}\, u'(vRv') \& \forall v'\, \epsilon^{g'}\, u' \exists v\, \epsilon^g\, u(vRv'). \tag{3}$$

Proposition 2. *The largest bisimulation relation between g and g' coincides with $\approx^{g,g'}$. Moreover, $\Rightarrow$ in (3) may be replaced by $\Leftrightarrow$ for the case of $\approx^{g,g'}$.*

□

Note that in the *finitely branching* case $\approx^{g,g'}$ may be obtained as intersection[4]

$$\approx^{g,g'} = \bigcap_{i=0}^{\infty} \approx_i^{g,g'} \quad \text{where} \tag{4}$$

---

[3] Other authors are using denotations HF$_0$ for HF, HF$_{1/2}$ for HFA and HF$_1$ for HFA$^\infty$.
[4] Here i ranges over natural numbers. In general case we should consider *ordinals*. Then e.g. $\approx_\omega^{g,g'} \rightleftharpoons \bigcap_{i\in\omega} \approx_i^{g,g'}$, and in finitely branching case $\approx^{g,g'} = \approx_\omega^{g,g'}$.

$$u \approx_0^{g,g'} u' \rightleftharpoons \mathbf{true} \quad \text{and}$$

$$u \approx_{i+1}^{g,g'} u' \rightleftharpoons \forall v \ \epsilon^g \ u \exists v' \ \epsilon^{g'} \ u'(v \approx_i^{g,g'} v') \& \forall v' \ \epsilon^{g'} \ u' \exists v \ \epsilon^g \ u(v \approx_i^{g,g'} v') \tag{5}$$

so that $\approx_0^{g,g'} \supseteq \approx_1^{g,g'} \supseteq \ldots \supseteq \approx^{g,g'}$. For the finite graphs this sequence evidently stabilizes on some *finite* step i: $\approx_i^{g,g'} = \approx^{g,g'}$, so that $\not\approx^{g,g'}$ and therefore $\approx^{g,g'}$ are definable in FO + LFP [5] and even polytime *computable*. For the finitely branching graphs with infinite number of vertices (4) also gives definability in FO + LFP (but only *semidecidability* wrt oracles g, g' [6] of $\not\approx^{g,g'}$).

The following two axioms with $\in$ interpreted as ϵ^g characterize *extensional* and, respectively, *strongly extensional*, graphs g ($\mathcal{EG}$ and $\mathcal{SEG}$):

$$\forall u' \in u \exists v' \in v(u' = v') \& \forall v' \in v \exists u' \in u(u' = v') \Rightarrow u = v$$
$$u \approx^g v \Rightarrow u = v.$$

For well-founded graphs these two versions of extensionality axiom are equivalent. Note, that HCA, HFA$^\infty$, HFA, HC and HF are strongly extensional.

According to a bijection between HF and the natural numbers [2], it was defined in [16, 18] corresponding (lexicographical) linear ordering on HF as the unique relation $< = <^{HF}$ satisfying the following equivalence (6). Actually,

Proposition 3. *The strict linear order $<$ and its non-strict version $\leq$ are defined uniquely on HF by each of the following equivalences. The same formulae define a linear ordering $<^g$ on the vertices of any extensional finitely branching well-founded graph ($\mathcal{EFBWG}$) [7] g with $\in$ interpreted as ϵ^g. In particular, $<^g$ is definable in FO + LFP, by positivity of $<$ in the right-hand-side of (7):*

$$x < y \Leftrightarrow \exists u \in y \setminus x \forall w \in x \cup y(u < w \Rightarrow w \in x \cap y), \tag{6}$$

$$x < y \Leftrightarrow \exists u \in y \setminus x \forall w \in x \dot- y(u \not< w),$$

$$x < y \Leftrightarrow \exists u \in y \setminus x \forall w \in x \dot- y(w < u \vee w = u), \tag{7}$$

$$x \leq y \Leftrightarrow \forall u \in x \setminus y \exists w \in x \dot- y(u \leq w \& u \neq w), \tag{8}$$

$$x \leq y \Leftrightarrow \forall u \in x \setminus y \exists w \in y \setminus x(u \leq w \& u \neq w),$$

$$x < y \Leftrightarrow \exists u \in y \setminus x \forall w \in x \setminus y(u \not< w).$$

$$x < y \Leftrightarrow \exists u \in y \setminus x \forall w \in x \setminus y(w < u).$$

□

Here $x \dot- y \rightleftharpoons (x \cup y) \setminus (x \cap y)$. Of course, $u \in y \setminus x$ is understood as abbreviation of $u \in y \& u \notin x$ and analogously for $w \in x \cap y$ so that the above equivalences

---

[5] We suppose that the extensions FO + LFP and FO + IFP of the language FO of the first-order logic, respectively, by the *least fixed point* and by the *inflationary fixed point operators* are known; cf. e.g. [14, 10, 11, 23, 8].

[6] Note, that a graph may be represented by a *computable* function $g : |g| \to \mathcal{P}(|g|)$. Let you imagine, e.g., that a vertex v is an address (URL) of a page of the World-Wide Web and invoking gv results in a set of "clickable" addresses on that page. Cf. also the corresponding brief note in the Introduction.

[7] Note, that any $\mathcal{EFBWG}$ is isomorphic to some transitive subclass of HF.

may be considered as first-order formulae in the signature $\{\epsilon\}$. Let us call this ordering $<^g$ on g (and $<^{HF}$ on HF) *canonical* one.

Define the following operations and relations for vertices of a graph g

$$[x]_g^0 \rightleftharpoons \emptyset, \quad [x]_g^{i+1} \rightleftharpoons \{[y]_g^i | y \,\epsilon^g\, x\},$$
$$x <_k^g y \rightleftharpoons [x]_g^k <^{HF} [y]_g^k. \tag{9}$$

Here all $[x]_g^n$ are HF-sets for *finite* n (independently on cardinality of $gx!$).[8]

Proposition 4.

$$[x]_g^{k+1} = [y]_g^{k+1} \Leftrightarrow \forall u\,\epsilon\,x \exists v\,\epsilon\,y([u]_g^k = [v]_g^k)\&\forall v\,\epsilon\,y \exists u\,\epsilon\,x([u]_g^k = [v]_g^k),$$

$$x \approx_k^g y \Leftrightarrow [x]_g^k = [y]_g^k \quad \text{(even for k any ordinal as in the footnote 8)},$$

$$x \approx_{k+1}^g y \Rightarrow x \approx_k^g y, \tag{10}$$

$$x \approx^g y \Leftrightarrow \forall i(x \approx_i^g y) \Leftrightarrow \forall i([x]_g^i = [y]_g^i) \quad \text{(}i\text{ is any ordinal if } g \notin \mathcal{FBG}\text{)},$$

$$[([x]_g^k)]_{HF}^n = [x]_g^{\min(k,n)},$$

$$x <_0^g y \Leftrightarrow \textbf{false},$$

$$<_k^g \text{ is a strict linear preorder},$$

$$x \approx_k^g y \Leftrightarrow x \not<_k^g y \& y \not<_k^g x,$$

$$x <_{k+1} y \Leftrightarrow \exists u\,\epsilon\,y[\forall v\,\epsilon\,x(u \not\approx_k v) \,\& \tag{11}$$
$$\forall w\,\epsilon\,x \cup y(u <_k w \to \exists p\,\epsilon\,x \exists q\,\epsilon\,y(p \approx_k w \,\& \, q \approx_k w))].$$

$\square$

Define *rank* of any vertex of a graph g by taking

$\text{rk}(x) = 0$ iff there are no edges $x \xrightarrow{3} \dots$,

$\text{rk}(x) = \sup\{\text{rk}(u) + 1 | u\,\epsilon\,x\}$ and

$\text{rk}(x) = \infty$ iff there exists an infinite chain $\dots \epsilon\, x_{i+1} \,\epsilon\, x_i \,\epsilon\, \dots \epsilon\, x_1 \,\epsilon\, x$.

The inequality $\text{rk}(x) \neq \infty$ characterizes vertices x from *well-founded part* of g.

Proposition 5. *For arbitrary graphs g*
(a) $[x]_g^i = [x]_g^{i+1}$ *iff* $\text{rk}(x) \leq i$ *and*
(b) $[x]_g^i \leq^{HF} [x]_g^{i+1}$.

Proof. (a&b): The base case $i = 0$ is trivial both for (a) and (b). For the induction step it suffices to show (for (a) and (b), respectively) that

$$[x]^{i+1} = [x]^{i+2} \Leftrightarrow \forall u\,\epsilon\,x([u]^i = [u]^{i+1}) \quad \text{and} \tag{12}$$

$$[x]^{i+1} < [x]^{i+2} \Leftrightarrow \exists u\,\epsilon\,x([u]^i < [u]^{i+1}), \tag{13}$$

---

[8] However this will not be the case if we extend the definition of $[x]_g^{\alpha+1}$ for arbitrary non-limit ordinals $\alpha + 1$ as above and take for limit ordinals $[x]_g^\delta \rightleftharpoons \langle [x]_g^0, [x]_g^1, \dots, [x]_g^\alpha, \dots \rangle_{\alpha < \delta}$.

HF-set of the form $[x]^i$ plays the role of i-th HF-approximation to $[x]$. This notion and some its version (for protosets) were considered by M. Boffa and, respectively, by M. Mislove, L. Moss and F. Oles; cf. [13].

assuming that (a&b) holds for i. The converse implication for (12) and the direct implication for (13) follow by the definition of $[x]^k$ and by induction hypothesis. For the rest, let us suppose that $[u]^i = [u]^{i+1}$ fails for some $u \in x$. Then, by induction hypothesis ($[u]^i \leq [u]^{i+1}$), we should have $[u]^i < [u]^{i+1}$. Take $u \in x$ with the largest such $[u]^{i+1}$ (in the sense of $<^{HF}$). This is possible because $[u]^{i+1} \in [x]^{i+2}$, for $x \in u$, and $[x]^{i+2} \in HF$. Then

$$[u]^{i+1} \in [x]^{i+2} \setminus [x]^{i+1}$$

because, otherwise, we would have $[u]^{i+1} \in [x]^{i+1}$, $[u]^i < [u]^{i+1} = [v]^i$ for some $v \in x$ and, by (10), $[u]^{i+1} \neq [v]^{i+1}$ and, therefore, $[u]^{i+1} = [v]^i < [v]^{i+1}$, by induction hypothesis ($[v]^i \leq [v]^{i+1}$), what contradicts to the choice of u. Moreover, $[u]^{i+1}$ witnesses that $[x]^{i+1} < [x]^{i+2}$ (cf. (6) and (11)):

$$\forall w \in x([u]^{i+1} < [w]^i \vee [u]^{i+1} < [w]^{i+1} \Rightarrow [w]^i = [w]^{i+1} \in [x]^{i+1} \cap [x]^{i+2})$$

by using induction hypothesis ($[w]^i \leq [w]^{i+1}$) and by the choice of u.

$\square$

3 Linear Ordering of Strongly Extensional Finitely Branching Graphs

Our goal is to define by reasonable means some global strict linear preordering on the vertices of arbitrary $\mathcal{FBG}$ g, up to bisimulation, and therefore linear ordering on the universes HFA and HFA$^\infty$.

A simple approach consists in defining for arbitrary vertices $x, y \in |g|$

$$x \lhd_k y \rightleftharpoons \exists i \leq k(x <_i y \& \forall j \leq i(x \approx_i y)) \quad \text{and} \tag{14}$$

$$x \lhd y \rightleftharpoons \exists k(x \lhd_k y) \Leftrightarrow \exists i(x <_i y \& \forall j \leq i(x \approx_i y))] \tag{15}$$

Evidently, $\lhd_k^g$ and $\lhd^g$ are "lexicographic" slp-o with the corresponding equivalence relations $\approx_k^g$ and $\approx_\omega^g$ ($\approx_\omega^g = \approx^g$ for $g \in \mathcal{FBG}$). We actually need to have a *uniform* definition of $\lhd^g$ (or any other appropriate) slp-o in terms of FO + LFP (or IFP) over $g \in \mathcal{FBG}$. Note, that this can be done for $g \in \mathcal{FG}$ by imitating an appropriate segment of the natural numbers (cf. the variables i, j in the definition of $\lhd$) with the help of *stage comparison theorem* of Y.N. Moschovakis [14]. It was this way how the main result of this paper (on definability of a global linear order in any $\mathcal{SEFG}$ and Theorem 7) was first obtained. However, in this paper we present more elegant solution in Theorem 6 by defining in FO + IFP (and even in FO + LFP, if g ranges over $\mathcal{FG}$ [8]) some global slp-o $\prec^g$ and showing in Proposition 10, that $\prec^g = \lhd^g$.

If used straightforwardly, the above approach to $<^{HF}$ via formulas (6,7) does not work to define a linear order on arbitrary strongly extensional finitely branching graph which may contain cycles/infinite paths. However, we may define the required strict linear preorder $\prec^g$ on any, even non-extensional $\mathcal{FBG}$ g, up to bisimulation $\approx^g$, by ordering step-by-step, for $k = 0, 1, \ldots$, the $\approx_k^g$-equivalence classes, called *k-classes*. (Actually, the following definition of $\prec_k^g$, $k = 0, 1, 2, \ldots$, works for *arbitrary* graphs.)

First note, that there exists only one 0-class, just the set of all vertices $|g|$, and only finite number of *k*-classes for each k. So, we let $x \prec_0 y \rightleftharpoons$ **false** for all $x, y \in |g|$. Let us suppose that we have defined a strict linear order on the *k*-classes and therefore corresponding strict linear preorder $\prec_k$ on $|g|$. Then we can order $k+1$-classes by a strict linear order $\prec_{k+1}$ inside any *k*-class *lexicographically* wrt previously defined order $\prec_k$. I.e. for any $x \approx_k y$ (inside one *k*-class) define recurrently linear *preorder* relation on the vertices (instead of classes)

$$x \prec_{k+1} y \rightleftharpoons \exists u \in y [\forall v \in x (u \not\approx_k v) \; \& $$
$$\forall w \in x \cup y (u \prec_k w \rightarrow \exists p \in x \exists q \in y (p \approx_k w \; \& \; q \approx_k w))].$$

In the case of $x \not\approx_k y$, i.e. of different *k*-classes, take $x \prec_{k+1} y \rightleftharpoons x \prec_k y$.

Evidently, $x \prec_{k+1} y$ does not depend on the choice of representatives x and y inside their $k+1$ classes so that this is indeed a strict linear order on these classes: $x \approx_k y \Leftrightarrow x \not\prec_k y \& y \not\prec_k x$. We also have $\approx_0 \supseteq \approx_1 \supseteq \ldots$ and $\prec_0 \subseteq \prec_1 \subseteq \ldots$. In the case of finite g this process stabilizes: *k*-classes = $k+1$-classes and $\prec_k = \prec_{k+1}$ for some k. This results in some slp-o $\prec^g$ on $|g|$ corresponding to the largest bisimulation $\approx^g$. In the case of arbitrary graph g take

$$\prec_\omega^g \rightleftharpoons \bigcup_{k=0}^\infty \prec_k^g \quad \text{i.e.} \quad x \prec_\omega^g y \rightleftharpoons \exists k (x \prec_k^g y).$$

It is easy to show that $\prec_\omega^g$ is slp-o such that $x \approx_\omega^g y \Leftrightarrow x \not\prec_\omega^g y \& y \not\prec_\omega^g x$. In the case of $\mathcal{FBG}$ we have $\approx^g = \approx_\omega^g$ and define $\prec^g \rightleftharpoons \prec_\omega^g$. We can resume the above iterative definition for *arbitrary* $x, y \in |g|$ (with the superscript g omitted) as

$$x \prec_{k+1} y \rightleftharpoons x \prec_k y \vee \{x \approx_k y \; \& \; \exists u \in y [\forall v \in x (u \not\approx_k v)$$
$$\& \; \forall w \in x \cup y (u \prec_k w \rightarrow \exists p \in x \exists q \in y (p \approx_k w \; \& \; q \approx_k w))]\}$$

where $x \approx_k y$ may be replaced by $x \not\prec_k y \& y \not\prec_k x$. It folows immediately

Theorem 6. *The strict linear preorder $\prec^g$ on any $\mathcal{FBG}$ g (whose equivalence relation is bisimulation $\approx^g$) is definable in FO + IFP, and even in FO + LFP in the case of finite g (cf. [8]).*

$\square$

It is clear that for any $\mathcal{SEFBG}$ g (i.e. such $\mathcal{FBG}$ that $\approx^g$ coincides with the equality on g) the relation $\prec^g$ is a strict linear order on $|g|$. In particular, we have defined strict linear orders $\prec^{HFA^\infty}$ and $\prec^{HFA}$ (= $\prec^{HFA^\infty}|_{HFA}$).

By description of PTIME in terms of FO + LFP + a linear order (cf. the Introduction), this implies positive answers to the open questions from [19, 20]:

Theorem 7. *(a) The class of global predicates definable in FO + LFP over strongly extensional finite graphs (SEFG) and therefore the corresponding class of graph transformers $SEFG \rightarrow FG$ coincide with those PTIME-computable.*

(b) It follows (from [19, 20]) *that the class of* Δ_D*-definable operations over* HFA *coincides with the class of all* PTIME-*computable operations with respect to the graph encoding of sets via* [·].

Here an operation q : HFA $\rightarrow$ HFA is called PTIME-computable if for some PTIME-computable graph transformer Q : $\mathcal{FG} \rightarrow \mathcal{FG}$ (or, equivalently, Q : $\mathcal{SEFG} \rightarrow \mathcal{SEFG}$) of graphs $\langle g, x \rangle$ with a distinguished vertex x such that $q([g, x]) = [Q(g, x)]$ where $[g, x] \rightleftharpoons [x]_g$.

Without going into details, we present only the syntax of Δ_D-*language*:

$$\Delta_D\text{-formulas} ::= a \in b \mid a = b \mid \varphi \& \psi \mid \varphi \vee \psi \mid \neg \varphi \mid \forall x \in a.\varphi \mid \exists x \in a.\varphi$$

$$\Delta_D\text{-terms} ::= \text{set-variables } p, q, x, y, \ldots \mid \{a, b\} \mid \bigcup a \mid$$
$$\text{Decoration}(t, s) \mid \text{TC}(s) \mid$$
$$\{t(x) \mid x \in a \& \varphi(x)\} \mid \text{the-least } p.(p = \{x \in a \mid \varphi(x, p)\})$$

where a, b, t, s are Δ_D-terms and φ, ψ are Δ_D-formulas, set-variables x, p are not free in a and all occurrences of set variable p in Δ_D-formula φ are only *of the form* '- $\in p$', *positive* and *not inside* of any complex subterm of φ. $\quad\square$

Proposition 8. *The relations* $\prec$ *and* $\approx$ *(as well as* $\approx_k$ *and* $\prec_k$*) are coherent for any two* $\mathcal{FBG}$ g, g' *and their vertices* $x, y \in |g|$, $x', y' \in |g'|$:

$$x \approx^{g,g'} x' \ \& \ y \approx^{g,g'} y' \Rightarrow (x \approx^g y \Leftrightarrow x' \approx^{g'} y') \ \& \ (x \prec^g y \Leftrightarrow x' \prec^{g'} y'),$$

$\quad\square$

Corollary 9. *The global slp-o* $\prec$ *over the class* $\mathcal{FG}$ *or countable* $\mathcal{FBG}$ *induces a unique slo on the entire universe* HFA *or, respectively,* HFA$^\infty$ *via decoration operation, which coincides with* $\prec^{\text{HFA}}$ *or, respectively,* $\prec^{\text{HFA}^\infty}$ *defined above:*

$$[x]_g \prec^{\text{HFA}^{(\infty)}} [y]_g \Leftrightarrow x \prec^g y, \quad \text{for any } g \in \mathcal{F(B)G}, \ x, y \in |g|).$$

$\quad\square$

Proposition 10. $x \prec_k y \Leftrightarrow x \lhd_k y$ *and therefore* $\prec = \lhd$ *in any* $\mathcal{FBG}$.

Proof. Let us define for each n slp-o $x \lhd'_{n+1} y$ as

$$x \lhd'_{n+1} y \rightleftharpoons \exists u \in y[\forall v \in x(u \not\approx_n v) \ \& \tag{16}$$
$$\forall w \in x \cup y(u \lhd_n w \rightarrow \exists p \in x \exists q \in y(p \approx_n w \ \& \ q \approx_n w))].$$

Note, that both $x \lhd_{n+1} y$ and $x \lhd'_{n+1} y$ generate the same equivalence relation $\approx_{n+1}$. Evidently, $x \lhd_n y \Rightarrow x \lhd_{n+1} y$ for all $x, y \in |g|$ and $n = 0, 1, \ldots$. Therefore, to prove Proposition we must show that for all $x \approx_n y$

$$x \lhd_{n+1} y \Leftrightarrow x \lhd'_{n+1} y \quad \text{i.e. that} \quad x <_{n+1} y \Leftrightarrow x \lhd'_{n+1} y.$$

Let, for the contrary, $x \lhd'_{n+1} y$, but $y <_{n+1} x$. Then (11) gives for *all* x, y

$$y <_{n+1} x \Leftrightarrow \exists v \in x[\forall u \in y(v \not\approx_n u) \ \& \tag{17}$$
$$\forall w \in x \cup y(v <_n w \rightarrow \exists p \in x \exists q \in y(p \approx_n w \ \& \ q \approx_n w))].$$

Then (16) and (17) give some $u \in y$ and $v \in x$ such that $u \not\approx_n v$, $\forall v \in x(u \not\approx_n v)$, $\forall u \in y(v \not\approx_n u)$, $\forall w \in x \cup y(v <_n w \rightarrow \exists p \in x \exists q \in y(p \approx_n w \,\&\, q \approx_n w))$, and $\forall w \in x \cup y(u \vartriangleleft_n w \rightarrow \exists p \in x \exists q \in y(p \approx_n w \,\&\, q \approx_n w))$.

It follows that $v \vartriangleleft_n u$ because, otherwise, $u \vartriangleleft_n v$ would imply $\exists q \in y(v \approx_n q)$ and contradict to $\forall u \in y(v \not\approx_n u)$. Analogously, $u <_n v$. The last two formulas imply $v \vartriangleleft_{n-1} u$ and, hence, $v \not\approx_{n-1} u$, what contradicts to $x \approx_n y$.

The case when $x <_{n+1} y$, but $y \vartriangleleft'_{n+1} x$ is reduced to a contradiction in the same way. Therefore, $x \vartriangleleft'_{n+1} y \Leftrightarrow x <_{n+1} y$ holds (for $x \approx_n y$), as required. $\square$

Example 1. Note, that $\emptyset < x$ and $\emptyset \prec x$ for $x \neq \emptyset$ in HF and HFA$^\infty$, respectibely. However, the orders $<$ and $\prec$ are different on the universe HF $\subseteq$ HFA$^\infty$. Take $x_0 \rightleftharpoons \{\emptyset, \{\emptyset\}\}$ and $x_{i+1} \rightleftharpoons \{x_i\}$. Then we have $x_i < x_{i+1}$, but $x_{i+1} \prec x_i$.

4 Application of DLW Approach to Strongly Extensional Graphs

The above representation of the linear order $\prec$ by lexicographical ordering of the refining equivalence classes was partially inspired by the paper [4] of A. Dawar, S. Lindell and S. Weinstein. As we will see, this 'DLW' approach itself may be used in our context of $\mathcal{SEFG}$ (and even $\mathcal{SEFBG}$), but leads to *non-coherent* linear orders in our sense, i.e. relatively to bisimulation $\approx^g$ (cf. Proposition 8). However, their orders are coherent wrt an equivalence relation $\equiv$ (cf. below) which they consider instead of $\approx$ used in this paper.

We begin a formal comparison with the following simple technical

Proposition 11. *Consider any binary relation $\bar{x} S \bar{y}$ between k-tuples $\bar{x}$ and $\bar{y}$ in arbitrary graph g which satisfies the following condition for all $\bar{x}, \bar{y} \in |g|$:*

$$\bar{x} S \bar{y} \Rightarrow \bar{x} \cong \bar{y} \,\&\, \bigwedge_{i=1}^{k} (\forall x_i \exists y_i (\bar{x} S \bar{y}) \,\&\, \forall y_i \exists x_i (\bar{x} S \bar{y})) \tag{18}$$

where $\bar{x} \cong \bar{y}$ means that $\bar{x}$ and $\bar{y}$ are "isomorphic", i.e. indistinguishable in g by atomic formulas in the language $\{\epsilon^g, =\}$. Then, for $k \geq 2$, $xRy \rightleftharpoons x \ldots x S \ldots y$ is a bisimulation relation on g.

$\square$

The largest S satisfying the condition (18) (for any fixed k) is an equivalence relation $\equiv^g$ on k-tuples.[9] Evidently $\equiv$ satisfies (18) even with $\Rightarrow$ replaced by $\Leftrightarrow$ and is definable in FO + LFP.

So, we have $xx \equiv^g yy \Rightarrow x \approx^g y$ (and also $xx \equiv^g_{2i} yy \Rightarrow x \approx^g_i y$). However, the inverse implication fails, as the following simple example of a graph

---

[9] Note, that there are two different equivalent reformulations of this relation. One is in terms of (k-) pebble games (with the quantifier alternations in (18) corresponding to moves of two players "$\forall$" and "$\exists$"), and another one is in terms of indistinguishability of the tuples $\bar{x}$ and $\bar{y}$ by the infinitary $L^k_{\infty\omega}$-formulas (or finitary L^k-formulas for the case $\mathcal{FG}$) involving only k different variables; cf. [4, 9].

$a \rightarrow \bullet \leftarrow b \leftarrow \bullet$ shows: $a \approx^g b$, but $aa \not\equiv^g bb$. It follows that the $\equiv^g$-classes (respectively, $\equiv^g_{2i}$-classes) on pairs xx are finer than those defined by $\approx^g$ (respectively, by $\approx^g_i$) on the corresponding single elements x. In particular,

Corollary 12. *Strongly extensional graphs g $\mathcal{SEG}$ are 2-rigid in the sense that $xx \equiv^g yy \Rightarrow x = y$ for all vertices x and y.*

$\square$

DLW approach [4, 9] allows to define uniformly in FO + IFP (and therefore in FO + LFP) a strict linear order on the equivalence classes of k-tuples relative to $\equiv^M$ for arbitrary finite relational structures M with finite signature. The idea of the step-by-step linear ordering of refining the corresponding equivalences classes $\equiv^M_i$ is used (which works also for infinite M and all finite i).

For the classes of finite k-rigid structures this gives global FO+LFP-definable linear order on the elements of the structures. Also, for $\mathcal{SEFBG}$ g a global slo $\ll^g$ is definable by this method in FO + IFP. This together with Corollary 12 also gives a positive answer to the question of our interest, on defining a linear order on arbitrary finite strongly extensional graphs.

In contrast to our approach, there are ≤ 18 of 0-classes of the corresponding equivalence relation $\equiv^g_0$. or even ≤ 4 classes if g is acyclic. Each ordering $\ll_0$ of these 0-classes generates corresponding orderings $\ll_k$ of $\equiv_k$-classes and eventually an ordering $\ll_\omega$ of $\equiv_\omega$-classes, as in Sect. 3.

The following theorem (which is actually quite expectable one) shows that the resulting global linear order $\ll^g$ on strongly extensional *finite* graphs is *not coherent* relatively to bisimulation and therefore does not induce in this way a strict linear order on the entire *infinite* universe HFA, and even on HF, in contrast to $\prec^g$.

Theorem 13. *For any ordering $\ll_0$ of the 0-classes, there exist two finite extensional acyclic graphs with two distinguished vertices in each $(g; x, y)$ and $(g'; x', y')$ such that $[x]_g = [x']_{g'}$ and $[y]_g = [y']_{g'}$, but $x \ll^g y$ and $y' \ll^{g'} x'$.*

$\square$

References

1. Aczel, P.: Non-Well-Founded Sets. CSLI Lecture Notes, No. 14, 1988
2. Barwise, J.K.: Admissible Sets and Structures. Springer, Berlin, 1975
3. Barwise, J.K., Moss, L.: Vicious Circles: on the mathematics of circular phenomena. CSLI Lecture Notes, 1996
4. Dawar, A., Lindell, S., Weinstein S.: Infinitary Logic and Inductive Definability over Finite Structures. Information and Computation, **119**, No. 2 (1995) 160–175
5. Fernando, T.: A Primitive Recursive Set Theory and AFA: On Logical Complexity of the Largest Bisimulation. Report CS-R9213 ISSN 0169-118XCWI P.O.Box 4079, 1009 AB Amsterdam, Netherlands
6. Fernando, R.T.P.: On Substitutional Recursion over Non-Well-Founded Sets. LICS'89 (1989) 273–282
7. Gurevich, Y.: Algebras of feasible functions. Proc. 24th IEEE Conf. on Foundations of Computer Science (1983) 210–214

8. Gurevich, Y., Shelah, S.: Fixed-point extensions of first-order logic. Annals of Pure and Applied Logic **32** (1986) 265-280
9. Hodkinson, I.: Finite variable logics. Bulletin of the EATCS, **51** (1993) 111-140
10. Immerman, N.: Relational queries computable in polynomial time. Proccedings of 14th ACM Symposium on Theory of Computation, (1982) 147-152
11. Immerman, N.: Relational queries computable in polynomial time. Information and Control **68** (1986) 86-104
12. Livchak, A.B.: Languages of polynomial queries. Raschet i optimizacija teplotehnicheskih ob'ektov s pomosh'ju EVM, Sverdlovsk, 1982, p. 41 (in Russian)
13. Mislove, M., Moss L., Oles, F.: Non-Well-Founded Sets Modeled as Ideal Fixed Points. Information and Computation, bf 93 (1991) 16-54
14. Moschovakis, Y.N.: Elementary Induction on Abstract Structures. Amsterdam, North-Holland, 1974.
15. Sazonov, V.Yu.: Polynomial computability and recursivity in finite domains. Elektronische Informationsverarbeitung und Kybernetik. **16**, N7 (1980) 319-323
16. Sazonov, V.Yu.: Bounded set theory, polynomial computability and Δ-programming. Application aspects of mathematical logic. Computing systems **122** (1987) 110-132 (In Russian) Cf. also a short English version of this paper in: Lect. Not. Comput. Sci. **278** Springer (1987) 391-397
17. Sazonov, V.Yu.: Bounded set theory and inductive definability. Abstracts of Logic Colloquium'90. JSL **56** No.3 (1991) 1141-1142
18. Sazonov, V.Yu.: Hereditarily-finite sets, data bases and polynomial-time computability. TCS **119** Elsevier (1993) 187-214
19. Sazonov, V.Yu.: A bounded set theory with anti-foundation axiom and inductive definability. Computer Science Logic, 8th Workshop, CSL'94 Kazimierz, Poland, September 1994, Selected Papers. Lecture Notes in Computer Science **933** Springer (1995) 527-541.
20. Sazonov, V.Yu.: On Bounded Set Theory (Invited talk). M.L. Dalla Chiara, et al. (eds.), Logic and Scientific Methods, Volume One of the Tenth International Congress of Logic, Methodology and Philosophy of Sciences, Florence, August 1995, Kluwer Academic Publishers, Dordrecht, 1997, 85-103.
21. Sazonov, V.Yu., Lisitsa, A.P.: Δ-languages for sets and sub-PTIME graph transformers. Database Theory – ICDT'95, 5th International Conference, Prague, Czech Republic, January 1995, Proceedings. Lecture Notes in Computer Science **893** Springer (1995) 125-138
22. Lisitsa, A.P., Sazonov, V.Yu.: Δ-languages for sets and LOGSPACE-computable graph transformers, Theoretical Computer Science, tentatively in Vol. 175, 1997
23. Vardi, M.: Complexity of relational query languages. Proceedings of 14th Symposium on Theory of Computation (1982) 137-146

Functions for the General Solution of Parametric Word Equations

G.S. Makanin[1] H. Abdulrab[2] P. Goralcik[3]

[1] Steklov Mathematical Institute, Vavilova 42, 117966, Moscow GSP-1, Russia
[2] PSI/LIRINSA, INSA de Rouen, BP 08, 76131 Mont Saint Aignan Cedex, France.
[3] LIR, Université de Rouen, BP 118, 76134 Mont Saint Aignan Cedex, France.

Abstract. In this article we introduce the functions $F^i\,(x_1, x_2)^{\,\lambda_1,\dots,\lambda_s}$ and $T^h\,(x_1, x_2, x_3)^{\,\lambda_1,\dots,\lambda_{2s}}$ $(i = 1, 2, 3)$, of the word variables x_i and of the natural number variables λ_i, where $s \geq 0$. By means of these functions, we give exactly the general solution (i.e., the set of all the solutions) of the first basic parametric equation: $x_1 x_2 x_3 x_4 = x_3 x_1^{\lambda} x_2 x_5$, in a free monoid.

1 Introduction

The following four parametric equations:

$$x_1 x_2 x_3 x_4 = x_3 x_1^{\lambda} x_2 x_5,$$
$$x_1 x_2 x_3 x_4 = x_2 x_3^{\lambda} x_1 x_5,$$
$$x_1 x_2^2 x_3 x_4 = x_3 x_1^2 x_2 x_5,$$
$$x_1 x_2^{\lambda+1} x_3 x_4 = x_3 x_2^{\mu+1} x_1 x_5,$$

in a free monoid, are called *basic* equations. They arise in the graph of the prefixe-equations in free monoid (cf. [2], [3]) and play an important role in the hierarchy of the parametric equations, in reason of the structures of their solutions. In particular, the general solution of any equation in a free monoid of the form $\Phi\,(x_1, x_2, x_3)\, x_4 = \Psi\,(x_1, x_2, x_3)\, x_5$, where $\Phi\,(x_1, x_2, x_3)$ and $\Psi\,(x_1, x_2, x_3)$ are any words on the alphabet $\{x_1, x_2, x_3\}$, is described by means of a finite number of parametric transformations and by the general solutions of the basic equations.

2 Definitions and notations

Let Π be a free monoid (a free semigroup with unit) with a countable alphabet of generators:

$$a_1, a_2, \dots, a_k, \dots \tag{1}$$

Let

$$x_1, x_2, \dots, x_n, \dots \tag{2}$$

be a countable alphabet of word variables, and let

$$\lambda_1, \lambda_2, \dots, \lambda_i, \dots \tag{3}$$

be a countable alphabet of natural number variables (called also natural parameters). Define inductively a parametric word as follows: - Any word on the alphabet (2) is a parametric word. - If P is a parametric word, λ is a natural parameter, then $(P)^\lambda$ is a parametric word. - If P and Q are two parametric words, then PQ is a parametric word.

A *parametric equation* in a free monoid is given by an equality of parametric words

$$\Phi(x_1, x_2, \ldots, x_n, \lambda_1, \lambda_2, \ldots, \lambda_t) = \Psi(x_1, x_2, \ldots, x_n, \lambda_1, \lambda_2, \ldots, \lambda_t) \quad (4)$$

If Φ and Ψ are empty words, the equation (4) is called the trivial equation, denoted by 1. We denote by Φ the set of all linear polynomials in naturals, of the form: $k_0 + \sum_{i=1}^{r} k_i \lambda_i$, where $r, k_0, k_1, \ldots, k_r$ are natural numbers, and $\lambda_1, \ldots, \lambda_r$ are natural parameters. A *parametric transformation* is defined by the application:

$$\begin{cases} x_1 \to W_1(x_1, x_2, \ldots, x_n, \lambda_1, \lambda_2, \ldots, \lambda_q) \\ \cdots\cdots\cdots\cdots\cdots\cdots\cdots\cdots\cdots\cdots\cdots\cdots \\ x_n \to W_n(x_1, x_2, \ldots, x_n, \lambda_1, \lambda_2, \ldots, \lambda_q) \\ \lambda_1 \to L_1(\lambda_1, \lambda_2, \ldots, \lambda_q) \\ \cdots\cdots\cdots\cdots\cdots\cdots\cdots\cdots\cdots\cdots \\ \lambda_t \to L_t(\lambda_1, \lambda_2, \ldots, \lambda_q) \end{cases} \quad (5)$$

where W_i is a parametric word, and L_i is a linear polynomial in naturals. The components of the form $x_i \to x_i$, and $\lambda_i \to \lambda_i$ are often omitted. The result of the application of the transformation (5) to the parametric equation (4) is the parametric equation given by the substitution of (5). A parametric transformation (5) is called a parametric solution of the equation (4), if the result of the application of the transformation (5) to the equation (4) is the trivial equation.

A *coefficient transformation* is defined by the application:

$$\begin{cases} x_1 \to X_1 \\ \cdots\cdots\cdots \\ x_n \to X_n \\ \lambda_1 \to \Lambda_1 \\ \cdots\cdots\cdots \\ \lambda_t \to \Lambda_t \end{cases} \quad (6)$$

where X_i are words on the alphabet (1), and Λ_i are natural numbers.

A coefficient transformation (6) is called a *solution* of the equation (4), if the words $\Phi(X_1, \ldots, X_n, \Lambda 1, \ldots, \Lambda_t)$ and $\Psi(X_1, \ldots, X_n, \Lambda_1, \ldots, \Lambda_n)$ coincide. A coefficient transformation

$$\begin{cases} x_1 \to X_1 \\ \cdots\cdots \\ x_n \to X_n \\ \lambda_1 \to \Lambda_1 \\ \cdots\cdots \\ \lambda_t \to \Lambda_t \end{cases}$$

is called an *extension* of the coefficient transformation (6), if $p \geq t$.

We will say that the parametric transformation T *contains* the coefficient transformation C, by means of the coefficient transformation C_1 if $TC_1 = C$.

A finite list of parametric solutions of the equation E will be called a general solution of E, if every solution of E is contained in some parametric solution of this list. The general solution of E will be denoted $\langle E \rangle$.

The length of a word A on the alphabet (1) will be denoted $|A|$.

Let us define parametric solutions which satisfy *conditions on the lengths* of solutions, and that which satisfy *conditions on natural parameters*.

A *condition on natural parameters* has the form:

$$L_1 (\lambda 1, \ldots, \lambda q) \; \#_i L_2 (\lambda_1, \ldots, \lambda_q), \tag{7}$$

where $\#_i \in \{<, \leq, =\}$, and $L1, L2$ are linear polynomials in naturals. A coefficient transformation (6) is called a *solution* of the equation E with a condition (7) on natural parameters, if (6) is a solution of E that satisfies:

$$L_1 (\Lambda_1, \ldots, \Lambda_q)) \; \#_i L_2 (\Lambda_1, \ldots, \Lambda_q),$$

where $\#_i$ is the relation determined by (7). A *condition on the length of a solution* has the form:

$$\partial (P_1 (x_1, \ldots, x_n, \lambda_1, \ldots, \lambda_q)) \; \#_i \; \partial (P_2 (x_1, \ldots, x_n, \lambda_1, \ldots, \lambda_q)) \tag{8}$$

where $\#_i \in \{<, \leq, =\}$ and $P1, P2$ are two parametric words. A coefficient transformation (6) is called a solution of the equation E with a condition (8) on the lengths of solutions, if (6) is a solution of E that satisfies:

$$|P_1 (X_1, \ldots, X_n, \Lambda_1, \ldots, \Lambda_q)| \; \#_i \; |P_1 (X_1, \ldots, X_n, \Lambda_1, \ldots, \Lambda_q)|,$$

where $\#_i$ is the relation determined by (8).

Let E be an equation with conditions and let $R_1, \ldots, R_m$ be the list of new conditions. By (E, R_i) we denote the equation E with the condition R_i. The equation E is said to be divided into a collection of equations $(E, R_1), \ldots, (E, R_m)$ if every solution S of E is a solution of some (E, R_i).

We say that the equation E_1 is *reduced by the parametric transformation* T to the equation E_2 if $E_1 T = E_2$, and for every solution S_1 of the equation E_1 there exists an extension S_2^* of a solution S_2 of the equation E_2 such that $S_1 = TS_2^*$. We say that S_2^* is the image of S_1 via the transformation T. The need of the extension S_2^* is justified by the fact that T could have some variables that are not in E_2.

A *parametric function* will be given by some expression of the form $f_{(x_1, \ldots, x_r)}\omega$ where $x_1, \ldots, x_r$ are letters of the alphabet (2) and ω is a finite sequence of natural number parameters. A *parametric functional transformation* is then defined by the application:

$$\begin{cases} x_1 \to {}^{f_1}(x_1, \ldots, x_r)^{\omega} x_1 \\ \cdots \cdots \cdots \cdots \cdots \\ x_r \to {}^{f_r}(x_1, \ldots, x_r)^{\omega} x_r \end{cases}$$

The notion of parametric transformation (5) can be then extended to include parametric functional transformation. And the notions of parametric solutions and general solution can be extended according to the extension of the notion of parametric transformation.

Lemma 1. *Let $E1$ be reduced by T to E_2. And let S_1 be a solution of E_1 and S_2^* its image via T. Then we have:*

- *If Q_2 is a parametric solution of E_2, then TQ_2 is a parametric solution of E_1.*

- *If the parametric solution Q_2 contains a solution S_2 of E_2, then the parametric solution TQ_2 of E_1 contains the solution S_1 of E_1.*

Proof.

- Since $E_2Q_2 = 1$ we have: $E_1(TQ_2) = (E_1T)Q_2 = E_2Q_2 = 1$.

- There exists a coefficient transformation C such that $Q_2C = S_2$. Let S_2^* be:
$$\begin{cases} S_2 \\ A \end{cases}$$

Define the coefficient transformation C^* by $\begin{cases} C \\ A \end{cases}$

Then, $Q_2C^* = \begin{cases} Q_2C \\ A \end{cases} = \begin{cases} S_2 \\ A \end{cases} = S_2^*$. Therefore $T(Q_2C^*) = TS_2^*$.
Thus we have: $(TQ_2)C^* = T(Q_2C^*) = TS_2^* = S_1$.

Theorem 2. *Let the equation E_1 be reduced by the parametric transformation T to the equation E_2. If the general solution $< E_2 >$ of E_2 is $Q_1, \ldots, Q_r$ then the general solution $< E_1 >$ of E_1 is $TQ_1, \ldots, TQ_r$.*

Proof.
By definition, $E_1T = E_2$, and for every $i = 1, \ldots, r$ we have $E_2Q_i = 1$. Therefore, for every i we have:

$$E_1(TQ_i) = (E_1T)Q_i = E_2Q_i = 1.$$

That is, the parametric solution TQ_i is a parametric solution of the equation E_1. Let S_1 be an arbitrary solution of E_1. Since E_1 is reduced by T to E_2, there exists an extension S_2^* of a solution S_2 of the equation E_2, such that $S_1 = TS_2^*$. Since $Q_1, \ldots, Q_r$ is $\langle E_2 \rangle$, some Q_i contains S_2. According to lemma 1, the parametric solution TQ_i contains the solution S_1 of E_1.

Theorem 3. *Let the equation E be divided into a collection of equations with conditions $(E, R_1), \ldots, (E, R_m)$, with the list of conditions $R_1, \ldots, R_m$. And let the general solution $\langle (E, R_i) \rangle$ of (E, R_i) be $Q_{i,1}, \ldots, Q_{i,r_i}$ Then, the general solution $\langle E \rangle$ of E is $Q_{1,1}, \ldots, Q_{1,r_1}, \ldots, Q_{m,1}, \ldots, Q_{m,r_m}$*

Proof.

By definition, every parametric transformation $Q_{i,j}$ of the joint lists $Q_{1,1}, \ldots, Q_{1,r_1}, \ldots, Q_{m,1}, \ldots, Q_{m,r_m}$ is a parametric solution of the equation E.

If S is an arbitrary solution of the equation E, then S is a solution of some equation (E, R_i). Since $Q_{i,1}, \ldots, Q_{i,r_i}$ is $< (E, R_i) >$, some parametric solution $Q_{i,j}$ contains S. Therefore, the joint lists is $\langle E \rangle$.

3 The function $F_i(x_1, x_2)^{\lambda_1, \ldots, \lambda_s}$

Define inductively the function $F_i(x_1, x_2)^{\lambda_1, \ldots, \lambda_s}$, for $s \geq 0$, x_1 and x_2 are two word variables, and $\lambda_1, \ldots, \lambda_s$ are natural parameters.

$$F_i(x_1, x_2) = 1$$
$$F_i(x_1, x_2)^{\lambda_1, \lambda_2, \ldots, \lambda_s} = \left(F_i(x_2, x_1)^{\lambda_2, \ldots, \lambda_s} x_2\right)^{\lambda_1} F_i(x_1, x_2)^{\lambda_3, \ldots, \lambda_s}; \quad (s \geq 1)$$

In particular:

$$F_i(x_1, x_2)^{\lambda_1} = (x_2)^{\lambda_1}$$
$$F_i(x_1, x_2)^{\lambda_1, \lambda_2} = \left((x_1)^{\lambda_2} x_2\right)^{\lambda_1}$$
$$F_i(x_1, x_2)^{\lambda_1, \lambda_2, \lambda_3} = \left(\left((x_2)^{\lambda_3} x_1\right)^{\lambda_2} x_2\right)^{\lambda_1} (x_2)^{\lambda_3}$$

Theorem 4 $F_i 1$. *The following identities hold for $k \geq 0$.*

$$F_i(x_1, x_2)^{\lambda_1, \ldots, \lambda_{2k}} = F_i(x_1, x_1^{\lambda_{2k}} x_2)^{\lambda_1, \ldots, \lambda_{2k-1}}.$$
$$F_i(x_1, x_2)^{\lambda_1, \ldots, \lambda_{2k+1}} = F_i(x_2^{\lambda_{2k+1}}, x_1, x_2)^{\lambda_1, \ldots, \lambda_{2k}} x_2^{\lambda_{2k+1}}.$$

Proof

By a joint induction on k.

− If $k = 0, 1$ the proof is obvious.

− Suppose that $k > 1$. By definition, $F_i(x_1, x_2)^{\lambda_1, \ldots, \lambda_{2k}}$ equals:

$$\left(F_i(x_2, x_1)^{\lambda_2, \ldots, \lambda_{2k}} x_2\right)^{\lambda_1} F_i(x_1, x_2)^{\lambda_3, \ldots, \lambda_{2k}}.$$

According to the inductions proposition (second identity), this last expression equals:

$$\left(F_i(x_1^{\lambda_{2k}} x_2, x_1)^{\lambda_2, \ldots, \lambda_{2k-1}} x_1^{\lambda_{2k}} x_2\right)^{\lambda_1} F_i(x_1, x_2)^{\lambda_3, \ldots, \lambda_{2k}}.$$

According to the inductions proposition (first identity), this last expression equals:

$$\left(F_i(x_1^{\lambda_{2k}} x_2, x_1)^{\lambda_2, \ldots, \lambda_{2k-1}} x_1^{\lambda_{2k}} x_2\right)^{\lambda_1} F_i(x_1, x_1^{\lambda_{2k}} x_2)^{\lambda_3, \ldots, \lambda_{2k-1}}.$$

This expression is equal (by definition) to $F_i(x_1, x_1^{\lambda_{2k}} x_2)^{\lambda_3, \ldots, \lambda_{2k-1}}$.

By definition, $F_i(x_1, x_2)^{\lambda_1, \ldots, \lambda_{2k+1}}$ equals

$$\left(F_i(x_2, x_1)^{\lambda_2, \ldots, \lambda_{2k+1}} x_2 \right)^{\lambda_1} F_i(x_1, x_2)^{\lambda_3, \ldots, \lambda_{2k+1}}.$$

According to the inductions proposition (first identity), this last expression equals:

$$\left(F_i(x_2, x_2^{\lambda_{2k+1}} x_1)^{\lambda_2, \ldots, \lambda_{2k}} x_2 \right)^{\lambda_1} F_i(x_1, x_2)^{\lambda_3, \ldots, \lambda_{2k+1}}.$$

According to the inductions proposition (second identity), this last expression equals:

$$\left(F_i(x_2, x_2^{\lambda_{2k+1}} x_1)^{\lambda_2, \ldots, \lambda_{2k}} x_2 \right)^{\lambda_1} F_i(x_2^{\lambda_{2k+1}} x_1, x_2)^{\lambda_3, \ldots, \lambda_{2k}} x_2^{\lambda_{2k+1}}.$$

This expression is equal (by definition) to $F_i(x_2^{\lambda_{2k+1}} x_1, x_2)^{\lambda_1, \ldots, \lambda_{2k}} x_2^{\lambda_{2k+1}}$.

$$\diamond$$

Consider an equation of the form $x_1 P x_2 \ldots = x_2 Q x_1 \ldots$, where P, Q are words on the alphabet $x_3, \ldots, x_n, \ldots$ and consider the sequence of parametric transformations:

$$
\begin{array}{lll}
1. & & x_1 \to (x_2 Q)^{\lambda_1} x_1 \\
2. & & x_2 \to (x_1 P)^{\lambda_2} x_2 \\
3. & & x_1 \to (x_2 Q)^{\lambda_3} x_1 \\
4. & & x_2 \to (x_1 P)^{\lambda_4} x_2 \\
\cdots\cdots & \cdots\cdots\cdots\cdots\cdots\cdots\cdots \\
2k-1. & & x_1 \to (x_2 Q)^{\lambda_{2k-1}} x_1 \\
2k. & & x_2 \to (x_1 P)^{\lambda_{2k}} x_2 \\
2k+1. & & x_1 \to (x_2 Q)^{\lambda_{2k+1}} x_1 \\
\cdots\cdots & \cdots\cdots\cdots\cdots\cdots\cdots\cdots
\end{array}
\tag{9}
$$

where $\lambda_1, \lambda_2, \ldots$ are natural parameters.

Theorem 5 $F_i 2$. *For every natural s, the sequence (9) of the first s parametric transformation can be collected by the following common transformation:*

$$
\begin{cases}
x_1 \to F_i (x_1 P, x_2 Q)^{\lambda_1, \lambda_2, \ldots, \lambda_s} x_1 \\
x_2 \to (x_2 Q, x_1 P)^{\lambda_2, \ldots, \lambda_s} x_2
\end{cases}
$$

Proof

If $s = 0$ the proposition holds obviously. Consider now two cases.

case 1) Suppose that the sequence of the first $2k - 1$ transformations can be collected by the common transformation:

$$
\begin{cases}
x_1 \to F_i (x_1 P, x_2 Q)^{\lambda_1, \lambda_2, \ldots, \lambda_{2k-1}} x_1 \\
x_2 \to F_i (x_2 Q, x_1 P)^{\lambda_2, \ldots, \lambda_{2k-1}} x_2
\end{cases}
$$

Let the $2k$-th transformation be of the form:

$$x_2 \rightarrow (x_1 P)^{\lambda_{2k}} x_2.$$

Then the sequence of the first $2k$ transformations can be collected by the common transformation:

$$\begin{cases} x_1 \rightarrow {}^{F_i}\left(x_1 P, (x_1 P)^{\lambda_{2k}} x_2 Q\right)^{\lambda_1, \lambda_2, \ldots, \lambda_{2k-1}} x_1 \\ x_2 \rightarrow {}^{F_i}\left((x_1 P)^{\lambda_{2k}} x_2 Q, x_1 P\right)^{\lambda_2, \ldots, \lambda_{2k-1}} (x_1 P)^{\lambda_{2k}} x_2 \end{cases}$$

According to Theorem 4 this transformation coincide with the transformation:

$$\begin{cases} x_1 \rightarrow {}^{F_i}\left(x_1 P, x_2 Q\right)^{\lambda_1, \lambda_2, \ldots, \lambda_{2k}} x_1 \\ x_2 \rightarrow {}^{F_i}\left(x_2 Q, x_1 P\right)^{\lambda_2, \ldots, \lambda_{2k}} x_2 \end{cases}$$

case 2) Suppose that the sequence of the first $2k$ transformations can be collected in the common transformation:

$$\begin{cases} x_1 \rightarrow {}^{F_i}\left(x_1 P, x_2 Q\right)^{\lambda_1, \lambda_2, \ldots, \lambda_{2k}} x_1 \\ x_2 \rightarrow {}^{F_i}\left(x_2 Q, x_1 P\right)^{\lambda_2, \ldots, \lambda_{2k}} x_2 \end{cases}$$

Let the $(2k+1)$–th transformation be of the form:

$$x_1 \rightarrow (x_2 Q)^{\lambda_{2k+1}} x_1.$$

Then the sequence of the first $2k+1$ transformations can be collected by the common transformation:

$$\begin{cases} x_1 \rightarrow {}^{F_i}\left((x_2 Q)^{\lambda_{2k+1}} x_1 P, x_2 Q\right)^{\lambda_1, \lambda_2, \ldots, \lambda_{2k}} (x_2 Q)^{\lambda_{2k+1}} x_1 \\ x_2 \rightarrow {}^{F_i}\left(x_2 Q, (x_2 Q)^{\lambda_{2k+1}} x_1 P\right)^{\lambda_2, \ldots, \lambda_{2k}} x_2 \end{cases}$$

According to Theorem 4 this transformation coincide with the transformation:

$$\begin{cases} x_1 \rightarrow {}^{F_i}\left(x_1 P, x_2 Q\right)^{\lambda_1, \lambda_2, \ldots, \lambda_{2k+1}} x_1 \\ x_2 \rightarrow {}^{F_i}\left(x_2 Q, x_1 P\right)^{\lambda_2, \ldots, \lambda_{2k+1}} x_2 \end{cases}$$

4 The function ${}^{Th}(x_1, x_2, x_3)_i{}^{\lambda_1, \lambda_2, \ldots, \lambda_{2s}}$

Define, by a joint induction, the functions
${}^{Th}(x_1, x_2, x_3)_1{}^{\lambda_1, \lambda_2, \ldots, \lambda_{2s}}, {}^{Th}(x_1, x_2, x_3)_2{}^{\lambda_1, \lambda_2, \ldots, \lambda_{2s}}, {}^{Th}(x_1, x_2, x_3)_3{}^{\lambda_1, \lambda_2, \ldots, \lambda_{2s}},$

for $s \geq 0$, x_1, x_2 and x_3 are three word variables, and $\lambda_1, \lambda_2, \ldots, \lambda_{2s}$ are natural parameters.

$${}^{Th}(x_1, x_2, x_3)_i = 1, \qquad (i = 1, 2, 3).$$

$${}^{Th}(x_1, x_2, x_3)_1 {}^{\lambda_1, \lambda_2, \ldots, \lambda_{2s}} =$$

$$\left({}^{Th}(x_1, x_2, x_3)_3 {}^{\lambda_1, \lambda_2, \ldots, \lambda_{2s}} x_3\right)^{\lambda_1} {}^{Th}(x_1, x_2, x_3)_1 {}^{\lambda_3, \ldots, \lambda_{2s}}$$

$${}^{Th}(x_1, x_2, x_3)_2 {}^{\lambda_1, \lambda_2, \ldots, \lambda_{2s}} =$$

$$\left({}^{Th}(x_1, x_2, x_3)_2 {}^{\lambda_3, \ldots, \lambda_{2s}} x_2 {}^{Th}(x_1, x_2, x_3)_3 {}^{\lambda_3, \ldots, \lambda_{2s}} x_3 {}^{Th}(x_1, x_2, x_3)_1 {}^{\lambda_3, \ldots, \lambda_{2s}} x_1\right.$$

$$\left. {}^{Th}(x_1, x_2, x_3)_1 {}^{\lambda_1, \lambda_2, \ldots, \lambda_{2s}} x_1\right)^{\lambda_2} {}^{Th}(x_1, x_2, x_3)_2 {}^{\lambda_3, \ldots, \lambda_{2s}}.$$

$${}^{Th}(x_1, x_2, x_3)_3 {}^{\lambda_1, \lambda_2, \ldots, \lambda_{2s}} =$$

$${}^{Th}(x_1, x_2, x_3)_1 {}^{\lambda_3, \ldots, \lambda_{2s}} x_1 {}^{Th}(x_1, x_2, x_3)_2 {}^{\lambda_3, \ldots, \lambda_{2s}} x_2 {}^{Th}(x_1, x_2, x_3)_3 {}^{\lambda_3, \ldots, \lambda_{2s}}.$$

And define inductively the auxiliary function

$${}^{Oc}(x_1, x_2, x_3)_i = 1.$$

$${}^{Oc}(x_1, x_2, x_3)^{\lambda_1, \lambda_2, \ldots, \lambda_{2s}} = {}^{Oc}(x_1, x_2, x_3)^{\lambda_3, \ldots, \lambda_{2s}} \left({}^{Th}(x_1, x_2, x_3)_3 {}^{\lambda_3, \ldots, \lambda_{2s}} x_3\right.$$

$$\left.{}^{Th}(x_1, x_2, x_3)_1 {}^{\lambda_3, \ldots, \lambda_{2s}} x_1 {}^{Th}(x_1, x_2, x_3)_2 {}^{\lambda_3, \ldots, \lambda_{2s}} x_2\right)^{\lambda_1}.$$

Theorem 6. : *The following identities hold for $s \geq 0$:*

$${}^{Th}(x_1, x_2, x_3)_1^{\lambda_1, \lambda_2, \ldots, \lambda_{2s}+2} = {}^{Th}((x_1 x_2 x_3)^{\lambda_{2s}+1} x_1,$$

$$(x_2 x_3 x_1 (x_1 x_2 x_3)^{\lambda_{2s}+1} x_1)^{\lambda_{2s}+2} x_2, x_1 x_2 x_3)_1^{\lambda_1, \lambda_2, \ldots, \lambda_{2s}} (x_1 x_2 x_3)^{\lambda_{2s}+1}$$

$${}^{Th}(x_1, x_2, x_3)_2^{\lambda_1, \lambda_2, \ldots, \lambda_{2s}+2} =$$

$${}^{Th}((x_1 x_2 x_3)^{\lambda_{2s}+1} x_1, (x_2 x_3 x_1 (x_1 x_2 x_3)^{\lambda_{2s}+1} x_1)^{\lambda_{2s}+2} x_2, x_1 x_2 x_3)_2^{\lambda_1, \lambda_2, \ldots, \lambda_{2s}}$$

$$(x_2 x_3 x_1 (x_1 x_2 x_3)^{\lambda_{2s}+1} x_1)^{\lambda_{2s}+2}$$

$${}^{Th}(x_1, x_2, x_3)_3^{\lambda_1, \lambda_2, \ldots, \lambda_{2s}+2} = {}^{Th}((x_1 x_2 x_3)^{\lambda_{2s}+1} x_1,$$

$$(x_2 x_3 x_1 (x_1 x_2 x_3)^{\lambda_{2s}+1} x_1)^{\lambda_{2s}+2} x_2, x_1 x_2 x_3)_3^{\lambda_1, \lambda_2, \ldots, \lambda_{2s}} x_1 x_2$$

$${}^{Oc}(x_1, x_2, x_3)^{\lambda_1, \lambda_2, \ldots, \lambda_{2s}+2} = (x_3 x_1 x_2)^{\lambda_{2s}+1} {}^{Oc}((x_1 x_2 x_3)^{\lambda_{2s}+1} x_1,$$

$$(x_2 x_3 x_1 (x_1 x_2 x_3)^{\lambda_{2s}+1})^{\lambda_{2s}+2} x_2, x_1 x_2 x_3)^{\lambda_1, \lambda_2, \ldots, \lambda_{2s}}.$$

$$\diamond$$

Consider the equation $x_1 x_2 x_3 x_4 = x_3 x_1^2 x_2$ and the sequence of joint parametric transformations: (10-13)

$$\begin{aligned}
x_1 &\to (x_1 x_2 x_3)^{\lambda_1} x_1 \\
x_2 &\to (x_2 x_3 x_1 (x_1 x_2 x_3)^{\lambda_1} x_1)^{\lambda_2} x_2 \\
x_3 &\to x_1 x_2 x_3 \\
x_4 &\to x_4 (x_3 x_1 x_2)^{\lambda_1}
\end{aligned} \tag{10}$$

$$\begin{aligned}
x_1 &\to (x_1 x_2 x_3)^{\lambda_3} x_1 \\
x_2 &\to (x_2 x_3 x_1 (x_1 x_2 x_3)^{\lambda_3} x_1)^{\lambda_4} x_2 \\
x_3 &\to x_1 x_2 x_3 \\
x_4 &\to x_4 (x_3 x_1 x_2)^{\lambda_3}
\end{aligned} \tag{11}$$

$$\cdots\cdots\cdots\cdots\cdots\cdots\cdots\cdots$$

$$\begin{aligned}
x_1 &\to (x_1 x_2 x_3)^{\lambda_{2s}-1} x_1 \\
x_2 &\to (x_2 x_3 x_1 (x_1 x_2 x_3)^{\lambda_{2s}-1} x_1)^{\lambda_{2s}} x_2 \\
x_3 &\to x_1 x_2 x_3 \\
x_4 &\to x_4 (x_3 x_1 x_2)^{\lambda_{2s}-1}
\end{aligned} \tag{12}$$

$$x_1 \to (x_1 x_2 x_3)^{\lambda_{2s+1}} x_1$$
$$x_2 \to (x_2 x_3 x_1 (x_1 x_2 x_3)^{\lambda_{2s+1}} x_1)^{\lambda_{2s+2}} x_2$$
$$x_3 \to x_1 x_2 x_3 \tag{13}$$
$$x_4 \to x_4 (x_3 x_1 x_2)^{\lambda_{2s+1}}$$

. . . .

Theorem 7 $^{Th}2$. *For every natural s, the sequence of the s joint parametric transformations (10-13) can be collected by the following common transformation:*

$$\begin{cases} x_1 \to {}^{Th}(x_1, x_2, x_3)_1^{\lambda_1, \lambda_2, \ldots, \lambda_{2s}} x_1 \\ x_2 \to {}^{Th}(x_1, x_2, x_3)_2^{\lambda_1, \lambda_2, \ldots, \lambda_{2s}} x_2 \\ x_3 \to {}^{Th}(x_1, x_2, x_3)_3^{\lambda_1, \lambda_2, \ldots, \lambda_{2s}} x_3 \\ x_4 \to x_4^{Oc}(x_1, x_2, x_3)^{\lambda_1, \lambda_2, \ldots, \lambda_{2s}} \end{cases}$$

5 Equations and Solutions

Proposition 8. *The general solution of the equation*

$$x_1 x_2 = x_2 x_1 \tag{14}$$

is described by the transformation:

$$\begin{cases} x_1 \to x_1^{\alpha} \\ x_2 \to x_1^{\beta} \end{cases}$$

where α and β are natural parameters.

Proof. See [1]

Proposition 9. *The general solution of the equation*

$$x_1 x_2 x_3 = x_3 x_1 x_2 \tag{15}$$

is described by the transformations:

$$\begin{cases} x_1 \to 1 \\ x_2 \to 1 \end{cases} \qquad \begin{cases} x_1 \to (x_1 x_2)^{\alpha} x_1 \\ x_2 \to (x_2 x_1)^{\beta} x_2 \\ x_3 \to (x_1 x_2)^{\gamma} \end{cases}$$

where α, β and γ are natural parameters.

Proof. See [1]

Proposition 10. *The general solution of the equation*

$$x_1 x_3 = x_2^\alpha x_1 \tag{16}$$

where a is natural parameter, is described by the transformations:

$$\text{t)} \begin{cases} x_2 \to 1 \\ x_3 \to 1 \end{cases} \qquad \text{tt)} \begin{cases} x_3 \to 1 \\ \alpha \to 0 \end{cases} \qquad \text{ttt)} \begin{cases} x_1 \to (x_1 x_2)^\beta \, x_1 \\ x_2 \to x_1 x_2 \\ x_3 \to (x_2 x_1)^\alpha \end{cases}$$

where β is a natural parameter.

Proposition 11. *The general solution of the equation*

$$x_1 x_2 x_3 x_4 = x_2 (x_1 x_2)^{\beta+1} x_3 \tag{17}$$

where β is a natural parameter, is described by the transformations:

$$\text{t)} \qquad \begin{cases} x_1 \to x_1^\gamma \\ x_2 \to x_1^\delta \end{cases}$$

$$\langle (16) \text{ with } \alpha = \delta + (\gamma + \delta)\beta \rangle$$

where γ and δ are natural parameters.

$\diamond$

By $^{Fi}(x_1, x_2)^{\lambda_1, \lambda_2, \ldots, \lambda_\sigma}$, where σ is a natural parameter, we denote the function depending on σ, whose values are the functions $^{Fi}(x_1, x_2)$, $^{Fi}(x_1, x_2)^{\lambda_1}$, $^{Fi}(x_1, x_2)^{\lambda_1, \lambda_2, \ldots}$

Proposition 12. *The general solution of the equation*

$$x_1 x_3 x_2 = x_2 x_1 x_4, \qquad \text{with } \partial(x_2) > 0, \partial(x_3) > 0 \tag{18}$$

is described by the transformations:

$$\text{t)} \qquad \begin{cases} x_1 \to {}^{Fi}(x_1 x_3, x_2)^{\lambda_1, \lambda_2, \ldots, \lambda_\sigma} x_1 \\ x_2 \to {}^{Fi}(x_2, x_1 x_3)^{\lambda_2, \ldots, \lambda_\sigma} x_2 \end{cases}$$

$$x_2 \to x_1 x_2$$

$$x_2 \to x_2 x_3$$

$$x_2 \to x_4 x$$

$$\langle (16) \text{ with } \alpha = 1 \rangle$$

By $^{Th}(x_1, x_2, x_3)_i^{\lambda_1, \lambda_2, \ldots, \lambda_{2\sigma}}$, where σ is a natural parameter, we denote the function depending on σ, whose values are the functions $^{Th}(x_1, x_2, x_3)_i$, $^{Th}(x_1, x_2, x_3)_i^{\lambda_1, \lambda_2}$, $^{Th}(x_1, x_2, x_3)_i^{\lambda_1, \lambda_2, \lambda_3, \lambda_4, \ldots}$

Proposition 13. *The general solution of the equation*

$$x_1 x_2 x_3 x_4 = x_3 x_1^2 x_2, \quad \text{with } \partial(x_3) > 0 \tag{19}$$

is described by the transformations **r)** *:*

$$
\begin{cases}
x_1 \to {}^{Th}(x_1, x_2, x_3)_1^{\lambda_1, \lambda_2, \ldots, \lambda_\sigma} x_1 \\
x_2 \to {}^{Th}(x_1, x_2, x_3)_2^{\lambda_1, \lambda_2, \ldots, \lambda_{2\sigma}} x_2 \\
x_3 \to {}^{Th}(x_1, x_2, x_3)_3^{\lambda_1, \lambda_2, \ldots, \lambda_{2\sigma}} x_3 \\
x_4 \to {}^{Oc}_{x_4}(x_1, x_2, x_3)^{\lambda_1, \lambda_2, \ldots, \lambda_{2\sigma}}
\end{cases}
$$

$$x_1 \to x_3^{\lambda_{2\sigma+1}} x_1,$$

$$x_3 \to x_1 x_3$$

$$x_2 \to \left(x_3 x_1^2 (x_3 x_1)^{\lambda_{2\sigma+1}}\right)^{\lambda_{2\sigma+2}} x_2$$

followed by one of the four transformations t), tt), ttt), tttt), where:

$$
\text{t)} \quad
\begin{array}{l}
x_2 \to x_3 x_2 \\
x_1 \to x_2 x_1 \\
x_4 \to x_4 x_2 (x_1 x_3 x_2)^{\lambda_{2\sigma+1}} \\
\langle (17)\text{with} \beta = 0 \rangle
\end{array}
$$

$$
\text{tt)} \quad
\begin{array}{l}
x_2 \to x_3 x_1 (x_1 x_3)^\gamma x_2, \text{with} \gamma < \lambda_{2\sigma+1} \\
x_1 \to x_2 x_1 \\
x_4 \to (x_1 x_3 x_2)^{\lambda_{2\sigma+1}-\gamma} x_1 x_2 (x_1 x_3 x_2)^\gamma \\
\langle (15) \rangle
\end{array}
$$

$$
\text{ttt)} \quad
\begin{array}{l}
x_2 \to x_3 x_1 (x_1 x_3)^\gamma x_1 x_2, \text{with} \gamma < \lambda_{2\sigma+1} \\
x_3 \to x_2 x_3 \\
x_4 \to (x_3 x_1 x_2)^{\lambda_{2\sigma+1}-\gamma-1} x_3 x_1^2 x_2 (x_1 x_3 x_2)^\gamma \\
\langle (15) \rangle
\end{array}
$$

$$
\text{tttt)} \quad
\begin{array}{l}
x_2 \to x_3 x_1 (x_1 x_3)^{\lambda_{2\sigma+1}} x_2 \\
x_1 \to x_2 x_1 \\
x_4 \to x_1 x_2 (x_1 x_3 x_2)^{\lambda_{2\sigma+1}} \\
\langle (15) \rangle
\end{array}
$$

Proposition 14. *The general solution of the equation*

$$x_1 x_2 x_3 x_4 = x_3 x_2 (x_2 (x_3 x_2)^{\lambda+1})^{\mu+1} x_1, \quad \text{with} \partial(x_2 x_3) > 0 \tag{20}$$

where λ, μ are natural parameters is described by the transformations:

$$x_1 \to (x_3 x_2 (x_2 (x_3 x_2)^{\lambda+1})^{\mu+1})^\rho x_1,$$

where ρ is a natural parameter, followed by one of the five transformations t),
tt), ttt), tttt), ttttt) where:

$$
\text{t)} \quad
\begin{array}{l}
x_3 \to x_1 x_3 \\
x_4 \to x_4 x_3 x_2 (x_1 x_3 x_2)^\lambda \left(x_2 (x_1 x_3 x_2)^{\lambda+1}\right)^\mu x_1 \\
\langle (19) \rangle
\end{array}
$$

$$x_3 \to x_3 x_1$$
$$x_1 \to x_1 x_2$$

tt)
$$x_4 \to x_4 x_1 x_2 \left(x_3 x_1 x_2\right)^{\lambda} \left(x_1 x_2 \left(x_3 x_1 x_2\right)^{\lambda+1}\right)^{q} x_3 x_1$$

$\langle (17) \text{with} \alpha = 1 \rangle$

$$x_1 \to x_3 x_2 \left(x_2 \left(x_3 x_2\right)^{\lambda+1}\right)^{\tau} x_1, \text{with} \tau < \mu + 1$$

ttt)
$$x_2 \to x_1 x_2$$
$$x_4 \to x_2 \left(x_3 x_1 x_2\right)^{\lambda} \left(x_1 x_2 \left(x_3 x_1 x_2\right)^{\lambda+1}\right)^{\mu-\tau} x_3 x_1 x_2 \left(x_1 x_2 \left(x_3 x_1 x_2\right)^{\lambda+1}\right)^{\tau} x_1$$

$\langle (15) \rangle$

tttt)
$$x_1 \to x_3 x_2 \left(x_2 \left(x_3 x_2\right)^{\lambda+1}\right)^{\tau} x_2 \left(x_3 x_2\right)^{\upsilon} x_1, \text{with } \tau < \mu + 1 \; rm, \; \upsilon < \lambda + 1$$
$$x_3 \to x_1 x_3$$

followed by one of the three transformations tttt1), tttt2), tttt3) where:

$$x_4 \to x_3 x_2 \left(x_1 x_3 x_2\right)^{\lambda-\upsilon-1} \left(x_2 \left(x_1 x_3 x_2\right)^{\lambda+1}\right)^{\mu-\tau} x_1 x_3 x_2$$

tttt1)
$$\left(x_2 \left(x_1 x_3 x_2\right)^{\lambda+1}\right)^{\tau} x_2 \left(x_1 x_3 x_2\right)^{\upsilon} x_1, \text{with } \lambda > \upsilon.$$

$\langle (15) \rangle$

tttt2)
$$x_4 \to x_3 x_2 \left(x_2 \left(x_1 x_3 x_2\right)^{\lambda+1}\right)^{\tau} x_2 \left(x_1 x_3 x_2\right)^{\upsilon} x_1, \text{with } \lambda = \upsilon, \mu = \tau$$

$\langle (15) \rangle$

tttt3)
$$x_4 \to x_4 x_3 x_2 \left(x_1 x_3 x_2\right)^{\lambda} \left(x_2 \left(x_1 x_3 x_2\right)^{\lambda+1}\right)^{\mu-\tau-1} x_1 x_3 x_2$$
$$\left(x_2 \left(x_1 x_3 x_2\right)^{\lambda+1}\right)^{\tau} x_2 \left(x_1 x_3 x_2\right)^{\lambda} x_1, \text{with} \lambda = \upsilon, \mu > \tau.$$

followed by the transformation $\langle (19) \rangle$ or $\langle < (16) \text{with} \alpha = 0 \rangle$.

ttttt)
$$x_1 \to x_3 x_2 \left(x_2 \left(x_1 x_3 x_2\right)^{\lambda+1}\right)^{\tau} x_2 \left(x_3 x_2\right)^{\upsilon} x_3 x_1, \text{with } \tau < \mu + 1, \upsilon < \lambda + 1$$
$$x_2 \to x_1 x_2$$

followed by one of the three transformations ttttt1), ttttt2), ttttt3) where:

$$x_4 \to x_2 \left(x_3 x_1 x_2\right)^{\lambda-\upsilon-1} \left(x_1 x_2 \left(x_3 x_1 x_2\right)^{\lambda+1}\right)^{\mu-\tau} x_3 x_1 x_2$$

ttttt1)
$$\left(x_1 x_2 \left(x_3 x_1 x_2\right)^{\lambda+1}\right)^{\tau} x_1 x_2 \left(x_3 x_1 x_2\right)^{\upsilon} x_3 x_1, \text{with } \lambda > \upsilon.$$

$\langle (15) \rangle$

ttttt2)
$$x_4 \to x_2 \left(x_1 x_2 \left(x_3 x_1 x_2\right)^{\lambda+1}\right)^{\tau} x_1 x_2 \left(x_3 x_1 x_2\right)^{\upsilon} x_3 x_1, \text{with } \lambda = \upsilon, \mu = \tau$$

$\langle (15) \rangle$

ttttt3)
$$x_4 \to x_1 x_2 \left(x_3 x_1 x_2\right)^{\lambda} \left(x_1 x_2 \left(x_3 x_1 x_2\right)^{\lambda+1}\right)^{\mu-\tau-1} x_3 x_1 x_2$$
$$\left(x_1 x_2 \left(x_3 x_1 x_2\right)^{\lambda+1}\right)^{\tau} x_1 x_2 \left(x_3 x_1 x_2\right)^{\upsilon} x_3 x_1$$

$\langle (17) \rangle, \text{ with } \alpha = 0, \lambda = \upsilon, \mu > \tau.$

Proposition 15. *The general solution of the equation*

$$x_1 x_2 x_3 x_4 = (x_3 x_1)^{\mu+2} x_2, \text{with} \partial(x_3) > 0 \tag{21}$$

where μ is a natural parameter is described by the transformations:

$$x_1 \to x_3^{\lambda} x_1$$
$$x_3 \to x_1 x_3$$
$$\langle(20)\rangle .$$

Proposition 16. *The general solution of the equation*

$$x_1 x_2 x_3 x_4 = x_2 (x_1 x_2)^{\tau} x_3 x_1 \Psi(x_1, x_2, x_3) \tag{22}$$

where τ is a natural parameter, $\Psi(x_1, x_2, x_3)$ is a parametric word, is described by the transformations:

$$\tau \to \alpha + 1 \qquad\qquad \tau \to 0$$
t) $x_4 \to x_4 x_1 \Psi(x_1, x_2, x_3),$ **tt)** $x_4 \to \Psi(x_1, x_2, x_3),$
$$\langle(17)\rangle \qquad\qquad\qquad \langle(15)\rangle$$

Proposition 17. *The general solution of the equation*

$$x_1 x_2 x_3 x_4 = x_3 x_2^{\mu+2} x_1, \text{with} \partial(x_2) > 0 \tag{23}$$

where μ is a natural parameter is described by the transformations:

$$\begin{cases} x_1 \to {}^{Fi}\left(x_1 x_2, x_3 x_2^{\mu+2}\right)^{\lambda_1, \lambda_2, \dots, \lambda_s} x_1 \\ x_3 \to {}^{Fi}\left(x_3 x_2^{\mu+2}, x_1 x_2\right)^{\lambda_2, \dots, \lambda_s} x_3 \end{cases}$$

followed by one of the two transformations ,t), tt) where:

$$x_3 \to x_1 x_3 \qquad x_1 \to x_3 x_2^{\upsilon} x_1 \quad (\upsilon < \mu + 2)$$
t) $x_2 \to x_3 x_2$ **tt)** $x_2 \to x_1 x_2$
$$\langle(21)\rangle \qquad\qquad \langle(22)\rangle$$

Proposition 18. *The general solution of the equation*

$$x_1 x_2 x_3 x_4 = x_2^{\mu+3} x_3 x_1, \text{with} \partial(x_2) > 0 \tag{24}$$

where μ is a natural parameter is described by the transformations:

$$x_1 \to x_2^{\upsilon} x_1 \quad (\upsilon < \mu + 3)$$
t) $\begin{array}{c} x_1 \to x_2^{\mu+3} x_1 \\ \langle(23)\rangle \end{array}$ **tt)** $x_2 \to x_1 x_2$
$$\langle(22)\rangle$$

Proposition 19. *The general solution of the equation*

$$x_1 x_2 x_3 x_4 = x_3 x_1^{\mu+3} x_2 \tag{25}$$

where μ is a natural parameter, is described by the transformations:

t) $\begin{array}{l} x_1 \to 1 \\ x_4 \to 1 \\ \langle(14)\rangle \end{array}$ tt) $\begin{array}{l} x_3 \to (x_1 x_2)^{\lambda} \\ \langle(16)\text{with}\alpha = \mu + 2\rangle \end{array}$

ttt) $\begin{array}{l} x_3 \to (x_1 x_2)^{\lambda} x_3 \\ x_1 \to x_3 x_1 \\ \langle(21)\rangle \end{array}$ tttt) $\begin{array}{l} x_3 \to (x_1 x_2)^{\lambda} x_1 x_3 \\ x_2 \to x_3 x_2 \\ \langle(24)\rangle \end{array}$

Proposition 20. *The general solution of the equation*

$$x_1 x_2 x_3 x_4 = x_3 x_1^{\lambda} x_2 x_5 \tag{26}$$

where λ is a natural parameter, is described by the transformations:

t1) $\begin{array}{l} \lambda \to 0 \\ x_1 \to x_3 x_1 \\ x_5 \to x_5 x_3 x_4 \\ \langle(16)\text{with}\alpha = 1\rangle \end{array}$ t2) $\begin{array}{l} \lambda \to 0 \\ x_3 \to x_1 x_3 \\ x_5 \to x_5 x_4 \\ \langle(18)\rangle \end{array}$ t3) $\begin{array}{l} \lambda \to 0 \\ x_1 \to 1 \\ x_5 \to x_4 \\ \langle(14)\rangle \end{array}$

t4) $\begin{array}{l} \lambda \to 1 \\ x_5 \to x_4 \\ \langle(15)\rangle \end{array}$ t5) $\begin{array}{l} \lambda \to 2 \\ x_4 \to x_4 x_5 \\ \langle(19)\rangle \end{array}$ t6) $\begin{array}{l} \lambda \to 2 \\ x_3 \to 1 \\ x_5 \to x_4 x_5 \\ \langle(16)\text{with}\alpha = 1\rangle \end{array}$

t7) $\begin{array}{l} \lambda \to \mu + 3 \\ x_4 \to x_4 x_5 \\ \langle(25)\rangle \end{array}$

References

1. Yu. I. Kmelevski. Equations in free semigroups, Trudy Mat. Inst. Steklov. 107 (1971); English transl. Proc. Steklov Inst. Math. 107 (1971). (1976).
2. G.S. Makanin. On general solution of equations in free semigroups, proceedings of IWWERT'91, LNCS n. 677. Edited by H. Abdulrab and J.P. Pcuchet. pp. 1-5.
3. G.S. Makanin, H. Abdulrab. On General Solution of Word Equations. Proceedings of "Important Results and Trends in Theoretical Computer Sciences", 9-10 June 1994, Graz, Autriche. LNCS, n: 812, p.p. 251-263.

A Proof Procedure for Hereditary Harrop Formulas with Free Equality

Evgeny Makarov

Department of Mathematical Logic
Faculty of Mathematics and Mechanics
Moscow State University
Moscow, 119899 Russia
e-mail: zhenya@lpcs.math.msu.ru

Abstract. We use a proof procedure for hereditary Harrop formulas to infer facts from programs containing Clark's Equational Theory (CET). In comparison with PROLOG, this allows to establish not only unifiability but also non-unifiability of terms. The described proof procedure is sound and complete w.r.t. minimal logic. As an interesting application, we translate program completion into hereditary Harrop formulas. SLDNF-resolution proves to be sound w.r.t. this translation. Since the described proof procedure treats negation as inconsistency, this kind of negation turns out to be a more general notion than negation as failure.

1 Introduction

One of the ways to generalize SLD-resolution is to extend the class of formulas permissible as clauses to first order hereditary Harrop formulas (hhf). The class of hereditary Harrop formulas (see [7]) properly includes Horn formulas. A hhf may contain implication in the clause body, and a goal $D \to A$ is deemed to succeed from a program P if A succeeds from $P \cup \{D\}$. In particular, $\neg A$, being defined by $A \to \bot$, is proved by inferring contradiction from a program augmented with A. Such an interpretation of negation was introduced in [2] and was called "negation as inconsistency".

The research presented in this paper originated with a question about the relation between negation as failure (NAF) and negation as inconsistency (NAI). We show that NAI is a more general notion. This is done by translating Clark's program completion $comp(P)$ into hhf (the result is denoted by $hcomp(P)$) and using a proof procedure for hhf suggested by G. Nadathur in [9]. The main result states that for a program P and a goal G (both of which can contain negative literals), if G returns a computed answer substitution (c.a.s.) θ via SLDNF-resolution then G^* returns θ from $hcomp(P)$ via this procedure for hhf, where G^* is obtained from G by substituting $C \to \bot$ for all occurrences of $\neg C$.

In order to model NAF, we must be able to prove the non-unifiability of two terms. We achieve this by including axioms of Clark's Equational Theory (CET) in $hcomp(P)$. Though $hcomp(P)$ becomes infinite, it does not make the implementation of the described proof procedure impossible since CET contains only a finite number of axiom schemes.

Our approach is similar to that of J. Harland [4] but our proof procedure is different. Harland's operational notion of provability $\vdash_s$ is not suitable for implementation (for example, it does not employ unification). We use Nadathur's proof procedure which can be efficiently implemented.

J. Harland proved completeness of NAF w.r.t. NAI as well. His main result states that for locally consistent programs P (the class of programs that includes locally stratified programs)

$$P \vdash_s G \Leftrightarrow hcomp(P) \vdash_s G^* .$$

Our result shows that local consistency is relevant only for ($\Leftarrow$).

The rest of the paper is organized as follows. Section 2 presents minimal logic. Section 3 introduces the definitions of hhf, programs and the proof procedure for hhf. It also states that this proof procedure is sound and complete w.r.t. minimal logic and that we can infer the necessary facts about unification. Section 4 describes SLDNF-resolution and contains our main result about the relation between NAF and NAI. Finally, section 5 is a conclusion.

2 Logical Preliminaries

This section describes minimal Gentzen's sequent calculus (MGC). We consider the first order language with a countable set of constants. The connectives are: $\wedge$, $\vee$ and $\rightarrow$ ($\neg$ is not used, $\neg A$ being interpreted as $A \rightarrow \perp$). There are also quantifiers $\forall$ and $\exists$.

The language has two special predicate symbols: $\perp$ of arity 0 and $=$ of arity 2. If F is a formula or some construct that includes formulas and terms then $Var(F)$ and $Const(F)$ denote respectively the set of free variables or constants in F or in formulas and terms occurring in F. As usually, if $Var(F) = \{x_1, \ldots, x_n\}$ then $\forall F$ ($\exists F$) denotes $\forall x_1 \ldots \forall x_n F$ (respectively $\exists x_1 \ldots \exists x_n F$).

A finite sequence of terms $t_1, \ldots, t_n$ is denoted by $\mathbf{t}$. Given two sequences of terms $\mathbf{t}$ and $\mathbf{s}$ of length n $\mathbf{t} = \mathbf{n}$ denotes the formula $t_1 = s_1 \wedge \ldots \wedge t_n = s_n$.

Sequents have the form $\Gamma \Rightarrow C$ where C is a formula and Γ is a (possibly empty) set of formulas. The union $\Gamma \cup \{A\}$ will be denoted by Γ, A. The only axiom scheme is $A \Rightarrow A$ where A is any formula. Rules of inference are shown in Fig. 1. In the last four rules w is a free variable in formula A and t is a term. In rules ($\Rightarrow\forall$) and ($\exists\Rightarrow$) their conclusion must not contain w.

If Γ is a set of formulas then we write $\Gamma \vdash_M C$ if for some finite subset Γ' of Γ the sequent $\Gamma' \Rightarrow C$ is derivable in MGC.

3 Proof Procedure for Hereditary Harrop Formulas

Definition 1. G- and D-formulas are defined by the following syntactic rules where A denotes atomic formulas:

$$G := A \mid G \wedge G \mid G \vee G \mid \exists x G \mid D \rightarrow G \mid \forall x G$$

$$\frac{\Gamma, A \Rightarrow C}{\Gamma, A \wedge B \Rightarrow C}(\wedge\Rightarrow) \qquad \frac{\Gamma, B \Rightarrow C}{\Gamma, A \wedge B \Rightarrow C}(\wedge\Rightarrow)$$

$$\frac{\Gamma \Rightarrow A \quad \Gamma \Rightarrow B}{\Gamma \Rightarrow A \wedge B}(\Rightarrow\wedge) \qquad \frac{\Gamma, A \Rightarrow C \quad \Gamma, B \Rightarrow C}{\Gamma, A \vee B \Rightarrow C}(\vee\Rightarrow)$$

$$\frac{\Gamma \Rightarrow A}{\Gamma \Rightarrow A \vee B}(\Rightarrow\vee) \qquad \frac{\Gamma \Rightarrow B}{\Gamma \Rightarrow A \vee B}(\Rightarrow\vee)$$

$$\frac{\Gamma \Rightarrow A \quad \Gamma, B \Rightarrow C}{\Gamma, A \to B \Rightarrow C}(\to\Rightarrow) \qquad \frac{\Gamma, A \Rightarrow B}{\Gamma \Rightarrow A \to B}(\Rightarrow\to)$$

$$\frac{\Gamma \Rightarrow A(w)}{\Gamma \Rightarrow \forall x A(x)}(\Rightarrow\forall) \qquad \frac{A(t), \Gamma \Rightarrow C}{\forall x A(x), \Gamma \Rightarrow C}(\forall\Rightarrow)$$

$$\frac{\Gamma \Rightarrow A(t)}{\Gamma \Rightarrow \exists x A(x)}(\Rightarrow\exists) \qquad \frac{A(w), \Gamma \Rightarrow C}{\exists x A(x), \Gamma \Rightarrow C}(\exists\Rightarrow)$$

Fig. 1. Inference rules of minimal logic

$$D := A \mid G \to A \mid D \wedge D \mid \forall x D.$$

D-formulas are called *first order hereditary Harrop formulas* (hhf for short). G-formulas are called *goals*.

Definition 2. A *clause* is a hhf of the form $\forall x_1 \ldots \forall x_n A$ or $\forall x_1 \ldots \forall x_n (G \to A)$, $n \geq 0$ where A is an atom and G is a goal. A *program* is a (possibly infinite) set of clauses.

Definition 3. The set of clauses $cl(D)$ contained in a hhf D is defined as follows.

1. If D is atomic or of the form $G \to A$ then $cl(D) = \{D\}$.
2. If D is $D_1 \wedge D_2$ then $cl(D) = cl(D_1) \cup cl(D_2)$.
3. If D is $\forall x D'$ then $cl(D) = \{\forall x D'' \mid D'' \in cl(D')\}$.

Let us consider how a goal can be proved from a program. The definition of a goal makes it possible to determine on each step how to resolve a given goal (cf. the notion of *uniform proof* in [7]). The only difficulty in actual implementation is encountered with universal quantifiers. It turns out that it is inadequate to merely substitute a new constant for the universally bound variable since this constant can be returned in computed answer substitution which is undesirable.

The solution is the following. First, constants and variables are partitioned into a countable collection of countable sets.

Definition 4. A mapping $\mathcal{L}$ from constants and variables to non-negative integers is called a labeling function if for all $i \geq 0$ the sets $\{c \mid c$ is a constant and $\mathcal{L}(c) = i\}$ and $\{c \mid x$ is a variable and $\mathcal{L}(x) = i\}$ are countable.

Then, the *current constant level l* is introduced. Suppose, a labeling function $\mathcal{L}$ is fixed. Each time a universal quantifier is encountered during computation l is increased by one and a new constant c such that $\mathcal{L}(c) = l$ is selected . Lastly, it is prohibited to assign a term t to a variable x during unification if t contains a constant c such that $\mathcal{L}(x) < \mathcal{L}(c)$. As a final remark, before assigning a term t to a variable x the value of $\mathcal{L}(y)$ should be set to $\mathcal{L}(x)$ for those variables y occurring in t that $\mathcal{L}(y) > \mathcal{L}(x)$. When existential quantifier is encountered during computation a new variable z is introduced and $\mathcal{L}(z)$ is set to the current constant level l.

Though the formal definition will be given below the following example, borrowed from [9], will clarify the matter.

Example 1. Consider the following goal $G = \exists x \forall y \exists z \, (p(x, f(z)) \wedge p(y, z))$ and a program $P = \{\forall x p(x, x)\}$. Let at the beginning the current constant level equal zero. One can see that G does not succeed from P though both formulas $G' = \exists x \exists z \, p(x, f(z))$ and $G'' = \forall y \exists z \, p(y, z)$ do.

In order to define the proof procedure for hhf we first give precise definition of the most general unifier (mgu) and the relevant notions in our setting.

The substitution $\theta = \{t_1/x_1, \ldots, t_n/x_n\}$ maps variables $x_1, \ldots, x_n$ to terms $t_1, \ldots, t_n$. If all t_i are ground then θ is called *ground*. ϵ denotes the identity substitution. The restriction of a substitution θ to a set V of variables is denoted by $\theta \,|\, V$. The result of application of a substitution to a term or a formula is defined as usual (this operation does not change bound variables).

Definition 5. Let $\theta = \{t_1/x_1, \ldots, t_n/x_n\}$ be a substitution. θ is *proper* w.r.t. to a labeling function $\mathcal{L}$ if, for $1 \leq i \leq n$, $\mathcal{L}(c) \leq \mathcal{L}(x_i)$ for every constant c appearing in t_i. The labeling *induced* by θ from $\mathcal{L}$ is the labeling function $\mathcal{L}'$ defined as follows:

$$\mathcal{L}'(x) = min(\{\mathcal{L}(x)\} \cup \{\mathcal{L}(x_i) \mid \{t_i/x_i\} \subseteq \theta \text{ and } x \text{ appears in } t_i\}).$$

The value of $\mathcal{L}'$ on constants coincides with that of $\mathcal{L}$.

The composition of two substitutions is defined as usual. We adopt the usual definition of mgu but introduce a new notion of *a most general unifier relative to a labeling function $\mathcal{L}$*.

Definition 6. A substitution θ_1 is more general than θ_2 relative to a labeling function $\mathcal{L}$ if θ_1 and θ_2 are proper w.r.t. $\mathcal{L}$ and there is a substitution σ that is proper w.r.t. the labeling induced by θ_1 from $\mathcal{L}$ such that $\theta_2 = \theta_1 \sigma$.

Definition 7. A pair of sequences of the same length consisting of terms or atomic formulas is called a *disagreement set*. A substitution θ is a *unifier* of the disagreement set $T = \langle t, s \rangle$ relative to $\mathcal{L}$ if θ is proper w.r.t. $\mathcal{L}$ and $t_i \theta = s_i \theta$ for each i. In this case T is called unifiable relative to $\mathcal{L}$. If $T = \langle t, s \rangle$ then θ is called the unifier of t and s relative to $\mathcal{L}$. A most general unifier for T relative to $\mathcal{L}$ is a unifier θ that is more general as a substitution than any other unifier relative to $\mathcal{L}$.

Theorem 8 (Nadathur [9]). *There is an algorithm which for any disagreement set T and labeling function $\mathcal{L}$ produces a most general unifier of T relative to $\mathcal{L}$ if T is unifiable relative to $\mathcal{L}$ and otherwise reports nonexistence of a unifier.*

We now describe the procedure which determines whether there is a proof of a goal from a program. A *state* S is a tuple

$$S = \langle \{\langle G_1, P_1, l_1 \rangle, \ldots, \langle G_k, P_k, l_k \rangle\}, \mathcal{C}, \mathcal{V}, \mathcal{L} \rangle . \tag{1}$$

Here for $i = 1, \ldots, k$ G_i are goals, P_i are programs, l_i are non-negative integers (representing current constants levels), $\mathcal{C}$ and $\mathcal{V}$ are the sets of constants and variables, respectively, and $\mathcal{L}$ is a labeling function.

Intuitively, $\mathcal{C}$ and $\mathcal{V}$ are sets of constant and variables that are prohibited to be picked as "fresh". A program P is kept with each goal because a program can be enlarged when resolving a goal of the form $D \to G'$. Finally, new triples $\langle G, P, l \rangle$ are added when a goal of the form $G_1 \wedge G_2$ is being resolved. For more discussion see [2], [3] and [9].

There are 6 (nondeterministic) computation rules that transform a state S into a state S' via some substitution σ (notation: $S \overset{\sigma}{\Rightarrow} S'$). Let S be as in (1) and let $1 \leq i \leq k$.

(R1) If G_i is atomic, there is a renaming substitution $\eta = \{w_1/x_1, \ldots, w_n/x_n\}$ such that, for $1 \leq j \leq n$, $w_j \notin \mathcal{V}$ and $\mathcal{L}(w_j) = l_i$, there is a clause $\forall x_1 \ldots \forall x_n (B \leftarrow G) \in P_i$ such that G_i and $B\eta$ have a most general unifier σ relative to $\mathcal{L}$ and $\mathcal{L}_1$ is induced from $\mathcal{L}$ by σ then $S \overset{\sigma}{\Rightarrow} S'$ where $S' = \langle \{\langle G_j\sigma, P_j\sigma, l_j \rangle \mid j \neq i\} \cup \{\langle G\sigma, P_i\sigma, l_i \rangle\}, \mathcal{C}, \mathcal{V} \cup \{w_1, \ldots, w_n\}, \mathcal{L}_1 \rangle$. As a special case, if the clause body G is empty, the tuple $\langle G\sigma, P_i\sigma, l_i \rangle$ is absent in S'.

(R2) If $G_i = G' \wedge G''$ then $S \overset{\epsilon}{\Rightarrow} \langle \{\langle G_j, P_j, l_j \rangle \mid j \neq i\} \cup \{\langle G', P_i, l_i \rangle, \langle G'', P_i, l_i \rangle\}, \mathcal{C}, \mathcal{V}, \mathcal{L} \rangle$.

(R3) If $G_i = G' \vee G''$ then $S \overset{\epsilon}{\Rightarrow} \langle \{\langle G_j, P_j, l_j \rangle \mid j \neq i\} \cup \{\langle G, P_i, l_i \rangle\}, \mathcal{C}, \mathcal{V}, \mathcal{L} \rangle$ where $G = G'$ or $G = G''$.

(R4) If $G_i = D \to G$ then $S \overset{\epsilon}{\Rightarrow} \langle \{\langle G_j, P_j, l_j \rangle \mid j \neq i\} \cup \{\langle G, P_i \cup cl(D), l_i \rangle\}, \mathcal{C}, \mathcal{V}, \mathcal{L} \rangle$.

(R5) If $G_i = \exists x G$ then $S \overset{\epsilon}{\Rightarrow} \langle \{\langle G_j, P_j, l_j \rangle \mid j \neq i\} \cup \{\langle G\{w/x\}, P_i, l_i \rangle\}, \mathcal{C}, \mathcal{V} \cup \{w\}, \mathcal{L} \rangle$ where $w \notin \mathcal{V}$ and $\mathcal{L}(w) = l_i$.

(R6) If $G_i = \forall x G$ then $S \overset{\epsilon}{\Rightarrow} \langle \{\langle G_j, P_j, l_j \rangle \mid j \neq i\} \cup \{\langle G\{c/x\}, P_i, l_i + 1 \rangle\}, \mathcal{C} \cup \{c\}, \mathcal{V}, \mathcal{L} \rangle$ where $c \notin \mathcal{C}$ and $\mathcal{L}(c) = l_i + 1$.

A possibly infinite sequence of states $S_0, S_1, \ldots$ together with a sequence of substitutions $\sigma_0, \sigma_1, \ldots$ such that for all $i = 0, 1, \ldots S_i \overset{\sigma_{i+1}}{\Rightarrow} S_{i+1}$ is called a *derivation*. A derivation is called *successful* if it is finite and the last state has the form $\langle \emptyset, \mathcal{C}, \mathcal{V}, \mathcal{L} \rangle$.

Below, if P is a program and G is a goal $P \cup \{G\}$ will be denoted by P, G. A program consisting of only closed clauses will be called *closed* program.

Consider a closed program P and a goal G. Let $\mathcal{C} = Const(P, G)$, $\mathcal{V} = Var(P, G)$ and $\mathcal{L}$ be a labeling function such that $\mathcal{L}$ equals zero over $\mathcal{C}$ and $\mathcal{V}$. Suppose there is a successful derivation $\mathbf{S} = S_0 \overset{\sigma_1}{\Rightarrow} \ldots \overset{\sigma_n}{\Rightarrow} S_n$ where $S_0 = \langle \{\langle G, P, 0 \rangle\}, \mathcal{C}, \mathcal{V}, \mathcal{L} \rangle$. Then $\mathbf{S}$ is called a derivation of G from P with c.a.s. $\theta = \sigma_1 \ldots \sigma_n | Var(G)$ (notation: $GR^{ni}(P)\theta$.

One can easily see that our proof procedure is monotonic, i.e. if $GR^{ni}(P)\theta$ then $GR^{ni}(P \cup P')\theta$.

The following are immediate corollaries of the results proved in [9].

Theorem 9 (Soundness). *Let P be a closed program and G a goal. If $GR^{ni}(P)\theta$ then $P \vdash_M G\theta$.*

Theorem 10 (Completeness). *Let P be a closed program, G a goal and σ a substitution acting only on $Var(G)$. Then if $P \vdash_M G\sigma$ then $GR^{ni}(P)\theta$ and $\sigma = \theta\eta$ for some η.*

We define $CET(F)$ for each set F of formulas or terms. The idea is to have axioms of the form $c = d \to \bot$ only for $c, d \in Const(F)$ because other constants can represent universally bound variables. The axioms of $CET(F)$ are:

1. $\forall x \; x = x$.
2. $\forall (\mathbf{x} = \mathbf{y} \to f(\mathbf{x}) = f(\mathbf{y}))$ for each function symbol f.
3. $\forall (\mathbf{x} = \mathbf{y} \to (p(\mathbf{x}) \to p(\mathbf{y})))$ for each predicate symbol p including $=$.
4. $c = d \to \bot$ for all distinct constants c and d such that $c, d \in Const(F)$.
5. $\forall (f(\mathbf{x}) = f(\mathbf{y}) \to x_i = y_i)$ for each n-ary function symbol f and for all $i = 1, \ldots, n$.
6. $\forall (f(\mathbf{x}) = g(\mathbf{y}) \to \bot)$ for all function symbols f and g such that $f \neq g$.
7. $\forall (x = t \to \bot)$ for each variable x and term t distinct from x such that x occurs in t.

Note that all these axioms are clauses. So we obtain the following corollary of soundness and completeness theorems.

Corollary 11. *Let P be a closed program and G a goal. Then*

$$GR^{ni}(P \cup CET(P, G))\epsilon \text{ iff } P \cup CET(P, G) \vdash_M G .$$

It is well-known that CET can prove necessary facts about unification (see [6]). Namely, the following statement holds.

Lemma 12. *Let $T = \langle \mathbf{s}, \mathbf{t} \rangle$ be a disagreement set. If T has a unifier (in an ordinary sense) then for some of its mgu $\{u_1/x_1, \ldots, u_k/x_k\}$ $CET(T) \vdash_M \forall (\mathbf{s} = \mathbf{t} \to \mathbf{x} = \mathbf{u})$ Otherwise $CET(T) \vdash_M \forall (\mathbf{s} = \mathbf{t} \to \bot)$.*

Corollary 13. *Let* $T = \langle \mathbf{s}, \mathbf{t} \rangle$ *be a disagreement set.*

1. *If* T *has a unifier then for some of its mgu* $\{u_1/x_1, \ldots, u_k/x_k\}$ $\forall(\mathbf{s} = \mathbf{t} \rightarrow \mathbf{x} = \mathbf{u})\mathbf{R}^{ni}(\mathrm{CET}(T))\epsilon.$
2. T *has no unifier iff* $\forall(\mathbf{s} = \mathbf{t} \rightarrow \perp)\mathbf{R}^{ni}(\mathrm{CET}(T))\epsilon.$

4 Soundness of NAF w.r.t. NAI

The definition of SLDNF-resolution and all related notions is customary (see [1], [5]). First, a new connective $\neg$ is added to the language. A *literal* is of the form A or $\neg A$ where A is an atom. A (*general*) *query* is a finite conjunction of literals. The empty general query is denoted by $\square$.

A *general clause*, or just clause, if there is no danger to confuse it with a hhf, is a formula of the form $A \leftarrow G$ where A is an atom and G is a query. A *general program* is a finite set of general clauses. We require that general programs and general goals do not contain equalities, i.e. atoms of the form $t = s$, and the atom $\perp$.

We present Kunen's definition of SLDNF-resolution.

Definition 14. Given a program P, the set $\mathbf{F}(P)$ of queries and the set $\mathbf{R}^{nf}(P)$ of pairs $\langle G, \theta \rangle$, where G is a query and θ is a substitution acting only on free variables of G, are defined by mutual induction as follows.

0) $\square \mathbf{R}^{nf}(P)\epsilon.$

R+) If $G = G^{(1)} \wedge A \wedge G^{(3)}$, there is a variant of a clause $B \leftarrow G^{(2)}$ from P with fresh variables such that A and B are unifiable with a most general unifier σ and $(G^{(1)} \wedge G^{(2)} \wedge G^{(3)})\sigma \mathbf{R}^{nf}(P)\theta$ then $G\mathbf{R}^{nf}(P)\sigma\theta|Var(G)$

R−) If $G = G^{(1)} \wedge \neg A \wedge G^{(2)}$, A is a ground atom in $\mathbf{F}(P)$ and $(G^{(1)} \wedge G^{(2)})\mathbf{R}^{nf}(P)\theta$ then $G\mathbf{R}^{nf}(P)\theta.$

F+) Suppose $G = G^{(1)} \wedge A \wedge G^{(3)}$ and for all variants of a clause $B \leftarrow G^{(2)}$ from P with fresh variables such that A and B are unifiable if σ is a most general unifier of A and B then $(G^{(1)} \wedge G^{(2)} \wedge G^{(3)})\sigma \in \mathbf{F}(P)$. Then $G \in \mathbf{F}(P)$.

F−) If $G = G^{(1)} \wedge \neg A \wedge G^{(2)}$, A is a ground atom such that $A\mathbf{R}^{nf}(P)\epsilon$ then $G \in \mathbf{F}(P)$.

$G\mathbf{R}^{nf}(P)\theta$ is read 'a query Q returns c.a.s. θ from P via SLDNF-resolution' and $G \in \mathbf{F}(P)$ is read 'a query G finitely fails from P'.

The declarative semantics of a general program can be described using Clark's completion which is defined in the following way. Let P be a general program. Suppose that for a predicate symbol p of arity n

$$p(\mathbf{t}_1) \leftarrow G_1, \ldots, p(\mathbf{t}_k) \leftarrow G_k \tag{2}$$

are all clauses from P that have p in their heads. Denote by $\mathrm{iff}(p)$ the following formula

$$\forall \mathbf{x} \left(p(\mathbf{x}) \leftrightarrow \bigvee_{i=1}^{k} \exists \mathbf{y}_i (\mathbf{x} = \mathbf{t}_i \wedge G_i) \right)$$

where **x** are new variables, i.e. they do not appear in any of clauses above, and y_i are original variables of the ith clause. If p is not encountered in the head of any clause from P then iff(p) is $\forall \mathbf{x} \neg p(\mathbf{x})$. The completion $comp(P)$ of P is $\{\text{iff}(p) \mid p \text{ is a predicate symbol}\} \cup \text{CET}(P)$.

We transform the completion of a general program P into a set of hhf, denoted by $hcomp(P)$, in a classically, though not intuitionistically, equivalent way. First, for each formula F we denote by F^* the formula obtained from F by replacing every occurrence of $\neg C$ by $C \to \bot$.

Let, as above, for a predicate symbol p (2) be all clauses from P with p in their heads. Then we denote by only_if(p) the following formula

$$\forall \mathbf{x} \left(\bot \leftarrow p(\mathbf{x}) \wedge \bigwedge_{i=1}^{k} \forall y_i (\mathbf{x} = t_i \wedge G_i^* \to \bot) \right)$$

where **x** and y_i $1 \le i \le k$ are as above.

If p is not encountered in the head of any general clause from P then only_if(p) is $\forall \mathbf{x} (\bot \leftarrow p(\mathbf{x}))$. Now

$$hcomp(P) = \{\forall (A \leftarrow G^*) \mid A \leftarrow G \in P\}$$
$$\cup \{\text{only\_if}(p) \mid p \text{ is a predicate symbol}\} \cup \text{CET}(P) \ .$$

Note that any formula from $hcomp(P)$ is a clause in the sense of hhf, so $hcomp(P)$ is a program. The relation between NAF and NAI is clarified by the following statement.

Theorem 15. *Let P be a general program and G a general query. Then*

1. *if $\mathbf{GR}^{nf}(P)\theta$ then $G^* \mathbf{R}^{ni}(hcomp(P))\theta$;*
2. *if $G \in \mathbf{F}(P)$ then $\forall (G^* \to \bot) \mathbf{R}^{ni}(hcomp(P))\epsilon$.*

To prove this theorem we need two technical lemmas. The first one is rather clear and we omit its proof.

Lemma 16. *Let P be a closed program (in the sense of hhf), G a closed goal and $\mathcal{L}$ a labeling function. Let also $l = \max\{\mathcal{L}(c) \mid c \in Const(P,G)\}$. Then the following are equivalent:*

1. *$\mathbf{GR}^{ni}(P)\epsilon$.*
2. *There is a successful derivation $S_0, S_1, \ldots$ such that $S_0 = \langle \{\langle G, P, l \rangle\}, Const(P,G), \emptyset, \mathcal{L} \rangle$.*

Lemma 17. *Let $F(\mathbf{x})$ be a formula and $T = \langle \mathbf{s}(\mathbf{x}), \mathbf{t}(\mathbf{x}) \rangle$ be a disagreement set ($\mathbf{s}$ and $\mathbf{t}$ are sequences of terms). Suppose all free variables of F, $\mathbf{s}$ and $\mathbf{t}$ are among $\mathbf{x} = x_1, \ldots, x_n$. Further, let $\sigma = \{u_1/x_1, \ldots, u_n/x_n\}$ be a mgu of T and $\eta = \{c_1/x_1, \ldots, c_n/x_n\}$ be a substitution such that each c_i is a distinct constant not in $Const(\{F, \mathbf{s}, \mathbf{t}\})$. Then $\text{CET}(\{F, \mathbf{s}, \mathbf{t}\}), \mathbf{s}\eta = \mathbf{t}\eta, F\eta \vdash_M F\sigma\eta$.*

Proof. Let $P = \text{CET}(\{F, s, t\})$. By Corollary 13 $\forall(s = t \rightarrow x = u)\mathbf{R}^{ni}(P)\epsilon$. Therefore, by Lemma 16 $(s\eta = t\eta \rightarrow c = u\eta)\mathbf{R}^{ni}(P)\epsilon$. By Soundness Theorem 9 $P, s\eta = t\eta \vdash_M c = u\eta$. Since $F\eta = F(c)$, $F\sigma\eta = F(u\eta)$ and P is a theory with equality (see, e.g., [8]) $P, c = u\eta, F\eta \vdash_M F\sigma\eta$; hence, the lemma holds. $\square$

Let us introduce a new notation. Let $\mathcal{G}_i = \{\langle G^i_j, P^i_j \rangle \mid 1 \leq j \leq k_i\}$, $i = 1, 2$. Then we write $\mathcal{G}_1 \Rightarrow \mathcal{G}_2$ if for some non-negative integers $l^i_j, 1 \leq j \leq k_i$, sets of constants and variables $\mathcal{C}_i$ and $\mathcal{V}_i$, labeling functions $\mathcal{L}_i, i = 1, 2$ and substitution σ $\langle \mathcal{G}_1, \mathcal{C}_1, \mathcal{V}_1, \mathcal{L}_1 \rangle \overset{\sigma}{\Rightarrow} \langle \mathcal{G}_2, \mathcal{C}_2, \mathcal{V}_2, \mathcal{L}_2 \rangle$.

Proof of Theorem 15. The proof is by induction on the definition of the sets $\mathbf{R}^{nf}(hcomp(P))$ and $\mathbf{F}(hcomp(P))$. The case 0) is clear.

Case R+). By induction hypothesis $(G^{(1)} \wedge G^{(2)} \wedge G^{(3)})^* \sigma \mathbf{R}^{ni}(hcomp(P))\theta$. But $\forall(B \leftarrow G) \in hcomp(P)$; therefore, $G^* \mathbf{R}^{ni}(hcomp(P))\sigma\theta|Var(G)$.

Case R-). According to induction hypothesis $(A \rightarrow \bot)\mathbf{R}^{ni}(hcomp(P))\epsilon$ and $(G^{(1)} \wedge G^{(2)})^* \mathbf{R}^{ni}(hcomp(P))\theta$. Therefore, $G^* \mathbf{R}^{ni}(hcomp(P))\theta$.

Case F+). Let $A = p(s)$. We have to show that $\forall(G^* \rightarrow \bot)\mathbf{R}^{ni}(hcomp(P))\epsilon$. One can easily verify that

$$\{\langle \forall(G^* \rightarrow \bot), hcomp(P)\rangle\} \Rightarrow \ldots \Rightarrow$$

$$\{\langle G^*\eta \rightarrow \bot, hcomp(P)\rangle\} \Rightarrow$$

$$\{\langle \bot, hcomp(P) \cup cl((G^{(1)}\eta \wedge G^{(2)}\eta)^*) \cup \{p(s)\eta\}\rangle\}$$

for some ground substitution η. Let $P' = hcomp(P) \cup cl((G^{(1)}\eta \wedge G^{(2)}\eta)^*) \cup \{p(s)\eta\}$ and $S = \{\langle \bot, P'\rangle\}$.

Suppose there are no clauses in P with p in their heads. Then $\forall(\bot \leftarrow p(x)) \in hcomp(P)$ and $S \Rightarrow \{\langle p(z), P'\rangle\} \Rightarrow \emptyset$ since $p(s)\eta \in P'$.

Let now (2) be all clauses in P with p in their heads. Then

$$\forall x \left(\bot \leftarrow p(x) \wedge \bigwedge_{i=1}^{k} \forall y_i(x = t_i \wedge G^*_i \rightarrow \bot) \right) \in P' .$$

Hence,

$$S \Rightarrow \ldots \Rightarrow \{\langle p(z), P'\rangle, \langle \forall y_i(z = t_i \wedge G^*_i \rightarrow \bot), P'\rangle \mid 1 \leq i \leq k\} \Rightarrow \ldots \Rightarrow$$

$$\{\langle p(z), P'\rangle\} \cup \{\langle (z = t_i\gamma \wedge G^*_i\gamma \rightarrow \bot), P'\rangle \mid 1 \leq i \leq k\} \Rightarrow \ldots \Rightarrow$$

$$\{\langle \bot, P' \cup cl(G^*_i\gamma) \cup cl(s\eta = t_i\gamma)\rangle \mid 1 \leq i \leq k\} \tag{3}$$

for some ground substitution γ.

Suppose that for some i s and t_i do not have a unifier. Then by Corollary 13

$$\forall(s = t \rightarrow \bot)\mathbf{R}^{ni}(\text{CET}(\{s, t_i\}))\epsilon .$$

Since $\text{CET}(\{\mathbf{s}, \mathbf{t}_i\}) \subseteq \text{CET}(P) \subseteq hcomp(P)$ one can consider that

$$\perp \mathbf{R}^{ni}(hcomp(P) \cup cl(\mathbf{s}\eta = \mathbf{t}_i\gamma))\epsilon \ .$$

By Lemma 16

$$\{\langle \perp, P' \cup cl(G_i^* \gamma) \cup cl(\mathbf{s}\eta = \mathbf{t}_i\gamma)\rangle\} \Rightarrow \ldots \Rightarrow \emptyset \ .$$

Suppose now that σ is a mgu of $\mathbf{s}$ and $\mathbf{t}_i$. It is clear from (3) and Completeness Theorem 10 that we have to show that

$$P', cl(G_i^* \gamma), cl(\mathbf{s}\eta = \mathbf{t}_i\gamma) \vdash_M \perp \ .$$

Note that η acts on $Var(G)$ and γ acts on $\mathbf{y}_i$. These sets of variables are disjoint; therefore, it follows from induction hypothesis that

$$\perp \mathbf{R}^{ni}(hcomp(P) \cup cl((G^{(1)} \wedge G_i \wedge G^{(2)})^* \sigma\eta\gamma))\epsilon \ .$$

By Soundness Theorem 9

$$hcomp(P), cl((G^{(1)} \wedge G^{(2)} \wedge G_i)^* \sigma\eta\gamma) \vdash_M \perp \ .$$

Then by Lemma 17 we obtain (3).

Case F−). We have that

$$\{\langle \forall \, ((G^{(1)})^* \wedge (A \to \perp) \wedge (G^{(2)})^* \to \perp), hcomp(P)\rangle\} \Rightarrow \ldots \Rightarrow$$

$$\{\langle \perp, hcomp(P) \cup \{A \to \perp\} \cup cl((G^{(1)}\eta)^*) \cup cl((G^{(2)}\eta)^*)\rangle\} \Rightarrow$$

$$\{\langle A, hcomp(P) \cup \{A \to \perp\} \cup cl((G^{(1)}\eta)^*) \cup cl((G^{(2)}\eta)^*)\rangle\} \Rightarrow \ldots \Rightarrow \emptyset \ .$$

because by induction hypothesis $A\mathbf{R}^{ni}(hcomp(P))\epsilon$. □

5 Conclusion

We described a proof procedure for hereditary Harrop formulas that can be efficiently implemented. Then, having noticed that the axioms of Clark's Equational Theory can be written as hhf, we used them to prove facts about term unification. An advantage we obtained in comparison with PROLOG is that we are capable to prove not only that two terms unify, but also that two terms do not unify.

The price we paid for that is, firstly, we consider much more broad class of program clauses than general clauses, namely, hhf (one can consider a general clause to be a hhf identifying $\neg A$ with $A \to \perp$) and, secondly, we adopted infinite programs (since CET is infinite). Nevertheless, all axioms of CET are instances of a small number of schemes and thus processing of equalities can be implemented.

We used the ability to prove the non-unifiability of terms in order to model negation as failure with negation as inconsistency. This was done by translating the completion of a general program into hhf. It was shown that if a goal succeeds from a program via SLDNF-resolution that the translation of a goal succeeds from the modified completion via the described procedure for hhf. Thus it was shown that NAI, which is embedded into this procedure, is a more general notion than NAF.

The obtained result can be shown to be trivial if another translation of program completion is used. For example, it is possible to include into $hcomp(P)$ clauses $\perp \leftarrow A$ for all (infinitely many) ground atoms A such that $A \in \mathbf{F}(P)$, but this translation would not be effective. Our approach has the advantage that $hcomp(P)$ is constructed from P using purely syntactic rules, without performing any computation.

Of course, completeness would also be of great interest. Completeness does not hold for all general programs. Consider, for example, $P = \{p \leftarrow \neg p\}$. Then $p\mathbf{R}^{ni}(hcomp(P))\epsilon$ though p does not succeed from P via SLDNF-resolution. Judging by the work of Harland, local stratification can be adopted as a sufficient condition that guarantees completeness.

References

1. K.R. Apt. Logic Programming and Negation: a Survey. Technical Report CS-R9402, CWI, Netherlands, 1994.
2. D.M. Gabbay and M.J. Sergot. Negation as Inconsistency. I *Journal of Logic Programming*, 3(1):1–36, 1986.
3. A. Gomolko. Negation as Inconsistency in Prolog via Intuitionistic Logic. LNCS 832, 1994.
4. J. Harland. A Clausal Form for the Completion of Logic Programs. In *Proceedings of the 8th International Conference on Logic Programming*, pages 711–725, Paris, France, 1991.
5. K. Kunen. Signed data dependencies in logic programs. *Journal of Logic Programming*, 7(4):231–246, 1989.
6. J.W. Lloyd. *Foundations of Logic Programming*. Springer-Verlag, Berlin, 1984.
7. D. Miller, G. Nadathur, F. Pfenning and A. Scedrov. Uniform proofs as a foundation for logic programming. *Annals of Pure and Applied Logic*, 51:125–157, 1991.
8. E. Mendelson. *Introduction to Mathematical Logic*. Van Nostrand, Princeton, NJ. 2nd ed., 1979.
9. G. Nadathur. A Proof Procedure for the Logic of Hereditary Harrop Formulas. *Journal of Automated Reasoning*, 11:115–145, 1993.

Basic Forward Chaining Construction for Logic Programs

V.W. Marek[1]*, A. Nerode[2]**, J.B. Remmel[3]***

[1] Department of Computer Science, University Kentucky, Lexington, KY 40506–0027.
[2] Mathematical Sciences Institute, Cornell University, Ithaca, NY 14853.
[3] Department of Mathematics, University of California at San Diego, La Jolla, CA 92903.

1 Introduction and Motivation

One of the problems which motivated this paper is how do we deal with inconsistent information. For example, suppose that we want to develop an expert system using logic programming with negation as failure. It may be the case that the knowledge engineer gathers facts, i.e. clauses of the form $p \leftarrow$, rules without exceptions, i.e. clauses of the form $p \leftarrow q_1, \ldots q_n$, and rules with exception or rules of thumb, i.e. clauses of the form $p \leftarrow q_1, \ldots q_n, \neg r_1, \ldots, \neg r_m$, from several experts. One problem is that the resulting program may be inconsistent in the sense that the program has no stable model. That is, the experts may not be consistent. The question then becomes how can we eliminate some of the clauses so that we can get a consistent program. That is, at a minimum, we would like to select a subprogram of the original program which has a stable model. Various schemes have been proposed in the literature to do this [GS92, KL89]. For example, we may throw away the rules which came from what we feel are the most unreliable experts until we get a consistent program. However even in the case when the knowledge engineer consults only a single expert, the rules that the knowledge engineer produces may be inconsistent because the rules that he or she abstracted are not specific enough or simply because the expert did not give us a consistent set of rules.

The above scenario is one practical reason that we would desire some procedure to construct, for a given program which has no stable model, a maximal subprogram that does have a stable model. Another practical reason occurs when we are using a logic program to control a plant in real time, see [KN93a] for examples. In this case, the program may have a stable model but that stable model may be very complicated and we do not have enough time to compute the full stable model. It has been shown [MT91] that the problem of determining whether a finite propositional logic program has a stable model is NP-complete. Moreover, the authors have shown [MNR92a] that there are finite predicate logic

* Research partially supported by NSF grant IRI-9400568.
** Research partially supported by ARO contract DAAL03-91-C-0027 and SDIO contract DAAH04-93-C-O113.
*** Research partially supported by NSF grant DMS-93064270.

programs which have stable models but which have no stable models which are hyperarithmetic so that there is no possible hope that one could compute the a stable model of the program no matter how much time one has. Thus if there are time problems, one may be satisfied by a procedure which would construct a subprogram of the original program and a stable model of the subprogram as long as both the subprogram and stable model of the subprogram can be computed rapidly, at the very least in polynomial time.

Indeed some see as a general problem with the stable model semantics the fact that there are many programs which have no stable models. For example, if we have any program P and p is new statement letter, the program P plus the clause $p \leftarrow \neg p$ has no stable model even if the original program P has a stable model. Thus a single superfluous clause which may have nothing to do with the rest of the program may completely destroy the possibility of the program possessing a stable model. This is one of the reasons that researchers have looked for alternatives to the stable model semantics such as the well-founded semantics [VGRS91].

In this paper, we shall present a basic Forward Chaining type construction which can be applied to any general logic program. The input of the construction will be any well-ordering of the non-Horn clauses of the program. The construction will then output a subprogram of the original program and a stable model of the subprogram. It will be the case that for any stable model M of the original program P, there will be a suitable ordering of the non-Horn clauses of the program so that the subprogram produced by our construction is just P itself and the stable model of subprogram produced by our construction will be M. Thus all stable models of the original program will be constructed by our Forward Chaining construction for suitable orderings. Moreover, we shall show that for finite propositional logic programs, our construction will run in polynomial time. That is, we shall prove that our Forward Chaining construction runs in order of the square of the length of the program.

We shall see that any stable model M of P can be produced via our Forward Chaining construction for some well-ordering $\prec$, i.e. every stable model of P is a stable submodel of P. In the case where our original program P is inconsistent in the sense that P has no stable models, we can view our Forward Chaining construction as a way of extracting a maximal consistent subset of clauses $C^\prec \subseteq P$ such that the system $C^\prec$ has stable model.

2 General logic programs

A *definite logic program* consists of clauses of the form

$$a \leftarrow a_1, \ldots, a_m$$

where $a, a_1, \ldots, a_m$ are atoms of some underlying language. We call such clauses *Horn program clauses* or simply Horn clauses. The set of atoms occurring in some clause of P is called the Herbrand base of P, and is denoted by H_P. We will be dealing here with the propositional case only.

A *general logic program* consists of clauses of the form

$$C = a \leftarrow a_1, \ldots, a_m, \neg b_1, \ldots, \neg b_n. \qquad (1)$$

where $a_1, \ldots, a_m, b_1, \ldots, b_n$ are atoms.. Here $a_1, \ldots, a_n$ are called the *premises* of clause C, $b_1, \ldots, b_m$ are called the *constraints* of clause C, and a is called the *conclusion* of clause C.

Each Horn program can be identified with the a general program in which every clause has an empty set of constraints.

Definition 1. A subset $M \subseteq H_P$ is called a *model* of P if for all $C = a \leftarrow a_1, \ldots, a_m \neg b_1, \ldots, \neg b_n \in P$, whenever all the premises $a_1, \ldots, a_n$ of C are in M and all the constraints $b_1, \ldots, b_m$ of C are not in M, then the conclusion a of C belongs to M.

Given sets $M \subseteq H_P$ and $I \subseteq H_P$, an M-*deduction* of c from I in P is a finite sequence $\langle c_1, \ldots, c_k \rangle$ such that $c_k = c$ and for all $i \leq k$, each c_i either
(1) belongs to I, or (2) is the conclusion of an axiom, or
(3) is the conclusion of a clause $C \in P$ such that all the premises of C are included in $\{c_1, \ldots, c_{i-1}\}$ and all constraints of C are in $H_P \setminus M$ (see [MT93], also [RDB89]).
An M-*consequence* of I is an element of H_P occurring in some M-deduction from I. Let $C_M(I)$ be the set of all M-consequences of I in P. Clearly I is a subset of $C_M(I)$. However note that M enters solely as a restraint on the use of the clauses which may be used in an M-deduction from I. M contributes no members directly to $C_M(I)$, although members of M may turn up in $C_M(I)$ by an application of a clause which happens to have its conclusion in M. For a fixed M, the operator $C_M(\cdot)$ is monotonic. That is, if $I \subseteq J$, then $C_S(M) \subseteq C_M(J)$. Also, $C_M(C_M(I)) = C_M(I)$. However, for fixed I, the operator $C_M(I)$ is anti-monotonic in the argument M. That is if $M' \subseteq M$, then $C_M(I) \subseteq C_{M'}(I)$.

We say that $M \subseteq H_P$ is *grounded* in I if $M \subseteq C_M(I)$. We say that $M \subseteq H_P$ is a *stable model of P over I* of I if $C_M(I) = M$.

With each clause C of form (1), we associate a Horn clause of form (2)

$$C' = a \leftarrow a_1, \ldots, a_m \qquad (2)$$

obtained from C by dropping all the constraints. The clause C' is called the *projection* of clause C. Let M be any subset of H_P and let $G(M, P)$ be the collection of all M-applicable clauses. That is, a clause C belongs to $G(M, P)$ if all the premises of C belong to M and all constraints of C are outside of M. We write $P|_M$ for the collection of all projections of all clauses from $G(M, P)$. The projection $P|_M$ is a Horn program. Our definition of stable model was different from but equivalent to that given by Gelfond and Lifschitz in [GL88].

3 The Forward Chaining Construction and Stable Submodels

Given a general program P, we then let $mon(P)$ denote the set of all Horn clauses of P and $nmon(P) = P \setminus mon(P)$. The elements of $nmon(P)$ will be called *nonmonotonic* clauses.

Our Forward Chaining construction will take as an input a program P and a well-ordering $\prec$ of $nmon(P)$. The principal output of the Forward Chaining construction will be a subset $D^{\prec}$ of H_P. Although such subset is not, necessarily, a stable model of P, it will be a stable model of $A^{\prec}$ for a subset $A^{\prec} \subseteq P$. This subset, $A^{\prec}$, will also be computed out of our construction and will be the maximal set of clauses of P for which $D^{\prec}$ is a stable model. We thus call $D^{\prec}$ a *stable submodel* of P.

The first feature of our construction is that in every stage of our construction we will close the sets we construct under $mon(P)$. The point is that stable models are always closed under the operator associated with the Horn part of the program, and the applicability of a clause from $mon(P)$ is not restricted. We shall denote by cl_{mon} the monotone operator of closure under the clauses in $mon(P)$. Thus $cl_{\text{mon}}(I) = T_{mon(P)} \uparrow \omega(I)$ is the least set Z of atoms from H_P such that $I \subseteq Z$ and Z is closed under every clause r of $mon(P)$. That is, if premises of such a clause are all in Z, then its conclusion also belongs to Z. The second important aspect of our construction is that when we inspect the clauses of $nmon(P)$ for a possible application, we look at the possible effect of their application on the applicability of those clauses which were previously applied. Rules that may invalidate applicability of previously used clauses are *not* used. The execution of this idea requires some book-keeping. Our Forward Chaining construction will define two sequences of subsets of H_P: $\langle D_\xi^{\prec} \rangle_{\xi \leq |P|^+}$ and $\langle R_\xi^{\prec} \rangle_{\xi \leq |P|^+}$. $D_\xi^{\prec}$ will be the set of *elements derived* by stage ξ. $R_\xi^{\prec}$ will be the set of *elements restrained* by stage ξ. Here and below α^+ is the least cardinal greater than α. Thus, if P is countable, then $|P|^+$ is either finite or the first uncountable ordinal. We shall prove, however, that if $|P|$ is countably infinite, then the construction actually stops *below* the first uncountable ordinal and therefore, for denumerable P, the use of nondenumerable cardinals can be eliminated.

In addition, we shall define two sets of clauses, $I^{\prec}$ (for "inconsistent clauses") and $A^{\prec}$ (for "acceptable" clauses). These sets of clauses will depend on previously defined hierarchies.

3.1 Forward Chaining Construction

Definition 2. Let P be a general program and let $\prec$ be a well-ordering of $nmon(P)$. We define two sequences of sets of atoms from H_P, $\langle D_\xi \rangle$ as well as $\langle R_\xi \rangle$. The set D_ξ is the set of atoms *derived* by stage ξ and R_ξ is the set of atoms *rejected* by the stage ξ.

1. $D_0^{\prec} = cl_{\text{mon}}(\emptyset)$, $R_0^{\prec} = \emptyset$;

2. If $\gamma = \beta + 1$ and there is a clause $C \in nmon(P)$ such that

$$prem(C) \subseteq D_\beta^{\prec}, \quad (\{c(C)\} \cup cons(C)) \cap D_\beta^{\prec} = \emptyset$$

and

$$cl_{\text{mon}}(D_\beta^{\prec} \cup \{c(C)\}) \cap (cons(C) \cup R_\beta^{\prec}) = \emptyset$$

(we call such clause *applicable clause*), then let C_γ be the $\prec$-first applicable clause and set

$$D_\gamma^\prec = cl_{\mathrm{mon}}(D_\beta^\prec \cup \{c(C_\gamma)\}) \quad R_\gamma^\prec = R_\beta^\prec \cup cons(C_\gamma).$$

If there is no C such that

$$prem(C) \subseteq D_\beta^\prec, \quad (\{c(C)\} \cup cons(C)) \cap D_\beta^\prec = \emptyset$$

and

$$cl_{\mathrm{mon}}(D_\beta^\prec \cup \{c(C)\}) \cap (cons(C) \cup R_\beta^\prec) = \emptyset,$$

then set

$$D_\gamma^\prec = D_\beta^\prec \quad \text{and} \quad R_\gamma^\prec = R_\beta^\prec$$

3. If γ is a limit ordinal, then

$$D_\gamma^\prec = \bigcup_{\xi < \gamma} D_\xi^\prec \quad \text{and} \quad R_\gamma^\prec = \bigcup_{\xi < \gamma} R_\xi^\prec.$$

4. Finally let

$$D^\prec = D_{|P|^+}^\prec = \bigcup_{\xi < |P|^+} D_\xi^\prec \quad \text{and} \quad R^\prec = R_{|P|^+}^\prec = \bigcup_{\xi < |P|^+} R_\xi^\prec.$$

Sets $D^\prec$ and $R^\prec$ are sets of atoms *derived* and *rejected* during the forward chaining construction along the well-ordering $\prec$.

We define the set of inconsistent clauses, $I^\prec$, and the set of consistent clauses, $A^\prec$, relative to ordering $\prec$ as follows:

5. C is *inconsistent with* $\prec$ (or simply *inconsistent* if $\prec$ is fixed) if $prem(C) \in D^\prec$, $(\{c(C)\} \cup cons(C)) \cap D^\prec = \emptyset$, but $cl_{\mathrm{mon}}(D^\prec \cup \{c(C)\}) \cap (cons(C) \cup R^\prec) \neq \emptyset$. $I^\prec = \{C \in P : C \text{ is inconsistent with } \prec\}$;

6. $A^\prec = P \setminus I^\prec$

We then say that a subset $D \subseteq H_P$ is a *stable submodel* of P, if there is a well-ordering $\prec$ of $nmon(P)$ such that $D = D^\prec$.

   The following observations should be clear: First, the clause that is used for construction of $D_{\gamma+1}^\prec$ from $D_\gamma^\prec$ is different from any clause used before in the construction. Therefore, by cardinality argument, the construction, eventually, stabilizes.

Next, both hierarchies $\langle D_\xi^\prec \rangle$ and $\langle R_\xi^\prec \rangle$ are increasing. Moreover, it is easy to prove by induction on ξ that $D_\xi^\prec \cap R_\xi^\prec = \emptyset$. Therefore $D^\prec \cap R^\prec = \emptyset$.

The sets $R_\xi^\prec$ accumulate the restraints of all clauses applied during the construction. Since $D^\prec \cap R^\prec = \emptyset$, the applicability of clauses applied during the construction is preserved at the end. This immediately implies the following result. First, let $P_\alpha = \{C_\xi : \xi < \alpha\}$, $P^* = \{C_\alpha : \alpha < |P|^+ \text{ and } C_\alpha \text{ is defined}\}$. We have

Proposition 3. $D_\xi^\prec$ is a stable model of P_ξ, and $D^p rec$ is a stable model of P^*.

We now have a result showing that the set $D^\prec$ we produced in the Forward Chaining construction behaves as promised:

Theorem 4. Let P be a general program. Let $\prec$ be a well-ordering of $nmon(P)$. Then $D^\prec$ is a stable model of $A^\prec$. Hence if $I^\prec = \emptyset$, then $D^\prec$ is a stable model of P.

We define the set of nonmonotonic generating clauses for a set $M \subseteq H_P$, $NG(M, P)$.

Definition 5. Let P be a general program. Let $M \subseteq H_P$.

$$NG(M, P) = \{C \in nmon(P) : prem(C) \subseteq M, cons(C) \cap M = \emptyset\}$$

Theorem 6. If P is a general program, then every stable model of P is a stable submodel of P. That is, if M is a stable model of P, then there exists a well-ordering $\prec$ of $nmon(P)$ such that $D^\prec = M$. In fact, for every well-ordering $\prec$ such that $NG(M, P)$ forms an initial segment of $\prec$, $D^\prec = M$.

While we stated Theorem 4 and Theorem 6 in full generality, we are most interested in the case when program P is finite or countable. In this case we can show that to construct stable models via forward chaining, one need consider orderings of type smaller or equal of order type ω.

Proposition 7. Let P be a program such that $|H_P| \leq \omega$ and let M be a stable model of P. There exists a well-ordering $\prec'$ of $nmon(P)$ in type $\leq \omega$ such that $D^{\prec'} = M$. Moreover the forward Chaining construction stabilizes in at most ω steps.

We note that Proposition 7 does not hold for all stable submodels. That is, the sets $D^\prec$ which are *not* stable models may have the property that they can only be obtained by means of orderings of the length $> \omega$.

Our construction of the set $D^\prec$ persists with respect to prolongation of the well-ordering (providing the Horn part is the same).

Proposition 8. Let $P \subset P'$ be two sets of clauses such that $mon(P) = mon(P')$. Let $\prec'$ be a well-ordering of $nmon(P')$ and let $nmon(P)$ be an initial segment in $\prec'$. Finally, let $\prec = \prec'|_P$. Then $D^\prec \subseteq D^{\prec'}$ and $R^\prec \subseteq R^{\prec'}$.

4 Complexity of Stable Submodels

4.1 Preliminaries

Let ω denote the set of natural numbers. The canonical index, $can(X)$, of finite set $X = \{x_1 < \ldots < x_n\} \subseteq \omega$ is defined as $2^{x_1} + \ldots + 2^{x_n}$ and the canonical index of $\emptyset$ is defined as 0. Let D_k be the finite set whose canonical index is k, i.e., $can(D_k) = k$.

We shall identify a clause r with a triple $\langle k, l, \varphi \rangle$ where $D_k = prem(r)$, and $D_l = cons(r)$, $\varphi = c(r)$. In this way, when $H_P \subseteq \omega$ we can think about P as a subset of ω as well. This given, we then say that a program P is *recursive* if H_P and P are recursive subsets of ω.

Next we shall define various types of recursive trees and Π_1^0 classes. Let $[,]: \omega \times \omega \to \omega$ be a fixed one-to-one and onto recursive pairing function such that the projection functions π_1 and π_2 defined by $\pi_1([x, y]) = x$ and $\pi_2([x, y]) = y$ are also recursive. Extend our pairing function to code n-tuples for $n > 2$ by the usual inductive definition, that is, let $[x_1, \ldots, x_n] = [x_1, [x_2, \ldots, x_n]]$ for $n \geq 3$. Let $\omega^{<\omega}$ be the set of all finite sequences from ω and let $2^{<\omega}$ be the set of all finite sequences of 0's and 1's. Given $\alpha = \langle \alpha_1, \ldots, \alpha_n \rangle$ and $\beta = \langle \beta_1, \ldots, \beta_k \rangle$ in $\omega^{<\omega}$, write $\alpha \sqsubseteq \beta$ if α is initial segment of β, i.e. , if $n \leq k$ and $\alpha_i = \beta_i$ for $i \leq n$. In this paper, we identify each finite sequence $\alpha = \langle \alpha_1, \ldots, \alpha_n \rangle$ with its code $c(\alpha) = [n, [\alpha_1, \ldots, \alpha_n]]$ in ω. Let 0 be the code of the empty sequence $\emptyset$. When we say that a set $S \subseteq \omega^{<\omega}$ is recursive, recursively enumerable, etc., what we mean is that the set $\{c(\alpha): \alpha \in S\}$ is recursive, recursively enumerable, etc. Define a *tree* T to be a nonempty subset of $\omega^{<\omega}$ such that T is closed under initial segments. Call a function $f: \omega \to \omega$ an infinite *path* through T provided that for all n, $\langle f(0), \ldots, f(n) \rangle \in T$. Let $[T]$ be the set of all infinite paths through T. Call a set A of functions a Π_1^0-class if there exists a recursive predicate R such that $A = \{f: \omega \to \omega : \forall n(R(n, [f(0), \ldots, f(n)]))\}$. Call a Π_1^0-class A *recursively bounded* if there exists a recursive function $g: \omega \to \omega$ such that $\forall f \in A \forall n(f(n) \leq g(n))$. It is not difficult to see that if A is a Π_1^0-class, then $A = [T]$ for some recursive tree $T \subseteq \omega^{<\omega}$. Say that a tree $T \subseteq \omega^{<\omega}$ is *highly recursive* if T is a recursive finitely branching tree and also there is a recursive procedure which, applied to $\alpha = \langle \alpha_1, \ldots, \alpha_n \rangle$ in T, produces a canonical index of the set of immediate successors of α in T. Then if A is a recursively bounded Π_1^0-class, it is easy to show that $A = [T]$ for some highly recursive tree $T \subseteq \omega^{<\omega}$, see [JS72b]. For any set $A \subseteq \omega$, let $A' = \{e: \{e\}^A(e) \text{ is defined}\}$ be the jump of A, let $\mathbf{0}'$ denote the jump of the empty set $\emptyset$. We write $A \leq_T B$ if A is Turing reducible to B and $A \equiv_T B$ if $A \leq_T B$ and $B \leq_T A$.

We say that there is an effective, one-to-one degree preserving correspondence between the set of stable models $Stab(P)$ of a recursive program P and the set of infinite paths $[T]$ through a recursive tree T if there are indices e_1 and e_2 of oracle Turing machines such that

(i) $\forall_{f \in [T]} \{e_1\}^{gr(f)} = M_f \in Stab(P)$,

(ii) $\forall_{M \in Stab(P)} \{e_2\}^M = f_M \in [T]$, and

(iii) $\forall_{f \in [T]} \forall_{M \in Stab(P)} (\{e_1\}^{gr(f)} = M \text{ if and only if } \{e_2\}^M = f)$.

where $\{e\}^B$ denotes the function computed by the eth oracle machine with oracle B. Also, write $\{e\}^B = A$ for a set A if $\{e\}^B$ is a characteristic function of A. For any function $f: \omega \to \omega$, $gr(f) = \{[x, f(x)]: x \in \omega\}$. Condition (i) says that the infinite paths of the tree T uniformly produce stable models via an algorithm with index e_1. Condition (ii) says that stable models of P uniformly produce infinite paths through T via an algorithm with index e_2. Condition (iii) asserts that if $\{e_1\}^{gr(f)} = M_f$, then f is Turing equivalent to M_f. In the sequel we shall not explicitly construct the indices e_1 and e_2, but it will be clear that such indices can be constructed in each case.

4.2 Complexity of the Forward Chaining Construction.

In this section we discuss complexity issues for sets of the form $D^\prec$, where P is a recursive program and $\prec$ is either some ordering of type ω or some finite ordering. First of all, recall that every stable model of P can be obtained as $D^\prec$ for a suitably chosen ordering $\prec$. This means that, since the stable models can be very complex, even if there is only one stable model, we cannot obtain results on complexity of $D^\prec$ without restricting the class of orderings. Our restriction is related to the fact that in any attempt to implement even a partial construction of $D^\prec$, we cannot go beyond ω. Moreover, ω (and finite ordinals) have the following property:

Lemma 9. *Let P be a program and let $\prec$ be a well-ordering of $nmon(P)$ of order type $\leq \omega$. Then the closure ordinal of the construction of the family $\langle D_\xi^\prec \rangle$ is at most ω.*

It is easy to see that the property indicated in Lemma 9 does not hold for ordinals greater than ω.
We shall restrict our attention now to the case when P is recursive and $\prec$ is a recursive well-ordering of type ω.

Proposition 10. *Let P be a recursive general program. Let $\prec$ be a recursive well-ordering of $nmon(P)$ of order type $\leq \omega$. Finally, let $D^\prec, R^\prec, I^\prec$, and $A^\prec$ be sets of atoms and of clauses defined in Definition 2. Then: $D^\prec$ is r.e. in $0'$, $R^\prec$ is r.e. in $0'$, $I^\prec$ is recursive in $0''$, and $A^\prec$ is recursive in $0''$.*

Corollary 11. *If P is a recursive program such that $nmon(P)$ is finite, then for any ordering $\prec$ of $nmon(P)$, $D^\prec$ is r.e., $R^\prec$ is finite, and $I^\prec$ is finite and $A^\prec$ is recursive.* $\qquad\square$

Now let us look at the case of finite P. In our complexity considerations, every atom a will have the cost $\|a\|$. Next, for a clause $r = c \leftarrow a_1, \ldots, a_n, \neg b_1, \ldots, \neg b_m$ we define $\|r\| = (\sum_{i \leq n} \|a_i\|) + (\sum_{i \leq m} \|b_j\|) + \|c\|$. Finally, for a set Q of clauses we define

$$\|Q\| = \sum_{r \in Q} \|r\|.$$

Theorem 12. *Suppose P is a finite general program and $\prec$ is some well-ordering of $nmon(P)$. Then $D^\prec, R^\prec, A^\prec$, and $I^\prec$ can be computed in time*

$$O(\|mon(P)\| \, \|nmon(P)\| + \|nmon(P)\|^2).$$

5 FC-Normal Programs

In this section we shall define FC-normal programs and state the basic results about such programs proved in [MNR93b]. We shall see that FC-normal programs have the property that the Forward Chaining construction always produces a stable model. In fact for FC-normal programs, one can drop the consistency check in the Forward Chaining construction and it will still always produce a stable model.

Definition 13. Let P be a program. We say that a subset $Con \subseteq \mathcal{P}(H_P)$ (where $\mathcal{P}(H_P)$ is the power set of H_P) is a *consistency property* over P if:
(1) $\emptyset \in Con$, (2) $\forall_{A,B \subseteq H_P}(A \subseteq B \quad \& \quad Con(B) \Rightarrow Con(A))$,
(3) $\forall_{A \subseteq H_P}(Con(A) \Rightarrow Con(cl_{mon}(A)))$, and
(4) whenever $\mathcal{A} \subseteq Con$ has the property that $A, B \in \mathcal{A} \to \exists_{C \in \mathcal{A}}(A \subseteq C \wedge B \subseteq C)$, then $Con(\bigcup \mathcal{A})$.

We note that conditions (1),(2), and (4) are Scott's conditions for information systems. Condition (3) connects "consistent" sets to the Horn part of the program; if A is consistent then adding elements derivable from A via Horn clauses preserves "consistency".

Definition 14. Let P be a program and let Con be a consistency property over P.

1. A clause $C = c \leftarrow a_1, \ldots, a_n, \neg b_1, \ldots, \neg b_k \in nmon(P)$ is *FC-normal* (with respect to Con) if $Con(V \cup \{c\})$ and not $Con(V \cup \{c, b_i\})$ for all $i \leq k$ whenever $V \subseteq H_P$ is such that $Con(V)$, $cl_{mon}(V) = V$, $a_1, \ldots, a_n \in V$, and $c, b_1, \ldots, b_k \notin V$.

2. P is a *FC-normal* (with respect to Con) program if all $r \in nmon(P)$ are FC-normal with respect to Con.

3. P is *FC-normal program* if for some consistency property $Con \subseteq \mathcal{P}(H_P)$, P is FC-normal with respect to Con.

FC-normal programs have all the desirable properties that are possessed by normal default theories as defined by Reiter in [Rei80]. In fact, it is shown in [MNR93b] that when one translates FC-normal programs back into the language of default logics than one obtains a class of default theories called *extended FC-normal* default theories which properly contains all normal default theories. We next shall state the basic results about FC-normal programs from [MNR93b].

Theorem 15. *Let P be a FC-normal program then there exists a stable model of P.*

Theorem 16. *Let P be a FC-normal program with respect to consistency property Con and let I be a subset of H_P such that $I \in Con$. Then there exists a stable model M of P such that $I \subseteq M$.*

In fact all stable models of FC-normal programs can be constructed via a slightly simplified version of the Forward Chaining construction which we shall call the Normal Forward Chaining construction. To this end, fix some well-ordering $\prec$ of $nmon(P)$. That is, the well-ordering $\prec$ determines some listing of the clauses of $nmon(P), \{r_\alpha : \alpha \in \gamma\}$ where γ is some ordinal. Let Θ_γ be the least cardinal such that $\gamma \leq \Theta_\gamma$. In what follows, we shall assume that the ordering among ordinals is given by $\in$. Our normal Forward Chaining construction will define an increasing sequence of sets $\{M_\alpha^\prec\}_{\alpha \in \Theta_\gamma}$. We will then define $M^\prec = \bigcup_{\alpha \in \Theta_\gamma} M_\alpha^\prec$. In [MNR93b] it is shown that $M^\prec$ is always an stable model of P.

The Normal Forward Chaining construction of $M^\prec$.

<u>Case 0.</u> Let $M_0^\prec = cl_{mon}(\emptyset)$.

<u>Case 1.</u> $\alpha = \eta + 1$ is a successor ordinal.
Given $M_\eta^\prec$, let $\ell(\alpha)$ be the least $\lambda \in \gamma$ such that

$$r_\lambda = s \leftarrow a_1, \ldots, a_p, \neg b_1, \ldots, \neg b_k$$

where $a_1, \ldots, a_p \in M_\eta^\prec$ and $b_1, \ldots, b_k, s \notin M_\eta^\prec$. If there is no such $\ell(\alpha)$, then let $M_{\eta+1}^\prec = M_\alpha^\prec = M_\eta^\prec$. Otherwise, let

$$M_{\eta+1}^\prec = M_\alpha^\prec = cl_{mon}(M_\eta^\prec \cup \{cln(r_{\ell(\alpha)})\}).$$

<u>Case 2.</u> α is a limit ordinal. Then let $M_\alpha^\prec = \bigcup_{\beta \in \alpha} M_\beta^\prec$.
This given, we have the following.

Corollary 17. *If P is a FC-normal program and $\prec$ is any well-ordering of $nmon(p)$, then*

1. *$M^\prec$ is a stable model of P.*

2. *(Completeness of the construction). Every stable model of P is of the form $M^\prec$ for a suitably chosen ordering $\prec$ of $nmon(P)$.*

It is quite straightforward to prove by induction that if P is FC-normal with respect to consistency property Con, then $M_\alpha^\prec \in Con$ for all α and hence $M^\prec \in Con$. Thus the following is an immediate consequence of Theorem 17(2).

Corollary 18. *Let P be a FC-normal program with respect to consistency property Con, then every stable model of P is in Con.*

We should also point out that if we restrict ourselves to countable programs P, i.e. if H_P is countable, then we can restrict ourselves to orderings of order type ω where ω is the order type of the natural numbers. That is, suppose we fix some well-ordering $\prec$ of $nmon(P)$ of order type ω. Thus, the well-ordering $\prec$ determines some listing of the clauses of $nmon(P), \{r_n : n \in \omega\}$. Our normal Forward Chaining construction can be presented in an even more straightforward manner in this case. Our construction again will define an increasing sequence of sets $\{M_n^\prec\}_{n \in \omega}$ in stages. This given, we will then define $M^\prec = \bigcup_{n \in \omega} M_n^\prec$. By the *Countable Normal Forward Chaining construction of $M^\prec$* we mean Normal Forward Chaining Construction restricted to orderings of type ω.

Theorem 19. *If P is a countable FC-normal program, then:*

1. $M^{\prec}$ is a stable model of P if $M^{\prec}$ is constructed via the Countable Normal Forward Chaining algorithm with respect to $\prec$, where $\prec$ is any well-ordering of $nmon(P)$ of order type ω.

2. Every stable model of P is of the form $M^{\prec}$ for a suitably chosen well-ordering $\prec$ of $nmon(P)$ of order type ω where $P^{\prec}$ is constructed via the Countable Normal Forward Chaining algorithm.

FC-normal programs also possess what Reiter terms the "semi-monotonicity" property.

Theorem 20. *Let P_1 and P_2 be two FC-normal program such that $P_1 \subseteq P_2$ but $mon(P_1) = mon(P_2)$ (that is, P_1, P_2 have the same Horn part). Assume, in addition, that both are FC-normal with respect to the same consistency property. Then for every stable model M_1 of P_1, there is a stable model M_2 of P_2 such that*

1. $M_1 \subseteq M_2$ and

2. $NG(M_1, P_1) \subseteq NG(M_2, P_2)$.

FC-normal programs also satisfy the *orthogonality of stable models* property with respect to their consistency property.

Theorem 21. *Let P be a FC-normal program with respect to a consistency property Con. Then if M_1 and M_2 are two distinct stable models of P, $M_1 \cup M_2 \notin Con$.*

We end this section with three more theorems which are analogues of results that hold for normal default theories.

Theorem 22. *Let P be a FC-normal program with respect to a consistency property Con. Suppose that $cl_{mon}\{cln(r) : r \in nmon(P)\}$ is in Con. Then P has a unique stable model.*

Theorem 23. *Suppose P is a FC-normal program and that $D \subseteq nmon(P)$. Suppose further that M_1' and M_2' are distinct stable models of $D \cup mon(P))$. Then P has distinct stable models M_1 and M_2 such that $M_1' \subseteq M_1$ and $M_2' \subseteq M_2$.*

References

[ABW87] K. Apt, H.A. Blair, and A. Walker. Towards a theory of declarative knowledge. *Foundations of Deductive Databases and Logic Programming*, pages 89–142, 1987.

[Apt90] K. Apt. Logic programming. *Handbook of Theoretical Computer Science*, pages 493–574, 1990.

[GL88] M. Gelfond and V. Lifschitz. The stable semantics for logic programs. *Proceedings of the 5th International Symposium on Logic Programming*, pages 1070–1080, 1988.

[GS92]   J. Grant and V.S. Subrahmanian. Reasoning about inconsistent knowledge bases. *IEEE Trans. on Knowledge and Data Engineering*, to appear.

[JS72b] C.G. Jockusch and R.I. Soare. π_1^0 classes and degrees of theories. *Transactions of American Mathematical Society*, 173:33–56, 1972.

[KL89]   M. Kifer and E. Lozinskii. RI: A logic for reasoning about inconsistency. TARK IV, pages 253-262, 1989.

[KN93a] W. Kohn and A. Nerode. Models for Hybrid Systems: Automata, Topologies, Controllability, Observability. In: *Hybrid Systems*, R.L. Grossman, A. Nerode, A.P. Ravn, H. Rischel, eds. Springer LN in CS 736, pages 317-356, 1993.

[MNR90] W. Marek, A. Nerode, and J.B. Remmel. Nonmonotonic rule systems I. *Annals of Mathematics and Artificial Intelligence*, 1:241–273, 1990.

[MNR92c] W. Marek, A. Nerode, and J.B. Remmel. Nonmonotonic rule systems II. *Annals of Mathematics and Artificial Intelligence*, 5:229–263, 1992.

[MNR92a] W. Marek, A. Nerode, and J. B. Remmel. The stable models of predicate logic programs. *Proceedings of International Joint Conference and Symposium on Logic Programming*, pages 446–460, Boston, MA, 1992. MIT Press.

[MNR95] W. Marek, A. Nerode, and J. B. Remmel. Complexity of Normal Default Logic and Related Modes of Nonmonotonic Reasoning, Proceedings of 10th Annual IEEE Symposium on Logic in Computer Science, pp. 178-187, 1995.

[MNR93b] W. Marek, A. Nerode, and J. B. Remmel. Context for Belief Revision: FC-Normal Nonmonotonic Rule Systems, Annals of Pure and Applied Logic 67(1994) pp. 269-324.

[MT91]   W. Marek and M. Truszczyński. Autoepistemic logic. *Journal of the ACM*, 38:588 – 619, 1991.

[MT93]   W. Marek and M. Truszczyński. *Nonmonotonic Logic – Context-dependent reasonings* 1993, Springer Verlag.

[Prz87] T. Przymusinski, On the declarative semantics of stratified deductive databases and logic programs, *Foundations of Deductive Databases and Logic Programming*, pages 193–216, 1987.

[RDB89] M. Reinfrank, O. Dressler, and G. Brewka. On the relation between truth maintenance and non-monotonic logics. *Proceedings of IJCAI-89*, pages 1206–1212.

[Rei80] R. Reiter. A logic for default reasoning. *Artificial Intelligence*, 13:81–132, 1980.

[VGRS91] A. Van Gelder, K.A. Ross and J.S. Schlipf. Unfounded sets and well-founded semantics for general logic programs. Journal of the ACM 38(1991).

Decidability and Undecidability of the Halting Problem on Turing Machines, a Survey

Maurice Margenstern

I.U.T. de Metz, Département d'Informatique,
Île du Saulcy,
57045 Metz Cedex, France,
L.I.A.F.A. (Université Paris 7),
L.R.I.M. (Université de Metz),
e-mail: margens@iut.univ-metz.fr

Abstract. The paper surveys the main results obtained for Turing machines about the frontier between a decidable halting problem and universality. The notion of *decidability criterion* is introduced. Techniques for decidability proofs and for contructing universal objects are sketchily explained. A new approach for finding very small universal Turing machines is considered in the last part of the paper.

1 Introduction

Let M be the set of deterministic Turing machines with one head and one tape, infinite on both sides. Considerations of our survey only take place in M. One knows that the halting problem for Turing machines in M is undecidable. This was settled by Turing, see [20], in the paper of 1936 in which he defined the machines later called after his name and in which he also proved the existence of universal Turing machines.

It should be noticed that Turing's undecidability theorem is proved by a simple diagonal argument, using the assumption that the set of *all* Turing machines in M is taken under consideration. What happens if only a *subset* of machines is considered?

The same problem is still undecidable when it is restricted to machines with alphabet $\{0, 1\}$ as the machine alphabet. In this case, the argument involves a universal Turing machine in the considered set of machines, say S, which allows to reduce the general halting problem to the halting problem on S.

1.1 A bit of history

Since the sixties and the early seventies, some research has been developed in the following direction, leading to partial answers which are far from closing the subject.

Identify any point $s \times l$ – take natural numbers s and l as coordinates – with the set of all Turing machines in M whose program contains s states, except the halting state(s), and uses l letters, including the blank symbol. Let us call such a point *decidable* if the halting problem is decidable for any Turing machine in the corresponding set, and *undecidable* if the halting problem restricted to this set of machines is undecidable. As a matter of fact, for large enough s and l, it is easy to construct a universal machine occurring in the set attached to the point $s \times l$. By restricting the number of states or the alphabet of the machine, it is easily seen that any universal machine belonging to $x \times y$ set implies that for any n and m, with $n \geq x$ and $m \geq y$, the point $n \times m$ is undecidable too. Then it comes as a natural thing to wonder what are the smallest values of s and l for which the point $s \times l$ is undecidable.

An account of the results obtained up to the last seventies can be found in [15]. However, the best results were established in 1982, see [16], by Yuri Rogozhin who was the only one to improve them later on, see [17] and [18]. Yet his paper of 1982 had remained unnoticed by the scientific community during ten years. The technique used in order to obtain these results will be explained in our survey, in subsection 4.1.

Let's have a look now in the opposite direction: it is trivial that Turing machines on a single letter alphabet have a decidable halting problem. It is not that trivial that the same property holds for machines with a single state, whatever the number of symbols is in the machine alphabet. This was definitely proved by Hermann in 1966, see [2].

For more than one state, one does not know much. As for proving the decidability of the halting problem, M. Minsky writes in his famous book of 1967 that he and one of his students "did this for all 2×2 machines [1961, unpublished] by a tedious reduction to thirty-odd cases (unpublishable)", see [10], p. 281, last two lines. Six years later, Liudmila Pavlotskaya proved the same theorem, see [11], in a compact, very short proof. This paper has also remained unnoticed for many years. A few years later, see [13], she proved that the point 3×2 is also decidable.

All these results about the decidability and undecidability of the halting problem for Turing machines can be represented on figure 1, below.

1.2 Decidability criteria

Let us now define the notion of *decidability criterion*. Let c be an integer valued function defined on a set M of Turing machines in $\mathcal{M}$ with the following property: there is an integer f such that the halting problem is decidable for any machine $T \in M$ such that $c(T) < f$, and for any $k \geq f$ a universal machine $U \in M$ such that $c(U) = k$ can always be constructed. In that case, c is called *decidability criterion* and f is called its *frontier value*, see [5].

Shannon's result, see [19], shows that starting from two letters for the machine alphabet, it is always possible to construct a universal Turing machine. Thus the number of symbols of the machine alphabet, whatever the number of states is, provides a simple example of a decidability criterion. The same paper by Shannon and paper [2] show that the number of states, whatever the number of symbols is, also provides a decidability criterion with again two as a frontier value.

If we look closer at figure 1, below, we see that for Turing machines with only two states, the number of symbols provides a function that has not yet been proved to be a decidability criterion. Of course, the frontier value does exist. But the actual value is still unknown. Figure 1 shows that it lies somewhere between 4 and 18.

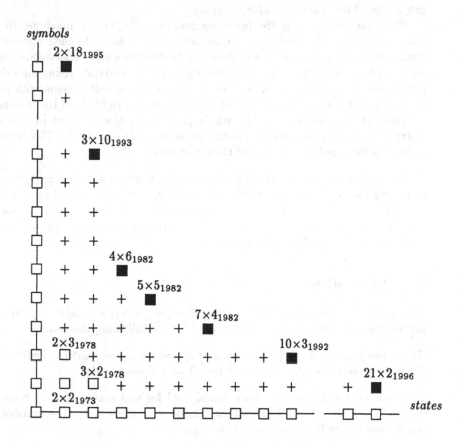

Fig. 1. The present state of the art

Black squares indicate undecidable points, white squares indicate decidable points, points marked with plus between the white squares lines and the line of black squares in the shape of a hyperbola correspond to sets for which the status of the halting problem is not known.

In our survey, we shall consider two other criteria, the only ones that have so far been established for Turing machines on alphabet $\{0, 1\}$.

One of them is the number of *colours* in the program of a Turing machine, see [4], where this number is defined as follows. The *colour* of an instruction of the program of Turing machine T is defined as the *projection* of the 5-tuple which encodes the instruction — input state, input symbol, move, output symbol, new state — on the triple which consists in the input symbol, the move and the output symbol. In other words, the colour is precisely what can be seen of the tape by an observer to whom the internal states of the machine are hidden when instructions are performed. The number of colours is the cardinality of the projection of the whole machine program.

The other criterion is the laterality number of a Turing machine. In this case, we consider the projection which associates its move to each instruction. Then it can easily be seen that stationnary instructions can be associated to an ultimate true move on the left or on the right, unless a trivial cycling on a single cell occurs. This allows to extend the projection to all instructions with range in $L, R - L$ for move to the left, R for move to the right. Call this extension *laterality* of the instruction. The whole program is thus divided in two sets: instructions with L as laterality, instructions with R as laterality. The *laterality number* is the smallest cardinal of these two sets.

In the first section, we introduce the results. Then, in a second part, we shall deal with the decidability part of the stated theorems. In a third one we shall deal with the techniques used for the universality part of the results. In the last part, we shall consider a new direction for founding out a new kind of frontier result about decidability and universality.

2 The results

The number of colours for a Turing machine program is a decidability criterion for Turing machines on $\{0, 1\}$. It is stated by the following theorem:

Theorem 1 ([11] and [12]). *The number of colours is a decidability criterion for Turing machines on alphabet $\{0, 1\}$ with 3 as a frontier value.*

A similar result has later been established for non-erasing Turing machines on $\{0, 1\}$, *i.e* Turing machines which can never replace 1 by 0 — symbol 0 is considered as the blank symbol of the tape:

Theorem 2 ([4]). *The number of colours is a decidability criterion for non-erasing Turing machines on alphabet $\{0, 1\}$ with 5 as a frontier value.*

The laterality number is also a criterion for Turing machines on $\{0, 1\}$ as it is now stated:

Theorem 3 ([11] and [8]). *The laterality number is a decidability criterion for Turing machines on alphabet $\{0, 1\}$ with 2 as a frontier value.*

This is also the case for non-erasing Turing machines with a different number as a frontier value:

Theorem 4 ([5] and [6]. *The laterality number is a decidability criterion for non-erasing Turing machines on alphabet $\{0, 1\}$ with 3 as a frontier value.*

In section 5 of our survey, we shall see a new direction jointly examined by Maurice Margenstern and Liudmila Pavlotskaya, generalizing the setting introduced in [14] by Liudmila Pavlotskaya. The direction consists in considering a new machine computation based on a Turing machine working with a finite automaton in the following conditions: the tape is initially restricted to the input word with the Turing machine head scanning some letter of the word and the automaton set on some state. The machine head is assumed to never go to the left of the leftmost symbol of the initial word. Each time when the Turing machine goes out from the word, the tape is extended by one cell. The content of this cell is determined by the finite automaton according to its own state and the state under which the Turing machine head exited. The following result holds:

Theorem 5 (unpublished). *The number of instructions of the Turing machine, whatever the number of states of the automaton is, is a decidability criterion for couples of Turing machines and automata associated as above indicated, with 5 as a frontier value.*

3 Decidability theorems

As pointed out in our introduction, the first proof of the decidability of the halting problem for 2×2 Turing machines was a tedious reduction by cases. Pavlotskaya's proof, a very short one, is based on a different idea. She introduced various integer valued *mathematical functions* connected with the behaviour of the head of the Turing machine on its tape. Technically, Pavlotskaya's proofs consist in proving that the functions she introduced in the general case, in an abstract way, turn out to be *recursive*, from which it ensues that the halting problem is decidable for the corresponding set of machines.

As indicated in our introduction, this proof remained unnoticed a long time. And so, another proof was found out by Volker Diekert and Manfred Kudlek, which takes place in a completely different setting. These authors succeeded in characterizing the *languages* constituted of the words on $\{0, 1\}$ on which the considered Turing machine halts. They proved that in almost all cases, such a set is a rationnal language, see [1] and [3]. In the remaining cases, it turns out to be a linear language. Here the proof is also a reduction by cases, strongly simplified by symmetry considerations and a general lemma based on the motion of the machine head which applies to a large number of machines. Details can be found in [3].

For what is the decidability criteria, the decidability part appears in Pavlotskaya's proof of the decidability of the point 2×2 in figure 1. Indeed, that result is an immediate corollary of the following lemma, which is also the decidability part of theorem 2.

Lemma 6 [11]. *If the program of a Turing machine T on $\{0, 1\}$ contains a single left instruction, then the halting problem is decidable for machine T.*

In the same paper, Liudmila Pavlotskaya proves the decidability part of theorem 1. Say that a Turing machine is *unilateral* if its program contains instructions with the same move except, possibly, stationnary instructions. It is clear that the halting problem is decidable for unilateral machines in $\mathcal{M}$. Combining that fact with lemma 6 reduces drastically the number of cases to be scrutinized.

3.1 A general approach

However, all the decidability parts of theorems 1 to 4 can be proved according to the same general plan clearly defined in [5] and [7].

The starting point consists in noticing that the decidability of the halting problem is obtained as soon as we get an algorithm for *recognizing* a non-halting computation. Indeed, a halting computation always eventually halts and so the algorithm is trivial: wait until the halting occurs.

Next, it can be noticed that non-halting computations, which, by definition, are computations in *infinite* time, may turn out to use either finite space or infinite space. The first case leads to the occurrence of two indentical configurations: this is a trivial case of non-halting, easy to recognize.

For conveniently analyzing the second case, we need accurate tools in order to describe the motion of the machine head on its tape. This is done in [4, 5] and mainly in [7]. In our settings, this makes use of *partial* recursive functions. We have to prove that they are *total* recursive. The corresponding proofs of [4, 5, 7] are constructive.

Let us informally summarize the notions introduced in these papers: at inital time, the *current bounds* of the configuration are the ends of the smallest segment containing both the initial word written on the tape and the machine head outside which all cells contain the blank symbol. Then, an *exit* is any time when the head goes out by one cell of the current bounds of the configuration. And for the next time, the current bound is moved to that latter position. We define *extremal exits* to be exits at which a half-turn occur, *i.e.* a change of direction in the motion.

Let us now turn back to the case when the machine motion is infinite in space. Trivial cases of unilateral motions put aside, it can easily be seen that either there are infinitely many extremal exits only on the right side, or there are infinitely many extremal exits on the other side, or there are infinitely many extremal exits on both sides. Say, respectively, right infinite case, left infinite case and traversal case.

First, consider the right infinite case. Between two consecutive extremal exits, there is a *leftmost* position for the head of the machine, we shall say an *lmp*.

This is below illustrated by figure 2.

It is now possible to formulate an important necessary and sufficient condition of ultimately periodic motion for the machine head of a Turing machine in the case it does not halt on the given data, see [4, 5, 7]:

Lemma 7 ([5]). *Let p_i be the sequence of absolute lmp's, S_i the extremal right exit which just occured before lmp time corresponding to p_i and l_i the distance* *from p_i to $v(S_i)$, position of the head at time S_i. The motion of the machine is utimately periodic if and only if:*

$$\liminf_{i \to \infty} l_i < +\infty,$$

and this condition is recursively enumerable,

where an *lmp* p_i is said to be *absolute* if the head never goes to the left of that position after the time it is reached as an *lmp*.

Notice that the lim inf condition precisely states that there is an integer L and there are infinitely many i's such that $\ell_i \leq L$ for all those selected i's.

By symmetry, analogous results hold for the left infinite case.

In [7], a more accurate necessary and sufficient condition of ultimately periodic motion is given. It intuitively says that when the head goes back to the left after an extremal right exit, the next *lmp* cannot be too often too far from the previous *lmp*.

Lemma 7 appears to solve the problem for most cases belonging to both right and left infinite cases. And so, proofs focus the attention on the 'difficult' cases where the necessary and sufficient conditions of ultimately periodic motion cannot be satisfied, in particular, in the traversal case.

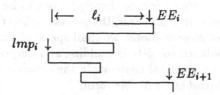

Fig. 2. On this figure, the horizontal pieces of line indicate the spaces scanned by the machine head and vertical lines indicate half-turns. This provides a condensed space-time diagram of that motion: moves in the same direction remain on the same horizontal line. EE_i is the ith extremal exit (here, on the right side).

The analyses given by the decidability proofs of [4, 7] show that 'difficult' cases do happen. Up to now, what happens is an ultimately periodic increasing of the ℓ_i's or the length of traversals. It is somehow difficult to prove that such a condition is a sufficient one. Generally speaking it is surely not the single possibility of 'difficult' motion in the case of infinite space.

3.2 Using the notion of colour

The decidability part of theorem 2 uses theorem 1. It also uses a precise characterization of what could be called 'decidable' sets of five colours, see [4]. In order to state the corresponding property, we say that a colour is of *pure move* if and only if its output symbol is the same as its input one: corresponding instructions perform their motion without altering the content of the tape cells. The following property holds:

Lemma 8 [4]. *If in the program of a non-erasing Turing machine on* $\{0,1\}$ *at least one of the pure move colour is missing, then the halting problem is decidable for that machine.*

The proof of lemma 8 is typically based on the above general considerations.

It should be noticed that lemma 8 appears not only in the proof of the decidability part of theorem 2 but also for a drastic reduction of the cases appearing in the decidability proof of theorem 4, allowing to lower the number of main cases to be examined from twenty one downto four. Moreover, with the help of particular considerations lemma 8 allows to discard three of the remaining four cases. And so, only in this last one the 'difficult' cases mentionned above do appear. See [7] for a thorough analysis.

4 Universality theorems

Various tools have been deviced for proving the universality of several computational models.

The main idea is to *simulate* a process already known to be universal. Two ways are used for this purpose. In some cases, the process is directly implemented in the considered system of computations: we shall speak of an *executive code*. In the other cases, the simulation consists in building an *interpreter* of a class of *processes* known to contain a universal element. In the simulating computation, simulated elements are *encoded* in a suitable form.

Notice that like russian dolls, simulations can be nested. This is often used for making *small* universal objects, small with respect to the size of the program.

4.1 Two register machines

The most popular tool of simulation is certainly the simulation of two register machines which are known to be universal, see for instance [10].

This is the case, in particular, for the universality parts of theorems 1 and 3. It is worth noticing that the proof of theorem 1, see [12] brings in a new ingredient which turns out to be very robust with respect to constraints.

Recall that for simulating a two register machine, it is enough to encode the content of its registers, say x and y, as a number $X = 2^x.3^y$. In this context, incrementation of one register is simulated by multiplication of X by 2, of the other one is simulated by multiplying X by 3. It is easy to see that for implementing decrementation, it is enough to simulate division by 6. But, for strong limitations on the number of colours or of laterality, it is not known how to divide by 6. Paper [12] shows that the situation is saved by simulating the following mapping: $x \mapsto x + \dfrac{x}{6}$, called *pseudo-division* by 6. This boils down to multiply

x by $\frac{7}{6}$. This introduces powers of 7 in the prime decomposition of X but this keeps the right value for the exponents of 2 and 3 in that decomposition.

4.2 Tag-systems

Another important tool, far less known than two register machines, but very powerful, are tag-systems. There are used, in particular, for making the smallest universal Turing machines: see [10, 16, 17, 18].

A p-tag-system − p a fixed positive integer − is a calculus which is associated to a mapping from alphabet A into A^*, the set of words on A. The image of $a \in A$ is its *production*. One step in the computation consists in performing the following three operations, illustrated, below on the right, in the case $p = 2$:

- a_i, first letter of the *tagged* word, *i.e.* submitted to the tag-system, is memorized ;
- first p letters of the word are erased ;
- P_i, production associated with a_i, is appended at the end of what remains from the tagged word.

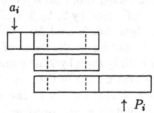

Repeat the process on the word thus obtained until either the tagged word has less than p letters, or the just obtained tagged word comes from a word, the first letter of which was a *halting* letter. By definition, halting letters are distinguished letters of A making the computation halt. A single halting letter is enough, which is assumed in most simulations.

Coke-Minsky theorem (see, for instance, [10]) states that any Turing machine in $\mathcal{M}$ can be simulated by a 2-tag-system. On another hand, Wang proved that the halting problem is decidable for any 1-tag-system ([21]).

This tool is used in the universality part of theorems 2 and 4. In theorem 4, three nested simulations are used: the 218 state non-erasing machine on $\{0,1\}$ simulates another Turing machine, say Z, on a richer alphabet that simulates a three head machine on $\{0,1\}$ which, on its turn, simulates an interpreter of 2-tag-systems. It is enough to put the encoding of a two-register machine simulating a universal Turing machine for obtaining the universality of the non-erasing machine. Notice that the non-erasing machine simulates machine Z by always 'refreshing' the current configuration: it endlessly replicates the content of the leftmost significant cell at the other end of the configuration.

5 A new direction

In section two, we described the computation associated to a couple constituted of a Turing machine and a finite automaton.

The decidability part of theorem 5 uses the general setting we indicated in section 3 which has to be extended due to the association with a finite automaton. It also uses the notion of colour but, mainly a new notion is taken into account for leading the splitting process into cases and for reducing the number of cases to consider. It is the notion of *phase* which is a dual notion to the notion of colour. The phase of an instruction is the projection constituted of the input state, the output state and the move of the instruction. The proof also introduces the notion of *connectivity* of a Turing machine program. Informally, a program is connected if and only if there are infinite computations which infinitely often make use of all the instructions of the machine. This allows to rule out a great number of cases. Then, inside each remaining case, colour type arguments allow to reduce again the number of subcases. Lemma 7 and its stronger version are also very useful in this proof.

The universality proofs of theorems 1 and 5 basically use simulations by a two register machine. They also use the pseudo-division introduced in [12]. But in both cases, more must be done. What is needed is a *uniform* simulation of mutiplication by 2, by 3 **and** of pseudo-division by 6. This means that the *same* algorithm is used for computing the operations, depending on a parameter. According to the value of the parameter, it is then possible to simulate now multiplication by 2, now multiplication by 3 and now, pseudo-division by 6. This allows to satisfy the constraint of two left instructions in the program for theorem 1, also see [8]. In theorem 5, a first machine is built which simulates the computation explained in [8] for proving theorem 1. It is a machine with 8 instructions, associated to a finite automaton which contains a great lot of the information needed for the computation, especially the connection between the instructions used by the simulated two register machine. Indeed, this couple works as an executive code. It is not an interpreter. Then, the couple constructed with a 5 instruction Turing machine simulates the couple with a machine with 8 instructions. Full details can be found in technical report [9].

Acknowledgements

The author is much indebt to S.I. Adian for very valuable comments. He also has to acknowledge both grant *HighTech. EV No*950525 given by NATO Division of Scientific Affairs and Metz University Institute of Technology for making possible Liudmila Pavlotskaya's visits in 1995 and 1996, thus providing the best conditions for common works with her.

References

1. Diekert V. and Kudlek M. Small Deterministic Turing Machines, Papers on Automata and Languages, Dep. of Math., Karl Marx Univ. of Economics, Budapest, 1988-4, p.77-87, (1989)
2. Hermann G.T. The uniform halting problem for generalised one state Turing machines, Proceedings of the Ninth Annual Switching and Automata Theory Symposium, (1968)
3. Kudlek M. Small Deterministic Turing Machines, Theoretical Computer Science, **168-2**, special issue on Universal Machines and Computations, 241-255, (1996)

4. Margenstern M. Turing machines: on the frontier between a decidable halting problem and universality. Lecture Notes in Computer Science, **710**, 375-385, in *Proceedings of FCT'93*, (1993)
5. Margenstern M. Non-erasing Turing machines: a new frontier between a decidable halting problem and universality. Lecture Notes in Computer Science, **911**, 386-397, in *Proceedings of LATIN'95*), (1995)
6. Margenstern M. Results on the halting problem, LMPS'95 International Conference, Florence, August 1995
7. Margenstern M. The laterality problem for non-erasing Turing machines on $\{0, 1\}$ is completely solved. RAIRO, 39p. (1997) (to appear)
8. Margenstern M. and Pavlotskaya L. Deux machines de Turing universelles à au plus deux instructions gauches. *C.R.A.S., Paris*, **320**, I, 1395-1400, (1995)
9. Margenstern M. and Pavlotskaïa L. Vers une nouvelle approche de l'universalité concernant les machines de Turing, LITP/IBP research report N°95.58 (1995)
10. Minsky M.L. Computation: Finite and Infinite Machines. Prentice Hall, Englewood Cliffs, N.J. (1967)
11. Pavlotskaya L.M. Razreshimost' problemy ostanovki dlja nekotorykh klassov mashin T'juringa. Matematicheskie Zametki, **13**, (6), 899-909, Ijun' 1973 899-909. (transl. Solvability of the halting problem for certain classes of Turing machines, Notes of the Acad. Sci. USSR, 13 (6) Nov.1973, 537-541)
12. Pavlotskaya L.M. O minimal'nom chisle razlichnykh kodov vershin v grafe universal'noj mashiny T'juringa. Diskretnyj analiz, Sbornik trudov instituta matematiki SO AN SSSR, **27**, 52-60, (1975) (On the minimal number of distinct codes for the vertices of the graph of a universal Turing machine) (in Russian)
13. Pavlotskaya L.M. Dostatochnye uslovija razreshimosti problemy ostanovki dlja mashin T'juring, Avtomaty i mashiny, 91-118, 1978 (Sufficient conditions for halting problem decidability of Turing machines) (in Russian)
14. Pavlotskaya L.M. On machines, universal by extensions, Theoretical Computer Science, **168-2**, special issue on Universal Machines and Computations, pp.257-266, (1996)
15. Priese L. Towards a precise characterization of the complexity of universal and nonuniversal Turing machines, SIAM Journal of Computation, **8**, 4, 508-523, (1979)
16. Rogozhin Ju.V. Sem' universal'nykh mashin T'juringa. Matematicheskie Issledovanija, **69**, 76-90, 1982 (Seven universal Turing machines) (in Russian)
17. Rogozhin Ju.V. Universal'naja mashina T'juringa s 10 sostojanijami i 3 simvolami. Matematicheskie Issledovanija, **69**, 76-90, 1992 (A universal Turing machine with 10 states and 3 symbols) (in Russian)
18. Rogozhin Ju.V. Small universal Turing machines, Theoretical Computer Science, **168-2**, special issue on Universal Machines and Computations, 215-240, (1996)
19. Shannon C.E. A universal Turing machine with two internal states. Ann. of Math. Studies, **34**, 157-165, (1956)
20. Turing A.M. On computable real numbers, with an application to the Entscheidungsproblem. Proc. Lond. Math. Soc., ser. 2, **42**, 230-265 (1936)
21. Wang H. Tag Systems and Lag Systems, Mat. Annalen, 152, (1963), 65-74

Case Study: Additive Linear Logic and Lattices

Jean-Yves Marion

Université Nancy 2, CRIN - CNRS & INRIA Lorraine,
Projet Calligramme,
Campus Scientifique - B.P. 239,
54506 Vandœuvre-lès-Nancy Cedex, France.
email : Jean-Yves.Marion@loria.fr

Abstract. We investigate sequent calculus where contexts, called additive contexts, are governed by the operations of a non-distributive lattice. We present a sequent calculus ALL_m with multiple antecedents and succedents. ALL_m is complete for non-distributive lattices and is equivalent to the additive fragment of linear logic. Weakenings and contractions are postulated for ALL_m and cut is redundant. We extend this construction in order to get a sequent calculus for propositional linear logic with both additive and multiplicative contexts.
Then we show that a bottom-up decision procedure based on the cut-free sequent calculi runs in exponential time. We provide a decision algorithm that exploits analytic cuts and whose runtime is polynomial.

1 Introduction

The morphology of statements of a deductive system is of the greatest importance. Take sequent calculi. There are two presentations. The additive one where rule pattern is thus

$$\frac{\Gamma \vdash \Delta, A \qquad \Gamma \vdash \Delta, B}{\Gamma \vdash \Delta, A\&B}$$

and the multiplicative one

$$\frac{\Gamma_1 \vdash \Delta_1, E_1 \qquad \Gamma_2 \vdash E_2, \Delta_2}{\Gamma_1, \Gamma_2 \vdash \Delta_1, E_1 \otimes E_2, \Delta_2}$$

It is well known that in the absence of weakening and contraction rules both formulations are not equivalent. So, for example in linear logic ([4, 10]), there are two distinct fragments, the additive one and the multiplicative one.

The sequent calculus associated with the additive fragment of linear logic ALL is a singular calculus. That is, there is only one formula on each side of the deducibility relation $\vdash$. In Section 2, we recall that ALL is complete for non-distributive lattices. In Section 3, we investigate sequent calculi when structural rules are governed by operations on non-distributive lattices. ¿From there, we introduce a multiple sequent calculus ALL_m. A sequent in ALL_m has multiple antecedents and succedents. We shall show that ALL_m-calculus is equivalent to ALL and therefore complete for non-distributive lattices. Then, Section 4 is devoted to decision procedures for non-distributive lattices. We show that in both cut-free sequent calculi, even if we delay choices, the number of steps is exponential. In the other hand, we present two decision procedures which run in quadratic time by maintaining partial results. In the last Section 5, we extend additive contexts to define a sequent calculus for the propositional fragment of linear logic. The two last sections are independent.

2 Non-distributive Lattices & Additives

The sequent calculus ALL, see Figure 1, is the standard calculus for the *additive* fragment of linear logic. The system ALL is complete for free non distributive lattices. Let $(\mathcal{L}, \leq)$ be the free non distributive lattice generated from $c_0, c_1, \ldots$ and closed by $\smile$ (join) and $\frown$ (meet). We refer to [1] and to [2] for definitions and details. An interpretation over $(\mathcal{L}, \leq)$ is given by a valuation ϕ which satisfies: $\phi(A)$ is some element of $\mathcal{L}$ when A is atomic; $\phi(E\&F) = \phi(E)\frown\phi(F)$; $\phi(E \oplus F) = \phi(E)\smile\phi(F)$.

Proposition 1. $E \vdash F$ *in* ALL *iff* $\phi(E) \leq \phi(F)$ *in* $(\mathcal{L}, \leq)$.

Proof. If we replace $\vdash$ by $\leq$ then each rule of ALL is valid in $\mathcal{L}$, and inversely. This result is quite immediate because it is merely two different notations. $\square$

$$\frac{}{E \vdash E}\ ax \qquad\qquad \frac{E_1 \vdash F \qquad F \vdash E_2}{E_1 \vdash E_2}\ cut$$

$$\frac{E_i \vdash F}{E_1 \& E_2 \vdash F}\ \&L_{i=1,2} \qquad\qquad \frac{E \vdash F_1 \qquad E \vdash F_2}{E \vdash F_1 \& F_2}\ \&R$$

$$\frac{E_1 \vdash F \qquad E_2 \vdash F}{E_1 \oplus E_2 \vdash F}\ \oplus L \qquad\qquad \frac{E \vdash F_i}{E \vdash F_1 \oplus F_2}\ \oplus R_{i=1,2}$$

Fig. 1. Rules of ALL

Theorem 2. *The cut elimination property holds in* ALL, *i.e. if* $E \vdash F$ *holds in* ALL *then there is a cut-free derivation of* $E \vdash F$.

We could have a negation, denoted $\perp$, which is involutive, $A^{\perp\perp} = A$, and which verifies De Morgan laws, $(A\&B)^{\perp} = A^{\perp} \oplus B^{\perp}$ and $(A \oplus B)^{\perp} = A^{\perp}\&B^{\perp}$. Here we restrict, without loss of generality, to sentences where negation is applied only to atoms. We do not consider constants here.

3 Multiple sequent calculus for additive logic

In sequent calculus, the meaning of the logical connectors is conveyed by both the operational rules and by the structural rules. For example, the intuitionistic logic is usually presented with one succedent sequent. This syntactic restriction implies that there is no contraction and no weakening on the right side of the sequent. In the other hand, there are intuitionistic sequent calculi with multiple succedents like the calculus LA_m of Curry [2], or yet the system $G4$ of Kleene. In those systems, $A_1, \ldots, A_p \vdash_{LI} B_1, \ldots, B_q$ means that the intuitionistic conjunction of $A_1, \ldots, A_p$ entails, in an intuitionistic way, the intuitionistic disjunction of the succedent $B_1, \ldots, B_q$. In those multiple systems, we can apply structural rules on both side of $\vdash_{LI}$, but there are restrictions on the implication, written below, and the negation, on the right.

3.1 Multiple Additive Sequent Calculus

We now formulate a multiple calculus ALL_m which is equivalent to ALL. For this, we define additive contexts. An additive context is a sequence of formulas separated by a semicolon (:). A ALL_m-sequent is a sequent where both the antecedent and the succedent are additive contexts. A ALL_m-sequent is called additive and is of the form $A_1: \cdots :A_p \vdash_m B_1: \cdots :B_q$. We interpret the antecedent by $\phi(A_1) \frown \cdots \frown \phi(A_p)$, and the succedent by $\phi(B_1) \smile \cdots \smile \phi(B_q)$. By setting $\Gamma = \{A_1, \ldots, A_p\}$ and $\Delta = \{B_1, \ldots, B_q\}$, we will often write $\Gamma \vdash_m \Delta$ for $A_1: \cdots :A_p \vdash_m B_1: \cdots :B_q$. Rules of ALL_m are written in Figure 2.

Theorem 3. $A_1: \cdots :A_p \vdash_m B_1: \cdots :B_q$ *holds in* ALL_m *if and only if the statement* $\phi(A_1) \frown \cdots \frown \phi(A_q) \le \phi(B_1) \smile \ldots \smile \phi(B_q)$ *is valid in* $(\mathcal{L}, \le)$.

We postpone the proof to Section 3.2 in order to give an example of proof technique and also to make some comments on ALL_m.

Example 1. We prove that $\oplus$ is distributive over &.

$$
\cfrac{
\cfrac{
\cfrac{
\cfrac{\cfrac{A \vdash_m A}{A:B \vdash_m A:C}\,w_L}{A\&B \vdash_m A:C}\&_L
}{A\&B \vdash_m A \oplus C}\oplus_R
\quad
\cfrac{\cfrac{\cfrac{C \vdash_m C}{C \vdash_m A:C}\,w_L}{C \vdash_m A \oplus C}\oplus_R}{}
}{(A\&B) \oplus C \vdash_m A \oplus C}\oplus_L
\quad
\cfrac{
\cfrac{
\cfrac{\cfrac{\cfrac{B \vdash_m B}{A:B \vdash_m B:C}\,w_L}{A\&B \vdash_m B:C}\&_L}{A\&B \vdash_m B \oplus C}\oplus_R
\quad
\cfrac{\cfrac{\cfrac{C \vdash_m C}{C \vdash_m B:C}\,w_L}{C \vdash_m B \oplus C}\oplus_R}{}
}{(A\&B) \oplus C \vdash_m B \oplus C}\oplus_L
}{}
}{(A\&B) \oplus C \vdash_m (A \oplus C)\&(B \oplus C)}\&_R
$$

The converse is not provable. □

We notice that contraction and weakening rules are postulated on both side of $\vdash_m$ [1]. The application of $\&_R$ (resp. $\oplus_L$) rule is restricted to right context (resp.

---

[1] So, we could also define a multiplicative version of ALL_m.

Axiom :

$$\frac{}{A \vdash_m A} \; ax$$

Logical Rules :

$$\frac{E_1 : E_2 : \Gamma \vdash_m \Delta}{E_1 \& E_2 : \Gamma \vdash_m \Delta} \; \&_L \qquad\qquad \frac{\Gamma \vdash_m F_1 \qquad \Gamma \vdash_m F_2}{\Gamma \vdash_m F_1 \& F_2} \; \&r$$

$$\frac{E_1 \vdash_m \Delta \qquad E_2 \vdash_m \Delta}{E_1 \oplus E_2 \vdash_m \Delta} \; \oplus_L \qquad\qquad \frac{\Gamma \vdash_m F_1 : F_2 : \Delta}{\Gamma \vdash_m F_1 \oplus F_2 : \Delta} \; \oplus_R$$

Cut :

$$\frac{\Gamma_1 \vdash_m \Delta_1 : E \qquad E \vdash_m \Delta_2}{\Gamma_1 \vdash_m \Delta_1 : \Delta_2} \; cut_L \qquad\qquad \frac{\Gamma_1 \vdash_m E \qquad E : \Gamma_2 \vdash_m \Delta_2}{\Gamma_1 : \Gamma_2 \vdash_m \Delta_2} \; cut_R$$

Structural Rules :
Weakening :

$$\frac{\Gamma \vdash_m \Delta}{E : \Gamma \vdash_m \Delta} \; w_L \qquad\qquad \frac{\Gamma \vdash_m \Delta}{\Gamma \vdash_m F : \Delta} \; w_R$$

Contraction :

$$\frac{E : E : \Gamma \vdash_m \Delta}{E : \Gamma \vdash_m \Delta} \; c_L \qquad\qquad \frac{\Gamma \vdash_m F : F : \Delta}{\Gamma \vdash_m F : \Delta} \; c_R$$

Exchanges :

$$\frac{\Gamma_1 : A : B : \Gamma_2 \vdash_m \Delta}{\Gamma_1 : B : A : \Gamma_2 \vdash_m \Delta} \; e_L \qquad\qquad \frac{\Gamma \vdash_m \Delta_1 : A : B : \Delta_2}{\Gamma \vdash_m \Delta_1 : B : A : \Delta_2} \; e_R$$

Fig. 2. ALL_m rules

left) containing only a single formula. Actually, this restriction on $\&_R$-rule is similar to the right implication rule in a multiple conclusion sequent calculus for intuitionistic logic, that we recall below

$$\frac{\Gamma, A \vdash_{LI} B}{\dfrac{\Gamma \vdash_{LI} A \to B}{\Gamma \vdash_{LI} A \to B, \Delta}} \; \to_R$$
$$w_R$$

In fact, by replacing $\&_R$ and $\oplus_L$ rules of ALL_m by these unrestricted versions

$$\frac{E_1 : \Gamma \Rightarrow \Delta \qquad E_2 : \Gamma \Rightarrow \Delta}{E_1 \oplus E_2 : \Gamma \Rightarrow \Delta} \; \oplus_L * \qquad\qquad \frac{\Gamma \Rightarrow F_1 : \Delta \qquad \Gamma \Rightarrow F_2 : \Delta}{\Gamma \Rightarrow F_1 \& F_2 : \Delta} \; \&_R *$$

we prove the distributive law: $(A \oplus B)\&C \Rightarrow A \oplus (B\&C)$. Such rules are valid in the extensionnal fragment of Relevance Logic, [3], where distributive law holds. In the same vein, if we substitute to both ALL_m-cut rules, the single cut rule :

$$\frac{\Gamma_1 \Rightarrow \Delta_1 : E \qquad E : \Gamma_2 \Rightarrow \Delta_2}{\Gamma_1 : \Gamma_2 \Rightarrow \Delta_1 : \Delta_2}$$

Then we can not eliminate cuts because the distributive law is again provable with cuts but not without.

3.2 Equivalence of the systems

Lemma 4. *Let E and F be two ALL formulas. If $E \vdash F$ holds in ALL then $E \vdash_m F$ holds in the contraction-free fragment of ALL_m. Moreover, a cut-free proof of ALL is mapped onto a cut-free proof of ALL_m.*

Proof. Assume that a proof in ALL ends with $(\oplus_{R_i})$ rule, then

$$\frac{\vdots}{\cfrac{E \vdash F_1}{E \vdash F_1 \oplus F_2}} \oplus_{R_i} \qquad \text{is transformed in } ALL_m \qquad \cfrac{\cfrac{\cfrac{\vdots}{E \vdash_m F_1}}{E \vdash_m F_1 : F_2} w_L}{E \vdash_m F_1 \oplus F_2} \oplus_R$$

All other right rules of ALL proof are valid in ALL_m. The left rules are symmetric. Observe that this homorphic translation do not increase the number of cut and do not introduce any contractions. □

It remains to show the converse. We illustrate how to map a ALL_m-proof onto a ALL-proof by an example. Take the following ALL_m-proof.

$$\cfrac{\cfrac{\cfrac{A \vdash_m A}{A:B:C \vdash_m A} w_L \quad \cfrac{B \vdash_m B}{A:B:C \vdash_m B} w_L}{A:B:C \vdash_m A\&B} \&_R}{A\&B:C \vdash_m A\&B} \&_L$$

We encode a ALL_m-sequence like $A\&B:C$ by $(A\&B)\&C$. Since $\&_L$ and $\&_R$ are not inversible in ALL, we apply first $\&_R$

$$\frac{(A\&B)\&C \vdash A \qquad (A\&B)\&C \vdash B}{(A\&B)\&C \vdash A\&B} \&_R$$

Next to simulate weakenings, we must, first, reorder formulas in $(A\&B)\&C$ on the right proof branch. This is performed by a cut.

$$\cfrac{\cfrac{A\vdash A}{\cfrac{A\&B\vdash A}{(A\&B)\&C\vdash A}\&L}\&L \qquad \cfrac{(A\&B)\&C\vdash A\&(C\&B) \quad \cfrac{\cfrac{B\vdash B}{C\&B\vdash B}\&L}{A\&(C\&B)\vdash B}\&L}{(A\&B)\&C\vdash B}cut}{(A\&B)\&C\vdash A\&B}\&R$$

Lemma 5. *Suppose that* $\Gamma = \{A_1, \ldots, A_p\}$ *and* $\Delta = \{B_1, \ldots, B_q\}$. *If* $\Gamma \vdash_m \Delta$ *holds in* ALL_m *then* $A_1 \& (\cdots \& A_p) \vdash B_1 \oplus (\cdots \oplus B_q)$ *holds in* ALL.

Combining Lemma 4 and 5,

Theorem 6. *A necessary and sufficient condition that a statement* $E \vdash F$ *hold in* ALL *is that* $E \vdash_m F$ *hold in* ALL_m.

The Theorem 6 and the remark 1 immediatly entails the Theorem 3.

Corollary 7. *The cut and contraction free fragment of* ALL_m *is equivalent to* ALL_m.

Proof. By Lemma 4, a cut free proof in ALL is mapped onto a cut free proof of ALL_m. Therefore, the cut elimination process is carried out thus. We transform a proof of ALL_m into a proof of ALL. Next, we eliminate cuts in ALL. Afterwards, we map back the equivalent cut free proof in ALL_m. □

4 Decision Procedure

4.1 Invertible rules

Assume that $\Gamma \vdash \Delta$ is valid and let E be a formula in $\Gamma \cup \Delta$. There is only one logical rule R which matches E. We wonder whether we can make a proof of $\Gamma \vdash \Delta$ which ends by introducing E by an instance of this rule R. If it is so, we say that the the rule R is invertible with respect to $\Gamma \vdash \Delta$. Inversion of inferences goes back to Kleene, cited in [2]. So if a rule is invertible, we can permute it with any other inferences in order to push it down.

Proposition 8. *The invertible rules of* ALL_m *are*

1. $\Gamma \vdash A\&B$ *iff* $\Gamma \vdash A$ *and* $\Gamma \vdash B$
2. $A \oplus B \vdash \Delta$ *iff* $A \vdash \Delta$ *and* $B \vdash \Delta$
3. $\Gamma \vdash A \oplus B:\Delta$ *iff* $\Gamma \vdash A:B:\Delta$
4. $A\&B:\Gamma \vdash \Delta$ *iff* $A:B:\Gamma \vdash \Delta$

and (1) and (2) in ALL *with the appropriate restriction.*

4.2   Bottom-up search in cut-free fragment

A naive bottom-up decision procedure will pick up a formula A from the input $\Gamma \vdash \Delta$ and will apply the operational rule that matches A. Then, the process is repeated until either a proof is found or all choices have been exhausted without success. For further background on backward-chaining see [6, 9], for instance. An upper-bound on the size of search space, which is a forest, is 4^n where n is the number of connectors in $\Gamma \cup \Delta$. (see [1] chapter II.11.)

But even, if we exploit inference permutations which reduce drastically the search space, there is a serious drawback.

Proposition 9. *A bottom-up decision procedure, based on ALL or ALL_m in cut-free sequent calculus, where proofs are trees, runs in more that 2^n steps in the worst case, where n is the number of connectors in the input formulas.*

Proof. We just make the proof for *ALL* because the proof for ALL_m is similar. Let HS_n^m be $A_0 \& \cdots \& A_m \vdash B_0 \oplus \cdots \oplus B_n$. and $s(m, n)$ be the number of possible proof attempts of HS_n^m. We show by induction that $s(m, n) > 2^{m+n}$. First, $s(1, 1) = 1$ since HS_1^1 is an instance of an axiom or not. Next, from the sequent HS_n^m, we have to deal with four choices which are $A_0 \vdash B_0 \oplus \cdots \oplus B_n$, $A_1 \& \cdots \& A_m \vdash B_0 \oplus \cdots \oplus B_n$, $A_0 \& \cdots \& A_m \vdash B_0$ and $A_0 \& \cdots \& A_m \vdash B_1 \oplus \cdots \oplus B_n$. This implies $s(m, n) = s(1, n) + s(m - 1, n) + s(m, 1) + s(m, n - 1)$. By assumption, we have $s(m - 1, n) = s(m, n - 1) > 2^{m+n-1}$, and so conclude that $s(m, n) > 2 \cdot 2^{m+n-1} = 2^{n+m}$. $\square$

For similar observations on classical propositional logic consult [11].

4.3   A top-down method

A top-down procedure enumerates valid sequents until the input is generated. We can restrict the sequent construction to subformulas of the inputs. For, we write $X \trianglelefteq E$ if X is a subformula of E. The process computes the smallest set $\text{PRF}(E, F)$ which satisfies:

$A \vdash A \in \text{PRF}(E, F)$ if $A \in \text{atomic}(E) \cap \text{atomic}(F)$
$X_1 \& X_2 \vdash Y \in \text{PRF}(E, F)$ if $X_i \vdash Y \in \text{PRF}(E, F)$ where $i \in \{1, 2\}$, $X_1 \& X_2 \trianglelefteq E$
$X_1 \oplus X_2 \vdash Y \in \text{PRF}(E, F)$ if $X_i \vdash Y \in \text{PRF}(E, F)$ for $i = 1, 2$, $X_1 \oplus X_2 \trianglelefteq E$
$X \vdash Y_1 \oplus Y_2 \in \text{PRF}(E, F)$ if $X \vdash Y_i \in \text{PRF}(E, F)$ where $i \in \{1, 2\}$, $Y_1 \oplus Y_2 \trianglelefteq F$
$X \vdash Y_1 \& Y_2 \in \text{PRF}(E, F)$ if $X \vdash Y_i \in \text{PRF}(E, F)$ for $i = 1, 2$, $Y_1 \& Y_2 \trianglelefteq F$.

Proposition 10. *The sequent $E \vdash F$ holds if and only if $E \vdash F \in \text{PRF}(E, E)$. The set $\text{PRF}(E, F)$ is computed in $O(|E| \cdot |F|)$ steps.*

Actually, the search space is a directed acyclic graph. This procedure turns out to be much more efficient that bottom-up searches in the cut-free fragment. See [8] for references on forward-chaining and resolution.

4.4 Proof search with analytic cuts

We investigate a decision algorithm which is polynomial time like the above top-down procedure and whose design is closed to a bottom-up procedure.

The idea is the following. We disregard permutation rules of the calculus. We repeatedly apply left operational rules. Then, we determine, by using right rules, analytic lemmas that we store in order to combine other analytic lemmas. Afterwards, we proceed by exploring the right side until an axiom or an analytic lemma is reached. (This pass is obsolete with the algorithm below.)

For this, we define a set $\mathrm{IMP}(E, F)$ such that if $X \in \mathrm{IMP}(E, F)$ so then $E \vdash X$ and also $X \trianglelefteq F$. That is, $\mathrm{IMP}(E, F)$ is an analytic initial section whose least element is E.

First, let $\mathrm{CL}(F, \Theta)$ be the smallest set satistfying:

- $A \in \mathrm{CL}(F, \Theta)$ if $A \trianglelefteq F$ and $A \in \Theta$,
- $X \& Y \in \mathrm{CL}(F, \Theta)$ if $X, Y \in \mathrm{CL}(F, \Theta)$ and $X \& Y \trianglelefteq F$,
- $X \oplus Y \in \mathrm{CL}(F, \Theta)$ if $X \in \mathrm{CL}(F, \Theta)$ or $Y \in \mathrm{CL}(F, \Theta)$ and $X \oplus Y \trianglelefteq F$,

The set $\mathrm{CL}(F, \Theta)$ verifies $\forall Y \in \mathrm{CL}(F, \Theta), \&_{X \in \Theta} X \vdash Y$, and is closed, that is $\mathrm{CL}(F, \Theta) = \mathrm{CL}(F, \mathrm{CL}(F, \Theta))$. Then, $\mathrm{IMP}(E, F)$ is computed thus:

- $\mathrm{IMP}(A, F) = \mathrm{CL}(F, \{A\})$ whenever A is an atom.
- $\mathrm{IMP}(E_1 \& E_2, F) = \mathrm{CL}(F, \mathrm{IMP}(E_1, F) \cup \mathrm{IMP}(E_2, F))$
- $\mathrm{IMP}(E_1 \oplus E_2, F) = \mathrm{CL}(F, \mathrm{IMP}(E_1, F) \cap \mathrm{IMP}(E_2, F))$

Proposition 11. *A neccessary and sufficient condition that $E \vdash F$ be a theorem of ALL is that $F \in \mathrm{IMP}(E, F)$. $\mathrm{IMP}(E, F)$ is computed in time $O(|E| \cdot |F|)$.*

Example 2. Let $E = (A \& B) \& (A \oplus C)$ and $F = (B \& A) \oplus C$. We seek a proof of $E \vdash F$. For this, we compute:

$$\mathrm{IMP}(A, F) = \{A\} \qquad \mathrm{IMP}(B, F) = \{B\} \qquad \mathrm{IMP}(C, F) = \{C, (B \& A) \oplus C\}$$
$$\mathrm{IMP}(A \& B, F) = \{A, B, B \& A, (B \& A) \oplus C\} \qquad \mathrm{IMP}(A \oplus C, F) = \emptyset$$
$$\mathrm{IMP}((A \& B) \& (A \oplus C), F) = \{A, B, B \& A, (B \& A) \oplus C\}$$

So $E \vdash F$ holds because $E = (B \& A) \oplus C \in \mathrm{IMP}(E, F)$ □

Observe that it is sufficient to maintain the greatest subformulas, with respect to $\trianglelefteq$, at each step.

5 Propositional Linear Logic

To conclude, we show how to combine additive contexts to propositional linear logic *MALL*. For this, we develop an equivalent sequent calculus *MALL*∗.

A sequent of *MALL*∗ is of the form $\Gamma \vdash_* \Delta$ where both Γ and Δ are multiplicative contexts that we define as follows. A multiplicative context is a sequence $\gamma_1, \cdots, \gamma_p$ of additive contexts separated by commas $(,)$. An additive context γ is a sequence $\Gamma_1 : \cdots : \Gamma_q$ where Γ_i is either a formula or a multiplicative context, separated by semicolons $(:)$. *MALL*∗-rules are displayed in Figure 5.

Theorem 12. *If $\Gamma \vdash \Delta$ holds in MALL then $\Gamma \vdash_* \Delta$ holds in MALL$_*$, and conversely.*

Proof. Since both systems are symmetric, we just consider right rules.

MALL-rules and MALL$_*$-rules are identical except for the the right-$\oplus$ rule. In order to simulate this former rule in MALL$_*$, we drop the extra formula by the right weakening rule, like in Lemma 4.

Conversely, we show that if $\Gamma \vdash_* \Delta$ then we can derive $v^-(\Gamma) \vdash v^+(\Delta)$ in MALL where $v^-(F) = v^+(F) = F$ when F is a formula, and

$$v^-(\gamma, \gamma') = v^-(\gamma) \otimes v^-(\gamma') \qquad v^+(\gamma, \gamma') = v^+(\gamma) \wp v^+(\gamma')$$
$$v^-(\Gamma{:}\Gamma') = v^-(\Gamma) \& v^-(\Gamma') \qquad v^+(\Gamma{:}\Gamma') = v^+(\Gamma) \oplus v^+(\Gamma')$$

To complete the proof, we observe that we can always conclude a proof by applying $\wp$ right rules, and also that $v^+(\Delta[\delta]) \vdash v^+(\Delta[\delta{:}B])$. □

Corollary 13. *MALL$_*$ permits cut-elimination.*

Proof. The proof is analogous to the proof of the Corollary 7. Since a cut-free MALL-proof is mapped onto a cut-free MALL$_*$-proof and since cut is redundant in MALL, we eliminate cut in MALL$_*$-proof. □

The sequent calculus **R** for Relevant Logic in [3] §3.7 had already the same features, but with no cut elimination property. Light Linear Logic [5] uses similar sequents where blocks are akin additive contexts. To conclude, we give an example that points out some interest in MALL$_*$ investigation.

Example 3. Suppose that we want to encode in linear logic a non-deterministic system like the one drawn below. (For further examples about Linear Logic and system behavior see [7].)

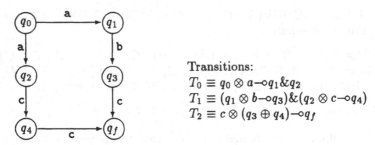

Transitions:
$$T_0 \equiv q_0 \otimes a \multimap q_1 \& q_2$$
$$T_1 \equiv (q_1 \otimes b \multimap q_3) \& (q_2 \otimes c \multimap q_4)$$
$$T_2 \equiv c \otimes (q_3 \oplus q_4) \multimap q_f$$

We encode an input word like abc by $a \otimes b \otimes c$. We prove that we reach the final state q_f from the initial configuration $(q_0, a \otimes b \otimes c)$.

$$
\cfrac{
\cfrac{
\cfrac{
\cfrac{T_0, q_1 \otimes b \multimap q_3 {:} q_2 \otimes c \multimap q_4, q_0, a, b \vdash_* q_3 {:} q_4}{T_0, T_1, q_0, a, b \vdash_* q_3 {:} q_4} \, \&_L \text{ on } T_1
}{T_0, T_1, q_0, a, b \vdash_* q_3 \oplus q_4} \, \oplus_R \qquad c \vdash_* c
}{T_0, T_1, q_0, a, b, c \vdash_* c \otimes (q_3 \oplus q_4)} \, \otimes_R \qquad q_f \vdash_* q_f
}{
\cfrac{T_0, T_1, T_2, q_0, a, b, c \vdash_* q_f}{T_0, T_1, T_2, q_0, a \otimes b \otimes c \vdash_* q_f} \, \otimes_L
} \, T_2
$$

Now, we choose either the transition starting with q_1 or the transition with q_2. In both cases, this choice is non-invertible. However, we delay the decision by using the following admissible rule:

$$\frac{\Gamma \vdash_* A_1: \cdots : A_k}{\Gamma, A_1 \multimap C_1: \cdots : A_k \multimap C_k \vdash_* C_1: \cdots : C_k} \; parallel \multimap$$

This rule means that we can fire both transitions simultaneously to get

$$\frac{T_0, q_0, a, b \vdash_* q_1 \otimes b{:}q_2 \otimes c}{T_0, q_1 \otimes b \multimap q_3{:}q_2 \otimes c \multimap q_4, q_0, a, b \vdash_* q_3{:}q_4}$$

Now, the proof keeps going straight.

$$\frac{\dfrac{\dfrac{}{q_0 \vdash_* q_0} \, ax \quad \dfrac{}{a \vdash_* a} \, ax}{q_0, a \vdash_* q_0 \otimes a} \otimes R \quad \dfrac{\dfrac{\overline{\overline{q_1{:}q_2 \vdash_* q_1}} \, w_L + ax \quad \dfrac{}{b \vdash_* b} \, ax}{q_1{:}q_2, b \vdash_* q_1 \otimes b{:}q_2 \otimes c} \otimes R}{q_1 \& q_2, b \vdash_* q_1 \otimes b{:}q_2 \otimes c} \& L}{T_0, q_0, a, b \vdash_* q_1 \otimes b{:}q_2 \otimes c} \, T_0$$

□

References

1. G. Birkhoff. *Lattice Theory*, volume XXV. American Mathematical Society, Second edition, 1948.
2. H.B. Curry. *Foundations of mathematical logic*. Dover, 1963 first edition, 1976.
3. M. Dunn. Relevance logic and entailment. In D. Gabbay and F.Guenthner, editors, *Handbook of philosophical logic*, chapter III.3, pages 117–224. D. Reidel Publishing Company, 1986.
4. J.-Y. Girard. Linear logic. *TCS*, 50:1–102, 1987.
5. J-Y Girard. Light linear logic. In D. Leivant, editor, *LCC'94*, pages 145–179. LNCS 960, 1995.
6. J.S. J odas und D. Miller. Logic programming in a fragment of linear logic. *Journal of Information and Computation*, 110(2):327–365, 1994.
7. M. Kanovich. Simulating guarded programs in linear logic. In T. Ito and A. Yonezawa, editors, *Int. Work. Theory and Practice of Parallel Programming*, pages 45–69. LNCS 907, 1994.
8. G. Mints. Resolution calculus for the first order linear logic. *Journal of Logic Languages and Information*, 2:59–93, 1993.
9. T. Tammet. Proof strategies in linear logic. *Journal of Automated Reasonning*, 12(3):273–304, 1994.
10. A. S. Troelstra. *Lectures on Linear Logic*, volume 29. CSLI, 1992.
11. A. Urquhart. Complexity of proofs in classical propositional logic. In Y.N. Moschovakis, editor, *Logic from computer science*, pages 597–608. Springer-Verlag, 1989.

Identity:

$$\frac{}{A \vdash_* A}\ \text{ax} \qquad\qquad \frac{\Gamma_1 \vdash_* \Delta_1, A \qquad A, \Gamma_2 \vdash_* \Delta_2}{\Gamma_1, \Gamma_2 \vdash_* \Delta_1, \Delta_2}\ \text{cut}$$

Additive Rules :

$$\frac{\Gamma[A\!:\!B] \vdash_* \Delta}{\Gamma[A\&B] \vdash_* \Delta}\ \&_L \qquad\qquad \frac{\Gamma \vdash_* \Delta, A \qquad \Gamma \vdash_* \Delta, B}{\Gamma \vdash_* \Delta, A\&B}\ \&_R$$

$$\frac{\Gamma, A \vdash_* \Delta \qquad \Gamma, B \vdash_* \Delta}{\Gamma, A \oplus B \vdash_* \Delta}\ \oplus_L \qquad\qquad \frac{\Gamma \vdash_* \Delta[A\!:\!B]}{\Gamma \vdash_* \Delta[A \oplus B]}\ \oplus_R$$

Multiplicative Rules:

$$\frac{\Gamma[A, B] \vdash_* \Delta}{\Gamma[A \otimes B] \vdash_* \Delta}\ \otimes_L \qquad\qquad \frac{\Gamma_1 \vdash_* \Delta_1, A \qquad \Gamma_2 \vdash_* B, \Delta_2}{\Gamma_1, \Gamma_2 \vdash_* \Delta_1, A \otimes B, \Delta_2}\ \otimes_R$$

$$\frac{\Gamma_1, A \vdash_* \Delta_1 \qquad B, \Gamma_2 \vdash_* \Delta_2}{\Gamma_1, A \wp B, \Gamma_2 \vdash_* \Delta_1, \Delta_2}\ \wp_L \qquad\qquad \frac{\Gamma \vdash_* \Delta[A, B]}{\Gamma \vdash_* \Delta[A\wp B]}\ \wp_R$$

$$\frac{\Gamma_1 \vdash_* A, \Delta_1 \qquad B, \Gamma_2 \vdash_* \Delta_2}{\Gamma_1, A \multimap B, \Gamma_2 \vdash_* \Delta_1, \Delta_2}\ \multimap_L \qquad\qquad \frac{\Gamma, A \vdash_* \Delta, B}{\Gamma \vdash_* \Delta, A \multimap B}\ \multimap_R$$

$$\frac{\Gamma \vdash_* A, \Delta}{\Gamma, A^\perp \vdash_* \Delta}\ \perp_L \qquad\qquad \frac{\Gamma, A \vdash_* \Delta}{\Gamma \vdash_* \Delta, A^\perp}\ \perp_R$$

Structural Rules :
  Weakening :

$$\frac{\Gamma[\gamma] \vdash_* \Delta}{\Gamma[\gamma\!:\!A] \vdash_* \Delta}\ w_L \qquad\qquad \frac{\Gamma \vdash_* \Delta[\delta]}{\Gamma \vdash_* \Delta[\delta\!:\!B]}\ w_R$$

Exchanges :

$$\frac{\Gamma_1 * \gamma' * \gamma * \Gamma \vdash_* \Delta}{\Gamma_1 * \gamma * \gamma' * \Gamma \vdash_* \Delta}\ e_l\ *\in\{:;,\} \qquad\qquad \frac{\Gamma \vdash_* \Delta_1 * \delta' * \delta * \Delta_2}{\Gamma \vdash_* \Delta_1 * \delta * \delta' * \Delta_2}\ e_r\ *\in\{:;,\}$$

Where we write $\Gamma[F]$ to say that F is a formula in Γ.

Fig. 3. *Rules of MALL*$_*$

Some Decision Problems for Traces

Yuri Matiyasevich

Steklov Institute of Mathematics at St.Petersburg,
27 Fontanka, St.Petersburg, 191011, Russia

Abstract. The notion of a word, considered as an element of a free monoïd, has been long ago generalized to the notion of a trace, an element of a partially commutative monoïd. Traces turned out to be useful tool for studying concurrency.
Every decision problem for words can be generalized to corresponding problem for traces. The main content of the paper is a proof of the decidability of trace equations, which is an extension of celebrated theorem of G. S. Makanin about decidability of word equations.
Several new results about decidable and undecidable cases of the code problem for traces are stated without proofs.

1 Introduction

1.1 The notion of trace

The notion of a trace can be viewed as a generalization of the notion of a word.

To be able to speak about words, we need first to fix some *alphabet*, the elements of which we call *generators*. The set of all words in an alphabet Σ will be denoted, as usual, by Σ^*.

To be able to speak about traces, we need to fix an *independence alphabet*. This is a couple $\langle \Sigma, I \rangle$ where Σ is an alphabet and I is an irreflexive symmetric binary *relation of independence* defined between the elements of Σ. Intuitively, if two generators a_p and a_q satisfy this relation (written $\langle a_p, a_q \rangle \in I$ or $a_q I a_p$), then we say that a_p and a_q are independent (in algebraic terminology, a_p and a_q commute) and consider strings $a_p a_q$ and $a_q a_p$ as equal. Formally, we say that two strings of the form

$$a_{i_1} \ldots a_{i_l} a_p a_q a_{j_1} \ldots a_{j_m} \tag{1}$$

and

$$a_{i_1} \ldots a_{i_l} a_q a_p a_{j_1} \ldots a_{j_m} \tag{2}$$

are *immediately equivalent* whenever $a_p I a_q$. By $\equiv_I$ we denote the reflexive transitive closure of the relation of immediate equivalence, that is, two strings are *equivalent* (satisfy relation $\equiv_I$) if one of them can be obtained from the other by a chain of changing the order of two neighbouring commutating generators. *Traces* in a given independence alphabet $\langle \Sigma, I \rangle$ are just classes of equivalent (with respect to $\equiv_I$) words from Σ^*.

Figuratively speaking, traces are strings of generators in which occurences of generators can "travel" along the string rather than occupy fixed positions. The limits of such "travelling" of a given occurence of a generator a_i are determined by the ends of the string and by the closest left and right occurences of generators which do not commute with a_i. In its turn, these left and right neighbours can "travel" themselves making the whole picture rather complicated.

Concatenation of strings, evidently, agrees with relation $\equiv_I$, so we can concatenate traces. In algebraic terminology, traces (together with concatenation as semigroup operation) form a *partially commutative monoïd* $\mathcal{M}(\Sigma, \mathcal{I}) = \Sigma^* / \equiv_\mathcal{I}$, also called *trace monoïd*.

1.2 Brief history

Systematic study of traces from combinatorial point of view was done by P. Cartier and D. Foata [1]. Later A. Mazurkiewicz [9] demonstrated that traces can be very useful in computer science, namely, as a tool for studying concurrency. The underlying idea can be explained very roughly as follows. Suppose, that we need to perform a sequence of actions, and two actions, say, A and B, are independent in the sense that they can be performed in parallel. Then they can be performed sequentially in both orders AB and BA. In this way a trace, as a class of equivalent strings, represents all possible manners to transform a concurrent program into an equivalent linear program.

The approach initiated by Mazurkiewicz has got further development in works of many researchers, and today the theory of traces is a well recognized area in the study of concurrency. The bibliography on traces has about 300 papers, and recent progress was summarized in "The Book of Traces" [3].

1.3 Decision problems

Any decision problem for words in some alphabet Σ can be naturally generalized to the case of traces in an independence alphabet $\langle \Sigma, I \rangle$. For some choice of the independence relation I the generalized problem about traces can turn out to be undecidable even if it was trivial for words. In such situation it is natural to look for a descriptions of those independence alphabets for which the generalized problem is decidable and of those for which the problem is undecidable. Often it is convenient to give such a classification in terms of the *graph of independence* in which the vertices are the generators, and the edges connect commuting generators.

The code problem. One can code words in some alphabet $\Sigma = \{a_1, \ldots, a_n\}$ by words in another alphabet Π by defining first a coding of generators, $\mu : a_k \mapsto A_k$ (where $A_k \in \Pi^*$), and then extending it in a natural way to a map $\mu^* : \Sigma^* \to \Pi^*$ which is usually required to be injective. The *code problem* is to decide, for given words $A_1, \ldots, A_n$, whether corresponding map μ^* is indeed injective or not.

The code problem for words was shown to be decidable by A. Sardinas and C. Patterson [10]. For traces the code problem turned out to be undecidable for some independence relation, for example, in the case when the independence graph is the circle C_4. This was shown by M. Chrobak and W. Rytter [2] who also established the decidability of the code problem for all other 4-generator independence alphabet with exception of the case of simple path P_4 as the independence graph.

For the code problem, we still are far from the complete classification, only a number of partial results is known. Recently, some progress was achieved by H. J. Hoogeboom and A. Muscholl [5] who showed the undecidability for one more graph and the decidabily in the case P_4. The latter result was also independently obtained by the author [7] and generalized to the case of any graph without circles. In the same paper the author established the undecidability for the case of two sufficiently long connected circles.

Trace equations. Solving equations is a standard occupation, and it is natural to consider string equations. The decidability of the basic case, word equations, was shown by G. S. Makanin [6]. His result was later extended in several directions, in particular, K. U. Schulz [11] showed that one can in addition restrict the admissible values of each unknown by corresponding regular set. Using this extension, we shall show in this paper the decidability of trace equations for arbitrary independence relation.

Remark. In fact, a bit more than the mere decidability is proved. Namely, the following relationship between trace and word equations has been established. For every trace equation in some unknowns $x_1, \ldots, x_m$ one can effectively construct a word equation in more unknowns $x_1, \ldots, x_m, \ldots, x_{m+l}$ (with restrictions on their admissible values) such that

- in every solution of the word equation, unknowns $x_1, \ldots, x_m$ produce a solution of the trace equation;
- every solution of the trace equation in which the values of $x_1, \ldots, x_m$ are in certain *normal form* can be extended (by the choice of values of $x_{m+1}, \ldots, x_l$) to a solution of the word equation.

This closer relationship between trace equations and word equations can be useful for reducing other problems about traces to corresponding problems for words. The crucial point for the success of such a reduction will be the above mentioned normal form. The proof below is based on the original normal form found by the author. Afterwards (see [4]) it was discovered that the classical least lexicographical form can also be used as normal. It seems interesting to search for other possible normal forms suitable for reducing trace equations to word equations.

2 Notation, definitions and main technical result

Notation 1. For a given word X from some alphabet Σ, by $\text{Alph}(X)$ we denote the set of all generators which do occur in X.

Definition 2. For given alphabet of generators Σ and alphabet of unknowns Ω such that $\Sigma \cap \Omega = \emptyset$, a *string equation* (with constants) is an expression of the form

$$L = R \tag{3}$$

where

$$L \in \{\Sigma \cup \Omega\}^*, \quad R \in \{\Sigma \cup \Omega\}^*. \tag{4}$$

Definition 3. An *assignment of value* to unknowns is any map

$$\tau : \Omega \to \Sigma^*. \tag{5}$$

Notation 4. For given assignment τ we will, by an abuse of notation, denote by $\tau(L)$ the result of replacing each occurence of each unknown x in L by its value $\tau(x)$.

Definition 5. For given independence relation I, $I \subseteq \Sigma \times \Sigma$, an *I-solution* of equation (3) is any assignment τ such that

$$\tau(L) \equiv_I \tau(R). \tag{6}$$

Definition 6. For given alphabet of generators Σ, an independence relation I, $I \subseteq \Sigma \times \Sigma$, and alphabet of unknowns Ω such that $\Sigma \cap \Omega = \emptyset$, a *commutativity equation* is an expression form

$$x\, I\, y \tag{7}$$

where

$$x \in \Omega, \quad y \in \Omega. \tag{8}$$

Definition 7. A *solution* of equation (7) is any assignment (5) such that

$$\mathrm{Alph}(\tau(x)) \times \mathrm{Alph}(\tau(y)) \subseteq I. \tag{9}$$

Definition 8. A *system of trace equations* is a quintuple

$$T = \langle \Sigma, \Omega, S, C, I \rangle \tag{10}$$

where

- $\Sigma = \{a_1, \ldots, a_n\}$ is an alphabet of generators;
- $\Omega = \{x_1, \ldots, x_m\}$ is an alphabet of unknowns, $\Sigma \cap \Omega = \emptyset$;
- S is a set of string equations in $\Sigma \cup \Omega$;
- C is a set of commutativity equations $\Sigma \cup \Omega$;
- I is an independence relation, $I \subseteq \Sigma \times \Sigma$;

Definition 9. A *solution of* a system of trace equations (10) is any assignment of values to the unknowns which is at the same time

- an I-solution of each of the string equations from S, and
- a solution of each of the commutativity equations from C.

Remark. In our terminology, classical *system of word equations* is just a system (10) with $C = \emptyset$, $I = \emptyset$.

Remark. For technical reasons, we work not with single equations but with systems of them. It is well known that a system of trace or word equations can be combined into single equivalent equation.

Definition 10. We say that an assignment

$$\tau' : \Omega' \to \Sigma^* \tag{11}$$

I-agrees with an assignment (5) if for every x from $\Omega \cap \Omega'$

$$\tau'(x) \equiv_I \tau(x). \tag{12}$$

Definition 11. System of trace equations $T_0 = \langle \Sigma, \Omega \cup \Omega_0, S_0, C_0, I_0 \rangle$ is a *detailization of* a system (10) if

- every solution of system T_0 is a solution of system T;
- for every solution τ of system T there is a solution τ_0 of system T_0 which I-agrees with τ.

Theorem 12. *Given a system of trace equations $T = \langle \Sigma, \Omega, S, C, I \rangle$, one can effectively construct an alphabet of additional unknowns Ω_0, a system S_0 of string equations and a system C_0 of commutativity equations, both in the alphabet $\Sigma \cup \Omega \cup \Omega_0$, such that the system of trace equations $T_0 = \langle \Sigma, \Omega \cup \Omega_0, S_0, C_0, \emptyset \rangle$ is a detailization of system T.*

Corollary 13. *There is an algorithm for deciding for given system of trace equations whether it has a solution.*

3 Proofs

Without loss of generality, we can assume that in our system T there are no constants, i.e., in each equation (3) from S both L and R are words from Ω^*. Moreover, we can assume that all equations of system S are of the form

$$vu = w \tag{13}$$

where $u, v, w \in \Omega$.

The required alphabet Ω_0 and systems S_0 and C_0 will be constructed in n steps. Namely, for $k = n, \ldots, 0$ let

$$\Sigma_k = \{a_1, \ldots, a_k\} \tag{14}$$

so that

$$\Sigma = \Sigma_n \supset \Sigma_{n-1} \supset \ldots \supset \Sigma_k \supset \Sigma_{k-1} \supset \ldots \supset \Sigma_1 \supset \Sigma_0 = \emptyset. \tag{15}$$

Respectively, let

$$I_k = I \cap (\Sigma_k \times \Sigma_k) \tag{16}$$

so that

$$I = I_n \supset I_{n-1} \supset \cdots \supset I_k \supset I_{k-1} \supset \cdots \supset I_1 \supset I_0 = \emptyset. \tag{17}$$

Let $\Omega_n = \Omega$, $S_n = S$ and $C_n = C$. We are to construct for $k = n-1, \ldots, 0$ an alphabet Ω_k, system of string equations S_k in alphabet $\Sigma \cup \Omega \cup \Omega_k$ and system of commutativity equations C_k such that the system of trace equations $T_k = \langle \Sigma, \Omega \cup \Omega_k, S_k, C_k, I_k \rangle$ is a detailization of system T_{k+1}. This will prove the theorem because detailization is clearly a transitive relation.

Let us fix some positive k and suppose that system T_k has already been found.

Definition 14. Generator a_k is called the *principal* generator (while constructing T_{k-1}).

Constructing T_{k-1} from T_k is based on a kind of normal form for a word representing given trace from $\mathcal{M}(\Sigma, \mathcal{I}_\parallel)$. Intuitively speaking, this is a form in which all occurences of the principal generator occupy the leftmost positions.

Definition 15. For a given word V from Σ^*, the *right wing of* an occurence

$$V = V'aV'' \tag{18}$$

of a non-principal generator a from Σ is (the occurence of) the longest word W such that

- $\{a\} \times \mathrm{Alph}(W) \subseteq I_k$;
- W is a prefix of V'';

in other words, $V = V'aWV'''$, a commutes with every generator from W but does not commute with the first generator of V''' (when this word is not empty).

Definition 16. An occurence (18) is *normal* if

- either $a = a_k$
- or $a \neq a_k$ and its right wing W does not contain a_k.

Definition 17. A word is *normal* if each occurence in it is normal.

The key role of normal words is expressed in the following two lemmas.

Lemma 18. *For every word V from Σ^* there is a normal word U such that $V \equiv_{I_k} U$.*

Lemma 19. *For every normal words V and W from Σ^**

$$V \equiv_{I_k} W \iff V \equiv_{I_{k-1}} W.$$

We need a bit more then the mere existence of an equivalent normal word stated in Lemma 18, namely, we are to fix a particular way of bringing an arbitrary word into normal form.

Definition 20. For every word V we define a unique word $\mathrm{Norm1}_k(V)$ in the following way:

- if V is empty then $\mathrm{Norm1}_k(V) = V$;
- if $V = a_k V''$ then $\mathrm{Norm1}_k(V) = a_k V''$;
- if $V = aV''$ and $a \neq a_k$, then $\mathrm{Norm1}_k(V) = WaV'''$ where and $aV'' = aWV'''$ is the right wing of the first occurence of generator a.

Lemma 21. *If a word V is normal, then for each generator a word $\mathrm{Norm1}_k(aV)$ is also normal.*

Definition 22. For every word V we define a unique word $\mathrm{Norm}_k(V)$ by induction on the length of V in the following way:

- if V is empty then $\mathrm{Norm}_k(V) = V$;
- if $V = aV''$ then $\mathrm{Norm}_k(V) = \mathrm{Norm1}_k(a\mathrm{Norm}_k(V''))$.

Lemma 23. *For every word V, word $\mathrm{Norm}_k(V)$ is normal and*

$$V \equiv_I \mathrm{Norm}_k(V). \tag{19}$$

Definition 24. We say that a word U is *normalized* if

$$U = \mathrm{Norm}_k(U). \tag{20}$$

Now we are to investigate what happens with normalized words during concatenation.

Definition 25. Let $a_{j_1}, \ldots, a_{j_l}$ be all generators occurring in a word U listed in the order of their first occurences in U, i.e., the leftmost occurence of generator a_{j_i} is situated to the left from the leftmost occurence of generator $a_{j_{i+1}}$. Then U has unique *canonical decomposition*

$$U = U_1 \ldots U_l \tag{21}$$

where

$$U_i \in a_{j_i}\{a_{j_1}, \ldots, a_{j_i}\}^*, \quad i = 1, \ldots, l. \tag{22}$$

Lemma 26. *For every word V from $\Sigma \backslash \{a_k\}$ and every normalized word U with canonical decomposition (21),*

$$\mathrm{Norm}_k(VU) = V_0 U_1 V_1 \ldots U_l V_l \tag{23}$$

for some words $V_0, \ldots, V_l$ such that

$$V \equiv_{I_{k-1}} V_0 V_1 \ldots V_l, \tag{24}$$
$$\tag{25}$$

and

$$U_i I_k V_j, \quad 1 \leq i \leq j \leq l. \tag{26}$$

Lemma 27. *If $V = V'a_k$ and U are normalized words, then word VU is also normalized.*

Lemma 28. *For every normalized word V and every normalized word U with canonical decomposition* (21),

$$\mathrm{Norm}_k(VU) = V_0 U_1 V_1 \ldots U_l V_l \tag{27}$$

for some words $V_0, \ldots, V_l$ such that

$$V \equiv_{I_{k-1}} V_0 V_1 \ldots V_l \tag{28}$$
$$\tag{29}$$

and

$$U_i I_k V_j, \quad 1 \le i \le j \le l. \tag{30}$$

Lemma 3 suggests that we can replace every string equation (13) to be solved in I_k by the system

$$v = v_0 \ldots v_n, \tag{31}$$
$$u = u_1 \ldots u_n, \tag{32}$$
$$w = v_0 u_1 v_1 \ldots u_n v_n, \tag{33}$$
$$u_i \, I_k \, v_j, \quad 1 \le i \le l \le n. \tag{34}$$

to be solved in I_{k-1}.

To construct T_{k-1}, we apply the above transformation to every string equation from T_k, selecting each time for $u_1, \ldots, u_n, v_0, \ldots, v_n$ new unknowns. Combining all resulting string equations of the form (31)-(33), we obtain T_{k-1}. To obtain C_{k-1}, we add the new commutativity equations of the form (34) to C_k.

In order to prove Corollary 13, it remains to transform commutativity equations into restrictions on admissible values of the unknowns of the form of regular sets and refer to [11]. This can be done in many ways, for example, we can introduce an extra generator a_0 and replace equation (7) by conditions

$$xa_0y = z, \quad z \in J \tag{35}$$

where z is a new unknown and

$$J = \{Xa_0Y \mid \mathrm{Alph}(\tau(X)) \times \mathrm{Alph}(\tau(Y)) \subseteq I\}, \tag{36}$$

which is, evidently, a regular set.

Remark. Mentioned in the Introduction normal form can be defined as

$$\mathrm{Norm}(V) = \mathrm{Norm}_1(\ldots \mathrm{Norm}_n(V)\ldots). \tag{37}$$

Remark. To simplify presentation, we constructed system T_0 by induction. If we used global normalization Norm from the above remark, we would be able to simplify system T_0 at the cost of making the proof more complicated. Namely, we could treat (32) and (33) (but not (31)) as word equations rather than equations in I_{k-1}. However, care should be taken when dealing with mixed systems containing both trace and word equations because in general they are undecidable.

This fact can be easily proved using the undecidability of Hilbert's tenth problem about Diophantine equations (see, e.g., [8]). Namely, we can treat natural numbers as words in one-generator alphabet $\Sigma_1 = \{a\}$, i.e., as solutions of word equation

$$xa = ax. \tag{38}$$

Under this representation addition of numbers corresponds to concatenation: if $x = a^k$, $y = a^l$ and $z = a^m$ then

$$k + l = m \iff xy = z. \tag{39}$$

Respectively, multiplication can be simulated by the following system of trace-word equations:

$$kl = m \iff \exists y_b y_{ab} z_{ab} \tag{40}$$

$$y_b b = b y_b \ \& \tag{41}$$

$$y_{ab} ab = ab y_{ab} \ \& \tag{42}$$

$$y_{ab} \equiv_I y y_b \ \& \tag{43}$$

$$z_{ab} x b = x b z_{ab} \ \& \tag{44}$$

$$z y_b \equiv_I z_{ab} \tag{45}$$

where $I = \{\langle a, b \rangle\}$.

References

1. P. Cartier and D. Floata. Problème combinatoires de commutation et réarrangements. *Lecture Notes in Mathematics*, 85, 1969.

2. M. Chrobak and W. Rytter. Unique decipherability for partially commutative alphabets. *Fundamenta Informaticae*, X:323–336, 1987.

3. V. Diekert and G. Rozenberg, editors. *The Book of Traces*. World Scientific, Singapoure a.o., 1995.

4. Volker Diekert, Yuri Matiyasevich, and Anca Muscholl. Solving trace equations using lexicographical normal forms. Report 1997/01, Universität Stuttgart, Fakultät Informatik, 1997.

5. Hendrik Jan Hoogeboom and Anca Muscholl. The code problem for traces—improving the boundaries. *Theoretical Computer Science*, 172:309–321, 1997.

6. G. S. Makanin. The problem of solvability of equations in a free semigroup. *Math. USSR Sbornik*, 32(2):129–198, 1977.

7. Youri Matiiassevitch. Mots et codes: Cas décidables et indécidables du problème du codage pour les monoïdes partiellement commutatifs. *Quadrature*, 27:23–33, 1997.

8. Yu. V. Matiyasevich. *Desyataya Problema Gilberta*. Nauka, Moscow, 1993. English translation: *Hilbert's tenth problem*. MIT Press, 1993. French translation: *Le dixime problème de Hilbert*. Masson, 1995.

9. A. Mazurkiewicz. Concurrent program schemes and their interpretations. DAIMI Rep. PB 78, Aarus University, Aaurus, 1977.

10. A. Sardinas and C. Patterson. A necessary and sufficient condition for the unique decomposition of coded messages. *IRE Intern. Conv. Record*, 8:104–108, 1953.

11. Klaus U. Schulz. Makanin's algorithm for word equations–Two improvements and a generalization. In K. U. Schulz, editor, *Word equations and related topics. 1st International Workshop, IWWERT'90, Tübingen, Germany, October 1990*, volume 572 of *Lecture Notes in Computer Science*, pages 85–150. Springer, 1992.

Existential Instantiation and Strong Normalization

G. Mints

Dept. of Philosophy, Stanford University, Stanford, CA 94305, USA,
mints@csli.stanford.edu

Abstract. We present a new manageable formulation of natural deduction with a rule of existential instantiation $\exists x A[x]/ A[b]$. It simplifies skolemizing devices of earlier formulations, includes a new rule for disjunction and admits a proof of strong normalization. This opens way to the treatment of new systems for which only normal form theorems are known.

1 Introduction

Natural deduction is used both for exact formalization of reasoning and as a tool for extraction of programs from proofs and normalization of proofs (deductions) which exactly corresponds under Curry-Howard isomorphism to computation by the corresponding programs. In both of these situations standard $\exists$-elimination rule

$$\frac{\Gamma \Rightarrow \exists x A \quad A[a], \Sigma \Rightarrow C}{\Gamma, \Sigma \Rightarrow C}$$

presents well-known difficulties. Interaction of immediate reduction of this rule (applicable when the first premise is deduced from $\Gamma \Rightarrow A[t]$) with permutative reductions, which prepare immediate reduction, is rather complicated, and makes a proof of strong normalization much more difficult than for negative fragment. The rule $\exists\,i$ of existential instantiation $\exists x A[x]/ A[a]$, which was introduced with the intension to bring formalization closer to human reasoning (the history is presented in [9, 7, 3]) simplifies deduction graph by replacing branching in the two-premise rule of $\exists$-elimination by direct flow. Moreover, it makes permutative reductions almost trivial. All previous formulations using this rule employed some kind of global restriction on variables in the whole deduction, or some skolemizing device to ensure correctness for eigenvariables (and for the assumptions in the intuitionistic case). This made the exact treatment of computational aspects very difficult. For example, no complete strong normalization proof for a system of this kind exists in the literature.

We propose a manageable formulation NJi of intuitionistic logic where restrictions on eigenvariables are made local and reduced to a minimum by a device from [4, sections 7,8], cf. also [6]. The only new syntactical objects are assumptions $< \Gamma >_{\exists x F, a}$, which cannot be combined into more complicated formulae, and are eventually interpreted as $F[x/a]$. System $M\epsilon$ of [3] is probably the closest to NJi in the literature. Main differences with $M\epsilon$ are the use of

ϵ-terms there and a different mechanism of restricting assumptions. Detailed comparison of $M\epsilon$ with other systems is presented in [3]. We present also a rule for disjunction which can play the same role for $\vee$ that $\exists i$ plays for $\exists$. Only intuitionistic predicate logic is considered here, but the extension to classical or higher order case is straightforward.

We describe the system NJi and define translation $*$ of NJi into intuitionistic natural deduction system NJ [1] and inverse translation, which proves soundness and completeness of NJi. Then we prove that $*$-translation preserves reductions (normalization steps) of proofs. This allows to derive strong normalization for NJi from strong normalization for NJ proved, for example, in [8].

It is plausible that a direct Tait-Girard-style [2] proof of strong normalization for NJi is feasible. In that case a perspicuous proof of strong normalization for NJ with permutative conversions would be obtained (via the translation NJ→NJi) and a way would be open to extensions to new systems (like [5]) for which no strong normalization proof is known.

2 Intuitionistic predicate logic NJi with existential instantiation

Formulas of first order predicate logic are constructed in a familiar way from atomic formulas (including propositional constant $\perp$) by the connectives $\&, \vee, \rightarrow, \forall, \exists$. Free and bound variables $FV(A), BV(A)$ of a formula A are defined as usual. There are no redundant quantifiers: $\exists x A, \forall x A$ are formulas only if $x \in FV(A)$.

Assumptions of the system NJi are formulas as well as expressions of the form

$$< \Gamma >_{\exists x A, a} \quad \text{and} \quad < \Gamma >_{A \vee B, C} \tag{1}$$

where Γ is a multiset of assumptions, A, B are formulas, C is A or B and a is an individual variable. By the definition,

$$FV(< \Gamma >_{\exists x A, b}) = FV(A[x/b]), \quad FV(< \Gamma >_{A \vee B, C}) = FV(C)$$

Assumptions (1) are called $<>$-*formulas*. Note that they cannot be subformulas of (real) formulas. Expression $E[x/t]$ (or $E[t]$ if x is known) stands for the result of substituting t for all free occurrences of x into E. *Sequent* of NJi is any expression

$$\Gamma \Rightarrow A$$

where Γ is a multiset of assumptions and A is a formula.

Axioms: $B \Rightarrow B$, $\perp \Rightarrow A$ where A is an atomic formula, B is an arbitrary formula.

Inference rules

Standard introduction and elimination rules for $\&, \rightarrow, \forall$,

$$\exists I \frac{\Gamma \Rightarrow A[x/t]}{\Gamma \Rightarrow \exists x A} \qquad \exists i \frac{\Gamma \Rightarrow \exists x A}{< \Gamma >_{\exists x A, b} \Rightarrow A[x/b]}$$

with a proviso for the *eigenvariable* b in the rule $\exists\, i$: $b \notin FV(\Gamma, \exists x A)$, i.e. b is not free in the premise of the rule (while the eigenvariable of $\forall\, I$ should not be free in the conclusion).

$$\diamond\, \exists\ \frac{<\Gamma>_{\exists x A, b},\, \Delta \Rightarrow C}{\Gamma, \Delta \Rightarrow C}$$

provided $b \notin FV(\Gamma, \Delta, C)$, i.e. b is not free in the conclusion.

$$\vee\, I\ \frac{\Gamma \Rightarrow A_i}{\Gamma \Rightarrow A_0 \vee A_1}\ i = 0, 1 \qquad\qquad \vee\, i\ \frac{\Gamma \to A_0 \vee A_1}{<\Gamma>_{A_0 \vee A_1, A_i} \Rightarrow A_i}\ i = 0, 1$$

$$\diamond\, \vee\ \frac{<\Gamma>_{A_0 \vee A_1, A_0},\, \Delta \Rightarrow C \qquad <\Gamma>_{A_0 \vee A_1, A_1},\, \Sigma \Rightarrow C}{\Gamma, \Delta, \Sigma \Rightarrow C}$$

Contraction rule has its standard form

$$\frac{A, A, \Gamma \Rightarrow C}{A, \Gamma \Rightarrow C}$$

where A is an assumption. Note that only identical assumptions can be contracted, not for example $<\Gamma>_{A,b}$ and $<\Gamma>_{A,c}$ with $b \neq c$. This concludes the description of the system NJi .

Examples of deductions in NJi .

1. Let $F = \exists x(Px \& q)$. We derive $\exists x(Px \& q) \Rightarrow \exists x Px$.

$$\frac{\dfrac{\dfrac{\dfrac{\exists x(Px \& q) \Rightarrow \exists x(Px \& q)}{<\exists x(Px \& q)>_{F,b} \Rightarrow Pb \& q}}{<\exists x(Px \& q)>_{F,b} \Rightarrow Pb}}{<\exists x(Px \& q)>_{F,b} \Rightarrow \exists x Px}}{\exists x(Px \& q) \Rightarrow \exists x Px}$$

or, without listing assumptions explicitly

$$\frac{\dfrac{\dfrac{\exists x(Px \& q)}{Pb \& q}}{Pb}}{\exists x Px}$$

2. We derive $\forall x \exists y(Px \& \neg Py) \Rightarrow \bot$ using abbreviations: $Fxy = (Px \& \neg Py)$, $Ax = \exists y Fxy$.

$$\frac{\dfrac{\dfrac{\dfrac{\forall x \exists y Fxy \Rightarrow \forall x \exists y Fxy}{\forall x \exists y Fxy \Rightarrow \exists y Fay}}{<\forall x \exists y Fxy>_{Aa,b} \Rightarrow Pa \& \neg Pb}}{<\forall x \exists y Fxy>_{Aa,b} \Rightarrow \neg Pb} \qquad \dfrac{\dfrac{\dfrac{\dfrac{\forall x \exists y Fxy \Rightarrow \forall x \exists y Fxy}{\forall x \exists y Fxy \Rightarrow \exists y Fby}}{<\forall x \exists y Fxy>_{Ab,c} \Rightarrow Pb \& \neg Pc}}{<\forall x \exists y Fxy>_{Ab,c} \Rightarrow Pb}}{\forall x \exists y Fxy \Rightarrow Pb}}{\dfrac{\dfrac{<\forall x \exists y Fxy>_{Aa,b}, \forall x \exists y Fxy \Rightarrow \bot}{\forall x \exists y Fxy, \forall x \exists y Fxy \Rightarrow \bot}}{\forall x \exists y Fxy \to \bot}}$$

or

$$\frac{\dfrac{\dfrac{\dfrac{\forall x \exists y Fxy}{\exists y Fay}}{Pa \& \neg Pb}}{\neg Pb} \qquad \dfrac{\dfrac{\dfrac{\forall x \exists y Fxy}{\exists y Fby}}{Pb \& \neg Pc}}{Pb}}{\bot}$$

2.1 System NJ+ . Soundness and completeness

To prove soundness we define translation of NJi into the standard natural deduction formulation NJ.

Lemma 1 *(a) Every deduction d in NJi of a sequent containing $< \Gamma >_{\exists x A,b}$ contains a subdeduction of $\Gamma \Rightarrow \exists x A$;*
    (b) Every deduction d in NJi of a sequent containing $< \Gamma >_{A \vee B,C}$ contains a subdeduction of $\Gamma \Rightarrow A \vee B$

Proof . Easy induction on d: when d is traced from the bottom up, assumptions with $<>$ disappear at i-rules furnishing necessary subdeductions.   □

Lemma 2 *If a sequent S in a deduction of a $<>$-free sequent contains assumptions $< \Gamma >_{F,a}$ and $< \Delta >_{G,a}$ with the same a, then these assumptions coincide, i.e. $\Gamma = \Delta$ (up to the order of elements) and $F = G$.*

Proof . Consider the uppermost sequent T below S where one of $< \Gamma >_{F,a}$, $< \Delta >_{G,a}$ (say the first) disappears. It is a premise of $<>$ $\exists$-rule and hence contains no occurrences of a except those shown explicitly, in particular no occurrences of $< \Delta >_{G,a}$. This is possible only if $< \Gamma >_{F,a}, < \Delta >_{G,a}$ were fused by a contraction between S and T, hence $\Gamma = \Delta$ and $F = G$.   □

$$S[< \Gamma >_{F,a}, < \Delta >_{G,a}]$$
$$|$$
$$\frac{< \Gamma >_{F,a}, < \Delta >_{G,a}, \Pi \Rightarrow E}{< \Gamma >_{F,a}, \Pi \Rightarrow E}$$
$$|$$
$$\frac{T = < \Gamma >_{F,a}, \Sigma \Rightarrow C}{\Gamma, \Sigma \Rightarrow C}$$

Now consider the union of NJ and NJi . The system NJ+ is obtained by adding standard elimination rules to NJi :

$$\exists E \; \frac{\Gamma \Rightarrow \exists x A \quad A[x/b], \Delta \Rightarrow C}{\Gamma, \Delta \Rightarrow C} \qquad \vee E \; \frac{\Gamma \Rightarrow A_0 \vee A_1 \quad A_0, \Delta \Rightarrow C \quad A_1, \Sigma \Rightarrow C}{\Gamma, \Delta, \Sigma \Rightarrow C}$$

Let us describe translations of NJ+ into NJ and NJi .
    The rules $(\exists E), (\vee E)$ are eliminated in favor of $\exists i, \vee i$ as follows:

$$\frac{\Gamma \Rightarrow \exists x A}{< \Gamma >_{\exists x A,b} \Rightarrow A[x/b]}$$

$$A[x/b] \Rightarrow A[x/b]$$
$$|$$
$$\frac{\Gamma \Rightarrow \exists x A \quad A[x/b], \Delta \Rightarrow C}{\Gamma, \Delta \Rightarrow C}$$

$$\frac{< \Gamma >_{\exists x A,b} \Rightarrow A[x/b]}{|}$$
$$\frac{< \Gamma >_{\exists x A,b}, \Delta \Rightarrow C}{\Gamma, \Delta \Rightarrow C}$$

$$A_0 \Rightarrow A_0 \qquad A_1 \Rightarrow A_1$$
$$| \qquad |$$
$$\frac{\Gamma \Rightarrow A_0 \vee A_1 \quad A_0, \Delta \Rightarrow C \quad A_1, \Sigma \Rightarrow C}{\Gamma, \Delta, \Sigma \Rightarrow C}$$

$$\frac{\Gamma \Rightarrow A_0 \vee A_1}{<\Gamma>_{A_0 \vee A_1, A_0} \Rightarrow A_0} \qquad \frac{\Gamma \Rightarrow A_0 \vee A_1}{<\Gamma>_{A_0 \vee A_1, A_1} \Rightarrow A_1}$$

$$\frac{<\Gamma>_{A_0 \vee A_1, A_0}, \Delta \Rightarrow C \quad <\Gamma>_{A_0 \vee A_1, A_1}, \Sigma \Rightarrow C}{\Gamma, \Delta, \Sigma \Rightarrow C}$$

Instantiation and $<>$-rules are eliminated as follows in favor of the standard rules of NJ:

$$\frac{\dfrac{\Gamma \Rightarrow \exists x A}{<\Gamma>_{\exists x A, b} \Rightarrow A[x/b]}}{\dfrac{<\Gamma>_{\exists x A, b}, \Delta \Rightarrow C}{\Gamma, \Delta \Rightarrow C}} \qquad \frac{\Gamma \Rightarrow \exists x A \quad \dfrac{A[x/b] \Rightarrow A[x/b]}{A[x/b], \Delta \Rightarrow C}}{\Gamma, \Delta \Rightarrow C},$$

and similarly for $\vee$-rules, i.e. by the transformations inverse to elimination of standard elimination rules.

If $S = \Delta \Rightarrow G$ is a sequent of NJi, then $S*$ denotes the result of replacing the outermost occurrences (in Δ) of $<\Gamma>_{\exists x A, a}$ by $A[a]$ and of $<\Gamma>_{A \vee B, C}$ by C.

Theorem 1 *If S is derivable in NJ+ then $S*$ is derivable in NJ and S is derivable in NJ. In particular NJ,NJi,NJ+ are equivalent for $<>$-free sequents.*

Proof . Apply translations described above.

3 Normalization

We define below conversion (one-step reduction) relation d *conv* d' for deductions in NJi. Reduction $d \succeq d'$ (by $n \geq 0$ conversions of subdeductions) and equality of deductions $d = d'$ up to reductions are defined in a standard way .

Two occurrences of sequents S, S' in a deduction d are *connected by a variable* b if there is a path from S to S' (going possibly up and down) in d, consisting of sequents containing b free. Similarly for two occurrences of a variable b itself.

A deduction in NJi is *pure* or has *pure variable property* if the eigenvariable of a $\forall$-introduction inference occurs only over the conclusion of that inference, and every occurrence of the eigenvariable of a $\exists i$-inference is connected with the conclusion of that inference.

One can prove in a standard way.

Lemma 3 *By renaming eigenvariables any deduction can be made pure*

Lemma 4 *In a pure deduction of a $<>$-free sequent any two assumptions $<\Gamma>_{F,a}$ and $<\Delta>_{F',a}$ with the same a coincide.*

A deduction is *uniform* if it is pure and all $\exists i$-inferences with the same eigenvariable have the same deduction of the premise (up to a suitable renaming).

Lemma 5 *Every deduction of a <>-free sequent can be made uniform by re-naming eigenvariables and identifying subdeductions.*

In the following we assume all deductions to be uniform. In most cases it is assumed also that the whole deduction (of which given subdeduction is a part) deduces a <>-free sequent.

3.1 Reductions for NJi

To simplify notation, we consider here the system without disjunction. This can be justified by the fact that $\vee$ is definable from $\exists, \&, \to, \neg$ in any system (say in arithmetic) with a constant 0 such that $x = 0$ is provably decidable:

$$A \vee B = \exists x (x = 0 \to A \& x \neq 0 \to B)$$

If d is a deduction in NJi , then d^+ stands for any deduction obtained from d by any number of <>-inferences and contractions:

$$\frac{d : \Gamma \Rightarrow C}{d^+ : \Gamma' \Rightarrow C} \tag{2}$$

Note that the conclusion C is the same. A figure like (2) will always stand for a sequence of <>-inferences and contractions. We define cut-conversions, and then extend the definition to achieve uniformity.

&-cut-conversion:

$$\frac{\dfrac{\Gamma_0 \Rightarrow A_0 \quad \Gamma_1 \Rightarrow A_1}{\Gamma_0, \Gamma_1 \Rightarrow A_0 \& A_1}}{\Gamma_0', \Gamma_1' \Rightarrow A_i} \qquad conv \qquad \frac{\Gamma_i \Rightarrow A_i}{\Gamma_i' \Rightarrow A_i}$$

with ususal treatment of redundant assumptions. $\forall$-cut-conversions are defined similarly, and $\neg, \to$-cut-conversions are defined in a standard way.

$\exists$-cut-conversion:

$$\frac{\dfrac{d' : \Gamma \Rightarrow A[t]}{d : \Gamma \Rightarrow \exists x A}}{< \Gamma' >_{\exists x A, b} \Rightarrow A[b]} \qquad conv \qquad \frac{\Gamma \Rightarrow A[t]}{\Gamma' \Rightarrow A[t]}$$
$$|/\, b \qquad\qquad\qquad |/\, t$$

where all occurrences of $b, < \Gamma' >_{\exists x A, b}$ connected with given occurrence are replaced by t, Γ', and all $\exists i$-inferences having such a b as an eigenvariable are replaced in the same way as $\exists i$ shown explicitly: each of them is inferred by the same deduction.

A *conversion* of a deduction in NJi is defined to be a series of cut conversions applied to a fixed cut **C** and to all other cuts identical with **C** and situated in the same way over the premises of $\exists i$-inferences with the same eigenvariable.

Lemma 6 *If d is uniform and d conv d' then d' is uniform*

3.2 Strong Normalization for NJi

For every deduction d of a sequent S in NJi let $d*$ stand for the deduction of the sequent $S*$ described in the section 2.1.

Lemma 7 *Let d be a deduction in NJi. If d conv d' then $d*$ reduces (in non-zero steps) to $d'*$.*

Proof . Check all conversions. To simplify notation, we consider only cut-conversions and ignore $<>$-rules and contractions which interfere inside the pair introduction/elimination to be converted.

&-conversion. We write down $d, d*$ and then $d', d'*$.

$$
\dfrac{\dfrac{d_0 : \ \Gamma \Rightarrow A \quad \Delta \Rightarrow B}{\Gamma, \Delta \Rightarrow A\&B}}{d : \ \Gamma, \Delta \Rightarrow A}
\qquad
\dfrac{\dfrac{d_0* : \ \Gamma* \Rightarrow A \quad \Delta* \Rightarrow B}{\Gamma*, \Delta* \Rightarrow A\&B}}{d* : \ \Gamma, \Delta \Rightarrow A}
$$

$$
d' = d_0 : \ \Gamma \Rightarrow A
\qquad\qquad
d_0* = d*' : \ \Gamma* \Rightarrow A
$$

$\forall$-conversion is treated similarly.

$\rightarrow$-conversion.

$$
\dfrac{\dfrac{\dfrac{\dfrac{\dfrac{A \Rightarrow A}{\ \big|\ }}{\dfrac{A, \Delta \Rightarrow \exists x C}{< A, \Delta > \Rightarrow C[b]}}}{\dfrac{\ \big|\ }{\dfrac{< A, \Delta >, A, \Pi \Rightarrow D}{A, \Delta, A, \Pi \Rightarrow D}}}}{\dfrac{\ \big|\ }{\dfrac{A, \Gamma \Rightarrow B}{\Gamma \Rightarrow A \rightarrow B} \quad \Sigma \Rightarrow A}}}{d : \ \Gamma, \Sigma \Rightarrow B}
\qquad
\dfrac{\dfrac{\dfrac{\dfrac{\dfrac{\Sigma \Rightarrow A}{\ \big|\ }}{\dfrac{\Sigma, \Delta \Rightarrow \exists x C}{< \Sigma, \Delta > \Rightarrow C[b]}}}{\dfrac{\ \big|\ }{\dfrac{< \Sigma, \Delta >, \Sigma, \Pi \Rightarrow D}{\Sigma, \Delta, \Sigma, \Pi \Rightarrow D}}}}{\ \big|\ }}{d' : \ \Sigma, \Gamma \Rightarrow B}
$$

$$
\dfrac{\dfrac{\dfrac{A \Rightarrow A \qquad C[b] \Rightarrow C[b]}{\ \big|\qquad\qquad\big|\ }}{\dfrac{A, \Delta* \Rightarrow \exists x C \quad C[b], A, \Pi* \Rightarrow D}{A, \Delta*, A, \Pi* \Rightarrow D}}}{\dfrac{\ \big|\ }{\dfrac{A, \Gamma* \Rightarrow B}{\Gamma* \Rightarrow A \rightarrow B} \quad \Sigma* \Rightarrow A}}{d : \ \Gamma*, \Sigma* \Rightarrow B}}
\qquad
\dfrac{\dfrac{\dfrac{\Sigma* \Rightarrow A \qquad C[b] \Rightarrow C[b]}{\ \big|\qquad\qquad\big|\ }}{\dfrac{\Sigma*, \Delta* \Rightarrow \exists x C \quad C[b], \Sigma*, \Pi* \Rightarrow D}{\Sigma*, \Delta*, \Sigma*, \Pi* \Rightarrow D}}}{\dfrac{\ \big|\ }{d*' : \ \Sigma*, \Gamma*, \Rightarrow B}}
$$

$\exists$-conversion.

$$
\dfrac{\dfrac{\dfrac{\dfrac{\Gamma \Rightarrow A[t]}{\Gamma \Rightarrow \exists x A}}{< \Gamma >_{\exists x A, a} \Rightarrow A[a]}}{\dfrac{a \big|}{< \Gamma >_{\exists x A, a}, \Sigma \Rightarrow C}}}{d : \ \Gamma, \Sigma \Rightarrow C}
\qquad
\dfrac{\dfrac{\Gamma \Rightarrow A[t]}{t \big|}}{d' : \ \Gamma, \Sigma \Rightarrow C}
$$

$$\frac{\Gamma* \Rightarrow A[t]}{\Gamma* \Rightarrow \exists x A} \quad \begin{array}{c} A[a] \Rightarrow A[a] \\ a| \\ \hline A[a], \Sigma* \Rightarrow C \end{array} \qquad \frac{\Gamma* \Rightarrow A[t]}{t|} $$

$$d* : \ \Gamma*, \Sigma* \Rightarrow C \qquad\qquad d'* : \ \Gamma*, \Sigma* \Rightarrow C$$

Theorem 2 *NJi is strongly normalizable for uniform deductions.*

Proof . Every sequence s of conversions of a deduction d in NJi induces by the previous Lemma a sequence of conversions s' of $d*$ with $length(s) \leq length(s')$. Since s' is finite, s is finite too.

References

1. Gentzen G.: Untersuchungen über das logische Schliessen. Mathematische Zeitschrift, **39** (1934) 176-210, 405-431
2. Girard J.-Y., Lafont Y., Taylor P.: Proofs and Types, Cambridge University Press, Cambridge (1988)
3. Leivant D.: Existential instantiation in a system of natural deduction, Mathematish Centrum-ZW **13-73** (1973)
4. Mints G.: Lewis Systems and the System T. In Mints G.: Selected Papers in Proof Theory, North-Holland-Bibliopolis (1993), 221-294 (Russian Original 1974)
5. Mints G.: A Normal Form Theorem for Second-order Classical Logic with an Axiom of Choice, Math. USSR Izvestiya **32** N3 (1989) 587-605
6. Mints G.: Normal Deduction in the Intuitionistic Linear Logic, CSLI report, (1996)
7. Prawitz D.: Natural Deduction, Almquist and Wiksell (1965)
8. Prawitz D.: Ideas and Results in Proof Theory, In Proc. 2-nd Scand.Logic Symp., North-Holland, (1972) 235-308
9. Quine W.V.: On natural deduction, Journal of Symbolic Logic **15** (1950) 93-102

Models for the Logic of Proofs

Alexey Mkrtychev

Moscow State University, Moscow 119899, Russia

Abstract. The operational logic of proofs $\mathcal{LP}$ was introduced by S. Artemov [1] as an operational version of $S4$. In this paper, we define a model for $\mathcal{LP}$ and prove the corresponding completeness theorem. Using this model, we prove the decidability of a variant of $\mathcal{LP}$ axiomatized by a finite set of schemes.

1 Introduction

The operational logic of proofs $\mathcal{LP}$ was introduced by S. Artemov in [1] for the description of the relation *"t is a proof of F"* in the presence of some computable operations over proofs. In addition to the usual propositional logic, it contains new formulas of the form $[t]F$ where t is a term generated inductively from *proof variables* and *axiom constants* by three function symbols: binary *"+", "×"* and monadic *"!"*. In the intended semantics, formulas are interpreted by arithmetical statements, and terms denote codes of proofs. Axiom constants denote trivial proofs of axioms and function symbols represent operations on proofs. Informally speaking *"×"* realizes the rule *modus ponens*, *"+"* is just a concatenation of two arbitrary proofs, and *"!"* represents a proof verifier, namely *"!t proves that t is a proof of F if it is the case".*

According to [1], the language of $\mathcal{LP}$ contains:

- Boolean constants $\top, \bot$, sentence letters $S_1, \ldots, S_n, \ldots$
- proof letters $p_1, \ldots, p_n, \ldots$
- proof axiom constants $a_1, \ldots, a_n, \ldots$
- Boolean connectives $\rightarrow, \wedge, \vee, \neg$
- functional symbols: monadic !, binary + and ×
- operator symbol $[\,](\,)$.

The sets Tm of terms and Fm of formulas are defined as follows. Any proof letter or axiom constant is in Tm; any sentence letter or Boolean constant is in Fm; whenever $s, t \in Tm$ then $!t, (s+t), (s \times t) \in Tm$; Boolean connectives behave conventionally, and if $F \in Fm$, $t \in Tm$ then $[t](F) \in Fm$. We shall write st instead of $s \times t$ and $[t]F$ instead of $[t](F)$ when convenient. Formulas $[t]F$ are called *quasiatomic* formulas (q-atomic, for short).

System $\mathcal{LP}_{AS}$ is given by the following axioms:

A0. Tautologies in the language of $\mathcal{LP}$

A1. $[t]F \to F$

A2. $[s](F \to G) \to ([t]F \to [st]G)$

A3. $[t]F \to [!t][t]F$

A4. $[s]F \vee [t]F \to [s+t]F$

AS. Axiom specification, that is a finite set of formulas of the form $[c]A$, where c is an axiom constant, and A is an axiom A0-A4.

The rule of inference: *modus ponens*.

Axiom A1 expresses the *weak reflexivity* of the proof predicate, while axioms A2-A4 describe the behavior of basic operations on proofs.

In [1], all the systems $\mathcal{LP}_{AS}$ are proved to be complete with respect to the intended arithmetical semantics . It is also established that modal logic $S4$ can be realized in operational modal logic. It means that $S4 \vdash F \Leftrightarrow \mathcal{LP}_{AS} \vdash F^r$ for some assignment r of $\mathcal{LP}$-terms to all occurrences of the modality $\Box$ in the formula F and some axiom specification AS. This result makes it possible to give an arithmetical semantics for $S4$ and to prove the completeness theorem with respect to this semantics.

In this paper, we allow axiom specifications to be infinite, so in what follows AS is an arbitrary set of formulas of the form $[c]A$, where c is an axiom constant and A is an axiom A0-A4. We give the definition of the model for $\mathcal{LP}_{AS}$ and prove the corresponding completeness theorem. Using this theorem, we establish the decidability of $\mathcal{LP}_{AS}$ with AS described by a finite set of schemes. Our result is a natural generalization of Artemov's result on the decidability of $\mathcal{LP}_{AS}$ with finite AS.

2 Semantics for operational modal logic

In this section we present semantics for $\mathcal{LP}_{AS}$ and formulate the completeness theorem with respect to it. Suppose $*(\cdot)$ is a function from Tm to 2^{Fm}. We call $*(\cdot)$ a *proof-theorem assignment* on Tm if it satisfies the following conditions:

a) $F \in *(t) \Rightarrow [t]F \in *(!t)$

b) $G \to F \in *(s),\ G \in *(t) \Rightarrow F \in *(st)$

c) $*(s) \cup *(t) \subseteq *(s+t)$

A *truth-assignment* is a mapping v with the domain the set of sentence letters and the range $\{True, False\}$. Let $v(\cdot)$ is a *truth-assignment* and $*(\cdot)$ is a *proof-theorem assignment*. We define an interpretation $\mathcal{I}$ of the language $\mathcal{LP}$ as a triple $(v, *, \models)$, where $\models$ is a truth relation on Fm:

0. for every sentence letter S: $\models S$ iff $v(S) = True$,
1. $\models$ commutes with Booleanconnectives,
2. $\models [t]F \Leftrightarrow F \in *(t)$.

We write "$\mathcal{I} \models F$" to denote that $\models F$ holds in interpretation $\mathcal{I}$, and "$\mathcal{I} \not\models F$" otherwise.

Definition 1. An interpretation $\mathcal{I}$ is called a *model* if $F \in *(t)$ implies $\mathcal{I} \models F$ for any formula F and any term t.

For the purpose of technical convenience we need to introduce the notion of *pre-model*. A pre-model is a triple $\mathcal{P} = (v, *, \models_p)$, where truth relation $\models_p$ defined in the following way:

0. for every sentence letter S: $\mathcal{P} \models_p S$ iff $v(S) = True$,
1. $\models_p$ commutes with Booleanconnectives,
2. $\models_p [t]F \Leftrightarrow F \in *(t)$ and $\models_p F$.

Also we write "$\mathcal{P} \models_p F$" to denote that $\models_p F$ holds in pre-model $\mathcal{P}$ and "$\mathcal{P} \not\models_p F$" otherwise.

Remark. It is clear that truth valuation in interpretation respects modus ponens i.e if $\mathcal{I} \models F$ and $\mathcal{I} \models F \to G$ then $\mathcal{I} \models G$. The same for pre-models.

Definition 2. An interpretation $\mathcal{I}$ and a pre-model $\mathcal{P}$ are called *equivalent* if they determine the same truth relation.

Lemma 3. *1) For any model $\mathcal{I} = (v, *, \models)$ there exists a pre-model $\mathcal{P} = (v', *', \models_p)$ equivalent to it;*
2) For any pre-model $\mathcal{P} = (v', *', \models_p)$ there exists a model $\mathcal{I} = (v, *, \models)$ equivalent to it.

Proof. 1) We put $v' = v$, $*' = *$. In order to show that $\mathcal{I} = (v, *, \models)$ and $\mathcal{P} = (v', *', \models_p)$ are equivalent we reason by induction on the length of a formula. The basis of induction is provided by the definition of v'. The case of Boolean-connectives is trivial. It remains to consider q-atomic formulas. If $\mathcal{P} \models_p [t]F$ then $F \in *'(t)$, whence $F \in *(t)$ and $\mathcal{I} \models [t]F$. If $\mathcal{I} \models [t]F$ then $F \in *(t)$, whence $\mathcal{I} \models F$ (since $\mathcal{I}$ is a model). By the induction hypothesis $\mathcal{P} \models_p F$. Thus $\mathcal{P} \models_p [t]F$.
2) We put $v = v'$ and define $*(\cdot)$ as follows: for an arbitrary term t and a formula F we put $F \in *(t)$ iff $F \in *'(t)$ and $\mathcal{P} \models_p F$. Let us show that $*(\cdot)$ is a proof-theorem assignment:

a) $F \in *(t) \Rightarrow [t]F \in *(!t)$:
   If $F \in *(t)$ then $\mathcal{P} \models_p F$ and $F \in *'(t)$. Hence $\mathcal{P} \models_p [t]F$ by the definition of a pre-model and $[t]F \in *'(!t)$ since $*'(\cdot)$ is a proof-theorem assignment. Therefore $[t]F \in *(!t)$.
b) $G \to F \in *(s)$, $G \in *(t) \Rightarrow F \in *(st)$:
   Suppose $G \to F \in *(s)$ and $G \in *(t)$. Then $\mathcal{P} \models_p G \to F$, $G \to F \in *'(s)$ and $\mathcal{P} \models_p G$, $G \in *'(t)$. Thus $\mathcal{P} \models_p F$ and $F \in *'(st)$ since $*'(\cdot)$ is a proof-theorem assignment. So we have $F \in *(st)$.
c) $*(s) \cup *(t) \subseteq *(s + t)$:
   If $F \in *(s) \cup *(t)$, then without loss of generality assume that $F \in *(s)$. Hence $\mathcal{P} \models_p F$ and $F \in *'(s)$. Since $*'(\cdot)$ is a proof-theorem assignment, we have $F \in *'(s + t)$. Therefore $F \in *(s + t)$.

Let us prove that the interpretation $\mathcal{I} = (v, *, \models)$ is equivalent to the pre-model $\mathcal{P} = (v', *', \models_p)$:

0. By definition of v, for every sentence letter S: $\mathcal{I} \models S$ iff $\mathcal{P} \models_p S$,
1. Both $\models$ and $\models_p$ commute with Booleanconnectives,
2. $\mathcal{I} \models [t]F \Leftrightarrow F \in *(t) \Leftrightarrow F \in *'(t)$ and $\mathcal{P} \models_p F \Leftrightarrow \mathcal{P} \models_p [t]F$.

It remains to show that interpretation $\mathcal{I} = (v, *, \models)$ is a model:
$\mathcal{I} \models [t]F \Rightarrow F \in *(t) \Rightarrow \mathcal{P} \models_p F$ (by definition of $*$) $\Rightarrow \mathcal{I} \models F$ (by equivalence of $\mathcal{I}$ and $\mathcal{P}$). $\qquad\square$

Formula F is called *valid* if $\mathcal{I} \models F$ for any model $\mathcal{I}$. Obviously, by lemma 3 formula F is valid if and only if $\mathcal{P} \models_p F$ for any pre-model $\mathcal{P}$.

Lemma 4. *Formulas A0-A4 are true in any pre-model* $\mathcal{P} = (v, *, \models_p)$.

Proof. **A0.** Case of tautologies is evident.
A1. $[t]F \to F$:
If $\mathcal{P} \not\models_p F$ then $\mathcal{P} \not\models_p [t]F$, so $\mathcal{P} \models_p [t]F \to F$.
A2. $[s](F \to G) \to ([t]F \to [st]G)$:
Suppose $\mathcal{P} \models_p [s](F \to G)$ and $\mathcal{P} \models_p [t]F$ then we have $\mathcal{P} \models_p F, F \to G$, (whence $\mathcal{P} \models_p G$) and $F \in *(t), F \to G \in *(s)$ (whence $G \in *(st)$). So one can conclude $\mathcal{P} \models_p [st]G$.
A3. $[t]F \to [!t][t]F$:
If $\mathcal{P} \models_p [t]F$, then $F \in *(t)$ and $[t]F \in *(!t)$, whence $\mathcal{P} \models_p [!t][t]F$.
A4. $[s]F \vee [t]F \to [s+t]F$:
If $\mathcal{P} \models_p [s]F \vee [t]F$ then $\mathcal{P} \models_p [s]F$ or $\mathcal{P} \models_p [t]F$. Assume that $\mathcal{P} \models_p [s]F$. Then $\mathcal{P} \models_p F$ and $F \in *(s)$. Thus we have $F \in *(s + t)$, whence $\mathcal{P} \models_p [s+t]F$.

$\qquad\square$

For a set $\mathcal{F}$ of formulas and an interpretation $\mathcal{I}$ (pre-model $\mathcal{P}$) we shall write $\mathcal{I} \models \mathcal{F}$ ($\mathcal{P} \models_p \mathcal{F}$) if any formula in $\mathcal{F}$ is true in $\mathcal{I}$ ($\mathcal{P}$).

Definition 5. A model $\mathcal{I}$ (pre-model $\mathcal{P}$) is an $AS - model$ ($AS - pre - model$) if $\mathcal{I} \models AS$ ($\mathcal{P} \models_p AS$).

Let us establish the existence of an AS-model for any axiom specification AS. According to lemma 3 we only need to present a pre-model $\mathcal{P} = (v, *, \models_p)$, such that $\mathcal{P} \models_p AS$. Let $\phi : Tm \mapsto 2^{Fm}$ be an arbitrary function, assigning sets of $\mathcal{LP}$-formulas to $\mathcal{LP}$-terms. Such a function we call an *evaluation*. Proof-theorem assignment is *based* on $\phi(\cdot)$ if $\phi(t) \subseteq *(t)$ holds for any term t. The following lemma states the existence of the proof-theorem assignment based on an arbitrary evaluation $\phi(\cdot)$.

Lemma 6. *For any evaluation* $\phi(\cdot)$ *there exists a proof-theorem assignment* $*(\cdot)$ *based on* $\phi(\cdot)$.

Proof. We define $*(t)$ by induction on the complexity of a term t: we put $*(a) := \phi(a)$ and $*(p) := \phi(p)$ for any axiom constant a and proof letter p. If $*(s)$ and $*(t)$ are already defined then:

a) $F \in *(!t)$ iff $F = [t]H$, where $H \in *(t)$ or $F \in \phi(!t)$,
b) $F \in *(st)$ iff there exists $G \in *(t)$, such that $G \to F \in *(s)$, or $F \in \phi(st)$,
c) $*(s+t) := *(s) \cup *(t) \cup \phi(s+t)$.

It is easily seen from the definition that $*(\cdot)$ the desired proof-theorem assignment. □

Obviously if $*(\cdot)$ is based on function $\phi(\cdot)$, such that $\phi(a) = \{F \in Fm \mid [a]F \in AS\}$ then $\mathcal{P} \models_p AS$.

Remark. One can easily see that $*(\cdot)$ constructed in the proof above is minimal in the following sense: if $*_1$ is any other proof-theorem assignment, based on the evaluation $\phi(\cdot)$, then $*(t) \subseteq *_1(t)$ for any term t.

Before the proof of completeness let us remind some well-known definitions.

Definition 7. A set $\mathcal{F}$ of formulas is called $AS - consistent$ if $\mathcal{LP}_{AS} \nvdash \neg(A_1 \wedge \ldots \wedge A_n)$ for any $A_1, \ldots, A_n \in \mathcal{F}$.

Definition 8. A set $\mathcal{M}$ of formulas is called *maximal AS-consistent* if $\mathcal{M}$ is AS-consistent and $A \in \mathcal{M}$ or $\neg A \in \mathcal{M}$ holds for any formula A.

The following lemma is standard.

Lemma 9. *1) If $\mathcal{M}$ is maximal AS-consistent then $\mathcal{LP}_{AS} \subseteq \mathcal{M}$; 2) Every maximal AS-consistent set is closed under modus ponens; 3) If set $\mathcal{F}$ is AS-consistent then there exists a maximal AS-consistent set $\mathcal{M}$, such that $\mathcal{F} \subseteq \mathcal{M}$.*

Lemma 10. *For any maximal AS-consistent set $\mathcal{M}$ there exists a pre-model $\mathcal{P}$, such that $\mathcal{P} \models_p \mathcal{M}$.*

Proof. First we construct $*(\cdot)$: for any term t put

$$*(t) = \{F \in Fm \mid [t]F \in \mathcal{M}\}$$

Let us establish that $*(\cdot)$ is a proof-theorem assignment:

a) $F \in *(t) \Rightarrow [t]F \in *(!t)$:
   If $F \in *(t)$, then $[t]F \in \mathcal{M}$, whence by A3 $[!t][t]F \in \mathcal{M}$. Therefore $[t]F \in *(!t)$.
b) $G \to F \in *(s)$, $G \in *(t) \Rightarrow F \in *(st)$:
   If $G \to F \in *(s), G \in *(t)$ then $[t]G, [s](G \to F) \in \mathcal{M}$. According to A2 we have $[st]F \in \mathcal{M}$, and hence $F \in *(st)$.
c) $*(s) \cup *(t) \subseteq *(s+t)$:
   If $F \in *(s) \cup *(t)$, then without loss of generality we assume that $F \in *(s)$. Therefore $[s]F \in \mathcal{M}$ and according to A4 $[s+t]F \in \mathcal{M}$. Thus we have $F \in *(s+t)$.

We define a truth-assignment v as follows: for any sentence letter S

$$v(S) = True \Leftrightarrow S \in \mathcal{M}.$$

Let $\mathcal{P} = (v, *, \models_p)$ denote the constructed pre-model. By induction on complexity of a formula F we prove that $\mathcal{P} \models_p F$ if and only if $F \in \mathcal{M}$. For any sentence letter, $\mathcal{P} \models_p S \Leftrightarrow S \in \mathcal{M}$ by definition of v. Case of Booleanconnectives is evident. Let $F = [t]H$, then: $\mathcal{P} \models_p [t]H \Rightarrow H \in *(t) \Rightarrow [t]H \in \mathcal{M}$. Conversely $[t]H \in \mathcal{M} \Rightarrow H \in *(t)$. By A1 we have $H \in \mathcal{M}$, whence $\mathcal{P} \models_p H$ by induction hypothesis. Thus $\mathcal{P} \models_p [t]H$.

Lemma 11. $\mathcal{LP}_{AS} \vdash F$ *if and only if* $\mathcal{P} \models_p F$ *for any AS-pre-model* $\mathcal{P}$.

Proof. If $\mathcal{LP}_{AS} \vdash F$ then, obviously $\mathcal{P} \models_p F$ for any AS-pre-model $\mathcal{P}$. Let $\mathcal{P} \models_p F$ for every AS-pre-model $\mathcal{P}$ and $\mathcal{LP}_{AS} \nvdash F$. Then the set $\{\neg F\}$ is AS-consistent. By lemma 9 there exists a maximal AS-consistent set $\mathcal{M}$, such that $\neg F \in \mathcal{M}$. A lemma 10 provides a pre-model $\mathcal{P}$, such that $\mathcal{P} \models_p \mathcal{M}$. Obviously, $\mathcal{P}$ is an AS-pre-model, and $\mathcal{P} \models_p \neg F$. Contradiction. □

Theorem 12. *(completeness)* $\mathcal{LP}_{AS} \vdash F \Leftrightarrow F$ *is true in all AS-models.*

Proof. By lemma 3 and lemma 11. □

3 Decidable fragments of $\mathcal{LP}$

In this section we prove the decidability of $\mathcal{LP}_{AS}$ for AS described by finite set of schemes. For this purpose we need some adaptation of the common unification technique for the language $\mathcal{LP}$. First of all we extend the language by a countable sets of new proof variables $q_1, q_2, \ldots$ and new sentence variables $T_1, T_2, \ldots$ called *special*. For the purpose of convenience we call an *expression* any formula or term in the extended language.

A *substitution* θ is a finite set of the form

$$(T_1 \leftarrow A_1, \ldots, T_n \leftarrow A_n, q_1 \leftarrow r_1, \ldots, q_m \leftarrow r_m),$$

where the T_i are distinct special sentence variables, the q_j are distinct special proof variables. Each A_i is an arbitrary formula and each r_j is an arbitrary term in the extended language. For an expression E we write $E\theta$ for the result of simultaneous replacing all occurrences of T_i in E by A_i and all occurrences of q_j in E by r_j for every $i \leq n, j \leq m$.

If X is a set of expressions then $X\theta = \{E\theta \mid E \in X\}$.

If τ and θ are substitutions, then $\tau = \theta$ iff $E\tau \equiv E\theta$ for every expression E. If $\theta = (T_1 \leftarrow A_1, \ldots, q_1 \leftarrow r_1, \ldots)$ and $\sigma = (T_1' \leftarrow A_1', \ldots, q_1' \leftarrow r_1', \ldots)$, then the *composition* $\theta \circ \sigma$ (or just $\theta\sigma$ for short) of θ and σ is

$$(T_1 \leftarrow A_1\sigma, \ldots, q_1 \leftarrow r_1\sigma, \ldots, T_1' \leftarrow A_1', \ldots, q_1' \leftarrow r_1', \ldots).$$

Let $\{E_1 = F_1, \ldots, E_n = F_n\}$ be a *set of equations*, where E_i and F_i are expressions. A substitution θ such that $E_1\theta \equiv F_1\theta, \ldots, E_n\theta \equiv F_n\theta$ is called a *unifier of the set of equations*. The set of equations is said to be unifiable if it has a unifier. If θ is a unifier of the set of equations, and for any other unifier σ of this set there exists a substitution δ such that $\sigma = \theta\delta$, then θ is called the *most general unifier (mgu)* of that set. A set of equations is called *solved* if it is of the form

$$T_1 = A_1, \ldots, T_n = A_n, q_1 = r_1, \ldots q_m = r_m,$$

where T_i and q_i are distinct special variables and none of them occurs in a right-hand side of any equation. Such solved set of equations determines the substitution

$$\theta = (T_1 \leftarrow A_1, \ldots, T_n \leftarrow A_n, q_1 \leftarrow r_1, \ldots, q_m \leftarrow r_m),$$

which is obviously a *mgu* of that solved set.

To obtain a mgu of the set of equations or to make sure that the set is nonunifiable, we run a standard unification algorithm [4].

Let X be the set of formulas in the extended language. Then $X\sigma$ is set of $\mathcal{LP}$-formulas if and only if σ has the form

$$(T_1 \leftarrow A_1, \ldots, T_n \leftarrow A_n, q_1 \leftarrow r_1, \ldots, q_m \leftarrow r_m),$$

where every A_i is an $\mathcal{LP}$-formula, every r_i is an $\mathcal{LP}$-term and $\{T_i, q_j \mid i \leq n, j \leq m\}$ is the list of the all special proof and sentence variables occurring in X. In what follows we denote $\bigcup_\sigma X\sigma$ the union of $X\sigma$ for all σ, described above.

Let us consider a function $\Phi : t \mapsto X^t$, that assigns finite set X^t of formulas in the extended language to any $\mathcal{LP}$-term t. We say that the evaluation $\phi(\cdot)$ *corresponds to* $\Phi(\cdot)$ if

$$\phi(t) = \bigcup_\sigma X^t\sigma$$

holds for any $\mathcal{LP}$-term t.

Lemma 13. *Let an evaluation $\phi(\cdot)$ correspond to function $\Phi(\cdot)$. Then there exists a function $U : t \mapsto Y^t$, where Y^t are finite sets of formulas in the extended language, such that minimal proof-theorem assignment $*(\cdot)$ based on evaluation $\phi(\cdot)$ corresponds to the function U i.e*

$$*(t) = \bigcup_\sigma Y^t\sigma$$

holds for any $\mathcal{LP}$-term t.

Proof. By induction on complexity of the term t we define $U(t)$. For any axiom constant a and proof variable p, put $U(a) = X^a$, $U(p) = X^p$. Suppose $U(t) = Y^t, U(s) = Y^s$ and $*(s) = \bigcup_\sigma Y^s\sigma$, $*(t) = \bigcup_\sigma Y^t\sigma$, then by lemma 6:

a) $F \in *(!t)$ iff $F = [t]H$, where $H \in *(t)$ or $F \in \phi(!t)$:
We have $*(!t) = \{[t]H \mid H \in \bigcup_\sigma Y^t\sigma\} \cup (\bigcup_\sigma X^{!t}\sigma)$. Let $[t]Y^t$ denote the finite set of formulas obtained by putting $[t]$ before each formula in Y^t. Then, obviously $\{[t]H \mid H \in \bigcup_\sigma Y^t\sigma\} = \bigcup_\sigma [t]Y^t\sigma$. Therefore $*(!t) = (\bigcup_\sigma [t]Y^t\sigma) \cup (\bigcup_\sigma X^{!t}\sigma) = \bigcup_\sigma ([t]Y^t \cup X^{!t})\sigma$. Thus we have $Y^{!t} = [t]Y^t \cup X^{!t}$.

b) $F \in *(st)$ iff there exists $G \in *(t)$, such that $G \to F \in *(s)$, or $F \in \phi(st)$. It means that $*(st) = \{F \mid exists\ G \in \bigcup_\sigma Y^t\sigma\ and\ G \to F \in \bigcup_\sigma Y^s\sigma\} \cup (\bigcup_\sigma X^{st}\sigma)$. Let $Y^t = \{H_1, \ldots, H_n\}$ and suppose $\{E_1 \to D_1, \ldots, E_m \to D_m\}$ are all the formulas of the form $E \to D$ in Y^s. We consequently apply an unification algorithm to each equation $E_i = H_j$, $i = 1, \ldots m, j = 1, \ldots n$. If for a pair (i, j) an equation is unifiable then we put $M_{ij} := D_i\theta$ where θ is a *mgu* of the equation. We put $M_{ij} := \emptyset$ otherwise. It is clear that $\{F \mid exists\ G \in \bigcup_\sigma Y^t\sigma\ and\ G \to F \in \bigcup_\sigma Y^s\sigma\} = \bigcup_\sigma (\bigcup_{i,j} M_{ij})\sigma$. Hence $*(st) = \bigcup_\sigma ((\bigcup_{i,j} M_{ij}) \cup X^{st})\sigma$. So we can define $Y^{st} = (\bigcup_{i,j} M_{ij}) \cup X^{st}$.

c) $*(s + t) := *(s) \cup *(t) \cup \phi(s + t)$, that is $*(s + t) = (\bigcup_\sigma Y^s\sigma) \cup (\bigcup_\sigma Y^t\sigma) \cup (\bigcup_\sigma X^{s+t}\sigma) = \bigcup_\sigma (Y^s \cup Y^t \cup X^{s+t})\sigma$. Thus we have $Y^{s+t} = Y^s \cup Y^t \cup X^{s+t}$.

One can easy see that Y^t is finite for any term t. □

Corollary 14. *If a function $\Phi : t \mapsto X^t$ is recursive then the function $U : t \mapsto Y^t$ constructed in the proof of lemma is recursive. Hence the relation "$F \in *(t)$" is recursive in F and t.*

Theorem 15. *Let axiom specification AS satisfy the following condition: for any axiom constant a*

$$\{F \in Fm \mid [a]F \in AS\} = \bigcup_\sigma X^a\sigma,$$

where X^a is a finite set of formulas in the extended language, and the function $\Psi : a \mapsto X^a$ is computable. Then the theory $\mathcal{LP}_{AS}$ is decidable.

Proof. Let F be an arbitrary $\mathcal{LP}$-formula. By lemma 11 the answer to the question if $\mathcal{LP}_{AS} \vdash F$ or $\mathcal{LP}_{AS} \nvdash F$ depends on the existence of an AS-pre-model $\mathcal{P}$, such that $\mathcal{P} \nvDash_p F$. Let us establish that the problem of existence of such $\mathcal{P}$ is decidable.

Restriction is a finite sequence of expressions of the form $\mathcal{P} \vDash_p A, \mathcal{P} \nvDash_p A, A \in *(t), A \notin *(t)$, where A is a $\mathcal{LP}$-formula and t is a $\mathcal{LP}$-term.

Restriction is *primitive* if it is a finite sequence of expressions of the form $\mathcal{P} \vDash_p S, \mathcal{P} \nvDash_p S, A \in *(t), A \notin *(t)$, where S is a sentence letter, A is a formula and t is a term.

Restriction is *inconsistent* if it contains $\mathcal{P} \vDash_p A$ and $\mathcal{P} \nvDash_p A$ or $A \in *(t)$ and $A \notin *(t)$ simultaneously. Now we describe the decision algorithm. The first stage of its computation consists in developing a tree whose nodes are labelled by *restrictions*. It starts with *restriction* $\mathcal{P} \nvDash_p F$ and proceeds accordingly to the following rules: (Γ and Δ are arbitrary *restrictions*.)

conjunction : $$\dfrac{\Gamma,\mathcal{P}\models_p A\wedge B,\Delta}{\Gamma,\mathcal{P}\models_p A,\mathcal{P}\models_p B,\Delta}\qquad\qquad \dfrac{\Gamma,\mathcal{P}\not\models_p A\wedge B,\Delta}{\Gamma,\mathcal{P}\not\models_p A,\Delta\quad\Gamma,\mathcal{P}\not\models_p B,\Delta}$$

disjunction : $$\dfrac{\Gamma,\mathcal{P}\not\models_p A\vee B,\Delta}{\Gamma,\mathcal{P}\not\models_p A,\mathcal{P}\not\models_p B,\Delta}\qquad\qquad \dfrac{\Gamma,\mathcal{P}\models_p A\vee B,\Delta}{\Gamma,\mathcal{P}\models_p A,\Delta\quad\Gamma,\mathcal{P}\models_p B,\Delta}$$

implication : $$\dfrac{\Gamma,\mathcal{P}\not\models_p A\rightarrow B,\Delta}{\Gamma,\mathcal{P}\models_p A,\mathcal{P}\not\models_p B,\Delta}\qquad\qquad \dfrac{\Gamma,\mathcal{P}\models_p A\rightarrow B,\Delta}{\Gamma,\mathcal{P}\not\models_p A,\Delta\quad\Gamma,\models_p B,\Delta}$$

negation : $$\dfrac{\Gamma,\mathcal{P}\models_p \neg A,\Delta}{\Gamma,\mathcal{P}\not\models_p A,\Delta}\qquad\qquad \dfrac{\Gamma,\mathcal{P}\not\models_p \neg A,\Delta}{\Gamma,\mathcal{P}\models_p A,\Delta}$$

[ ]( ) : $$\dfrac{\Gamma,\mathcal{P}\models_p [t]A,\Delta}{\Gamma,\mathcal{P}\models_p A,A\in *(t),\Delta}\qquad\qquad \dfrac{\Gamma,\mathcal{P}\not\models_p [t]A,\Delta}{\Gamma,\mathcal{P}\not\models_p A,\Delta\quad\Gamma,A\notin *(t),\Delta}$$

Obviously, if AS-pre-model $\mathcal{P}$ satisfies the *restriction* Γ_0 then it satisfies some restriction obtained from Γ_0 by any developing rule.

A branch of a tree, computed by this algorithm is declared *closed* as soon as that branch is labelled by inconsistent *restriction*. Otherwise branch is declared *open*. We place a cross "×" at the base of each closed branch.

The algorithm choses an open branch nondeterministically and develops it according to the rules above. The first stage terminates when no rule can be applied to any open branch. If it terminates with all branches closed, then we can surely see, that there is no AS-pre-model to disprove formula F, and thus $\mathcal{LP}_{AS} \vdash F$. Otherwise we have open branches with primitive *restrictions* in them. The desired AS-pre-model satisfies at least one of these *restrictions*. Here the second stage of the algorithm begins. For every restriction we try to construct an AS-pre-model satisfying it. In order to do it for a primitive *restriction* Γ we define $Y^t = \{H \in Fm \mid H \in *(t) \text{ in } \Gamma\}$ for any term t. We construct a function $\Phi(\cdot)$, such that $\Phi(a) = X^a \cup Y^a$ for any axiom constant a, and $\Phi(t) = Y^t$ for any other term. Obviously Φ is computable. Let an evaluation $\phi(\cdot)$ correspond to Φ. We construct a minimal proof-theorem assignment based on the evaluation $\phi(\cdot)$ and define truth-assignment v as follows: $v(S) = True$ iff there is expression $\mathcal{P} \models_p S$ in Γ. It is clear that $\mathcal{P} = (v, *, \models_p)$ is an AS-pre-model. According to corollary 14 relation "$A \in *(t)$" is recursive in A and t. Then we check $*(\cdot)$ to satisfy the expressions of the form $A \notin *(t)$ in Γ. If $*(\cdot)$ satisfies those expressions then $\mathcal{P} = (v, *, \models_p)$ disproves the formula F. If not then no other proof-theorem assignment based on $\phi(\cdot)$ satisfies Γ, since $*(\cdot)$ is minimal. In this case *restriction* Γ fails to provide us with AS-pre-model. Thus, trying each primitive *restriction* at the open branches we can verify if there exists desired AS-pre-model or not.

□

Acknowledgments

I would like to thank professor S. Artemov, who advised me on this work. I am also indebted to V. Krupski and T. Sidon for careful reading of this paper which led to valuable improvements.

References

1. S. Artemov, Operational modal logic, Tech. Rep. 95-29, Mathematical Sciences Institute, Cornell University, December 1995.
2. S. Artemov, Logic of proofs, Annals of pure and applied logic 67 (1994) 29-59
3. A. Nerode, "Some Lectures on Modal Logic", Tech. Rep 90-25, Mathematical Sciences Institute, Cornell University, April 1990.
4. F. Baader and J. Siekman, Unification Theory, in D. M. Gabbay, C. J. Hogger, and J. A. Robinson (ed.) Handbook of Logic in Artificial Intelligence and Logic Programming, Oxford University Press.

Interpretation of the Full Computation Tree Logic CTL* on Sets of Infinite Sequences

Ulrich Nitsche[*]

University of Zurich
Department of Computer Science
Winterthurerstr. 190
CH-8057 Zurich
Switzerland
email: nitsche@ifi.unizh.ch

Abstract. Formulae of the full computation tree logic CTL* are usually interpreted on infinite trees. In this paper, we propose how to interpret them on ω-languages, i.e. on sets of infinite sequences, even if the ω-language does not have a one-to-one corresponding tree representation.

Keywords. CTL*, ω-Languages, Verification.

1 Introduction

The semantics of the branching-time temporal logic CTL* (*full computation tree logic*) is defined originally on trees of infinite depth. To interpret a CTL*-formula on an ω-language, i.e. a set of infinite sequences, one cannot, in general, represent the ω-language by an infinite tree. Acceptance conditions needed for representation of an ω-language may prevent it from being representable by a tree. If the ω-language can be represented as the Eilenberg-limit of a *prefix-closed* language, no acceptance conditions are needed for its representation and, consequently, these ω-languages can be represented easily as an infinite tree.

We start with this tree representation of Eilenberg-limits of prefix-closed regular languages, defining the semantics of CTL*-formulae on the tree representation as usual. Further on, we relax the definition by defining the semantics of CTL* on ω-languages in general. To do so, the key notion is the one of an ω-language's leftquotient by a word. In this relaxed setting, words correspond to nodes in the tree and leftquotients correspond to subtrees. A leftquotient represents possible suffixes of a word in a given ω-language. Thus we establish a slightly more general semantics definition for CTL* than the one on infinite trees, taking into account that for an ω-language which can be represented by an infinite tree, interpreting a CTL*-formula directly on the ω-language is equivalent to interpreting it on the ω-language's tree representation, which proves the new semantics definition to be reasonable.

---
[*] Former address: GMD, Rheinstr. 75, D-64295 Darmstadt, Germany. Ulrich Nitsche is supported by the Swiss National Science Foundation.

¿From the introduction above it is not clear why, if at all, one may be interested in considering such a type of semantics definition. To say a few words about motivation, we look at a class of properties which is of practical interest for the verification of reactive concurrent systems: *eventuality properties* [10], also called *always possibly properties* in [8]. In logical terms, eventuality properties are represented by the $\mathcal{AGEF}$-fragment of CTL* [9, 12, 13]. Considering these properties by means of language theory as in [10], one finds that the language theoretic point of view slightly differs from the usual logical view. The present paper takes this observation on the $\mathcal{AGEF}$-fragment of CTL* up to the level of whole CTL*.

2 Preliminaries

Regular languages and the basic properties of regular languages are assumed to be known as well as their representation using finite automata [1, 6, 7].

Let Σ be a finite set, called an *alphabet*. The set of all finite sequences over Σ is denoted by Σ^* and the set of all infinite sequences over Σ is denoted by Σ^ω. A subset L of Σ^* is called a *(finitary) language* over Σ and a subset L_ω of Σ^ω is called an *ω-language* over Σ [14].

Let Σ be an alphabet, and let $L \subseteq \Sigma^*$ and $L_\omega \subseteq \Sigma^\omega$ be a finitary language and an ω-language over Σ respectively.

The set of *prefixes* of L is $pre(L) = \{v \in \Sigma^* \mid \exists w \in \Sigma^* : vw \in L\}$. The set of prefixes of L_ω is $pre(L_\omega) = \{w \in \Sigma^* \mid \exists x \in \Sigma^\omega : wx \in L_\omega\}$. We call L *prefix-closed* if and only if $pre(L) = L$.

We define the *limit* of L [4, 14] to be

$$lim(L) = \{x \in \Sigma^\omega \mid \text{infinitely many different finite prefixes of } x \text{ are in } L.\}.$$

If an ω-language is the limit of a prefix-closed language, then the ω-language will be called *limit-closed*.

For an *ω-word* $x \in \Sigma^\omega$, x_i designates the ith letter of x and $x_{(i...)}$ designates the suffix of x starting with the ith letter: $\forall j \in \mathbb{N} : x_{(i...)j} = x_{i+j-1}$. Let w be a word in Σ^*. Then the *leftquotient* of L_ω by w is $w^{-1}(L_\omega) = \{v \in \Sigma^\omega \mid wv \in L_\omega\}$ [1, 6]. We can consider the leftquotient $w^{-1}(L_\omega)$ being the *continuation* of w in L_ω. The leftquotient of an ω-word $x \in \Sigma^\omega$ by a word $w \in \Sigma^*$ is defined by the equation $w \cdot w^{-1}(x) = x$.

3 Tree-Representation of Limit-Closed ω-Languages

Let $L \subseteq \Sigma^*$ be a prefix-closed language. We represent $lim(L)$ by a finitely branching tree $T_{lim(L)}$ of infinite depth (if $lim(L)$ is not empty). We define $T_{lim(L)}$ in such a way that all finitely long sequences of labels of nodes along a finitely long path in $T_{lim(L)}$ starting at $T_{lim(L)}$'s root are in L. All infinitely long paths of labels of nodes along infinitely long paths starting at $T_{lim(L)}$'s root are in $lim(L)$.

Definition 1. For all $w \in L$, let $succ(w) = w^{-1}(L) \cap \Sigma$, i.e. $succ(w)$ is the set of all letters in Σ that can follow w in L. We define $T_{lim(L)}$ recursively:

- The root of $T_{lim(L)}$, denoted by $root(T_{lim(L)})$ is labelled with ε.
- If ν is a node in $T_{lim(L)}$ such that the path from the root is labelled with w, then ν has $|succ(w)|$ many child nodes and each child node is labelled with an element of $succ(w)$ respectively. The set of all child nodes of a node ν is designated $children(\nu)$.
- $T_{lim(L)}$ does not contain any other nodes than the ones described.

For a node ν in $T_{lim(L)}$, $label(\nu)$ designates the label of ν. □

To be precise, $T_{lim(L)}$ is in fact a forest of $|succ(\varepsilon)|$ many trees. Each of these trees represents all ω-words in $lim(L)$ that start with the same letter. To have a single representation of $lim(L)$, we have linked these trees to a particular node ($T_{lim(L)}$'s root) which has to be labelled with the empty word ε for not changing the fact that path labels correspond to words in the language.

4 Defining CTL*with respect to $T_{lim(L)}$

We define the syntax and semantics of the full computation tree logic CTL* in this section. We take the usual semantics of CTL* as presented in [2, 3, 5] and modify it to fit the structure of $T_{lim(L)}$.

Definition 2. We define the syntax of CTL* with respect to a set AP of atomic propositions [5].

- All atomic propositions are state formulae.
- If ξ and ζ are state formulae then so are $(\xi) \wedge (\zeta)$ and $\neg(\xi)$.
- If ξ is a path formula then $\mathcal{A}(\xi)$ and $\mathcal{E}(\xi)$ are state formulae.
- Each state formula is also a path formula.
- If ξ and ζ are path formulae then so are $(\xi) \wedge (\zeta)$ and $\neg(\xi)$.
- If ξ and ζ are path formulae then so are $(\xi)\mathcal{U}(\zeta)$ and $\mathcal{X}(\xi)$.
- CTL*-formulae are exactly the state formulae.

□

For defining the semantics of CTL* on $T_{lim(L)}$, we have to take care of $T_{lim(L)}$'s root ρ, because it is the only node with $label(\rho) \notin \Sigma$. We define the satisfaction relation "$\models$" between nodes of $T_{lim(L)}$ and CTL*-formulae such that $T_{lim(L)}$ satisfies a CTL*-formula if and only if its root does (initial satisfaction). The satisfaction relation is defined with respect to a labelling function $\lambda : \Sigma \to 2^{AP}$ which takes node-labels to sets of atomic propositions. λ determines which atomic propositions are satisfied by a node with a particular node-label.

Definition 3. Let η, ξ and ζ be CTL*-formulae or path formulae and let ν be a node in $T_{lim(L)}$. Let, for all nodes γ of $T_{lim(L)}$, $path(\gamma)$ designate the set of all infinite paths in $T_{lim(L)}$ that start at γ.

1. If $\eta \in AP$ and $label(\nu) = \varepsilon$, then $\nu \models \eta$ if and only if, for all $\gamma \in children(\nu)$, we have $\eta \in \lambda(label(\gamma))$.
2. If $\eta \in AP$ and $label(\nu) \neq \varepsilon$, then $\nu \models \eta$ if and only if $\eta \in \lambda(label(\nu))$.
3. $\nu \models \neg(\xi)$ if and only if it is not the case that $\nu \models \xi$.
4. $\nu \models (\xi) \wedge (\zeta)$ if and only if $\nu \models \xi$ and $\nu \models \zeta$.
5. If $label(\nu) = \varepsilon$, then $\nu \models \mathcal{E}(\xi)$ if and only if there exists an infinite path $\sigma \in \bigcup_{\gamma \in children(\nu)} path(\gamma)$ such that $\sigma \models \xi$.
6. If $label(\nu) \neq \varepsilon$, then $\nu \models \mathcal{E}(\xi)$ if and only if there exists an infinite path $\sigma \in path(\nu)$ such that $\sigma \models \xi$.
7. If $label(\nu) = \varepsilon$, then $\nu \models \mathcal{A}(\xi)$ if and only if, for all infinite paths $\sigma \in \bigcup_{\gamma \in children(\nu)} path(\gamma)$, we have $\sigma \models \xi$.
8. If $label(\nu) \neq \varepsilon$, then $\nu \models \mathcal{A}(\xi)$ if and only if, for all infinite paths $\sigma \in path(\nu)$, we have $\sigma \models \xi$.

Let $\sigma = s_1 s_2 s_3 \ldots$ be an infinite path in $T_{lim(L)}$ starting at a node s_1 such that $label(s_1) \neq \varepsilon$, i.e. s_1 is not the root of $T_{lim(L)}$.

9. If $\eta \in AP$, then $\sigma \models \eta$ if and only if $\eta \in \lambda(label(s_1))$.
10. $\sigma \models \neg(\xi)$ if and only if it is not the case that $\sigma \models \xi$.
11. $\sigma \models (\xi) \wedge (\zeta)$ if and only if $\sigma \models \xi$ and $\sigma \models \zeta$.
12. $\sigma \models \mathcal{X}(\xi)$ if and only if $\sigma_{(2\ldots)} \models \xi$.
13. $\sigma \models (\xi)\mathcal{U}(\zeta)$ if and only if there exist an $i \in \mathbb{N}$ such that $\sigma_{(i\ldots)} \models \zeta$ and, for all $j < i$, we have $\sigma_{(j\ldots)} \models \xi$.
14. $\sigma \models \mathcal{E}(\xi)$ if and only if there exists and infinite path $\sigma' \in path(s_1)$ such that $\sigma' \models \xi$.
15. $\sigma \models \mathcal{A}(\xi)$ if and only if, for all infinite paths $\sigma' \in path(s_1)$, we have $\sigma' \models \xi$.
□

We use this satisfaction relation on nodes of $T_{lim(L)}$ to define the satisfaction relation on $T_{lim(L)}$ as satisfaction in $T_{lim(L)}$'s root (intial satisfaction).

Definition4. Let η be a CTL*-formula and let $\lambda : \Sigma \to 2^{AP}$ be a labelling function of node-labels with atomic propositions. Then $T_{lim(L)} \models \eta$ (with respect to λ) if and only if $root(T_{lim(L)}) \models \eta$ (with respect to λ). □

In Definition 3, we have adapted usual CTL* semantics to fit the structure of $T_{lim(L)}$. Based on this definition we define a CTL* semantics on ω-languages. Before we do so, we first give an example why we cannot always represent ω-languages by trees like in Section 3.

5 Not All ω-Languages Have a Tree Representation

If we consider the ω-language $lim(\{a,b\}^*)$ which is $\{a,b\}^\omega$, then $T_{lim(\{a,b\}^*)}$ is the *complete* binary tree with nodes labelled with a and b respectively except for its root which is labelled with ε. Obviously, $\{a,b\}^*$ is prefix-closed. $lim(\{a,b\}^*)$ is obtained by the set of all infinite sequences of node-labels along infinite paths in $T_{lim(\{a,b\}^*)}$. Hence we have a nice one-to-one correspondence between $lim(\{a,b\}^*)$ and $T_{lim(\{a,b\}^*)}$.

If we consider the ω-language $lim((\{a,b\}^* \cdot a)^*)$, then $(\{a,b\}^* \cdot a)^*$ is not prefix-closed anymore. Nevertheless $lim((\{a,b\}^* \cdot a)^*)$ is "not *very* different" from $lim(\{a,b\}^*)$, since $pre(\{a,b\}^*) = pre((\{a,b\}^* \cdot a)^*)$. If we want to construct a tree corresponding to Definition 1 for $lim((\{a,b\}^* \cdot a)^*)$, we would have to replace "L" by "$pre(L)$" in Definition 1. But according to $pre((\{a,b\}^* \cdot a)^*) = pre(\{a,b\}^*)$, $lim((\{a,b\}^* \cdot a)^*)$ would be presented by the same binary tree as $lim(\{a,b\}^*)$. Thus the resulting tree does not correspond precisely to $lim((\{a,b\}^* \cdot a)^*)$ and we cannot therefore evaluate a CTL*-formula on $lim((\{a,b\}^* \cdot a)^*)$ by using this tree representation.[2] Using the notion of leftquotients of ω-languages by words, we can make CTL* semantics definition applicable to all ω-languages.

6 Satisfaction of CTL*-Formulae with Respect to ω-Languages

We can use the fact that the leftquotient of an ω-language by a word extracts exactly the information about suffixes of the word in the ω-language to define a CTL* semantics on ω-languages.

Definition 5. Let $L_\omega \subseteq \Sigma^\omega$ be an ω-language. Let η, ξ and ζ be CTL*-formulae or path formulae. Let $\lambda : \Sigma \to 2^{AP}$ be a labelling function. Let w be a (finitary) word in $pre(L_\omega)$. If $w \neq \varepsilon$, let $first(w)$ designate the first letter in w, let $last(w)$ designate the last letter in w, and let $short(w)$ be the word that we obtain by removing w's last letter. For an ω-word $x \in \Sigma^\omega$, let $first(x)$ designate x's first letter.[3] We define the satisfaction relation with respect to L_ω and λ:

1. If $\eta \in AP$ and $w = \varepsilon$, then $w \models \eta$ if and only if, for all $x \in L_\omega$, we have $\eta \in \lambda(first(x))$.
2. If $\eta \in AP$ and $w \neq \varepsilon$, then $w \models \eta$ if and only if $\eta \in \lambda(first(w))$.
3. $w \models \neg(\xi)$ if and only if it is not the case that $w \models \xi$.
4. $w \models (\xi) \wedge (\zeta)$ if and only if $w \models \xi$ and $w \models \zeta$.
5. If $w = \varepsilon$, then $w \models \mathcal{E}(\xi)$ if and only if there exists $x \in L_\omega$ such that $(w, x) \models \xi$.
6. If $w \neq \varepsilon$, then $w \models \mathcal{E}(\xi)$ if and only if there exists $x \in w^{-1}(L_\omega)$ such that $(short(w), last(w)x) \models \xi$.
7. If $w = \varepsilon$, then $w \models \mathcal{A}(\xi)$ if and only if, for all $x \in L_\omega$, we have $(w, x) \models \xi$.
8. If $w \neq \varepsilon$, then $w \models \mathcal{A}(\xi)$ if and only if, for all $x \in w^{-1}(L_\omega)$, we have $(short(w), last(w)x) \models \xi$.

Let $u \in \Sigma^*$ be a word and let $x \in \Sigma^\omega$ be an ω-word over Σ.

9. If $\eta \in AP$, then $(u, x) \models \eta$ if and only if $\eta \in \lambda(first(x))$.

---

[2] In general, languages L that require accepting states for there representation (i.e. that are not prefix-closed) have a tree representation equivalent to $pre(L)$.

[3] $first(x) = x_1$, $first(w) = w_1$, $last(w) = w_{|w|}$, and $short(w) = w_1 w_2 \ldots w_{|w|-1}$.

10. $(u, x) \models \neg(\xi)$ if and only if it is not the case that $(u, x) \models \xi$.

11. $(u, x) \models (\xi) \wedge (\zeta)$ if and only if $(u, x) \models \xi$ and $(u, x) \models \zeta$.

12. $(u, x) \models \mathcal{X}(\xi)$ if and only if $(u\, first(x), first(x)^{-1}(x)) \models \xi$.

13. $(u, x) \models (\xi)\mathcal{U}(\zeta)$ if and only if there exist a word $w \in pre(x)$ such that $(uw, w^{-1}(x)) \models \zeta$ and such that, for all $v \in pre(x)$ such that $|v| < |w|$, we have $(uv, v^{-1}(x)) \models \xi$.

14. $(u, x) \models \mathcal{E}(\xi)$ if and only if there exists $x' \in (u\, first(x))^{-1}(L_\omega)$ such that $(u, first(x)x') \models \xi$.

15. $(u, x) \models \mathcal{A}(\xi)$ if and only if, for all $x' \in (u\, first(x))^{-1}(L_\omega)$, we have $(u, first(x)x') \models \xi$.

L_ω satisfies a CTL*-formulae η with respect to λ if and only if the empty word ε satisfies η with respect to L_ω and λ (the modified version of initial satisfaction). $\qquad\qquad\square$

If in Definition 5, $L_\omega = lim(L)$ for a prefix-closed language $L \subseteq \Sigma^*$, equally numbered parts of Definition 3 and Definition 5 become equivalent. We thus obtain that Definition 5 is reasonable, since for tree-representable ω-languages, it corresponds to the usual CTL* semantics:

Lemma 6. *Let $L \subseteq \Sigma^*$ be a prefix-closed language, let η be a CTL*-formula, and let $\lambda : \Sigma \to 2^{AP}$ be a labelling function. Then*

$$lim(L) \models \eta \quad \text{if and only if} \quad T_{lim(L)} \models \eta.$$

7 Conclusions

We have defined a semantics for CTL* on ω-languages instead of infinite trees. The definition is a slight modification of CTL*'s usual semantics definition, using the notion of an ω-language's leftquotient by a word to adapt the logical terminology on branching processes to the notations of formal language theory.

The way we have looked at CTL* in this paper has the advantage to relate property classes defined in terms of formal language theory to logical notions. Using, for instance, the presented "ω-language semantics" for CTL* enables us to identify *always possibly properties* of [8] with *eventuality properties* in [10], which can be represented by the $\mathcal{AGEF}$-fragment of CTL*.

Bringing CTL* to a formal language theory basis may also result in possibly using a "verification and abstraction"-concept for CTL*-properties as presented for the $\mathcal{AGEF}$-fragment of CTL* in [13], based on results in [11]. It is a topic for further study whether homomorphic abstractions are compatible with the ω-language semantics.

The topic of decidability of the satisfaction of a CTL*-formula on a given ω-language is not discussed in this paper. It is another topic for further study. Nevertheless it is rather probable that this problem is decidable for regular ω-languages, which may require adapting the usual model checking algorithm of CTL* [2, 3, 5].

References

1. J. Berstel. *Transductions and Context-Free Languages*. Studienbücher Informatik. Teubner Verlag, Stuttgart, first edition, 1979.
2. E. M. Clarke and E. A. Emerson. Design and synthesis of synchronization skeletons using branching-time temporal logic. In D. Kozen, editor, *Logic of Programs 1981*, volume 131 of *Lecture Notes in Computer Science*, pages 52–71. Springer Verlag, 1982.
3. E. M. Clarke, E. A. Emerson, and A. P. Sistla. Automatic verification of finite-state concurrent systems using temporal logic specifications. *ACM Transactions on Programming Languages and Systems*, 8(2):244–263, 1986.
4. S. Eilenberg. *Automata, Languages and Machines*, volume A. Academic Press, New York, 1974.
5. E. A. Emerson. Temporal and modal logic. In van Leeuwen [15], pages 995–1072.
6. M. A. Harrison. *Introduction to Formal Language Theory*. Addison-Wesley, Reading, Mass., first edition, 1978.
7. J. E. Hopcroft and J. D. Ullman. *Introduction to Automata Theory, Languages and Computation*. Addison-Wesley, Reading, Mass., first edition, 1979.
8. O. Kupferman and M. Y. Vardi. Module checking. In R. Alur and T. A. Henzinger, editors, *CAV'96*, volume 1102 of *Lecture Notes in Computer Science*, pages 75–86, New Brunswick, N.J., 1996. Springer Verlag.
9. R. P. Kurshan. *Computer-Aided Verification of Coordinating Processes*. Princeton University Press, Princeton, New Jersey, first edition, 1994.
10. U. Nitsche. *Verification of Co-Operating Systems and Behaviour Abstraction*. PhD thesis, University of Frankfurt, Germany. handed in 1996.
11. U. Nitsche. Propositional linear temporal logic and language homomorphisms. In A. Nerode and Y. V. Matiyasevich, editors, *Logical Foundations of Computer Science '94, St. Petersburg*, volume 813 of *Lecture Notes in Computer Science*, pages 265–277. Springer Verlag, 1994.
12. U. Nitsche. A verification method based on homomorphic model abstraction. In *Proceedings of the 13th Annual ACM Symposium on Principles of Distributed Computing*, page 393, Los Angeles, 1994. ACM Press.
13. U. Nitsche. Verification and behavior abstraction – towards a tractable verification technique for large distributed systems. *Journal of Systems and Software*, 33(3):273–285, June 1996.
14. W. Thomas. Automata on infinite objects. In van Leeuwen [15], pages 133–191.
15. J. van Leeuwen, editor. *Formal Models and Semantics*, volume B of *Handbook of Theoretical Computer Science*. Elsevier, 1990.

Type Introduction for Equational Rewriting

Hitoshi Ohsaki and Aart Middeldorp

Institute of Information Sciences and Electronics
University of Tsukuba
Tsukuba 305, Japan
{hitoshi,ami}@score.is.tsukuba.ac.jp

Abstract. Type introduction is a useful technique for simplifying the task of proving properties of rewrite systems by restricting the set of terms that have to be considered to the well-typed terms according to any many-sorted type discipline which is compatible with the rewrite system under consideration. A property of rewrite systems for which type introduction is correct is called persistent. Zantema showed that termination is a persistent property of non-collapsing rewrite systems and non-duplicating rewrite systems. We extend his result to the more complicated case of equational rewriting. As a simple application we prove the undecidability of AC-termination for terminating rewrite systems. We also present sufficient conditions for the persistence of acyclicity and non-loopingness, two properties which guarantee the absence of certain kinds of infinite rewrite sequences.

1 Introduction

Term rewriting is an important method for equational reasoning. In term rewriting the axioms of the equational system under consideration are used in one direction only. Since in the presence of axioms like commutativity, a common situation in equational reasoning, rewriting is non-terminating, the framework of equational term rewriting has been proposed. Equational term rewriting is an extension of rewriting in which certain axioms are used bidirectionally, more precisely, an equational rewrite system $\mathcal{R}/\mathcal{E}$ consists of a term rewriting system $\mathcal{R}$ and an equational system $\mathcal{E}$ and a term s rewrites in one step to a term t if there exists a rewrite rule $l \to r$ in $\mathcal{R}$ and a substitution σ such that s is equivalent (in the equational theory generated by $\mathcal{E}$) to a term s' which contains $l\sigma$ and t is equivalent to the term t' obtained from s' by replacing $l\sigma$ by $r\sigma$.

Here we are interested in termination of equational rewrite systems. An early paper on termination of equational rewriting is Jouannaud and Muñoz [7]. In that paper sufficient conditions are given for reducing (equational) termination of $\mathcal{R}/\mathcal{E}$ to termination of $\mathcal{R}$. In more recent papers [14, 13, 8] methods like the well-known recursive path order for proving termination of rewriting are extended to AC-termination, i.e., termination of equational rewrite systems $\mathcal{R}/\mathcal{E}$ where $\mathcal{E}$ consists of the associativity and commutativity axioms $f(f(x,y),z) \approx f(x, f(y, z))$ and $f(x,y) \approx f(y, x)$ for (some of) the binary function symbols in $\mathcal{R}$. Another recent paper is Ferreira [4] where the dummy elimination technique of [5] for proving termination is extended to equational rewriting.

In this paper we extend the type introduction technique of Zantema [15] for proving properties of rewriting to equational rewriting. More precisely, we show that termination is a persistent property of equational rewrite systems $\mathcal{R}/\mathcal{E}$ such that $\mathcal{R}$ does not contain both collapsing and duplicating rules and $\mathcal{E}$ is variable preserving and does not contain collapsing axioms. Type introduction is known to be useful for proving undecidability results for termination of rewriting [10], and in this paper we give a simple proof of the undecidability of AC-termination for terminating rewrite systems using type introduction. This result, which appears to be new, clearly shows that equational termination is a much harder problem than termination. We also show that, under the same conditions as for termination, acyclicity and non-loopingness are persistent properties of equational rewrite systems. The last result enables us to simplify several proofs of non-loopingness that can be found in the literature.

This paper is organized as follows. In the next section we briefly define equational rewriting and we recall the results of Zantema [15] on type introduction. In Section 3 we generalize these results to equational rewriting. In Section 4 the usefulness of the results of Section 3 is illustrated by showing the undecidability of AC-termination for terminating rewrite systems and in Section 5 we address persistence of acyclicity and non-loopingness. Persistence is closely related ([12, 15]) to modularity, a property which has been thoroughly investigated in the term rewriting literature. Along this line we obtained several new modularity results. Due to space restrictions however these cannot be described here. For the same reason proofs of several of our results have been omitted.

2 Preliminaries

Familiarity with the basic notions of term rewriting (as expounded in e.g. [2, 9]) will be helpful in the following. We start this preliminary section with a very brief introduction to *many-sorted* equational reasoning and term rewriting.

Let S be a set of sorts. An S-sorted signature is a set $\mathcal{F}$ of function symbols together with a sort declaration $\alpha_1 \times \cdots \times \alpha_n \to \alpha$ for every $f \in \mathcal{F}$. Here $\alpha_1, \ldots, \alpha_n, \alpha \in S$ and n is called the arity of f. Function symbols of arity 0 are called constants. We assume the existence of pairwise disjoint countably infinite sets of variables $\mathcal{V}_\alpha$ for every sort $\alpha \in S$. The union of all $\mathcal{V}_\alpha$ is denoted by $\mathcal{V}$. The set $\mathcal{T}(\mathcal{F}, \mathcal{V})$ of well-typed terms is the union of the sets $\mathcal{T}_\alpha(\mathcal{F}, \mathcal{V})$ for $\alpha \in S$ that are inductively defined as follows: $\mathcal{V}_\alpha \subseteq \mathcal{T}_\alpha(\mathcal{F}, \mathcal{V})$ and $f(t_1, \ldots, t_n) \in \mathcal{T}_\alpha(\mathcal{F}, \mathcal{V})$ whenever $f \in \mathcal{F}$ has sort declaration $\alpha_1 \times \cdots \times \alpha_n \to \alpha$ and $t_i \in \mathcal{T}_{\alpha_i}(\mathcal{F}, \mathcal{V})$ for all $1 \leqslant i \leqslant n$. If $t \in \mathcal{T}_\alpha(\mathcal{F}, \mathcal{V})$ for some $\alpha \in S$ then we say that t has sort α and we write $\mathrm{sort}(t) = \alpha$. For every $\alpha \in S$, let $\square_\alpha$ be a fresh constant, named hole, of sort α. Elements of $\mathcal{T}(\mathcal{F} \cup \{\square_\alpha \mid \alpha \in S\}, \mathcal{V})$ are called contexts. An empty context is a hole. If C is a context with n holes $\square_{\alpha_1}, \ldots, \square_{\alpha_n}$ (from left to right) and $t_1, \ldots, t_n$ are terms with $\mathrm{sort}(t_i) = \alpha_i$ then $C[t_1, \ldots, t_n]$ denotes the term obtained from C by replacing the holes by $t_1, \ldots, t_n$. A substitution is a mapping σ from $\mathcal{V}$ to $\mathcal{T}(\mathcal{F}, \mathcal{V})$ such that $\mathrm{sort}(\sigma(x)) = \alpha$ if $x \in \mathcal{V}_\alpha$. We write $t\sigma$ for the result of applying σ to a term t.

An S-sorted *equational system* (ES for short) consists of an S-sorted signature $\mathcal{F}$ and a set $\mathcal{E}$ of equations between terms in $\mathcal{T}(\mathcal{F}, \mathcal{V})$ such that $\mathsf{sort}(l) = \mathsf{sort}(r)$ for every equation $l \approx r \in \mathcal{E}$. We write $s \to_{\mathcal{E}} t$ if there exist an equation $l \approx r$ in $\mathcal{E}$, a substitution σ, and a context C such that $s = C[l\sigma]$ and $t = C[r\sigma]$. The symmetric closure of $\to_{\mathcal{E}}$ is denoted by $\vdash_{\mathcal{E}}$ and the transitive reflexive closure of $\vdash_{\mathcal{E}}$ by $\sim_{\mathcal{E}}$. Note that $\mathsf{sort}(s) = \mathsf{sort}(t)$ whenever $s \sim_{\mathcal{E}} t$. An equation $l \approx r$ is called *non-erasing* if the sets of variables in l and r are the same. We say that $l \approx r$ is *variable-preserving* if the multisets of variable occurrences in l and r are the same. The equation $l \approx r$ is called *collapsing* if l or r is a variable. An (S-sorted) ES is non-erasing (variable-preserving, collapsing) if all its equations are so.

A rewrite rule is an equation $l \approx r$ such that l is a not a variable and variables which occur in r also occur in l. Rewrite rules $l \approx r$ are written as $l \to r$. An S-sorted *term rewriting system* (TRS for short) is an S-sorted ES all of whose equations are rewrite rules. A rewrite rule $l \to r$ is *duplicating* if some variable occurs more often in r than in l. An S-sorted TRS is duplicating if it has a duplicating rewrite rule. An S-sorted *equational term rewriting system* (ETRS for short) $\mathcal{R}/\mathcal{E}$ consists of an S-sorted TRS $\mathcal{R}$ and an S-sorted ES $\mathcal{E}$ over the same signature. We write $s \to_{\mathcal{R}/\mathcal{E}} t$ if there exist terms s' and t' such that $s \sim_{\mathcal{E}} s' \to_{\mathcal{R}} t' \sim_{\mathcal{E}} t$.

An ES (TRS, ETRS) is an S-sorted ES (TRS, ETRS) with S a singleton set. This is equivalent to the usual (unsorted) definition found in the literature. The underlying ES $\Theta(\mathcal{E})$ of an S-sorted ES $\mathcal{E}$ is obtained by simply dropping all sort declarations; likewise for TRSs and ETRSs. The term rewriting literature is mainly concerned with unsorted (E)TRSs. In this paper we show how many-sorted ETRSs can help to simplify the task of proving properties of unsorted ETRSs. A property P of (many-sorted) ETRS is called *persistent* if the following equivalence holds for every many-sorted ETRS $\mathcal{R}/\mathcal{E}$: $\mathcal{R}/\mathcal{E}$ has the property P if and only if $\Theta(\mathcal{R}/\mathcal{E})$ has the property P. For most properties the "only if" direction is trivial; we are interested in the "if" direction. In order to show that a given ETRS $\mathcal{R}/\mathcal{E}$ has a certain property P, which is known to be persistent, it is sufficient to find suitable S and sort declarations such that the S-sorted ETRS $\mathcal{R}/\mathcal{E}$ has the property P. The latter is often easier to prove since only well-typed terms have to be considered. Hence persistence facilitates proving properties of ETRSs by type introduction. In this paper we are mainly concerned with the termination property. An ETRS $\mathcal{R}/\mathcal{E}$ is called *terminating* if there are no infinite $\mathcal{R}/\mathcal{E}$-rewrite sequences.

Zantema [15] obtained the following result. In the next section we generalize it to ETRSs.

Theorem 1. *Termination is persistent for TRSs that do not contain both collapsing and duplicating rules.* □

3 Persistence of Termination for Equational Rewriting

In the following few definitions and lemmata $\mathcal{R}$ is an S-sorted TRS and $\mathcal{E}$ an S-sorted ES. Terms in $\Theta(\mathcal{R})$ need not be well-typed (with respect to $\mathcal{R}$), but they can be partitioned into well-typed components. This yields a natural layered structure, which is formalized below.

Definition 2. We write $t = C[\![t_1, \ldots, t_n]\!]$ if $t = C[t_1, \ldots, t_n]$ with C non-empty and maximal well-typed. Note that every term can be uniquely written as $C[\![t_1, \ldots, t_n]\!]$. We write $\mathrm{top}(t) = C$. The subterms $t_1, \ldots, t_n$ of t are called *aliens* and we denote the multiset $\{t_1, \ldots, t_n\}$ by $\mathrm{alien}(t)$. The *rank* of a term is the maximum number of type-clashes along any of its paths:

$$\mathrm{rank}(t) = \begin{cases} 0 & \text{if } t \text{ is well-typed,} \\ 1 + \max\{\mathrm{rank}(s) \mid s \in \mathrm{alien}(t)\} & \text{otherwise.} \end{cases}$$

The rank of a $\Theta(\mathcal{R}/\mathcal{E})$-rewrite sequence is the rank of its initial term. We extend the definition of sort in Section 2 to arbitrary (non-well-typed) terms by letting $\mathrm{sort}(t) = \alpha$ if $t = f(t_1, \ldots, t_n)$ with $f: \alpha_1 \times \cdots \times \alpha_n \to \alpha$.

Let us illustrate these concepts on a small example. Consider $S = \{\alpha, \beta\}$ with sort declarations $f: \alpha \times \beta \to \beta$, $g: \beta \to \beta$, $a: \alpha$, and $b: \beta$. For the term $t = f(f(b, b), f(a, g(a)))$ we have $\mathrm{sort}(t) = \beta$, $\mathrm{top}(t) = f(\Box_\alpha, f(a, g(\Box_\beta)))$, $\mathrm{alien}(t) = \{f(b, b), a\}$, and $\mathrm{rank}(t) = 2$.

The following result is well-known.

Lemma 3. *If* $s \to_{\Theta(\mathcal{R})} t$ *then* $\mathrm{rank}(s) \geqslant \mathrm{rank}(t)$. $\qquad\square$

Definition 4. A rewrite step $s \to_{\Theta(\mathcal{R})} t$ is called *inner* if it takes place in one of the aliens of s. Non-inner steps are called *outer*. An outer step $s \to_{\Theta(\mathcal{R})} t$ is called *collapsing* if $\mathrm{sort}(s) \neq \mathrm{sort}(t)$. An inner step $s \to_{\Theta(\mathcal{R})} t$ is called *collapsing* if $\mathrm{top}(s) \neq \mathrm{top}(t)$.

Note that collapsing rewrite steps necessarily employ collapsing rewrite rules, but not every (outer) step using a collapsing rewrite rule is collapsing. The next two lemmata express well-known facts in the context of modularity.

Lemma 5. *Suppose* $s \to_{\Theta(\mathcal{R})} t$ *is non-collapsing. If* $s \to_{\Theta(\mathcal{R})} t$ *is outer then* $\mathrm{top}(s) \to_{\mathcal{R}} \mathrm{top}(t)$, *otherwise* $\mathrm{top}(s) = \mathrm{top}(t)$. $\qquad\square$

Lemma 6. *If* $s \to_{\Theta(\mathcal{R})} t$ *is outer, non-collapsing, and non-duplicating then* $\mathrm{alien}(t) \subseteq \mathrm{alien}(s)$. *If* $s = C[\![s_1, \ldots, s_i, \ldots, s_n]\!] \to_{\Theta(\mathcal{R})} C[\![s_1, \ldots, t_i, \ldots, s_n]\!] = t$ *with* $s_i \to_{\Theta(\mathcal{R})} t_i$ *is collapsing then* $\mathrm{alien}(t) = (\mathrm{alien}(s) - \{s_i\}) \uplus \mathrm{alien}(t_i)$, *otherwise* $\mathrm{alien}(t) = (\mathrm{alien}(s) - \{s_i\}) \uplus \{t_i\}$. $\qquad\square$

Next we consider how $\vdash_{\Theta(\mathcal{E})}$ and $\sim_{\Theta(\mathcal{E})}$-steps affect the layered structure of terms. Because $\mathcal{E}$ is assumed to be non-collapsing and non-erasing, $\mathcal{E} \cup \mathcal{E}^{-1}$ is a TRS and the relation $\vdash_{\Theta(\mathcal{E})}$ coincides with $\to_{\Theta(\mathcal{E} \cup \mathcal{E}^{-1})}$. Hence we can reuse the above results when reasoning about $\vdash_{\Theta(\mathcal{E})}$.

Lemma 7. *Let $\mathcal{E}$ be non-erasing and non-collapsing. If $s \vdash_{\Theta(\mathcal{E})} t$ then $\text{rank}(s) = \text{rank}(t)$.*

Proof. We have $s \rightarrow_{\Theta(\mathcal{E} \cup \mathcal{E}^{-1})} t$ and thus $\text{rank}(s) \geqslant \text{rank}(t)$ by Lemma 3. Symmetry yields $\text{rank}(t) \geqslant \text{rank}(s)$ and hence $\text{rank}(s) = \text{rank}(t)$. □

The next lemma expresses that the layered structure of terms is essentially preserved by $\Theta(\mathcal{E})$-steps. For the second part it is essential that $\mathcal{E}$ is variable-preserving.

Lemma 8. *Let $\mathcal{E}$ be variable-preserving and non-collapsing. Suppose $\text{alien}(s) = \{s_1, \ldots, s_n\}$ and $\text{alien}(t) = \{t_1, \ldots t_m\}$. If $s \sim_{\Theta(\mathcal{E})} t$ then $\text{top}(s) \sim_{\mathcal{E}} \text{top}(t)$, $m = n$, and there exists a permutation π such that $s_i \sim_{\Theta(\mathcal{E})} t_{\pi(i)}$ for all $1 \leqslant i \leqslant n$.* □

Using all of the preceding lemmata, the following result can now be proved by a routine 'minimal counterexample' argument (cf. Ohlebusch [11]).

Lemma 9. *Let $\mathcal{R}/\mathcal{E}$ be a terminating S-sorted ETRS with $\mathcal{E}$ variable-preserving and non-collapsing. If $\Theta(\mathcal{R}/\mathcal{E})$ is not terminating then there exists an infinite rewrite sequence which contains an outer duplicating and an inner collapsing $\Theta(\mathcal{R})$-step.*

Proof. Let $\mathcal{A}$ be an infinite rewrite sequence of minimal rank. According to Lemmata 3 and 7 this implies that all terms in $\mathcal{A}$ have the same rank and thus $\mathcal{A}$ contains no outer collapsing $\Theta(\mathcal{R})$-steps. For a proof by contradiction suppose that $\mathcal{A}$ lacks outer duplicating or inner collapsing $\Theta(\mathcal{R})$-steps.

First we show that there exists an infinite tail $\mathcal{B}$ of $\mathcal{A}$ with the property that all $\Theta(\mathcal{R})$-steps in $\mathcal{B}$ are either outer or inner non-collapsing. If $\mathcal{A}$ lacks inner collapsing $\Theta(\mathcal{R})$-steps then we can take $\mathcal{B} = \mathcal{A}$. If $\mathcal{A}$ lacks outer duplicating $\Theta(\mathcal{R})$-steps, we reason as follows. Associate with every term s the multiset $\sharp(t) = \{\text{rank}(t') \mid t' \in \text{alien}(t)\}$. As a consequence of Lemmata 7 and 8 we have $\sharp(u) = \sharp(v)$ for every $\Theta(\mathcal{E})$-step $u \sim v$ in $\mathcal{A}$. Let $u \rightarrow v$ be a $\Theta(\mathcal{R})$-step in $\mathcal{A}$. From Lemma 6 we infer that

1. $\sharp(u) \geqslant_{\text{mul}} \sharp(v)$ if $u \rightarrow v$ is outer (and thus non-duplicating),
2. $\sharp(u) >_{\text{mul}} \sharp(v)$ if $u \rightarrow v$ is inner collapsing, and
3. $\sharp(u) \geqslant_{\text{mul}} \sharp(v)$ if $u \rightarrow v$ is inner non-collapsing.

Since $>_{\text{mul}}$ is a well-founded order on multisets (Dershowitz and Manna [3]), the second alternative occurs finitely often. Hence there exists an index $n \geqslant 1$ such that all $\Theta(\mathcal{R})$-steps in the subsequence $t_n \rightarrow_{\Theta(\mathcal{R}/\mathcal{E})} t_{n+1} \rightarrow_{\Theta(\mathcal{R}/\mathcal{E})} \cdots$ of $\mathcal{A}$ are either outer or inner non-collapsing.

If $\mathcal{B}$ contains infinitely many outer $\Theta(\mathcal{R})$-steps then by applying top to every term in $\mathcal{B}$ we obtain an infinite $\mathcal{R}/\mathcal{E}$-rewrite sequence as a consequence of Lemmata 5 and 8, contradicting the terminating of $\mathcal{R}/\mathcal{E}$. Hence there exists an infinite tail $\mathcal{C}$ of $\mathcal{B}$ such that all $\Theta(\mathcal{R})$-steps in $\mathcal{C}$ are inner non-collapsing. With help of Lemma 8 and the pigeon-hole principle, we obtain an infinite $\Theta(\mathcal{R}/\mathcal{E})$-rewrite sequence starting from one of the aliens of $\mathcal{C}$. This contradicts the minimality of $\mathcal{A}$. □

Corollary 10. *Termination is persistent for many-sorted ETRSs $\mathcal{R}/\mathcal{E}$ such that $\mathcal{R}$ is non-collapsing or non-duplicating and $\mathcal{E}$ is non-collapsing and variable-preserving.* □

Variable-preservingness of $\mathcal{E}$ cannot be weakened to non-erasingness. Consider for instance the $\{\alpha, \beta\}$-sorted ETRS $\mathcal{R}/\mathcal{E}$ with $\mathcal{R} = \{a \rightarrow b\}$, $\mathcal{E} = \{f(x, x, y) \approx f(y, x, y)\}$, and sort declarations $f \colon \alpha \times \alpha \times \alpha \rightarrow \beta$ and $a, b \colon \beta$. The ETRS $\mathcal{R}/\mathcal{E}$ is terminating since the only reducible well-typed term is a, but in $\Theta(\mathcal{R}/\mathcal{E})$ we have the following infinite sequence: $f(a, b, a) \rightarrow_{\mathcal{R}} f(b, b, a) \vdash_{\mathcal{E}} f(a, b, a) \rightarrow_{\mathcal{R}} \cdots$. Note that $f(a, b, a)$ is not well-typed.

At present it is unclear whether Corollary 10 holds for collapsing $\mathcal{E}$. Our proof doesn't allow for collapsing $\mathcal{E}$ since Lemmata 7 and 8 no longer hold. Note that $\mathcal{R}/\mathcal{E}$ cannot be terminating if $\mathcal{E}$ contains a collapsing equation $l \approx x$ such that x has more than one occurrence in l.

4 Undecidability of AC-Termination

We start this section by showing the undecidability of termination modulo commutativity for terminating TRSs. To this end we make use of the following well-known result (e.g. [10]).

Lemma 11. *It is undecidable whether a TRS admits an infinite rewrite sequence in which all steps take place at the root position.* □

Theorem 12. *It is undecidable whether a terminating TRS is C-terminating.*

Proof. Let $\mathcal{R}$ be an arbitrary TRS. Define

$$\mathcal{R}' = \{f(l, a) \rightarrow f(a, r) \mid l \rightarrow r \in \mathcal{R}\}$$

with f and a are fresh symbols. Termination of $\mathcal{R}'$ can be shown by the lexicographic path order with any total precedence in which f is maximum and a a minimum. We show that $\mathcal{R}'$ is C-terminating (i.e., $\mathcal{R}'/\mathcal{E}$ with $\mathcal{E} = \{f(x, y) \approx f(y, x)\}$ is terminating) if and only if $\mathcal{R}$ doesn't admit an infinite rewrite sequence in which all steps take place at the root position.

Let $l_1\sigma_1 \rightarrow r_1\sigma_1 = l_2\sigma_2 \rightarrow r_2\sigma_2 = \cdots$ with $l_i \rightarrow r_i \in \mathcal{R}$ for $i \geqslant 1$ be an infinite $\mathcal{R}$-rewrite sequence in which all steps take place at the root position. This sequence can be transformed into the following infinite $\mathcal{R}'/\mathcal{E}$-rewrite sequence: $f(l_1\sigma_1, a) \rightarrow_{\mathcal{R}'} f(a, r_1\sigma_1) \vdash_{\mathcal{E}} f(r_1\sigma_1, a) = f(l_2\sigma_2, a) \rightarrow_{\mathcal{R}'} f(a, r_2\sigma_2) \vdash_{\mathcal{E}} \cdots$.

For the other direction we reason as follows. Since $\mathcal{R}'$ is non-collapsing and $\mathcal{E}$ trivially non-collapsing and variable-preserving, we can apply Corollary 10. To this end we consider the sort declarations $a \colon \alpha$, $f \colon \alpha \times \alpha \rightarrow \beta$, and $g \colon \alpha \times \cdots \times \alpha \rightarrow \alpha$ for all function symbols g of $\mathcal{R}$. In order to show that $\mathcal{R}'$ is C-terminating, it is sufficient to prove termination of all well-typed terms. Terms of sort α are trivially terminating. An infinite $\mathcal{R}'/\mathcal{E}$-rewrite sequence starting from a well-typed term of sort β must have the form $f(l_1\sigma_1, a) \rightarrow_{\mathcal{R}'} f(a, r_1\sigma_1) \vdash_{\mathcal{E}} f(r_1\sigma_1, a) = f(l_2\sigma_2, a) \rightarrow_{\mathcal{R}'} f(a, r_2\sigma_2) \vdash_{\mathcal{E}} \cdots$ with $l_i \rightarrow r_i \in \mathcal{R}$ for $i \geqslant 1$. This gives rise to an infinite rewrite sequence $l_1\sigma_1 \to_\mathcal{R} r_1\sigma_1 = l_2\sigma_2 \to_\mathcal{R} r_2\sigma_2 = \cdots$ in which all steps take place at the root position, contradicting the assumption. Hence $\mathcal{R}'$ is C-terminating.

The desired result follows from the previous lemma. □

Next we show the undecidability of termination modulo associativity for terminating TRSs.

Theorem 13. *It is undecidable whether a terminating TRS is A-terminating.*

Proof. Let $\mathcal{R}$ be an arbitrary TRS. Define

$$\mathcal{R}' = \{f(f(e(l), a), a) \to f(e(r), f(a, a)) \mid l \to r \in \mathcal{R}\}.$$

Termination of $\mathcal{R}'$ is easily shown by the lexicographic path order. We can show that $\mathcal{R}'$ is A-terminating if and only if $\mathcal{R}$ doesn't admit an infinite rewrite sequence in which all steps take place at the root position, similar to the preceding proof. Let $\mathcal{F}$ be the signature of $\mathcal{R}$. For the "if" direction we use sort declarations $a: \beta$, $e: \alpha \to \beta$, $f: \beta \times \beta \to \beta$, and $g: \alpha \times \cdots \times \alpha \to \alpha$ for all function symbols $g \in \mathcal{F}$. Every well-typed term t of sort β can be (uniquely) written as $C[t_1, \ldots, t_n]$ such that C contains only f and a symbols and for every $1 \leqslant i \leqslant n$ we have either $t_i \in \mathcal{T}(\mathcal{F}, \mathcal{V})$ or $t_i = e(t_i')$ with $t_i' \in \mathcal{T}(\mathcal{F}, \mathcal{V})$. Let us denote the sequence $(t_1, \ldots, t_n)$ by $\phi(t)$. If $t \to_{\mathcal{R}'} t'$ then there exist an i, a rewrite rule $l \to r \in \mathcal{R}$, and a substitution σ such that $t_i = e(l\sigma)$ and $\phi(t') = (t_1, \ldots, t_i', \ldots t_n)$ with $t_i' = e(r\sigma)$. If $t \sim_\mathcal{E} t'$ then $\phi(t) = \phi(t')$. Using the pigeon-hole principle it follows that an infinite $\mathcal{R}'/\mathcal{E}$-rewrite sequence gives rise to infinite $\mathcal{R}$-rewrite sequence in which all steps take place at the root position, contradicting the assumption. □

Note that taking $\mathcal{R}' = \{f(f(l, a), a) \to f(r, f(a, a)) \mid l \to r \in \mathcal{R}\}$ in the above proof precludes the (good) use of type introduction as there can be only one sort.

Theorem 14. *It is undecidable whether a terminating TRS is AC-terminating.*

Proof sketch. Replace $\mathcal{R}'$ in the previous proof by $\mathcal{R}' = \{f(f(e(l), a), b) \to f(a, f(b, e(r))) \mid l \to r \in \mathcal{R}\}$ and use similar arguments. □

Since it is not difficult to show that $\mathcal{R}'$ in the above proof is both A and C-terminating, we can strengthen Theorem 14 to the undecidability of AC-termination for A and C-terminating TRSs. (Note that identifying the constants a and b in $\mathcal{R}'$ would result in a TRS that is not C-terminating.)

5 Persistence of Acyclicity and Non-Loopingness

An ETRS $\mathcal{R}/\mathcal{E}$ is *cyclic* if it admits a sequence of the form $t \to_{\mathcal{R}/\mathcal{E}}^+ t$. We say that $\mathcal{R}/\mathcal{E}$ is *looping* if there exist a term t, context C, and substitution σ such that $t \to_{\mathcal{R}/\mathcal{E}}^+ C[t\sigma]$. Terminating ETRSs are non-looping and non-looping

ETRSs are acyclic, but the reverse statements do not hold. A recent study of non-loopingness for TRSs is performed in Zantema and Geser [16].

By a straightforward modification of the proof of Lemma 9 we obtain the following result.

Theorem 15. *Acyclicity is persistent for many-sorted ETRSs $\mathcal{R}/\mathcal{E}$ such that $\mathcal{R}$ is non-collapsing or non-duplicating and $\mathcal{E}$ is non-collapsing and variable-preserving.* □

The proof of the analogous result for non-loopingness is quite a bit more involved. The reason is that because the involved substitution may substitute a term of sort β for a variable of sort α we don't obtain a contradiction by considering a loop of minimal rank.

Definition 16. A substitution σ is called *consistent* if $\text{sort}(x) = \text{sort}(x\sigma)$ for all $x \in \mathcal{V}$.

We show that every looping $\Theta(\mathcal{R}/\mathcal{E})$ admits a loop with consistent substitution. Most of the work is done in the following lemma.

Lemma 17. *For every substitution σ and finite set of variables V with $\text{dom}(\sigma) \subseteq V$ there exist a consistent substitution σ' and a variable substitution τ such that $\sigma\tau = \tau\sigma'$ $[V]$.*

Proof. The desired substitutions σ' and τ are computed by the following algorithm:

$$W := V;$$
$$\sigma' := \sigma;$$
$$\tau := \varepsilon;$$
$$\textbf{while } W \neq \varnothing \textbf{ do}$$
$$\quad \textbf{if } \exists x \in W \text{ with } x\sigma' \notin W \textbf{ then}$$
$$\quad\quad \tau' := \{x \mapsto x'\} \text{ with } x' \text{ a fresh variable of sort } \text{sort}(x\sigma');$$
$$\quad\quad \sigma' := \sigma'\tau'\lceil_{\text{dom}(\sigma')\setminus\{x\}} \cup \{x' \mapsto x\sigma'\tau'\}$$
$$\quad \textbf{else}$$
$$\quad\quad \tau' := \{x \mapsto \xi \mid x \in W\} \text{ with } \xi \text{ a fresh variable (of arbitrary sort)};$$
$$\quad\quad \sigma' := \sigma'\tau'\lceil_{\text{dom}(\sigma')\setminus W}$$
$$\quad \textbf{fi};$$
$$\quad \tau := \tau\tau';$$
$$\quad W := W \setminus \text{dom}(\tau')$$
$$\textbf{od}$$

It is not difficult to prove that the statements

1. τ is a variable substitution, i.e., a mapping from $\mathcal{V}$ to $\mathcal{V}$,
2. $\text{var}(\tau) \cap W = \varnothing$,
3. $\sigma\tau = \tau\sigma'$ $[V]$,
4. $\text{ran}(\sigma'\lceil_{\mathcal{V}\setminus W}) \cap W = \varnothing$, and
5. $\sigma'\lceil_{\mathcal{V}\setminus W}$ is consistent

are invariants of the while-loop which hold after the first three assignments. Here $\mathsf{var}(\tau)$ denotes the union of $\mathsf{dom}(\tau)$ and $\bigcup_{x \in \mathsf{dom}(\tau)} \mathsf{var}(x\tau)$. (Statements 2 and 4 are needed to show 3 and 5.) Termination of the while-loop is obvious since in each iteration at least one element of W is removed and initially $W = V$ is finite by assumption. Upon termination we have $W = \varnothing$ and thus $\sigma' {\restriction}_V = \sigma'$ is consistent. $\qquad\square$

Lemma 18. *Let $\mathcal{R}/\mathcal{E}$ be an S-sorted ETRS. If $\Theta(\mathcal{R}/\mathcal{E})$ is looping then there exists a loop $t \to^+ C[t\sigma]$ with consistent σ.*

Proof. Let $t \to^+ C[t\sigma]$ be a loop in $\Theta(\mathcal{R}/\mathcal{E})$. Without loss of generality we assume that $\mathsf{dom}(\sigma) \subseteq \mathsf{var}(t)$. Let $V = \mathsf{var}(t)$. According to the previous lemma there exist a consistent substitution σ' and a variable substitution τ such that $\sigma\tau = \tau\sigma'$ $[V]$. Let $t' = t\tau$ and $C' = C\tau$. Since (equational) rewriting is closed under substitutions we obtain $t' = t\tau \to^+ C\tau[t\sigma\tau] = C'[t\tau\sigma'] = C'[t'\sigma']$, which shows that $\Theta(\mathcal{R}/\mathcal{E})$ admits a loop with consistent substitution. $\qquad\square$

Lemma 19. *Let $\mathcal{R}/\mathcal{E}$ be a non-looping S-sorted ETRS with $\mathcal{E}$ variable-preserving and non-collapsing. If $\Theta(\mathcal{R}/\mathcal{E})$ is looping then there exists a loop which contains an outer duplicating and an inner collapsing $\Theta(\mathcal{R})$-step.*

Proof. Let $\mathcal{A} \colon t \to^+_{\Theta(\mathcal{R}/\mathcal{E})} C[t\sigma]$ be a loop with consistent σ of minimal rank, the existence of which is guaranteed by Lemma 18. (The rank of $\mathcal{A}$ may be greater that the minimal rank of a loop because the construction in the proof of Lemma 17 may increase the rank by one.) Because σ is consistent, $\mathsf{rank}(t) \leqslant \mathsf{rank}(t\sigma)$ and thus $\mathsf{rank}(t) \leqslant \mathsf{rank}(C[t\sigma])$. From Lemmata 3 and 7 we obtain $\mathsf{rank}(t) \geqslant \mathsf{rank}(C[t\sigma])$ and therefore $\mathsf{rank}(t) = \mathsf{rank}(C[t\sigma])$. Hence $\mathcal{A}$ contains no outer collapsing $\Theta(\mathcal{R})$-steps.

We show that $\mathcal{A}$ contains an inner collapsing $\Theta(\mathcal{R})$-step. Because $\mathsf{rank}(t) = \mathsf{rank}(C[t\sigma])$ the displayed occurrence of $t\sigma$ is not a subterm of an alien subterm of $C[t\sigma]$ and thus $\mathsf{top}(C[t\sigma]) = C'[\square_{\alpha_1}, \ldots, \mathsf{top}(t\sigma), \ldots, \square_{\alpha_m}]$ for some context C' with $m > 0$. Let $\sigma' = \mathsf{top} \circ \sigma$. An easy induction on the structure of t yields $\mathsf{top}(t\sigma) = \mathsf{top}(t)\sigma'$. Here we make use of the consistency of σ. Now suppose for a proof by contradiction that $\mathcal{A}$ contains no inner collapsing $\Theta(\mathcal{R})$-steps. According to Lemma 5 we have $\mathsf{top}(s) \to_\mathcal{R} \mathsf{top}(s')$ for every outer step $s \to_{\Theta(\mathcal{R})} s'$ in $\mathcal{A}$ and $\mathsf{top}(s) = \mathsf{top}(s')$ for every inner step $s \to_{\Theta(\mathcal{R})} s'$ in $\mathcal{A}$. Furthermore, $\mathsf{top}(s) \sim_\mathcal{E} \mathsf{top}(s')$ for every $s \sim_{\Theta(\mathcal{E})} s'$ in $\mathcal{A}$ by Lemma 8. Hence if $\mathcal{A}$ contains an outer $\Theta(\mathcal{R})$-step then

$$\mathsf{top}(t) \to^+_{\mathcal{R}/\mathcal{E}} \mathsf{top}(C[t\sigma]) = C'[\square_{\alpha_1}, \ldots, \mathsf{top}(t)\sigma', \ldots, \square_{\alpha_m}],$$

contradicting the non-loopingness of $\mathcal{R}/\mathcal{E}$. Consequently, $\mathsf{top}(t) \sim_\mathcal{E} \mathsf{top}(C[t\sigma])$. Because $\mathcal{E}$ is variable preserving and non-collapsing this implies that $\mathsf{top}(t)$ and $\mathsf{top}(C[t\sigma])$ have the same number of holes and thus the context C must be well-typed. Let $\mathsf{alien}(t) = \{t_1, \ldots, t_n\}$. Using the consistency of σ we obtain

$\mathsf{alien}(C[t\sigma])) = \mathsf{alien}(t\sigma) = \{t_1\sigma, \ldots, t_n\sigma\}$. With help of Lemma 8 we obtain a permutation π such that for all $1 \leqslant i \leqslant n$ either $t_i \sim_{\Theta(\mathcal{E})} t_{\pi(i)}\sigma$ or $t_i \to^+_{\Theta(\mathcal{R}/\mathcal{E})} t_{\pi(i)}\sigma$. Since there are inner $\Theta(\mathcal{R})$-steps in $\mathcal{A}$, the latter alternative must occur for some j. Let $k > 0$ satisfy $\pi^k(j) = j$. We obtain $t_j \to^+_{\Theta(\mathcal{R}/\mathcal{E})} t_j\sigma^k$ where σ^k denotes the k-fold composition of σ. Since $\mathsf{rank}(t_j) < \mathsf{rank}(t)$ and σ^k inherits consistency from σ, this contradicts the minimality of $\mathcal{A}$. We conclude that $\mathcal{A}$ contains an inner collapsing $\Theta(\mathcal{R})$-step.

It remains to show that $\mathcal{A}$ contains an outer duplicating $\Theta(\mathcal{R})$-step. Suppose to the contrary that there are no outer duplicating $\Theta(\mathcal{R})$-steps in $\mathcal{A}$. Consider the mapping $\natural$ defined in the proof of Lemma 9. There we noticed that $\natural(u) = \natural(v)$ for every $\Theta(\mathcal{E})$-step $u \sim v$ in $\mathcal{A}$ and if $u \to v$ is a $\Theta(\mathcal{R})$-step in $\mathcal{A}$ then

1. $\natural(u) \geqslant_{\mathrm{mul}} \natural(v)$ if $u \to v$ is outer,
2. $\natural(u) >_{\mathrm{mul}} \natural(v)$ if $u \to v$ is inner collapsing, and
3. $\natural(u) \geqslant_{\mathrm{mul}} \natural(v)$ if $u \to v$ is inner non-collapsing.

Since we know that the second alternative occurs at least once, we obtain $\natural(t) >_{\mathrm{mul}} \natural(C[t\sigma])$. However, using the consistency of σ, one easily verifies that $\natural(C[t\sigma]) \geqslant_{\mathrm{mul}} \natural(t\sigma) \geqslant_{\mathrm{mul}} \natural(t)$, yielding the desired contradiction. We conclude that $\mathcal{A}$ contains an outer duplicating $\Theta(\mathcal{R})$-step. $\qquad\Box$

Corollary 20. *Non-loopingness is persistent for many-sorted ETRSs $\mathcal{R}/\mathcal{E}$ such that $\mathcal{R}$ is non-collapsing or non-duplicating and $\mathcal{E}$ is non-collapsing and variable-preserving.* $\qquad\Box$

We illustrate the usefulness of the above theorem by giving a simple proof of non-loopingness for the following TRS from [6], depending on arbitrary instance P of Post's Correspondence Problem over the alphabet Γ:

$$\mathcal{R} = \begin{cases} h(F(c, c, a(z))) \to g(F(a(z), a(z), a(z))) & \text{for all } a \in \Gamma \\ F(\alpha(x), \beta(y), z) \to F(x, y, z) & \text{for all } (\alpha, \beta) \in P \\ h(g(x)) \to g(h(x)) \\ f(g(x)) \to f(h(h(x))) \end{cases}$$

In [6] this TRS is used to show that termination is an undecidable property of non-looping TRSs. Note that $\mathcal{R}$ is non-collapsing. Hence we can use type introduction to prove its non-loopingness. Consider $\mathcal{S} = \{\alpha, \beta, \gamma\}$ with sort declarations $F: \alpha \times \alpha \times \alpha \to \beta$, $c: \alpha$, $a: \alpha \to \alpha$ for all $a \in \Gamma$, $g, h: \beta \to \beta$, and $f: \beta \to \gamma$. Terms of sort α are in normal form, hence trivially non-looping. For terms of sort β we note that the rule $f(g(x)) \to f(h(h(x)))$ can never be applied, but since $\mathcal{R}$ minus this rule is terminating (by lexicographic path order) it follows that those terms are non-looping. So if $\mathcal{R}$ admits a loop $t \to^+ C[t\sigma]$ then $\mathsf{sort}(t) = \gamma$ and the rule $f(g(x)) \to f(h(h(x)))$ must be used. From $\mathsf{sort}(t) = \gamma$ we immediately infer that the root symbol of t is f and that C is empty. Hence $t \to^+ C[t\sigma]$ must be of the form

$$\begin{aligned} t = f(C_1[F(s_1, s_2, s_3)]) &\to^* f(g(C_2[F(t_1, t_2, t_3)])) \\ &\to f(h(h(C_2[F(t_1, t_2, t_3)]))) \\ &\to^* f(C_1[F(s_1\sigma, s_2\sigma, s_3\sigma)]) = t\sigma \end{aligned}$$

with C_1 and C_2 only containing g and h symbols. From the form of the rewrite rules of $\mathcal{R}$ we get the contradictory $|C_1| \leqslant |g(C_2)| = |C_2| + 1$ and $|C_2| + 2 = |h(h(C_2))| \leqslant |C_1|$. Hence also all terms of sort γ are non-looping. In [6] non-loopingness of $\mathcal{R}$ is shown by a more complicated ad-hoc argument.

We conclude by remarking that the proofs of non-loopingness of several of the examples in [16] can be simplified by an appeal to Corollary 20.

References

1. N. Dershowitz, *Termination of Rewriting*, Journal of Symbolic Computation **3** (1987) 69–116.
2. N. Dershowitz and J.-P. Jouannaud, *Rewrite Systems*, in: Handbook of Theoretical Computer Science, Vol. B (ed. J. van Leeuwen), North-Holland (1990) 243–320
3. N. Dershowitz and Z. Manna, *Proving Termination with Multiset Orderings*, Communications of the ACM **22** (1979) 465–476.
4. M.C.F. Ferreira, *Dummy Elimination in Equational Rewriting*, Proc. 7th RTA, New Brunswick, LNCS **1103** (1996) 78–92.
5. M.C.F. Ferreira and H. Zantema, *Dummy Elimination: Making Termination Easier*, Proc. 10th FCT, Dresden, LNCS **965** (1995) 243–252.
6. A. Geser, A. Middeldorp, E. Ohlebusch, and H. Zantema, *Relative Undecidability in Term Rewriting*, Proc. CSL, Utrecht, LNCS (1996). To appear. Available at http://www.score.is.tsukuba.ac.jp/~ami/papers/csl96.dvi.
7. J.-P. Jouannaud and M. Muñoz, *Termination of a Set of Rules Modulo a Set of Equations*, Proc. 7th CADE, Napa, LNCS **170** (1984) 175–193.
8. D. Kapur and G. Sivakumar, *A Total, Ground Path Ordering for Proving Termination of AC-Rewrite Systems*, Proc. 8th RTA, Sitges, LNCS (1997). To appear.
9. J.W. Klop, *Term Rewriting Systems*, in: Handbook of Logic in Computer Science, Vol. 2 (eds. S. Abramsky, D. Gabbay and T. Maibaum), Oxford University Press (1992) 1–116.
10. A. Middeldorp and B. Gramlich, *Simple Termination is Difficult*, Applicable Algebra in Engineering, Communication and Computing **6** (1995) 115–128.
11. E. Ohlebusch, *A Simple Proof of Sufficient Conditions for the Termination of the Disjoint Union of Term Rewriting Systems*, Bulletin of the EATCS **49** (1993) 178–183.
12. J. van de Pol, *Modularity in Many-Sorted Term Rewriting Systems*, Master's thesis, report INF/SCR-92-37, Utrecht University (1992).
13. A. Rubio and R. Nieuwenhuis, *A Total AC-Compatible Ordering Based on RPO*, Theoretical Computer Science **142** (1995) 209–227.
14. J. Steinbach, *Termination of Rewriting: Extensions, Comparison and Automatic Generation of Simplification Orderings*, Ph.D. thesis, Universität Kaiserslautern (1994).
15. H. Zantema, *Termination of Term Rewriting: Interpretation and Type Elimination*, Journal of Symbolic Computation **17** (1994) 23–50.
16. H. Zantema and A. Geser, *Non-Looping Rewriting*, report UU-CS-1996-03, Utrecht University, Department of Computer Science (1996). Available at ftp://ftp.cs.ruu.nl/pub/RUU/CS/techreps/CS-1996/1996-03.ps.gz.

Capturing Bisimulation-Invariant Ptime

Martin Otto

RWTH Aachen, Germany
otto@informatik.rwth-aachen.de

Abstract. Consider the class of all those properties of worlds in finite Kripke structures, that are recognisable in polynomial time *and* closed under bisimulation equivalence. It is shown that the class of these *bisimulation-invariant* PTIME *queries* has a natural logical characterisation. It is captured, in the sense of descriptive complexity theory, by the straightforward extension of propositional μ-calculus to arbitrary finite dimension. Bisimulation-invariant PTIME thus proves to be one of the very rare cases in which a logical characterisation is known in a setting of unordered structures.

0   Introduction

An outstanding issue in the study of the relation between computational complexity and logical definability concerns the search for exact matches. Paradigmatic results in this area are, for instance, Fagin's Theorem (the NP-recognisable properties of finite structures are exactly those that can be formalised in existential second-order logic), the Büchi-Elgot-Trakhtenbrot Theorem (the automaton-recognisable properties of finite words are those that are definable in monadic second-order logic), or the Immerman-Vardi Theorem (the PTIME properties of finite linearly ordered structures are exactly those that are definable in least fixed-point logic). It is a characteristic feature in these examples that they either concern complexity classes beyond PTIME or else concern classes of linearly ordered structures. Indeed no logical characterisation has been found for any of the standard complexity classes below NP, that would cover arbitrary rather than linearly ordered structures. In particular, the question whether PTIME itself – regarded as the class of all those properties of finite structures that can be recognised by PTIME algorithms – admits a logical characterisation, is a central open problem in finite model theory. This fundamental issue was raised by Chandra and Harel [6] and more rigorously formalised by Gurevich [9], cf. [7].

The present investigation deals with, and offers a positive solution for, a semantically defined fragment of PTIME concerning finite Kripke structures. Kripke structures not only form the natural models for *modal logics* but also play an important role as formalisations of *transition systems*. Under both aspects, *bisimulation equivalence* is the natural notion of indistinguishability. It is therefore natural in this framework to consider the class of those PTIME properties of finite Kripke structures, that are preserved under bisimulation. It turns out that this class possesses an exact logical match in higher-dimensional μ-calculus, which is here introduced as the obvious extension of ordinary propositional μ-calculus to arbitrary arities.

1 Preliminaries, basic definitions, and the theorem

We deal with *Kripke structures* that form the appropriate models for propositional modal logic ML, its infinitary variant ML_∞, and the propositional μ-calculus L_μ. We fix a finite set of basic propositions (propositional constants) $\overline{P} = P_1, \ldots, P_l$. A Kripke structure for $\overline{P}$ is a structure $\mathfrak{A} = (A, E^{\mathfrak{A}}, P_1^{\mathfrak{A}}, \ldots, P_l^{\mathfrak{A}})$:
- A is the universe or *set of worlds* of $\mathfrak{A}$,
- $E^{\mathfrak{A}} \subseteq A^2$ is the binary relation of *accessibility between worlds*,
- each $P_i^{\mathfrak{A}} \subseteq A$ interprets the set of worlds at which P_i *holds true*.

For the standard semantics of modal logics one usually deals with Kripke structures $(\mathfrak{A}, a)$ in which one element is designated: $a \in A$ is a *distinguished world*.

1.1 Bisimulation equivalence, modal logic and the μ-calculus

Bisimulation A fundamental notion of equivalence between Kripke structures with distinguished worlds is *bisimulation equivalence*, which we denote by $\sim$. This equivalence has a natural motivation as a notion of behavioural indistinguishability, if Kripke structures are taken as descriptions of transition systems [13, 16]. There is also a very elegant Ehrenfeucht-Fraïssé style characterisation of bisimulation equivalence due to van Benthem [4, 5]. We here work with the following inductive characterisation of bisimulation *inequivalence* $\not\sim$ as a least fixed point. Let $\mathfrak{A} = (A, E, P_1, \ldots, P_l)$ and $\mathfrak{A}' = (A', E', P_1', \ldots, P_l')$ be two Kripke structures:

$(\mathfrak{A}, a) \not\sim_0 (\mathfrak{A}', a')$ if $\bigvee_{i=1}^{l} \neg(a \in P_i \leftrightarrow a' \in P_i')$

$(\mathfrak{A}, a) \not\sim_{\alpha+1} (\mathfrak{A}', a')$ if $(\mathfrak{A}, a) \not\sim_\alpha (\mathfrak{A}', a')$

or $\exists b \in A \left[(a, b) \in E \wedge \forall b' \in A' ((a', b') \in E' \rightarrow (\mathfrak{A}, b) \not\sim_\alpha (\mathfrak{A}', b')) \right]$ (1)

or $\exists b' \in A' \left[(a', b') \in E' \wedge \forall b \in A ((a, b) \in E \rightarrow (\mathfrak{A}, b) \not\sim_\alpha (\mathfrak{A}', b')) \right]$

$(\mathfrak{A}, a) \not\sim_\lambda (\mathfrak{A}', a')$ if $\exists \alpha < \lambda \ (\mathfrak{A}, a) \not\sim_\alpha (\mathfrak{A}', a')$ for limits λ.

$(\mathfrak{A}, a) \not\sim (\mathfrak{A}', a')$ if $(\mathfrak{A}, a) \not\sim_\alpha (\mathfrak{A}', a')$ for some α. For cardinality reasons, the sequence of the $\not\sim_\alpha$ eventually becomes stationary in restriction to any given pair of structures $\mathfrak{A}$ and $\mathfrak{A}'$. Thus $(\mathfrak{A}, a) \not\sim (\mathfrak{A}', a')$ if $(\mathfrak{A}, a) \not\sim_\alpha (\mathfrak{A}', a')$ for the least α such that, for all $a \in A$ and $a' \in A'$, $(\mathfrak{A}, a) \not\sim_{\alpha+1} (\mathfrak{A}', a')$ implies $(\mathfrak{A}, a) \not\sim_\alpha (\mathfrak{A}', a')$. The least such α is bounded by $|A| \cdot |A'| + 1$ if $\mathfrak{A}$ and $\mathfrak{A}'$ are finite, whence bisimulation equivalence over finite Kripke structures is in PTIME (in fact PTIME-complete [1]).

There is an Ehrenfeucht-Fraïssé type theorem associated with bisimulation equivalence which involves the infinitary variant ML_∞ of ordinary propositional modal logic ML, see Theorem 1.1.

Modal logic Recall that ML for $\overline{P} = P_1, \ldots, P_l$ has atomic formulae $P_i x$, is closed under boolean operations, and under the modal constructors $\Diamond$ and $\Box$, whose semantics is given by $(\mathfrak{A}, a) \models \Diamond\varphi$ if $\exists b \in A((a, b) \in E^{\mathfrak{A}} \wedge (\mathfrak{A}, b) \models \varphi)$ and, dually, $(\mathfrak{A}, a) \models \Box\varphi$ if $\forall b \in A((a, b) \in E^{\mathfrak{A}} \rightarrow (\mathfrak{A}, b) \models \varphi)$.

ML_∞ further enriches the syntax and semantics of ML through closure under conjunctions and disjunctions over arbitrary sets of formulae.

Clearly $\Diamond$ and $\Box$ may be pictured as existential and universal first-order quantifications along accessibility edges. It is therefore straightforward that $\text{ML} \subsetneq L^2_{\omega\omega}$ and $\text{ML}_\infty \subsetneq L^2_{\infty\omega}$, where $L^2_{\omega\omega}$ is first-order logic with only two variable symbols, $L^2_{\infty\omega}$ its infinitary variant.

Theorem 1.1 (Barwise, Moss, van Benthem [2, 3])

$(\mathfrak{A}, a) \sim (\mathfrak{A}', a')$ *iff* $(\mathfrak{A}, a)$ *and* $(\mathfrak{A}', a')$ *satisfy exactly the same formulae of* ML_∞. *For finite (in fact even for finitely branching) Kripke structures:* $(\mathfrak{A}, a) \sim (\mathfrak{A}', a')$ *iff* $(\mathfrak{A}, a)$ *and* $(\mathfrak{A}', a')$ *satisfy exactly the same formulae of* ML.

Propositional μ-calculus Propositional μ-calculus $\mathbf{L}_\mu$, as introduced in [12], augments the syntax and semantics of ML by constructors for least fixed points. To this end one firstly admits propositional variables $X, Y, Z, \ldots$ in formulae, with corresponding new atomic assertions Xx etc. Free propositional variables get interpreted by subsets of the universe like the propositional constants.

If φ is a formula of $\mathbf{L}_\mu$ that is positive in X (meaning that X does not occur free in the scope of an odd number of negations), then $\psi = \mu_X\varphi$ is also a formula of $\mathbf{L}_\mu$ (in which X no longer occurs free). The semantics can without loss of generality be explained in the case that no propositional variable apart from X is free in φ. Then φ induces a monotone operator on subsets of A according to $F^{\mathfrak{A}}_\varphi : P \subseteq A \longmapsto \{a \in A | (\mathfrak{A}, a) \models \varphi[P/X]\}$, where $\varphi[P/X]$ denotes φ under the interpretation that assigns P to X. Being a monotone operator, $F^{\mathfrak{A}}_\varphi$ possesses a least fixed point $\text{LFP}(F^{\mathfrak{A}}_\varphi)$. Now $(\mathfrak{A}, a) \models \mu_X\varphi$ if $a \in \text{LFP}(F^{\mathfrak{A}}_\varphi)$.

It is not hard to see that $\mathbf{L}_\mu$ is preserved under bisimulation: if $(\mathfrak{A}, a) \sim (\mathfrak{A}', a')$ and $\varphi \in \mathbf{L}_\mu$, then $(\mathfrak{A}, a) \models \varphi$ iff $(\mathfrak{A}', a') \models \varphi$. Over finite Kripke structures, in particular, $\mathbf{L}_\mu \subseteq \text{ML}_\infty$ (the inclusion is strict as $\mathbf{L}_\mu \subseteq \text{PTIME}$). More background on $\mathbf{L}_\mu$, its variants, and its role as a process logic can be found in [8]. It is customary to introduce $\mathbf{L}_\mu$ in a multi-modal framework, i.e. with several accessibility relations and corresponding modalities rather than just one. The results presented here have straightforward extensions to that scenario.

Remark We write all modal formulae φ as formulae $\varphi(x)$ in a single formal element variable x which is ultimately interpreted by the distinguished world in a Kripke structure. The semantics of φ may thus be associated with the monadic predicate defined by $\varphi(x)$ over the $\mathfrak{A}$: $\varphi[\mathfrak{A}] = \{a \in A \mid \mathfrak{A} \models \varphi[a]\}$, whereby formulae of ML or ML_∞ define *monadic global relations* over Kripke structures.

1.2 k-dimensional μ-calculus

We introduce extensions of L_μ which roughly correspond to the expressive power of L_μ over the k-th Cartesian power of the given Kripke structures. The elements of this power are k-tuples of worlds, $\bar{a} \in A^k$, and there are k different accessibility relations E_j corresponding to E-accessibility in the j-th component:

$$(\bar{a}, \bar{a}') \in E_j^{\mathfrak{A}} \quad \text{iff} \quad a_i = a_i' \text{ for } i \neq j, \text{ and } (a_j, a_j') \in E^{\mathfrak{A}}.$$

We write formulae of L_μ^k in a k-tuple $\bar{x} = (x_1, \ldots, x_k)$ of element variables. The syntax is governed by the following clauses:

atomic formulae: for $1 \leqslant i \leqslant l$ and $1 \leqslant j \leqslant k$, $P_i x_j$ is an atomic formula of L_μ^k. For propositional variables X, Y, etc. L_μ^k has atomic formulae $X\bar{x}$, $\ldots$
booleans: L_μ^k is closed under (finitary) boolean operations.
modalities: L_μ^k is closed under modalities $\Diamond_j$ and $\Box_j$ for $1 \leqslant j \leqslant k$.
variable substitutions: L_μ^k is closed w.r.t variable substitutions $\sigma:\{1, \ldots, k\} \rightarrow \{1, \ldots, k\}$, replacing $(x_1, \ldots, x_k)$ by $(x_{\sigma(1)}, \ldots, x_{\sigma(k)})$.
least fixed points: L_μ^k is closed under applications of the μ_X-operator to φ whenever X occurs positively. We stress the fact that X plays the role of a k-ary relation variable by speaking of a μ^k-operation.

Semantically we associate with a formula $\varphi \in L_\mu^k$ a k-ary global relation according to $\varphi[\mathfrak{A}] = \{\bar{a} = (a_1, \ldots, a_k) \in A^k \mid \mathfrak{A} \models \varphi[\bar{a}]\}$. The atomic and boolean cases are obvious. $\Diamond_j$ and $\Box_j$ are treated as modal operators for the accessibility relations $E_j^{\mathfrak{A}}$ over A^k. A substitution σ operates according to $\varphi^\sigma[\mathfrak{A}] = \{(a_1, \ldots, a_k) \in A^k \mid \mathfrak{A} \models \varphi[a_{\sigma(1)}, \ldots, a_{\sigma(k)}]\}$. The fixed-point operator μ_X corresponds to least fixed points of the induced monotone operation that sends $P \subseteq A^k$ to $\{\bar{a} \in A^k \mid \mathfrak{A} \models \varphi[P/X, \bar{a}]\}$. For ease of notation we shall also just write $\varphi(x_{\sigma(1)}, \ldots, x_{\sigma(k)})$ for φ^σ, or indicate simple substitutions as in $\varphi(x_k/x_i)$ for φ^σ if σ maps i to k and fixes all other indices.

Remark For a substitution that reduces the number of variables that occur free in φ, we may either think of the result as still defining a k-ary global relation (with trivial factors in those components that are not free any more), or as defining a global relation of correspondingly reduced arity. For instance a formula $\varphi(x_1, x_1) \in L_\mu^2$ could alternatively be associated with the monadic global relation given by $Q_1^{\mathfrak{A}} = \{a \in A \mid \mathfrak{A} \models \varphi[a, a]\}$, or with the binary global relation according to $Q_2^{\mathfrak{A}} = \{(a_1, a_2) \in A^2 \mid \mathfrak{A} \models \varphi[a_1, a_1]\}$. Clearly $Q_2^{\mathfrak{A}} = Q_1^{\mathfrak{A}} \times A$ and we shall often increase the arity of relations to some uniform common value by just this type of *padding* that occurs in the translation from Q_1 to Q_2.

Standard semantics Towards a standard *monadic* semantics of formulae of L_μ^k over Kripke structures $(\mathfrak{A}, a)$ with the single distinguished world a, we ultimately pass to substitution instances $\varphi(x_1, x_1, \ldots, x_1)$ (i.e. σ identically 1).

With these conventions it is clear that $L_\mu \equiv L_\mu^1$ and $L_\mu^k \subseteq L_\mu^{k+r}$ for definability of monadic global relations (and, up to the necessary padding, also for definability of k-ary global relations). It can also be shown that there are monadic global relations that are definable in L_μ^2 but not in L_μ. The following is also not hard to show.

Lemma 1.2 *Let* $\varphi(\overline{x}) \in L_\mu^k$, $\mathfrak{A}$ *and* $\mathfrak{A}'$ *Kripke structures. If* $\overline{a} \in A^k$ *and* $\overline{a}' \in A'^k$ *are such that for* $j = 1, \ldots, k$ $(\mathfrak{A}, a_j) \sim (\mathfrak{A}', a_j')$, *then* $\overline{a} \in \varphi[\mathfrak{A}]$ *if and only if* $\overline{a}' \in \varphi[\mathfrak{A}']$. *In particular any monadic global relation that is definable in* L_μ^k *is bisimulation-invariant.*

A crucial example of the expressive power of the L_μ^k is the following:

Lemma 1.3 *Consider bisimulation equivalence* $\sim$ *as a binary global relation on individual Kripke structures, according to* $(a, a') \in \sim^{\mathfrak{A}}$ *if* $(\mathfrak{A}, a) \sim (\mathfrak{A}, a')$. *Then* $\sim$ *is definable in* L_μ^k *for all* $k \geqslant 2$.

In fact, a comparison with the inductive generation of the $\not\sim_\alpha$ in (1) shows that $\not\sim$ is defined as a global relation in variables x_1 and x_2 in this sense by the formula $\varphi_{\not\sim} = \mu_X \left(\bigvee_{i=1}^l \neg(P_i x_1 \leftrightarrow P_i x_2) \vee \Diamond_1 \Box_2 X \overline{x} \vee \Diamond_2 \Box_1 X \overline{x} \right)$.

1.3 The capturing result

We now consider monadic queries or global relations on finite Kripke structures. Formally such a query Q is a mapping that associates with each finite Kripke structure $\mathfrak{A}$ a subset $Q^{\mathfrak{A}} \subseteq A$ such that any isomorphism $f: \mathfrak{A} \simeq \mathfrak{A}'$ is also an isomorphism of the expansions $(\mathfrak{A}, Q^{\mathfrak{A}})$ and $(\mathfrak{A}', Q^{\mathfrak{A}'})$. We are here interested in the stronger invariance condition imposed by bisimulation equivalence.

Definition 1.4 A monadic query Q on finite Kripke structures is
(i) *bisimulation-invariant* ($\sim$-invariant) if $(\mathfrak{A}, a) \sim (\mathfrak{A}', a')$ implies that $a \in Q^{\mathfrak{A}}$ iff $a' \in Q^{\mathfrak{A}'}$.
(ii) *polynomial time computable* (in PTIME) if there is a PTIME algorithm for deciding on input $(\mathfrak{A}, a)$ whether $a \in Q^{\mathfrak{A}}$ or not.
We denote by PTIME $\cap$ ML$_\infty$ the class of all those monadic PTIME queries on finite Kripke structures that also are $\sim$-invariant.

The notation PTIME $\cap$ ML$_\infty$ is motivated by the observation that (by a standard argument using the boundedness of the class of finite Kripke structures with designated worlds and ML$_\infty$-Scott sentences for these) a monadic query on finite Kripke structures is $\sim$-invariant if and only if it is definable in ML$_\infty$.

A monadic query Q on finite Kripke structures is definable in L_μ^k if there is a formula $\varphi(x_1, \ldots, x_1) \in L_\mu^k$ such that for all finite $\mathfrak{A}$: $\varphi[\mathfrak{A}] = Q^{\mathfrak{A}}$. As stated above, any L_μ^k-definable query is bisimulation invariant; clearly it is also in PTIME. We now state the main theorem as follows.

Theorem 1.5 *A monadic query* Q *on finite Kripke structures is in* PTIME *and bisimulation invariant if and only if it is definable in* L_μ^k *for some* k. *In formulae:*

$$\text{PTIME} \cap \text{ML}_\infty \equiv \bigcup_k L_\mu^k.$$

Corollary 1.6 *In particular* PTIME ∩ ML$_\infty$ *admits a recursive presentation in the sense of descriptive complexity: there is a language with recursive syntax and with* PTIME *semantics which is semantically complete for the class of all* PTIME *bisimulation-invariant queries over finite Kripke structures.*

For background on the underlying notion of *capturing complexity classes* compare [7]. The rest of this sketch is devoted to the proof of the non-trivial inclusion PTIME ∩ ML$_\infty$ ⊆ $\bigcup_k L_\mu^k$.

2 Towards the proof of the main theorem

2.1 Canonical structures and a normal form

Let for a Kripke structure $\mathfrak{A} = (A, E, P_1, \ldots, P_l)$ and an element $a \in A$, $\langle a \rangle^E$ denote the *forward E-closure of* a:

$$\langle a \rangle^E = \{ b \in A \mid \exists n \, \exists a_1 \ldots a_n \text{ such that } a_1 = a, a_n = b \text{ and } (a_i, a_{i+1}) \in E^{\mathfrak{A}} \}.$$

Clearly $(\mathfrak{A}, a) \sim (\mathfrak{A}, a) \restriction \langle a \rangle^E$. Bisimulation equivalence over $\mathfrak{A}$ may be factored out to obtain a quotient Kripke structure which is a minimal representative for the $\sim$-class of the given $(\mathfrak{A}, a)$. We want to denote the result of applying this process to $(\mathfrak{A}, a) \restriction \langle a \rangle^E$ by can$(\mathfrak{A}, a)$ and call it the *canonical structure for* $(\mathfrak{A}, a)$ (cf. the *strongly extensional quotients* in [3]). Explicitly,

$$\text{can}(\mathfrak{A}, a) = (\langle a \rangle^E / \sim, E^\sim, P_1 / \sim, \ldots, P_l / \sim, [a]),$$

where $\sim$ is bisimulation equivalence (viewed as a binary relation over A), $[b]$ denotes the $\sim$-equivalence class of b, and the predicates $E^\sim$ and $P_i / \sim$ are defined as follows: $P_i / \sim = \{ [b] \mid b \in P_i^{\mathfrak{A}} \cap \langle a \rangle^E \}$, and $([b], [b']) \in E^\sim$ if there is some some $b'' \sim b'$ such that $(b, b'') \in E^{\mathfrak{A}}$ (this is independent of the choice of b within its equivalence class, even though $\sim$ is not a congruence with respect to E). It is obvious that can$(\mathfrak{A}, a)$ is computable from $(\mathfrak{A}, a)$ in PTIME. It is also straightforward to verify that for all $(\mathfrak{A}, a)$ we have $(\mathfrak{A}, a) \sim \text{can}(\mathfrak{A}, a)$, and that $(\mathfrak{A}, a) \sim (\mathfrak{A}', a')$ if and only if can$(\mathfrak{A}, a) \simeq \text{can}(\mathfrak{A}', a')$. In the special case that $\mathfrak{A} = \mathfrak{A}'$ this becomes

$$a \sim^{\mathfrak{A}} a' \quad \text{iff} \quad \text{can}(\mathfrak{A}, a) = \text{can}(\mathfrak{A}, a'), \tag{2}$$

since $a \sim a'$ in particular also implies that $\langle a \rangle^E / \sim = \langle a' \rangle^E / \sim$ (even though not necessarily $\langle a \rangle^E = \langle a' \rangle^E$).

Passage to can$(\mathfrak{A}, a)$ may be seen as a canonization procedure that (almost) picks unique representatives from bisimulation equivalence classes. [1]

Passage to can$(\mathfrak{A}, a)$ may also be used as a filter to enforce bisimulation invariance of queries. Let $\mathcal{A}$ be an algorithm that recognises some monadic query Q_0 on finite Kripke structures and identify Q_0 with the class of those $(\mathfrak{A}, a)$ for

---

[1] '*almost*' because in general the result is really only determined up to isomorphism. But here we could do better, see Proposition 2.10.

which $a \in Q^{\mathfrak{A}}$. Let $\mathcal{A} \circ$ can denote the algorithm which on input $(\mathfrak{A}, a)$ first computes $\text{can}(\mathfrak{A}, a)$ and then applies $\mathcal{A}$ to the outcome. Then clearly $\mathcal{A} \circ$ can recognises the $\sim$-invariant query $Q = \{(\mathfrak{A}, a) \mid \text{can}(\mathfrak{A}, a) \in Q_0\}$. Furthermore, if $\mathcal{A}$ is in PTIME then so is $\mathcal{A} \circ$ can, and Q_0 is itself $\sim$-invariant if and only if $Q_0 = Q$. In other words we have the following.

Proposition 2.1 *For any monadic query Q on finite Kripke structures: Q is $\sim$-invariant if and only if $Q = \{(\mathfrak{A}, a) \mid \text{can}(\mathfrak{A}, a) \in Q\}$. Note that in any case $\{(\mathfrak{A}, a) \mid \text{can}(\mathfrak{A}, a) \in Q\}$ is in PTIME if Q itself is.*

Corollary 2.2 *The queries in* PTIME∩ML$_\infty$ *are those obtained as compositions of arbitrary* PTIME *queries with* can: PTIME ∩ ML$_\infty \equiv$ PTIME ∘ can.

The notorious difficulties of capturing complexity classes – e.g. of finding a recursive syntax with PTIME semantics for the class of all PTIME queries on finite graphs – has to do with the implicit isomorphism invariance that is required of queries. For Kripke structures we have so far established a smooth passage from arbitrary queries (characterised by isomorphism-invariance) to bisimulation-invariant queries. This passage lends itself to a translation of the capturing issue. But why should it help? The answer is, that the domain to which we have reduced the original problem through Proposition 2.1 has *the* crucial advantage on which all known capturing results ultimately rely: over the can$(\mathfrak{A}, a)$, there is a definable and PTIME computable, in fact LFP-definable, global linear ordering. This allows us to reduce the present capturing issue to the well known Immerman-Vardi Theorem, that least fixed-point logic LFP captures PTIME over linearly ordered finite structures.

2.2 Order and the Immerman-Vardi Theorem

Among the prominent logics in finite model theory are various fixed-point extensions of first-order logic, most notably *least fixed-point logic* LFP. We here only sketch the definitions of LFP and the related *inductive fixed-point logic* IFP to make them available for technical applications. For more background the reader should see for instance [7]. LFP is the extension of first-order logic that is obtained through closure under the formation of least fixed points of positively defined operations on predicates. If φ is a formula in which the second-order variable X occurs positively (no free occurrence in the scope of an odd number of negations), X of arity r and $\bar{x}$ a tuple of r distinct first-order variables, $\bar{y}$ any tuple of r first-order variables, then $\psi = [\text{LFP}_{X, \bar{x}} \varphi](\bar{y})$ is also a formula (in which the $\bar{y}$ are free and X is not). ψ asserts of $\bar{y}$ that it is contained in the least fixed point of the monotone operation $P \mapsto \{\bar{x} \mid \varphi[P/X]\}$. Least-fixed point logic LFP is the smallest extension of first-order logic that is closed under first-order operations and the LFP-constructor.

Inductive fixed-point logic IFP similarly extends first-order by the formation of *inductive fixed points* of operations on predicates. For φ and X, $\bar{x}$ and $\bar{y}$ as above, but φ not necessarily positive in X, $\psi = [\text{IFP}_{X, \bar{x}} \varphi](\bar{y})$ is also a formula which asserts that $\overline{y}$ is in the limit of the inductive sequence of r-ary relations X_α generated through $X_0 = \emptyset$, $X_{\alpha+1} = X_\alpha \cup \{\overline{x} \mid \varphi[X_\alpha/X]\}$, and $X_\lambda = \bigcup_{\alpha<\lambda} X_\alpha$ for limits λ.

By a result of Gurevich and Shelah [10], LFP and IFP have the same expressive power over finite structures.

A class of finite structures admits an LFP-definable global ordering if there is an LFP-definable binary query which on all members of that class evaluates to a total linear ordering.

Theorem 2.3 (Immerman [11], Vardi [17]) *A query over linearly ordered structures is in* PTIME *if and only if it is definable in LFP. The same holds of* PTIME *queries in restriction to classes of structures that admit an LFP-definable global linear ordering.*

Let $\mathcal{CAN}$ stand for the class of all $\mathrm{can}(\mathfrak{A}, a)$ where $(\mathfrak{A}, a)$ is any finite Kripke structure (of our fixed type $\overline{P}$).

Proposition 2.4 $\mathcal{CAN}$ *admits an LFP-definable global ordering.*

Proof. In view of the Gurevich-Shelah Theorem we need only show that a linear order can uniformly be obtained through an IFP-like inductive process over all $\mathrm{can}(\mathfrak{A}, a)$. The idea is the same as in the colour refinement for finite graphs (cf. the detailed treatment in [15]). We know that $\sim$ (or rather its complement $\not\sim$) is definable in a fixed-point process, and the elements of $\mathrm{can}(\mathfrak{A}, a)$ just are the $\sim$-classes. The refinement steps in the generation of $\not\sim$ as a limit of successive stages $\not\sim_i$ may be adapted to generate an ordered representation of the $\sim_i$-classes in each stage. The resulting limit will be a linear ordering of the $\sim$-classes.

Recall that $\sim_0$ is atomic equivalence with respect to $\overline{P} = (P_1, \ldots, P_l)$. Enumerate the 2^l atomic $\overline{P}$-types in some fixed order, and let $\prec_0$ be the strict pre-ordering induced on $\mathrm{can}(\mathfrak{A}, a)$ by this enumeration: $[b] \prec_0 [b']$ if the atomic type of b precedes that of b' in this fixed enumeration. Clearly $\prec_0$ is a first-order definable global relation on $\mathcal{CAN}$.

Note that the elements of $\mathrm{can}(\mathfrak{A}, a)$ are $\sim$-classes, and as $\sim$ is the common refinement of the $\sim_i$, the $\sim_i$-classes are represented over $A/\sim$ by sets of $\sim$-classes. In this picture, the limit $\sim$ is reached over $\mathfrak{A}$ with that $\sim_i$ whose classes are singleton subsets of $A/\sim$. We shall inductively define the $\prec_i$ so that $\prec_i$ induces a linear ordering of the $\sim_i$-classes, i.e. such that each $(A/\sim, \prec_i) / \sim_i$ is a well defined strict linear ordering. The limit $\prec := \bigcup_i \prec_i$ then provides a linear ordering on $A/\sim$ and similarly on the universes $\langle a \rangle^E/\sim$ of the $\mathrm{can}(\mathfrak{A}, a)$ as desired. We define $\prec_{i+1}$ in terms of $\prec_i$ through

$$[b] \prec_{i+1} [b'] \quad \text{iff} \quad [b] \prec_i [b'] \quad \text{or} \quad [b]\sim_i[b'] \text{ but } \Gamma([b], \prec_i) <_{\mathrm{lex}} \Gamma([b'], \prec_i),$$

where $<_{\mathrm{lex}}$ is lexicographic comparison, applied to boolean tuples $\Gamma([b], \prec_i)$ which are defined as follows. Let $\gamma_1, \ldots, \gamma_t$ be the enumeration of $\sim_i$-classes in $\prec_i$-increasing order. Then $\Gamma([b], \prec_i) = (i_1, \ldots, i_t) \in \{0,1\}^t$ where $i_s = 1$ if $([b], [c]) \in E^\sim$ for some $[c] \in \gamma_s$.

In comparison with the inductive generation of the $\not\prec_\alpha$ in (1) it is shown inductively that $\prec_i$ is a linear ordering of the $\sim_i$-classes: $[b] \not\prec_i [b']$ if and only if $[b] \prec_i [b']$ or $[b'] \prec_i [b]$. It follows that the limit $\prec := \bigcup_i \prec_i$ does indeed define a linear ordering of the $\sim$-classes, i.e. a linear ordering on the universe of $\operatorname{can}(\mathfrak{A}, a)$.

It is also not hard to see that the crucial lexicographic comparison in the refinement step is first-order definable in the sense that there is a first-order formula that defines $\prec_{i+1}$ in terms of $\prec_i$ and $E^\sim$. Altogether this shows that the global ordering $\prec$ is IFP-definable, hence LFP-definable, over $\mathcal{CAN}$. $\qquad\square$

Corollary 2.5 *A bisimulation-invariant monadic query Q over finite Kripke structures is in* PTIME *if and only if* $\operatorname{can}(Q) = \big\{ \operatorname{can}(\mathfrak{A}, a) \mid a \in Q^{\mathfrak{A}} \big\}$ *is definable in* LFP *over* $\mathcal{CAN}$. *In formulae:* PTIME $\cap$ ML$_\infty$ $\equiv$ LFP $\circ$ can.

2.3 Getting it all into L_μ^k

Consider L_μ^k-definability over Kripke structures $\mathfrak{A}$. From Lemma 1.2 we know that bisimulation equivalence is a congruence for any L_μ^k-definable relation, whence an r-ary L_μ^k-definable relation over $\mathfrak{A}$ has a faithful representation as an r-ary relation over the quotient $A/\sim$. In the converse direction we want to establish a translation for LFP-definable relations over the $\operatorname{can}(\mathfrak{A}, a)$ to L_μ^k-definable relations over the $\mathfrak{A}$ themselves. It will be convenient to fix a sufficiently large value of r and to regard all global relations under consideration as r-ary (through appropriate padding).

Definition 2.6 Let R be of arity r over $\operatorname{can}(\mathfrak{A}, a)$ and let $k = r+1$. The *pull-back* of R to $\mathfrak{A}$ is defined to be the following k-ary relation R^* on $\mathfrak{A}$

$$R^* = \big\{ (a_1, \ldots, a_k) \in A^k \mid ([a_1], \ldots, [a_r]) \in R^{\operatorname{can}(\mathfrak{A}, a_k)} \big\}.$$

This notion of pull-back naturally extends to lead from global relations over the $\operatorname{can}(\mathfrak{A}, a)$ to global relations over the $\mathfrak{A}$. Note that R^* is invariant under $\sim$: if $a_j \sim a_j'$ for $j = 1, \ldots, k$, then $(a_1, \ldots, a_k) \in R^*$ if and only if $(a_1', \ldots, a_k') \in R^*$. This is obvious for the components $j = 1, \ldots, r$. For the k-th component it follows from the fact that for $a_k \sim a_k'$ even $\operatorname{can}(\mathfrak{A}, a_k) = \operatorname{can}(\mathfrak{A}, a_k')$, cf. (2).

Lemma 2.7 *Let $r \geqslant 1$ and put $k = r + 1$. For pull-backs of global relations of arity r we have the following:*

(i) *U^*, the pull-back of the full r-ary relation U over the $\operatorname{can}(\mathfrak{A}, a)$ is L_μ^k-definable.*

(ii) *The pull-backs of the basic relations in the vocabulary of the $\operatorname{can}(\mathfrak{A}, a)$ and of equality (between variables or with $[a]$) over $\operatorname{can}(\mathfrak{A}, a)$ are L_μ^k-definable.*

(iii) *Pull-back commutes with variable substitutions.*

(iv) *In restriction to U^*, pull-back commutes with boolean combinations.*

(v) *If S is defined by $\exists x_j R x_1 \ldots x_r$ over the $\operatorname{can}(\mathfrak{A}, a)$, then S^* is uniformly L_μ^k-definable in restriction to U^* from R^* by $\big[\mu_X \big(R^* \overline{x} \vee \Diamond_j X \overline{x} \big) \big] (x_k / x_j)$.*

Sketch of proof. Recall from Lemma 1.3 that $\sim$ is L_μ^k-definable for each $k \geqslant 2$. We shall simply write $x_i \sim x_j$ as shorthand for the formula defining $\sim$ in the i-th and j-th component, $i \neq j$.

(i) Let $i, j \in \{1, \ldots, k\}$, $i \neq j$. The formula $\chi_{ij}(\overline{x}) = \mu_X (x_j \sim x_i \vee \Diamond_j X\overline{x})(\overline{x})$ defines the relation $\{\overline{x} \mid \exists x_j' \in \langle x_j \rangle^E (x_j' \sim x_i)\} = \{\overline{x} \mid [x_i] \cap \langle x_j \rangle^E \neq \emptyset\}$. Therefore $\chi(\overline{x}) = \bigwedge_{i \neq j} \chi_{ij}(\overline{x})(x_k/x_j)$ defines U^*, the pull-back of the full r-ary relation over $\mathrm{can}(\mathfrak{A}, a)$.

(ii) Consider equalities first. For $R = \{(x_1, \ldots, x_r) \mid x_i = x_j\}$, $i \neq j$, use the formula $\chi(\overline{x}) \wedge x_i \sim x_j$. For equality with $[a]$, $R = \{(x_1, \ldots, x_r) \mid x_j = c\}$ (where c is the constant symbol for $[a]$ in $\mathrm{can}(\mathfrak{A}, a)$), R^* is definable by $\chi(\overline{x}) \wedge x_j \sim x_k$. The monadic basic predicates $R = \{(x_1, \ldots, x_r) \mid P_i x_j\}$ are similarly dealt with. For $R = \{(x_1, \ldots, x_r) \mid E^\sim x_i x_j\}$ finally, R^* is definable by $\chi(\overline{x}) \wedge \Diamond_i (x_i \sim x_j)(\overline{x})$.

(iii) and (iv) are straightforward.

(v) Suppose $S = \{(x_1, \ldots, x_r) \mid \exists x_j (x_1, \ldots, x_r) \in R\}$. The crucial observation is that the existential quantifier $\exists x_j$ ranges over $\langle [a] \rangle^{E^\sim} = \langle a \rangle^E / \sim$ in $\mathrm{can}(\mathfrak{A}, a)$. While L_μ^k does not allow global existential quantification, it can express just this relativised quantification. If φ defines R^*, then S^* is definable by $\chi(\overline{x}) \wedge [\mu_X (\varphi(\overline{x}) \vee \Diamond_j X\overline{x})](x_k/x_j)$, since the least fixed point $[\mu_X (\varphi(\overline{x}) \vee \Diamond_j X\overline{x})]$ exactly comprises those tuples of worlds for which a positive instance for R^* can be reached on a forward E-path in the j-th component. The substitution of x_k in the j-th component thus relativises the existential quantification over $\mathfrak{A}$ to $\langle x_k \rangle^E$, which is just as desired. $\qquad \square$

The following weak normal-form for LFP will be useful.

Lemma 2.8 *Any formula of LFP is logically equivalent with one in which – for some suitable value of r – all first-order variables that occur (free or bound) are among $x_1, \ldots, x_r$ and all applications of the fixed-point operator are of the form $[\mathrm{LFP}_{X, \overline{x}} \psi]\overline{x}'$, where the arity of X is r and $\overline{x} = (x_1, \ldots, x_r)$.*

We turn to the translation of fixed-point constructions over $\mathrm{can}(\mathfrak{A}, a)$ to μ^k-constructions over $\mathfrak{A}$. Let $\varphi(\overline{x}, \overline{X}) = [\mathrm{LFP}_{X, \overline{x}} \psi(\overline{x}, X, \overline{X})]\overline{y}$ be a fixed-point application that conforms to the normal form pattern of Lemma 2.8. Let X, $\overline{x}$, $\overline{y}$ and the extra second-order parameters $\overline{X}$ be of arity r and put $k = r + 1$. Up to a variable substitution in the desired formula φ^*, we may assume that $\overline{y} = \overline{x} = (x_1, \ldots, x_r)$. Assume that $\psi^*(\overline{x}, X^*, \overline{X}^*)$ is such that for all interpretations R, $\overline{R}$ over any $\mathrm{can}(\mathfrak{A}, a)$ and for $R' := \{(x_1, \ldots, x_r) \mid \psi(\overline{x}, X, \overline{X})\}$, $(R')^*$ is definable over $(\mathfrak{A}, R^*, \overline{R}^*)$ by $\psi^*(\overline{x}, X^*, \overline{X}^*)$. Let $S = \{(x_1, \ldots, x_r) \mid \varphi(\overline{x}, \overline{X})\}$. Then the following L_μ^k-formula φ^* defines S^* as a global relation over the $\mathfrak{A}$: $\varphi^*(\overline{x}, \overline{X}^*) = \mu_X \cdot (\psi^*(\overline{x}, X^*, \overline{X}^*))$.

Corollary 2.9 *Given any LFP-definable query over the $\mathrm{can}(\mathfrak{A}, a)$, this query may be padded to some sufficiently large arity r so that the pull-back of this r-ary query becomes L_μ^k-definable over the $\mathfrak{A}$, for $k = r + 1$.*

This finally yields a proof of the main theorem. Let Q be a monadic query on finite Kripke structures that is in PTIME and bisimulation-invariant. Then by

Corollary 2.5 there is some sentence φ in LFP (in the language of the $\operatorname{can}(\mathfrak{A}, a)$) such that $\{(\mathfrak{A}, a) \mid a \in Q^{\mathfrak{A}}\} = \{(\mathfrak{A}, a) \mid \operatorname{can}(\mathfrak{A}, a) \models \varphi\}$. Viewing this sentence as defining an r-ary global relation $\{\overline{x} \mid \varphi(\overline{x})\}$ (which is either full or empty, since there are no *free* variables in φ) for suitably large r, we get a formula φ^* of L_μ^k for $k = r + 1$ such that

$$
\begin{aligned}
a \in Q^{\mathfrak{A}} \;&\Leftrightarrow\; \varphi[\operatorname{can}(\mathfrak{A}, a)] = (\langle a\rangle/\sim)^r \;\Leftrightarrow\; \varphi[\operatorname{can}(\mathfrak{A}, a)] \neq \emptyset \\
&\Leftrightarrow\; \{(a_1, \ldots, a_r) \in A^r \mid \mathfrak{A} \models \varphi^*[a_1, \ldots, a_r, a]\} \neq \emptyset \\
&\Leftrightarrow\; \mathfrak{A} \models \varphi^*[a, \ldots, a] \;\Leftrightarrow\; (\mathfrak{A}, a) \models \varphi^*(x_1, \ldots, x_1).
\end{aligned}
$$

2.4 Some remarks

It should be noted that the *existence* of a logic for bisimulation-invariant PTIME, in the sense of descriptive complexity, follows directly from Corollary 2.5 which may be looked at as a normal form theorem or as a logical characterisation. Putting the reasoning that leads to Corollary 2.5 in different perspective, the abstract capturing result we have here is due to the following.

Proposition 2.10 *There is a PTIME functor H defined on finite Kripke structures with distinguished worlds, for canonization up to bisimulation equivalence:*

$$
\forall (\mathfrak{A}, a) \quad (\mathfrak{A}, a) \sim H(\mathfrak{A}, a),
$$
$$
\forall (\mathfrak{A}, a) \forall (\mathfrak{A}', a') \quad (\mathfrak{A}, a) \sim (\mathfrak{A}', a') \Rightarrow H(\mathfrak{A}, a) = H(\mathfrak{A}', a').
$$

Such H is directly gained from $(\mathfrak{A}, a) \overset{\mathrm{can}}{\longmapsto} \operatorname{can}(\mathfrak{A}, a) \overset{\mathrm{stan}}{\longmapsto} H(\mathfrak{A}, a)$, where the functor can is composed with a functor stan which maps $\operatorname{can}(\mathfrak{A}, a)$ to its isomorphic standard representation over an initial segment of the natural numbers, naturally ordered by the PTIME computable global ordering $\prec$ (according to Proposition 2.4). PTIME canonization functors generally induce capturing results for corresponding fragments of PTIME, as outlined in [15]. Indeed, two other interesting capturing results for fragments of PTIME were obtained along these lines [14]. These concern the full two-variable fragments of infinitary first-order logic and its extension by counting quantifiers. In those cases, canonization requires far more elaboration, since the passage to quotients (as in the formation of $\operatorname{can}(\mathfrak{A}, a)$) does not lead to structures of the original kind. The reconstruction of a standard representative from its concise quotient description becomes an essential, and technically involved, step in those arguments. It just so happens that for bisimulation invariance the quotient structures themselves are canonical representatives of their equivalence class. It should also be stressed that the present capturing result is not a consequence of the capturing result for the two-variable fragment of PTIME , PTIME $\cap L^2_{\infty\omega}$, in [14], although the latter result is strictly stronger in the sense that PTIME $\cap$ ML$_\infty \subsetneq$ PTIME $\cap L^2_{\infty\omega}$.

Beyond the abstract capturing result, however, the major point of the present result is the close link with L_μ, a logic that is well-known for many other reasons. Even though it is not L_μ itself but rather its 'vectorisations' L_μ^k that come up here, this capturing result still illustrates the naturalness of L_μ from yet another angle.

References

1. J. BALCAZAR, J. GABARRO AND M. SANTHA, *Deciding bisimilarity is P-complete*, Formal Aspects of Computing, 4, 1992, pp. 638–648.
2. J. BARWISE AND J. VAN BENTHEM, *Interpolation, preservation, and pebble games*, preprint, 1996.
3. J. BARWISE AND L. MOSS, *Modal correspondence for models*, preprint, 1996.
4. J. VAN BENTHEM, *Modal correspondence theory*, PhD thesis, University of Amsterdam, 1976.
5. ——, *Modal Logic and Classical Logic*, Bibliopolis, 1985.
6. A. CHANDRA AND D. HAREL, *Structure and complexity of relational queries*, Journal of Computer and System Sciences, 25, 1982, pp. 99–128.
7. H.-D. EBBINGHAUS AND J. FLUM, *Finite Model Theory*, Springer, 1995.
8. E. EMERSON, *Temporal and modal logic*, in J. van Leeuwen (ed.), *Handbook of Theoretical Computer Science*, vol. B: Formal Models and Semantics, Elsevier, 1990, pp. 995–1072.
9. Y. GUREVICH, *Logic and the challenge of computer science*, in E. Börger (ed.), *Current Trends in Theoretical Computer Science*, Computer Science Press, 1988, pp. 1–57.
10. Y. GUREVICH AND S. SHELAH, *Fixed-point extensions of first-order logic*, Annals of Pure and Applied Logic, 32 (1986), pp. 265–280.
11. N. IMMERMAN, *Relational queries computable in polynomial time*, Information and Control, 68, 1986, pp. 86–104.
12. D. KOZEN, *Results on the propositional μ-calculus*, Theoretical Computer Science, 27, 1983, pp. 333–354.
13. R. MILNER, *A Calculus of Communicating Systems*, Lecture Notes in Computer Science, vol. 92, Springer, 1980.
14. M. OTTO, *Canonization for two variables and puzzles on the square*, to appear in Annals of Pure and Applied Logic, 1997.
15. ——, *Bounded Variable Logics and Counting — A Study in Finite Models*, Lecture Notes in Logic, vol. 9, Springer, 1997.
16. D. PARK, *Concurrency and automata on infinite sequences*, in: P. Deussen (ed.) *Proc. 5th GI-Conf. on Theoretical Computer Science*, Lecture Notes in Computer Science, vol. 104, Springer, 1981, pp. 167–183.
17. M. VARDI, *The complexity of relational query languages*, in Proc. 14th ACM Symp. on Theory of Computing, 1982, pp. 137–146.

Equivalence of Multiplicative Fragments of Cyclic Linear Logic and Noncommutative Linear Logic

Mati Pentus*

Department of Mathematical Logic, Faculty of Mechanics and Mathematics, Moscow State University, Moscow 119899, Russia

Abstract. In this paper we consider two associative noncommutative analogs of Girard's linear logic. These are Abrusci's noncommutative linear logic and Yetter's cyclic linear logic.

We give a linear-length translation from the multiplicative fragment of Abrusci's noncommutative linear logic into the multiplicative fragment of Yetter's cyclic linear logic and vice versa. As a corollary we obtain that the decidability problems for derivability in these two calculi are in polynomial time reducible to each other.

1 Introduction

In this paper we consider two associative noncommutative analogs of Girard's linear logic [2] (which is associative and commutative). The *cyclic linear logic* was introduced in 1990 by D. N. Yetter [3]. In 1991 V. M. Abrusci [1] introduced another noncommutative linear analog of the original linear logic PNCL and called it *pure noncommutative classical linear propositional logic*.

Here we give a linear-length translation from the multiplicative fragment of Abrusci's PNCL into the multiplicative fragment of Yetter's cyclic linear logic and vice versa.

2 The calculus MNCL

In [1] V. M. Abrusci introduced a sequent calculus PNCL for the pure noncommutative classical linear propositional logic. In the same paper two one-sided sequent calculi SPNCL and SPNCL$'$ were introduced and it was proved that they are equivalent to PNCL.

We shall use a slightly modified (but equivalent) version of the multiplicative fragment of SPNCL$'$ and denote it by MNCL.

We assume that an enumerable set of *variables* $\mathrm{Var_{MNCL}}$ is given. We introduce the set of formal symbols called *atoms*

$$\mathrm{At_{MNCL}} \rightleftharpoons \{p^{\perp n} \mid p \in \mathrm{Var_{MNCL}},\ n \in \mathbf{Z}\}.$$

* The research described in this publication was made possible in part by the Russian Foundation for Basic Research (grant 96-01-01395).

We denote the set of all integers by $\mathbf{Z}$ and the set of all natural numbers by $\mathbf{N}$.

Intuitively, if $n \geq 0$, then $p^{\perp n}$ means 'p with n right negations' and $p^{\perp(-n)}$ means 'p with n left negations'.

Definition 1. The set of MNCL-*formulas* is defined to be the smallest set Fm_{MNCL} satisfying the following conditions:

1. $\text{At}_{\text{MNCL}} \subset \text{Fm}_{\text{MNCL}}$;
2. $\perp \in \text{Fm}_{\text{MNCL}}$;
3. $\mathbf{1} \in \text{Fm}_{\text{MNCL}}$;
4. if $A \in \text{Fm}_{\text{MNCL}}$ and $B \in \text{Fm}_{\text{MNCL}}$, then $(A \otimes B) \in \text{Fm}_{\text{MNCL}}$ and $(A \wp B) \in \text{Fm}_{\text{MNCL}}$.

Here $\otimes$ is the multiplicative conjunction, called 'tensor', and $\wp$ is the multiplicative disjunction, called 'par'. The constants $\perp$ and $\mathbf{1}$ are multiplicative falsity and multiplicative truth respectively.

By $\text{Fm}^*_{\text{MNCL}}$ we denote the set of all finite sequences of normal formulas. The empty sequence is denoted by Λ.

Definition 2. We define the right negation $(\ )^{\perp} : \text{Fm}_{\text{MNCL}} \to \text{Fm}_{\text{MNCL}}$ and the left negation $^{\perp}(\ ) : \text{Fm}_{\text{MNCL}} \to \text{Fm}_{\text{MNCL}}$.

$$(p^{\perp n})^{\perp} \rightleftharpoons p^{\perp(n+1)}$$
$$\perp^{\perp} \rightleftharpoons \mathbf{1}$$
$$\mathbf{1}^{\perp} \rightleftharpoons \perp$$
$$(A \otimes B)^{\perp} \rightleftharpoons (B^{\perp}) \wp (A^{\perp})$$
$$(A \wp B)^{\perp} \rightleftharpoons (B^{\perp}) \otimes (A^{\perp})$$

$$^{\perp}(p^{\perp n}) \rightleftharpoons p^{\perp(n-1)}$$
$$^{\perp}\perp \rightleftharpoons \mathbf{1}$$
$$^{\perp}\mathbf{1} \rightleftharpoons \perp$$
$$^{\perp}(A \otimes B) \rightleftharpoons (^{\perp}B) \wp (^{\perp}A)$$
$$^{\perp}(A \wp B) \rightleftharpoons (^{\perp}B) \otimes (^{\perp}A)$$

The sequents of this calculus are of the form $\to \Gamma$, where $\Gamma \in \text{Fm}^*_{\text{MNCL}}$. The calculus MNCL has the following axioms and rules.

$$\frac{}{\to (p^{\perp(n+1)})(p^{\perp n})} \ (id)$$

$$\frac{}{\to \mathbf{1}} \ (\mathbf{1}) \qquad\qquad \frac{\to \Gamma \Delta}{\to \Gamma \perp \Delta} \ (\perp)$$

$$\frac{\to \Gamma A B \Delta}{\to \Gamma (A \wp B) \Delta} \ (\wp) \qquad\qquad \frac{\to \Gamma A \quad \to B \Delta}{\to \Gamma (A \otimes B) \Delta} \ (\otimes)$$

$$\frac{\to A \Gamma}{\to \Gamma(^{\perp \perp}A)} \ (^{\perp \perp}(\ )) \qquad\qquad \frac{\to \Gamma A}{\to (A^{\perp \perp})\Gamma} \ ((\ )^{\perp \perp})$$

Here capital letters A, B, ... stand for formulas, capital Greek letters denote finite (possibly empty) sequences of formulas, p ranges over Var_{MNCL}, and n ranges over $\mathbf{Z}$.

3   The calculus MCLL

We denote by MCLL the multiplicative fragment of the cyclic linear logic introduced by D. N. Yetter [3].

The formulas of MCLL are defined similarly to those of MNCL except that the set of atoms is defined as follows

$$At_{MCLL} \rightleftharpoons Var_{MCLL} \cup \{p^\perp \mid p \in Var_{MCLL}\}.$$

We shall denote the set of all MCLL-formulas by Fm_{MCLL}.

There is only one negation operator $(\ )^\perp$ in MCLL.

Definition 3.

$$(p)^\perp \rightleftharpoons p^\perp$$
$$(p^\perp)^\perp \rightleftharpoons p$$
$$\perp^\perp \rightleftharpoons 1$$
$$1^\perp \rightleftharpoons \perp$$
$$(A \otimes B)^\perp \rightleftharpoons (B^\perp) \wp (A^\perp)$$
$$(A \wp B)^\perp \rightleftharpoons (B^\perp) \otimes (A^\perp)$$

The calculus MCLL has the following axioms and rules.

$$\frac{}{\rightarrow p^\perp \ p} \ (id_1) \qquad\qquad \frac{}{\rightarrow p \ p^\perp} \ (id_2)$$

$$\frac{}{\rightarrow 1} \ (1) \qquad\qquad \frac{\rightarrow \Gamma \Delta}{\rightarrow \Gamma \perp \Delta} \ (\perp)$$

$$\frac{\rightarrow \Gamma A B \Delta}{\rightarrow \Gamma (A \wp B) \Delta} \ (\wp) \qquad\qquad \frac{\rightarrow \Gamma A \quad \rightarrow B \Delta}{\rightarrow \Gamma (A \otimes B) \Delta} \ (\otimes)$$

$$\frac{\rightarrow \Delta \Gamma}{\rightarrow \Gamma \Delta} \ (rotate)$$

4   The invariant function $\flat$

Definition 4. We define the function $\flat \colon Fm_{MNCL} \rightarrow \mathbb{Z}$ by induction as follows.

$$\flat(p^{\perp n}) \rightleftharpoons 1 \qquad \flat(1) \rightleftharpoons 2 \qquad \flat(\perp) \rightleftharpoons 0$$

$$\flat(A \wp B) \rightleftharpoons \flat(A) + \flat(B) \qquad \flat(A \otimes B) \rightleftharpoons \flat(A) + \flat(B) - 2$$

For sequences of formulas we put

$$\flat(A_1 \ldots A_n) \rightleftharpoons \flat(A_1) + \ldots + \flat(A_n).$$

Lemma 5. *For every formula* $A \in Fm_{MNCL}$ *the equality*

$$\flat(A^{\perp \perp}) = \flat(A) = \flat(^{\perp \perp} A)$$

holds true.

Proof. Straightforward induction on the complexity of A. $\qquad\square$

Lemma 6. *If* $\mathrm{MNCL} \vdash \to \Gamma$, *then* $\flat(\Gamma) = 2$.

Proof. Easy induction on the length of the MNCL-derivation of Γ. To check the rules $(\perp\perp(\,))$ and $((\,)^{\perp\perp})$ we use Lemma 5. $\qquad\square$

Next we define a very similar function that maps formulas of MCLL to integers. We shall denote this function too by $\flat$.

Definition 7. The function $\flat\colon \mathrm{Fm}_{\mathrm{MCLL}} \to \mathbf{Z}$ is defined by induction as follows.

$$\flat(p^{\perp}) \rightleftharpoons 1 \qquad \flat(p) \rightleftharpoons 1 \qquad \flat(1) \rightleftharpoons 2 \qquad \flat(\perp) \rightleftharpoons 0$$

$$\flat(A \wp B) \rightleftharpoons \flat(A) + \flat(B) \qquad \flat(A \otimes B) \rightleftharpoons \flat(A) + \flat(B) - 2$$

Again, for sequences of formulas we put

$$\flat(A_1 \ldots A_n) \rightleftharpoons \flat(A_1) + \ldots + \flat(A_n).$$

Lemma 8. *If* $\mathrm{MCLL} \vdash \to \Gamma$, *then* $\flat(\Gamma) = 2$.

Proof. Induction on the length of the MCLL-derivation of Γ. $\qquad\square$

5 Translation

We assume now that

$$\mathrm{Var}_{\mathrm{MNCL}} = \{p_{i,j} \mid i \in \mathbf{N},\ j \in \{0,1\}\}$$

and

$$\mathrm{Var}_{\mathrm{MCLL}} = \{q_{i,k} \mid i \in \mathbf{N},\ k \in \mathbf{Z}\}.$$

Definition 9. For every integer m we define a one-to-one translation α_m mapping formulas of MNCL to formulas of MCLL.

$$\alpha_m(1) \rightleftharpoons 1$$
$$\alpha_m(\perp) \rightleftharpoons \perp$$
$$\alpha_m(p_{i,j}^{\perp n}) \rightleftharpoons \begin{cases} q_{i,n+m} & \text{if } m \equiv j \pmod 2 \\ q_{i,n+m}^{\perp} & \text{if } m \not\equiv j \pmod 2 \end{cases}$$
$$\alpha_m(A \wp B) \rightleftharpoons \alpha_m(A) \wp \alpha_{m+\flat A}(B)$$
$$\alpha_m(A \otimes B) \rightleftharpoons \alpha_m(A) \otimes \alpha_{m+\flat A-2}(B)$$

For sequences of formulas we put

$$\alpha_m(\Lambda) \rightleftharpoons \Lambda \qquad \alpha_m(\Gamma A) \rightleftharpoons \alpha_m(\Gamma)\, \alpha_{m+\flat\Gamma}(A).$$

Lemma 10. *For every formula* $A \in \mathrm{Fm}_{\mathrm{MNCL}}$ *and every integer* $m \in \mathbf{Z}$ *the following equalities hold.*

1. $b(\alpha_m A) = b(A)$
2. $\alpha_m(A^{\perp\,\perp}) = \alpha_{m+2}(A)$
3. $\alpha_m(^{\perp\,\perp}A) = \alpha_{m-2}(A)$

Proof. Induction on the complexity of A. □

The converse translations $\beta_m : \mathrm{Fm}_{\mathrm{MCLL}} \to \mathrm{Fm}_{\mathrm{MNCL}}$ can be explicitly defined as follows.

Definition 11.

$$\beta_m(1) \rightleftharpoons 1$$
$$\beta_m(\perp) \rightleftharpoons \perp$$
$$\beta_m(q_{i,k}) \rightleftharpoons p_{i,m\,\mathrm{mod}\,2}^{\perp(k-m)}$$
$$\beta_m(q_{i,k}^{\perp}) \rightleftharpoons p_{i,m+1\,\mathrm{mod}\,2}^{\perp(k-m)}$$
$$\beta_m(A \,\wp\, B) \rightleftharpoons \beta_m(A) \,\wp\, \beta_{m+\flat A}(B)$$
$$\beta_m(A \otimes B) \rightleftharpoons \beta_m(A) \otimes \beta_{m+\flat A-2}(B)$$

For sequences of formulas we put

$$\beta_m(\Lambda) \rightleftharpoons \Lambda \qquad \beta_m(\Gamma A) \rightleftharpoons \beta_m(\Gamma)\,\beta_{m+\flat\Gamma}(A).$$

Lemma 12.

1. *For every formula $A \in \mathrm{Fm}_{\mathrm{MNCL}}$ and every integer $m \in \mathbf{Z}$, $\beta_m(\alpha_m(A)) = A$.*
2. *For every formula $A \in \mathrm{Fm}_{\mathrm{MCLL}}$ and every integer $m \in \mathbf{Z}$, $\alpha_m(\beta_m(A)) = A$.*

Proof. Straightforward induction on the complexity of A. □

Theorem 13. *For every sequent $\to\Gamma$ in the language of MNCL the sequent $\to\Gamma$ is derivable in MNCL if and only if $\to\alpha_0(\Gamma)$ is derivable in MCLL.*

Proof. First we prove by induction on MNCL-derivations that if $\mathrm{MNCL} \vdash \to\Gamma$, then $\mathrm{MCLL} \vdash \to\alpha_m(\Gamma)$ for every $m \in \mathbf{Z}$.

The case of the axiom (id) is checked as follows. Let $\Gamma = (p^{\perp(n+1)})\,(p^{\perp n})$. Then $\alpha_m(\Gamma) = \alpha_m(p^{\perp(n+1)})\,\alpha_{m+1}(p^{\perp n})$. If $m \equiv j \pmod 2$, then $\alpha_m(\Gamma) = q_{i,n+m+1}\,q_{i,n+m+1}^{\perp}$, else $\alpha_m(\Gamma) = q_{i,n+m+1}^{\perp}\,q_{i,n+m+1}$. In both cases $\alpha_m(\Gamma)$ is an axiom of MCLL.

To check the rules $(^{\perp\,\perp}(\;))$ and $((\;)^{\perp\,\perp})$ we use Lemma 6 and Lemma 10. Cases of other rules are easy.

For the converse, by induction on MCLL-derivations it can be proved that if $\mathrm{MCLL} \vdash \to\Gamma$, then $\mathrm{MNCL} \vdash \to\beta_m(\Gamma)$ for every $m \in \mathbf{Z}$.

To check the rule

$$\frac{\to\Delta\Gamma}{\to\Gamma\Delta} \;\; (rotate)$$

we need to prove that MNCL $\vdash \rightarrow \beta_m(\Gamma)\beta_{m+b\Gamma}(\Delta)$. According to induction hypothesis, MNCL $\vdash \rightarrow \beta_k(\Delta)\beta_{k+b\Delta}(\Gamma)$ for every $k \in \mathbf{Z}$. We put $k \rightleftharpoons m + b\Gamma$ and recall that $b\Delta + b\Gamma = 2$ according to Lemma 8. Thus we have MNCL $\vdash \rightarrow \beta_{m+b\Gamma}(\Delta)\beta_{m+2}(\Gamma)$. Applying the rule $(( )^{\perp \perp})$ for each formula of $\beta_{m+2}(\Gamma)$ in turn, we obtain MNCL $\vdash \rightarrow \beta_{m+2}(\Gamma)^{\perp \perp} \beta_{m+b\Gamma}(\Delta)$. But this is the desired sequent, because $\beta_{m+2}(\Gamma)^{\perp \perp} = \beta_m \alpha_m (\beta_{m+2}(\Gamma)^{\perp \perp}) = \beta_m \alpha_{m+2} \beta_{m+2}(\Gamma) = \beta_m(\Gamma)$. $\qquad \Box$

Corollary 14. *The decidability problem for derivability in* MNCL *is in polynomial time reducible to the decidability problem for derivability in* MCLL *and vice versa.*

Remark. It is well-known that both these decidability problems belong to NP. Whether they are NP-complete is still an open question.

References

1. Abrusci, V. M.: Phase semantics and sequent calculus for pure noncommutative classical linear propositional calculus. Journal of Symbolic Logic, **56** (1991) 1403–1451
2. Girard, J.-Y.: Linear logic. Theoretical Computer Science, **50** (1987) 1–102
3. Yetter, D. N.: Quantales and noncommutative linear logic. Journal of Symbolic Logic, **55** (1990) 41–64

A Decidable Fragment of Second Order Linear Logic

G. Perrier

CRIN-CNRS & INRIA Lorraine
Campus Scientifique -B.P. 239
54506 Vandœuvre-les-Nancy Cedex
France
e-mail: perrier@loria.fr

Abstract. Existentially quantified variables are the source of non decidability for second order linear logic without exponentials (MALL2). We propose a decision procedure for a fragment of MALL2 based on a canonical instantiation of these variables and using inference permutability in proofs.

Introduction

The decision problem for second order linear logic without exponentials (MALL2) has given rise recently to many papers and results. P. Lincoln, A. Scedrov and N. Shankar have shown that the multiplicative fragment of second order intuitionistic linear logic (IMLL2) is undecidable by encoding second order intuitionistic propositional logic — known to be undecidable — into IMLL2 [10]. M. Emms has extended this strategy to prove the undecidability of second order Lambek Calculus (L2) [4], which can be viewed as the multiplicative fragment of intuitionistic non-commutative linear logic. The undecidability of MALL2 has been shown by Y. Lafont [7] and the undecidability of the multiplicative fragment of second order propositional linear logic (MLL2) by Y. Lafont and A. Scedrov [8],in both cases with an encoding of two-counter machines.

From linguistic motivations, M. Emms has shown the decidability of a fragment of L2 [3]. In L2, undecidability comes from universally quantified formulas on the left hand side of sequents. M. Emms limits such formulas to five formulas that represent all polymorphic categories according to his view of the Categorial Grammar theory for the syntax of natural languages. For these five formulas, he proposes a canonical method for eliminating their variable and the source of undecidability at the same time. When such a formula has to be decomposed in a bottom-up proof search, the propositional variable is not instantiated immediately. The decomposition of the formula goes on until all occurrences of the variable emerge. At this moment and because of the particular syntax of the decomposed formula, a canonical instantiation of the variable can be found.

We propose to generalise the result of M. Emms to the larger framework of MALL2. With respect to our problem of decidability, the main difference between L2 and MALL2 comes from the presence of additives in MALL2, the non-commutativity or the commutativity of the logic being irrelevant. If we present

MALL2 in the formalism of one-sided sequent calculus, undecidability comes from the existentially quantified formulas. We exhibit general criteria so that the method of M. Emms could be applied to such formulas. A crucial point is to have the possibility of decomposing these formulas at one go in order to make the occurrences of the variable emerge. That means the possibility to apply the inference rules in a particular order. That is why in Section 1 we begin by a preliminary study of inference permutability in MALL2. This study will guide us in Section 2 in order to define the syntax of the existentially quantified formulas that will belong to the fragment for which a decision procedure will be defined.

1 Deduction procedures and inference permutability in MALL2

1.1 Deduction procedures in MALL2

We present the inference system of MALL2 in the framework of the one-sided sequent calculus. Sequents have the form $\vdash \Delta$ where Δ is a finite multi-set of MALL2 formulas.

To define MALL2 formulas, we assume a countably infinite set V of propositional variables and a countably infinite set C of propositional constants. X, Y, Z range over propositional variables and a, b, c, d, ... over propositional constants.

MALL2 formulas F are built recursively from V and C by the following grammar:

$$F ::= X \mid a \mid X^{\perp} \mid a^{\perp} \mid 1 \mid \perp \mid \top \mid 0 \mid F \otimes F \mid FOF \mid FNF \mid F \oplus F \mid \forall X F \mid \exists X F$$

As usual, the negation of a MALL2 formula is defined recursively by using involutivity of negation and the de Morgan laws.

General formulas will be referenced by the capital letters F, G, H and atomic formulas by the capital letters A, B, C. Multi-sets of formulas will be referenced by the Greek letters Δ, Γ.

The inference rules for MALL2 are given in Figure 1. In the application of rule $\exists$, we use the notation $\overline{X}$ of M. Emms [3] for the formula G that instantiates the quantified variable X when this formula is not determined immediately.

Before we get to the heart of the matter, we need to define the vocabulary related to bottom-up proof search in MALL2 we use in the rest of the article.

Definition 1. A *goal* is a finite set of MALL2 sequents, which are its subgoals. A *deduction* is a finite or infinite sequence of goals (G_n) such that for any n, G_{n+1} is obtained from G_n by replacing a subgoal $\vdash \Delta$ with a finite (possibly empty) set of new subgoals $\vdash \Delta_1, \cdots, \vdash \Delta_p$. such that, if $\vdash \Delta_1, \cdots, \vdash \Delta_p$ are provable in MALL2, then $\vdash \Delta$ also.

A deduction is *successful* if it is finite and its last goal is empty.

Definition 2. A *deduction procedure* is a set of deductions.

A deduction procedure is *complete* if, for any provable sequent $\vdash \Delta$, there exists a successful deduction that belongs to this procedure and that begins with the goal $\{\vdash \Delta\}$.

Fig. 1. The rules of MALL2 sequent calculus

identity group

$$\frac{}{\vdash A, A^\perp}\ \text{id} \qquad\qquad \frac{\vdash \Delta_1, F \quad \vdash F^\perp, \Delta_2}{\vdash \Delta_1, \Delta_2}\ \text{cut}$$

logical group

  multiplicatives

$$\frac{\vdash \Delta_1, F_1 \quad \vdash F_2, \Delta_2}{\vdash \Delta_1, F_1 \otimes F_2, \Delta_2}\ \otimes \qquad\qquad \frac{\vdash F_1, F_2, \Delta}{\vdash F_1 O F_2, \Delta}\ O$$

$$\frac{}{\vdash 1}\ 1 \qquad\qquad \frac{\vdash \Delta}{\vdash \perp, \Delta}\ \perp$$

  additives

$$\frac{\vdash F_1, \Delta \quad \vdash F_2, \Delta}{\vdash F_1 N F_2, \Delta}\ N \qquad \frac{\vdash F_1, \Delta}{\vdash F_1 \oplus F_2, \Delta}\ \oplus_1 \qquad \frac{\vdash F_2, \Delta}{\vdash F_1 \oplus F_2, \Delta}\ \oplus_2$$

$$\frac{}{\vdash T, \Delta}\ T$$

second-order quantifiers

$$\frac{\vdash F[Y/X], \Delta}{\vdash \forall X F, \Delta}\ \forall \qquad\qquad \frac{\vdash F[G/X], \Delta}{\vdash \exists X F, \Delta}\ \exists$$

with Y not free in the conclusion

A deduction procedure *terminates* if, for any sequent $\vdash \Delta$, all deductions that belong to this procedure and that begin with the goal $\{\vdash \Delta\}$ are finite and their number is also finite.

A deduction procedure is a *decision procedure* if it is complete and it terminates.

The more general deduction procedure in MALL2, we denote by $\mathcal{P}_0$, consists in deductions where, at each step, we apply a rule of MALL2 sequent calculus from the conclusion to the premises. We drop the cut rule from $\mathcal{P}_0$ and from all deduction procedures because it is highly non-deterministic in bottom-up proof search. Despite this restriction, because of cut elimination, $\mathcal{P}_0$ remains complete but obviously it does not terminate. Our aim is to restrict $\mathcal{P}_0$ as much as possible while keeping completeness. As MALL2 is undecidable, we cannot hope to obtain a procedure for whole MALL2 that terminates. A way of restricting the procedure is to start from the following observation: because we use the formalism of sequent calculus, an important part of the order between the rules that are applied during a deduction is irrelevant and it can be fixed so that non-determinism of the procedure is reduced. To distinguish what is relevant and what is irrelevant in the order of applied rules, we have to make a static analysis of inference permutability in MALL2 proofs. Then we will transpose the conclusions of such an analysis into the definition of deduction procedures.

1.2   Inference permutability in MALL2

The property of inference permutability in a proof means the possibility for two consecutive inferences of inverting them without disturbing the rest of the proof. This property is defined precisely in [6] where it is studied for first order linear logic. The results that conclude this study can be easily transposed to MALL2. So, we can divide the inference types into two classes $T \downarrow$ and $T \uparrow$ according to the direction in which they can move in a proof the most easily.

$$T \downarrow = \{N, \perp, O, \forall\} \quad \text{and} \quad T \uparrow = \{\otimes, \oplus, \exists, cut\}$$

Using this, we normalise proofs by moving their inferences of a type $\in T \downarrow$ as far down and moving their inferences of a type $\in T \uparrow$ as far up as possible. This normalisation includes cut elimination but it goes further in ordering inferences in a particular way. As the resulting class of normal proofs is complete, we can restrict the proof search to normal proofs in order to establish the provability of sequents. This restricts deduction procedures in two ways:

- As soon as a formula of a type $\in T \downarrow$ [1] emerges in a subgoal, we decompose it. We have described this principle in [5] as the principle of *immediate decomposition*. This corresponds to the fact that formulas of a type $\in T \downarrow$ are introduced as low as possible in normal proofs. J.-M. Andreoli expresses that in [1] with the notion of invertibility.
- As soon as a formula of a type $\in T \uparrow$ begins to be decomposed, the decomposition goes on as long as the components are of a type $\in T \uparrow$. We have described this principle in [5] as the principle of *chaining decomposition*. This corresponds to the fact that formulas of a type $\in T \uparrow$ are introduced as high as possible in normal proofs. This corresponds to the notion of focusing introduced in [1].

We can embed both principles into the procedure $\mathcal{P}_0$ in a formal way to obtain the procedure $\mathcal{P}_1$.

Definition 3. The procedure $\mathcal{P}_1$ is the set of deductions $(G_n) \in \mathcal{P}_0$ such that every deduction step $G_n \to G_{n+1}$ has the following properties:

- if G_n contains a formula of a type $\in T \downarrow$, then $G_n \to G_{n+1}$ consists in decomposing such a formula (immediate decomposition principle);
- if $G_n \to G_{n+1}$ consists in decomposing a formula F of a type $\in T \uparrow$, then all following steps consist in decomposing subformulas of F as long as they have a type $\in T \uparrow$ (chaining decomposition principle).

---

[1] The type of a complex formula is its head connective.

The interest of the procedure $\mathcal{P}_1$ is that it reduces non-determinism in proof search while keeping completeness [5]. Unfortunately, $\mathcal{P}_1$ does not terminate.

2 A decision procedure for a fragment of MALL2

The $\exists$-rule is the source of non-termination for every proof procedure that aims to be complete for all sequents of MALL2 as this is shown in [7]. The problem comes from the fact that, in a deduction, a formula of the form $\exists X F$ can be decomposed into $F[G/X]$ where all occurrences of X are replaced by any formula G.

In one case, this difficulty can be easily overcome : when all occurrences of X are either positive or negative. If they are positive, X can be instantiated by the constant $\top$ and if they are negative, X can be instantiated by the constant 0. Then, all subgoals where the formula $\overline{X}$ will surface, will be satisfied automatically by means of the $\top$-axiom.

In the general case, we try to delay the instantiation of X as late as possible until we have enough information to do it. For this, we use two principles:

One-piece decomposition As soon as we start to decompose a formula $\exists X F$, we go on with the decomposition until all occurrences of $\overline{X}$ surface in the goal.

Elimination of unknowns When $\overline{X}$ is only present in subgoals that have the form $\vdash \overline{X}, \Delta_1 \cdots \vdash \overline{X}, \Delta_n \vdash \overline{X}^\perp, \Delta'_1 \cdots \vdash \overline{X}^\perp, \Delta'_m$, we replace these subgoals by the subgoals $\vdash \Delta_i, \Delta'_j$ where (i,j) is any element of $[1, n] \times [1, m]$. This amounts to replacing $\overline{X}$ by $O(\Delta'_1)N \cdots NO(\Delta'_m)$ [2].

The principle of elimination of unknowns is justified by the following lemma which is a generalisation of the lemma of elimination of unknowns introduced by M. Emms in [3].

Lemma 4. *There exists a MALL2 formula F such that the sequents* $\vdash F, \Delta_1 \cdots \vdash F, \Delta_n \vdash F^\perp, \Delta'_1 \cdots \vdash F^\perp, \Delta'_m$ *are provable if and only if the sequents* $\vdash \Delta_i, \Delta'_j$ *where (i,j) is any element of* $[1, n] \times [1, m]$, *are provable.*

Proof. The left to right direction follows immediately from the cut rule.
For the right to left direction, we choose F to be the formula $O(\Delta'_1)N \cdots N O(\Delta'_m)$. The provability of $\vdash F, \Delta_i$ follows from m-1 applications of the N-rule. The formula $F^\perp$ has the form $O(\Delta'_1)^\perp \oplus \cdots \oplus O(\Delta'_m)^\perp$. To prove $\vdash F^\perp, \Delta'_j$ we start from the provable sequent $\vdash O(\Delta'_j)^\perp, \Delta'_j$ and we apply the rules $\oplus_1$ or $\oplus_2$ m-1 times. □

---

[2] $O(\Delta)$ represents the unique formula that is obtained by linking all formulas of Δ with the connective O.

2.1 Completeness conditions

The one-piece decomposition principle is interesting insofar as it preserves the completeness of the deduction procedure. This holds if it consists in applying the chaining decomposition principle and then the immediate decomposition principle. For this, the formula F which is existentially quantified must have a particular syntax: F must be *decomposable at one go*.

Definition 5. A formula $\exists X F$ is *decomposable at one go* if each positive and negative occurrence L of X in F has the following property: if L belongs to a subformula F' of F with the type O, N or $\forall$, no subformula of F' that contains L has the type $\otimes$, $\oplus$ or $\exists$.

If we decompose such a formula $\exists X F$ to make all occurrences of X emerge, we begin with applying rules of $T \uparrow$. So, we can use the chaining decomposition principle. We terminate by applying rules of $T \downarrow$ and then we can use the immediate decomposition principle.

After application of the one-piece decomposition principle, the resulting goal must have the appropriate form so that the principle of elimination of unknowns will be applied: all generated subgoals must contain one occurrence of $\overline{X}$ at most. This implies that two occurrences of X cannot belong to two distinct components of a subformula of F of type O.

2.2 Termination conditions

We have to make sure that the procedure terminates. If we choose the number of connectives as measure of the size of the sequents, we remark that at every step of a deduction, the size of the active subgoal is strictly greater than the size of each subgoal that replaces it except when we apply the principle of elimination of unknowns. In this case, we compare the size of each generated subgoal $\vdash \Delta_i, \Delta_j$ with the size of the unique subgoal $\vdash \exists X F, \Delta$ from which they come. It can be greater because, in the decomposition of F, an application of the N-rule gives rise to a duplication of context. To avoid this problem, we restrict the syntax of each formula $\exists X F$ as follows: a positive occurrence and a negative occurrence of X cannot belong to two distinct components of a subformula of F that has O as head connective.

2.3 Soundness conditions

As the instantiation of an existentially quantified variable X is delayed until the phase of elimination of unknowns, this instantiation can violate the side condition of the $\forall$-rule which could have been applied during the phase of one-piece decomposition. We avoid the presence of fresh variables in the formula that instantiates the unknown X in the phase of elimination of unknowns by constraining the syntax of formulas $\exists X F$ as follows: for every subformula $\forall Y G$ of F, X and Y do not belong to two distinct components of a subformula of G of type O.

2.4 The syntax of a decidable fragment of MALL2

Now, we are in a position to make precise the syntax of the fragment of MALL2 for which it will be possible to define a decision procedure. This fragment is denoted MALL2'. In the definition, we use an abbreviation for formulas that are existentially quantified and that have both positive and negative occurrences of their bound variable : we call them *balanced formulas*.

Definition 6. A MALL2' formula is a MALL2 formula F that has the following properties:

1. all balanced subformulas of F are decomposable at one go;
2. a balanced subformula $\exists XG$ of F cannot have occurences of X in both components A and B of a common subformula AOB of G;
3. a balanced subformula $\exists XG$ of F cannot have a positive and a negative occurrence of X, one appearing in A and the other in B in a subformula ANB of G;
4. a balanced subformula $\exists XG$ of F and a subformula $\forall YH$ of G cannot have their variables X and Y one appearing in A and the other in B in a subformula AOB of H.

2.5 The decision procedure

The decision procedure for MALL2' we define now consists essentially in constraining the $\exists$-rule with the principles of one-piece decomposition and elimination of unknowns.

Definition 7. The procedure $\mathcal{P}_2$ is a set of deductions (G_n) such that any deduction step $G_n \to G_{n+1}$ verifies the following properties:

1. if G_n does not contain any undetermined formula, G_n is deduced from G_{n+1} by application of a MALL2 inference rule.
2. if G_n contains a complex formula with an occurrence of an undetermined formula, the step $G_n \to G_{n+1}$ consists in decomposing such a formula by using a MALL2 inference rule (one-piece decomposition principle);
3. In the application of the $\exists$-rule with $\exists XF$ as the formula to be decomposed, three cases are possible:
   - if F contains only positive occurrences of X, then X is instantiated in G_{n+1} with the constant $\top$;
   - if F contains only negative occurrences of X, then X is instantiated in G_{n+1} with the constant 0;
   - in the other cases, X is instantiated with the undetermined formula $\overline{X}$.
4. if all subgoals of G_n that contain occurrences of a formula $\overline{X}$ have the form $\vdash \overline{X}, \Delta_i$ or $\vdash \overline{X}^\perp, \Delta'_j$, we replace them with all possible combinations $\vdash \Delta_i, \Delta'_j$ (principle of elimination of unknowns) [3].

---

[3] In this case, we bend Definition 1 slightly: many subgoals are active in the deduction step but such a step can be viewed as replacing many elementary standard steps.

Now, we come to the central theorem, which is a natural consequence of the four conditions that define the syntax of MALL2' formulas.

Theorem 8. *The procedure P_2 is a deduction procedure which is a decision procedure when restricted to MALL2'.*

Proof. Here is only a sketch of the proof, which is developed in [12]. We have first to prove that P_2 is a deduction procedure, i.e. all its elements are deductions in the sense given in Definition 1. For this, we have to show that all steps of any element $\in P_2$ are sound. For steps verifying Condition 2, the only critical case is when the applied rule is ∀. In this case, soundness is guaranteed because of Condition 4 in Definition 6. Steps produced by Condition 4 are sound because of Lemma 4.

Now, we consider only MALL2' deductions and we will show that P_2 terminates and is complete for such deductions. Proving termination for P_2 consists first in proving that every deduction is finite. This follows immediately from this lemma:

Lemma 9. *For every goal G_n of a MALL2' deduction $\in P_2$ without undetermined formulas, there is a finite number of deduction steps that follow G_n or there is a goal G_m without undetermined formulas that follows G_n and that is obtained from G_n by replacing a subgoal with a finite number of new subgoals that have a lesser size.*

Now, let us prove completeness of P_2. We have to show that, for every goal G that is constituted of provable sequents, there is a successful deduction $\in P_2$. For this, we use induction over the maximum length of deductions $\in P_2$ that start from a goal constituted of provable sequents. Since P_1 is complete, there exists a successful deduction $D \in P_1$ that starts from G. If D does not contain applications of the ∃-rule, it is a deduction $\in P_2$, otherwise we consider the first application of the ∃-rule. Let $\exists X F$ be the formula that is decomposed in this application. We must distinguish two cases according to whether F contains both positive or negative occurrences of X, or not. In both cases, we prove there is a deduction that has length one at least and that starts with G and ends with a goal G' which is constituted of provable sequents and does not contain undetermined variables. The deductions of P_2 that start from G' have a maximum length that is strictly less than those starting from G and as G' is a set of provable sequents, we can apply to it the induction hypothesis: there is a successful deduction $\in P_2$ starting from G'. By completing it, we obtain a successful deduction $\in P_2$ starting from G. □

3   Transposition to Intuitionistic Linear Logic and Lambek Calculus

This decidability result can be easily transposed to the multiplicative and additive fragment of second order intuitionistic linear logic (IMALL2).
For linguistic applications, it is also interesting to transpose our result to second

order Lambek Calculus with additives (AL2). In the framework of the one-sided sequent calculus, AL2 is obtained from MALL2 by two modifications:

- sequents are considered as lists and not as multi-sets of formulas and a rule of circular permutation is added as a structural rule;
- the syntax of formulas is constrained by means of polarities [9]; in this way, formulas are considered either as output formulas or input formulas; all sequents must include one output formula exactly.

In this way, AL2 can be viewed as a fragment of the Yetter calculus [13] and the formulas that constitute the decidable fragment of AL2, are the formulas of MALL2' that respect the rules of polarity. In particular, they include the five formulas of M.Emms [3] but an infinite number of others too. Let us illustrate the execution of our decision procedure in L2 with additives by taking an example borrowed from computational linguistics [2].

Example 1. Parsing the German sentence *er findet und hilft Frauen* presents a difficulty caused by a coordination of unlike categories. The verb *findet* requires an accusative complement and the verb *hilft* a dative complement so that the conjunction *findet und hilft* requires a complement that can be used in both accusative and dative contexts : *Frauen* has this property. In the theory of Categorial Grammar [11], each grammatical category is represented by a logical formula, which usually comes from the Lambek Calculus. A lexicon gives the categories associated with the words of a natural language. For example, such a lexicon can give the following entries for the words of the sentence *er findet und hilft Frauen*:

$$er : np(n)/(np(n)\backslash s)$$
$$findet : np(n)\backslash s/np(a)$$
$$und : \forall X \forall Y (X\backslash (X \oplus Y)/Y)$$
$$hilft : np(n)\backslash s/np(d)$$
$$Frauen : np(n) \mathrm{N} np(a) \mathrm{N} np(d)$$

The atomic formulas np and s stand respectively for the categories *noun phrase* and *sentence*. The argument of np stands for the case of the noun phrase: nominative, accusative or dative. The connectives $\backslash$ and $/$ represent the left and right implications of the Lambek Calculus. The polymorphic category of the conjunction is represented by a second order formula.

To establish the wellformedness of the sentence *er findet und hilft Frauen* amounts to prove the sequent $np(n)/(np(n)\backslash s)$, $np(n)\backslash s/np(a)$, $\forall X \forall Y (X\backslash (X \oplus Y)/Y)$, $np(n)\backslash s/np(d)$, $np(n) \mathrm{N} np(a) \mathrm{N} np(d) \vdash s$ in the Lambek Calculus. In the framework of the one-sided sequent calculus, it amounts to proving the sequent $\vdash s$, $np(d)^{\perp} \oplus np(a)^{\perp} \oplus np(n)^{\perp}$, $np(d) \otimes s^{\perp} \otimes np(n)$, $\exists X \exists Y (Y \otimes (Y^{\perp} \mathrm{N} X^{\perp}) \otimes X$, $np(a) \otimes s^{\perp} \otimes np(n)$, $(np(n)^{\perp} O s) \otimes s^{\perp}$ in non-commutative linear logic. All formulas of this sequent are MALL2' formulas. So, we can apply the decision procedure $\mathcal{P}_2$ in the context of non-commutative linear logic. In all deduction steps, we use the following abbreviations for the inactive formulas:

$$F_{Frauen} = np(d)^{\perp} \oplus np(a)^{\perp} \oplus np(n)^{\perp}$$
$$F_{hilft} = np(d) \otimes s^{\perp} \otimes np(n)$$
$$F_{findet} = np(a) \otimes s^{\perp} \otimes np(n)$$
$$F_{er} = (np(n)^{\perp} O s) \otimes s^{\perp}$$

After a phase of one-piece decomposition, we obtain the following derivation:

$$
\cfrac{
  \cfrac{
    \vdash F_{hilft}, \overline{Y} \qquad
    \cfrac{
      \vdash s, F_{Frauen}, \overline{Y}^{\perp}, F_{er} \qquad \vdash s, F_{Frauen}, \overline{X}^{\perp}, F_{er}
    }{
      \vdash s, F_{Frauen}, \overline{Y}^{\perp} N \overline{X}^{\perp}, F_{er}
    } N
  }{
    \vdash s, F_{Frauen}, F_{hilft}, \overline{Y} \otimes (\overline{Y}^{\perp} N \overline{X}^{\perp}), F_{er}
  } \otimes \qquad \vdash \overline{X}, F_{findet}
}{
  \cfrac{
    \cfrac{
      \vdash s, F_{Frauen}, F_{hilft}, \overline{Y} \otimes (\overline{Y}^{\perp} N \overline{X}^{\perp}) \otimes \overline{X}, F_{findet}, F_{er}
    }{
      \vdash s, F_{Frauen}, F_{hilft}, \exists Y (Y \otimes (Y^{\perp} N \overline{X}^{\perp}) \otimes \overline{X}), F_{findet}, F_{er}
    } \exists
  }{
    \vdash s, F_{Frauen}, F_{hilft}, \exists X \exists Y (Y \otimes (Y^{\perp} N X^{\perp}) \otimes X), F_{findet}, F_{er}
  } \exists
} \otimes
$$

Then, by application of the principle of elimination of unknowns to the four resulting subgoals, we replace them with the following subgoals:

$$\vdash s, F_{Frauen}, F_{hilft}, F_{er} \qquad \vdash s, F_{Frauen}, F_{findet}, F_{er}$$

which correspond respectively to the sentences *er findet Frauen* and *er hilft Frauen*. Using the procedure $\mathcal{P}_2$, we have shown that the wellformedness of the initial sentence is a consequence of the wellformedness of these ones.

Conclusion

By generalising the result of M. Emms, this study can help us in understanding the foundations of undecidability problems in second order linear logic. Moreover, as the last example illustrates it, both additives and second order formulas can provide us with tools to express some subtle linguistic phenomena in the theory of Categorial Grammar. From this, the decision procedure can be used to parse natural languages.

A first direction for future research would be to extend the decidable fragment of MALL2 while keeping a simple enough definition of its syntax. Another direction would be to use the principles that ground the decision procedure $\mathcal{P}_2$ (immediate and chaining decomposition, delayed variable instantiation ...) to design deduction procedures for whole MALL2. Such procedures cannot be obviously decision procedures, however they can reduce non-determinism in proof search significantly.

Acknowledgments

Thanks to François Lamarche who reread this article.

References

1. J. M. Andreoli. Logic programming with focusing proofs in linear logic. *Journal of Logic and Computation*, 2(3):297–347, 1992.
2. S. Bayer and M. Johnson. Features and agreement. In *33rd Meeting of the Association for Computational Linguistics, San Francisco*, pages 70–76, 1995.
3. M. Emms. Parsing with Polymorphism. In *6th Conference of the European Asociation for Computational Linguistics, Utrecht*, pages 120–129, 1993.
4. M. Emms. An undecidibility result for polymorphic lambek calculus. Manuscript, 1995.
5. D. Galmiche and G. Perrier. Foundations of proof search strategies design in linear logic. In A. Nerode and Yu.V. Matiyasevich, editors, *Proceedings of Logical Foundations of Computer Science, St Petersburg, Russia, July 1994*, volume 813 of *Lecture Notes in Computer Science*, pages 101–113. Springer Verlag, 1994.
6. D. Galmiche and G. Perrier. On proof normalisation in linear logic. *Theoretical Computer Science*, 135(1):67–110, December 1994.
7. Y. Lafont. The undecidability of second order linear logic without exponentials. *Journal of Symbolic Logic*, to appear. Also available as preprint 95-06, Laboratoire de Mathématiques discrètes, University of Marseille.
8. Y. Lafont and A. Scedrov. The undecidability of second order multiplicative linear logic. Preprint 95-17, Laboratoire de Mathématiques discrètes, University of Marseille, 1995.
9. F. Lamarche. From proof nets to games. *Electronic Notes in Theoretical Computer Science*, 3, 1996. Special Issue of Linear Logic'96, Tokyo Meeting, march 1996.
10. Patrick Lincoln, Andre Scedrov, and Natarajan Shankar. Decision problems for second order linear logic. In D. Kozen, editor, *Tenth Annual IEEE Symposium on Logic in Computer Science*, pages 476–485, San Diego, California, June 1995.
11. M. Moortgart. Categorial Type Logics. In J. van Benthem and A. ter Meulen, editors, *Handbook of Logic and Language*, chapter 2. Elsevier, 1996.
12. G. Perrier. A decidable fragment of Second Order Linear Logic. Research Report 97-R-007, CRIN-CNRS, Nancy, January 1997. Available at *http://www.loria.fr/~perrier/papers.html*.
13. D. N. Yetter. Quantales and (noncommutative) linear logic. *Journal of Symbolic Logic*, 55(1):41–64, March 1990.

Some Results on Propositional Dynamic Logic with Fixed Points

Igor Rents[1] and Nikolaj Shilov[2]

[1] Institute of Computational Technologies, Novosibirsk-90, Lavrentjev ave., 6, Russia, 630090, e-mail: ir@net.ict.nsk.su
[2] Institute of Informatics Systems, Novosibirsk-90, Lavrentjev ave., 6, Russia, 630090, e-mail: shilov@iis.nsk.su

Abstract. We present a Gentzen-style cut-free sound and complete axiomatization for Propositional Dynamic Logic (PDL). The axiomatization exploits the conservative extension of PDL by means of a new program constructor for v-times iteration of a program, where v has a natural number value. Then we expand our axiomatization PDL to cover the extension of PDL by the least and the greatest fixed points (PDL+MuC).

1 Introduction

Propositional Dynamic Logic (PDL) [1, 2] is a polymodal program logic with program constructs for sequential composition ;, non-deterministic choice ∪, non-deterministic iteration ⋆ and test ?. For PDL, some proof formalization methods such as the tableux proof method and Hilbert-type axiomatizations are known [2]. Gentzen-style sequent calculus with cut for PDL without iteration has been developed in [8]. And quite recently, a Gentzen-style cut-free sequent calculus for PDL without iteration has been developed in [7].

In the second part of the paper we propose a Gentzen-style cut-free calculus for PDL and present the soundness and completeness theorem. This calculus deals with a so-called extended PDL and i-sequents. The extended PDL has one additional program constructor for v-times iteration, where v has a natural number value. An i-sequent is a sequent with induction assumptions. The soundness and completeness theorem is based on the Small Model Property of PDL [2].

Mu-Calculus (MuC) is a propositional polymodal logic with least (μ) and greatest (ν) fixed point operators [3, 4, 5]. Propositional Dynamic Logic with fixed points (PDL+MuC) is the extension of PDL by the same fixed points μ and ν [9]. Second Order Propositional Dynamic Logic of program schemata (SO-PDL) is a variant of Propositional Dynamic Logic (PDL) with second order quantifiers. Unfortunately SO-PDL is undecidable, but the validity problem in Herbrand Models (HM) is decidable with a single exponential upper bound [10].

Another algorithmic problem for PDL+MuC is axiomatization. In the original paper [3], Kozen proposed a very natural sound axiomatization for MuC, but the completeness of the axiomatization was proved for a fragment of MuC - for conjunctive formulae only. A complete axiomatization of MuC remained an unsolved problem for 10 years. The problem was solved by I. Walukiewicz in

1993 [6]. Since MuC is more expressive than PDL and there exists the standard translation of PDL into MuC [3, 5] then an axiomatization of MuC can be extended to an axiomatization of PDL+MuC by presentating the translation rules as new axioms. In the framework of an axiomatization of PDL+MuC of such this kind, formulae of PDL+MuC will be deductively equivalent to formulae of MuC and the main steps of a deduction of a formula will be done in terms of MuC.

Our approach to the axiomatization of PDL+MuC is different. We ensure that under our axiomatization of PDL+MuC, any formula of PDL+MuC will be deductively equivalent to a formula of PDL and the main steps of a deduction of the formula will be done in terms of PDL.

The third part of the paper starts with the syntax, semantics and some expressability results for PDL, MuC, PDL+MuC, SO-PDL and a special fragment of PDL+MuC — the so-called μ-formulae. We demonstrate that μ-formulae are more expressive then PDL and also design a Hilbert-style axiomatization of PDL+MuC. This axiomatization is an extension of the Hilbert-style sound and complete axiomatization of PDL [2]. In the framework of this axiomatization a simple rewriting strategy leads from a μ-formula to a proper deductively equivalent PDL formula. We also present the soundness of this axiomatization for PDL+MuC and completeness for μ-formulae. The proof of the soundness exploits some features of SO-PDL. It turns out that in this axiomatization, μ-formulae are deductively equivalent to PDL formulae. At the end of the third part, we design a sound Gentzen-style axiomatization for PDL+MuC as the conservative extension of the Gentzen-style i-sequent calculus of PDL and the Hilbert-type axiomatization of PDL+MuC developed in the paper.

This work is supported by the Russian Foundation for Basic Research (RFBR) and the International Association for the Promotion of Cooperation with Scientists from the Independent States of Former Soviet Union (INTAS), contract INTAS-RFBR 95-0378.

2   Gentzen-type I-Sequent Calculus for PDL

Propositional Dynamic Logic (PDL) [1, 2] is a polymodal program logic with program constructs. The syntax of PDL is constructed from a countable alphabet of (program) symbols and a countable alphabet of (propositional) variables. The syntax consists of programs (structured program schemata) and (logical) formulae which are defined by mutual induction as usual [2]. The semantics of PDL is defined in Kripke structures $M = < D, I >$ where program symbols are interpreted as binary relations over a domain D and propositional variables - as unary predicates on D in the usual manner [2].

Remark. We suppose that basic propositional operations in all variants of PDL are negation ¬, conjunction ∧ and disjunction ∨. The implication → and equivalence ↔ are usual abbreviations only.

Let us extend the syntax of PDL with a countable alphabet of natural value variables and the set of PDL programs by means of new program construct: for any natural value variable v and any program α let α^v be a program too. Semantics of this new construction with an evaluation $E(v) \in Nat$ is $\alpha^{E(v)}$ where α^0 is $true?$ and α^{i+1} is $(\alpha^i; \alpha)$ for any $i \geq 0$. In the extended PDL we will consider the formulae without double instances of any natural value variable.

Definition 1. A sequent is an ordered pair $\Gamma \vdash \Delta$ where Γ, Δ are finite sets of formulae. If S is a finite set of sequents then we will denote by $Var(S)$ a set of all natural value variables from S and for any evaluation $E : Var(S) \to Nat$ we will denote by $E(S)$ the substitution of the values of the variables into S.

Definition 2. The sequent with induction assumptions (in short i-sequent) is an ordered pair $\Sigma \| \Pi$, where Π is a sequent and Σ is a finite set of sequents (the induction assumptions of the sequent Π).

Definition 3. Let $M = < D, I >$ be a Kripke structure. The sequent $\Gamma \vdash \Delta$ is said to be valid in a state $s \in D$ (written $s \models (\Gamma \vdash \Delta)$) iff (for all $A \in \Gamma$)($s \models A$) $\Longrightarrow$ (there is some $B \in \Delta$) ($s \models B$).

Definition 4. A sequent $\Gamma \vdash \Delta$ is said to be valid (written $\models (\Gamma \vdash \Delta)$) if it is valid in all Kripke structures in all states. An i-sequent $\Sigma \| \Gamma \vdash \Delta$ is said to be valid iff for any evaluation $E : Var(\Sigma \| \Gamma \vdash \Delta) \to Nat$ the following holds: for all $(\Gamma' \vdash \Delta') \in \Sigma$ ($\models E(\Gamma' \vdash \Delta')$) $\Longrightarrow \models E(\Gamma \vdash \Delta)$).

Remark. Let us remark that a formula A is valid iff the sequent $\emptyset \vdash A$ is valid and a sequent $\Gamma \vdash \Delta$ is valid iff the i-sequent $\emptyset \| \Gamma \vdash \Delta$ is valid.

Definition 5. Gentzen type i-sequent calculus G(PDL) for the extended PDL is a set of axioms and a set of rules. An i-sequent Θ is said to be provable (written $\vdash_{G(PDL)} \Theta$) if there is a proof of the i- sequent. Below we will use the following notation:
$\Gamma^{[\alpha]} = \{A | [\alpha]A \in \Gamma\}$, $\Gamma^{<\alpha>} = \{A | < \alpha > A \in \Gamma\}$,
$< \alpha > \Gamma = \{< \alpha > A | < \alpha > A \in \Gamma\}$, $[\alpha]\Gamma = \{[\alpha]A | [\alpha]A \in \Gamma\}$,
$\Gamma_0 =$ the subset of all propositional formulae of Γ,
where Γ is a set of formulae and α is a program. For Gentzen-type cut-free axiomatization G(PDL) of the extended PDL — see the Appendix A. The rules of this system fall into three parts: modal rules $((<>\vdash), (\vdash [\,]))$, induction rules $((< * >\vdash), (\vdash [*]))$ and all the other rules are simplifying ones.

Lemma 6. Let $p_1, \ldots, p_n$ be program symbols,
$\Gamma = \Gamma_0 \cup < p_1 > \Gamma \cup \ldots \cup < p_n > \Gamma \cup [p_1]\Gamma \cup \ldots \cup [p_n]\Gamma$, and
$\Delta = \Delta_0 \cup < p_1 > \Delta \cup \ldots \cup < p_n > \Delta \cup [p_1]\Delta \cup \ldots \cup [p_n]\Delta$.
Then $\Gamma \models \Delta$ iff at least one of the following condition holds:
1. $\Gamma_0 \cap \Delta_0 \neq \emptyset$;
2. there is some $i \in \{1, \ldots, n\}$ and there is some $< p_i > A \in \Gamma$ such that $A, \Gamma^{[p_i]} \models \Delta^{<p_i>}$;
3. there is some $i \in \{1, \ldots, n\}$ and there is some $[p_i]A \in \Delta$ such that $\Gamma^{[p_i]} \models A, \Delta^{<p_i>}$.

Remark. Lemma 1 implies that if the conclusion of the modal rules is valid then it is an axiom or there is some instance of the rule such that premise is valid.

Lemma 7. *Gentzen-type i-sequent calculus G(PDL) is sound for i-sequents of the extended PDL.*

The above lemmas and the Small Model Property of PDL [2] lead to

Theorem 8. *For any PDL i-sequent* $\Sigma\| \ \Gamma \vdash \Delta$ *the following holds:* $\vdash_{G(PDL)} \ \Sigma\| \ \Gamma \vdash \Delta$ *iff* $\models \ \Sigma\| \ \Gamma \vdash \Delta$.

3 Syntax, Semantics, Expressability and Axiomatizations of PDL variants

The Propositional Mu-Calculus (MuC) is a propositional program- polymodal logic with constructions for the least and the greatest fixed points [3, 4]. The syntax of MuC is constructed from a countable alphabet of (program) symbols and a countable alphabet of (propositional) variables similar to PDL but with elementary programs (program symbols) only. Fixed point constructions are associated with propositional variables: If p is a variable then $\mu p.$ and $\nu p.$ is the pair of fixed point constructions associated with this variable and they are applicable to formulae with positive instances of the variable only. (– The notion of positive and negative instances of a subformula in a formula is standard [3, 4, 5] and correspond to even or odd number of negations over the instance in the formulae.) The semantics of MuC is defined in Kripke structures where fixed point constructions are the least and the greatest fixed points with respect to the inclusion on interpretations of propositional variables. This semantics can be defined constructively in accordance with the Tarski-Knaster theorem for fixed points of a monotonic function over subsets of a set.

Propositional Dynamic Logic with fixed points (PDL+MuC) is the combination of two logics PDL and MuC with united syntax and semantics.

The Second Order Propositional Dynamic Logic (SO-PDL) [10] is an extension of Propositional Dynamic Logic (PDL) with the pair of new modalities $\square$, $\lozenge$ and quantifiers $\forall$ and $\exists$ over propositional variables. The semantics of SO-PDL is defined in Kripke structures where modalities $\square$ and $\lozenge$ are the usual S5-modalities and the semantics of quantifiers is straightforward from their names - for all/some interpretation of a propositional variable as a unary predicate.

Remark. 1. For avoiding collisions of free and bound variable instances let us suppose that all bound variables differ from each other and from all free variables. 2. If A is a formula with a subformula B then we will use notation $A(B)$. In this context we will use notation $A(C)$ for result of the substitution on place of all instances of B in $A(B)$ instances of a formula C.

It is known that PDL is decidable with a one-exponential lower [1] and upper [2] time bounds as well as MuC and PDL+MuC [10], but PDL is less expressive then MuC [2]. Really, there exists a translation procedure of PDL to MuC [3, 5]. At the same time the following predicate Δ divides this expressive powers [2]: for any program α let $\nabla(\alpha)$ be a formula which is valid in a state s of a Kripke structure iff it is impossible to iterate α infinitely after starting from s, this predicate is not expressible in PDL but is expressible in MuC as $\mu p.(p \vee [\alpha]p)$.

In [10] is proved that SO-PDL is undecidable and that the expressive power of MuC is less than the expressive power of SO-PDL. It is obvious that the fixed points are expressible in terms of second order quantification. Simultaneously the following SO-PDL formula $\forall p. (< a > p \leftrightarrow < b > p)$, where a and b are program symbols, is not expressible in MuC and in PDL+MuC.

Finally let us define a μ-formula as a formula of PDL+MuC that each instance of ν is negative. Since the μ-formula $\mu p.(p \vee [\alpha]p)$ is equivalent to $\nabla(\alpha)$ then we can summarize expressability results as

Lemma 9. *The following expressability inequalities hold:*
$PDL < \mu$-*formulae* $\leq PDL+MuC = MuC < SO$-*PDL.*

Let us turn to an axiomatization of PDL+MuC. The idea of our approach consists in the design of a sound axiomatic system and a simple deductive strategy which establish the deductive equivalence of PDL and some fragments of PDL+MuC. Let us accept as the start point the sound and complete Hilbert-type axiomatization AS of PDL [2] with 8 Axiom Schemata and two Inference Rules: (MP) Modus Ponens $\frac{A,\ A\rightarrow B}{B}$, (G) Generalization $\frac{A}{[\alpha]A}$. We would like add to AS the new axiom scheme and two new inference rules. The axiom scheme is the following equivalence: (A9) $\neg\mu p.A(p) \leftrightarrow \nu p.\neg A(\neg p)$.

The first inference rule is a PDL+MuC version of the second order equivalence $B(\mu p.C) \leftrightarrow \forall p.(\Box(C \rightarrow p) \rightarrow B(p))$ which holds in the combined logic PDL+MuC+SO-PDL for normal formulae and an outer the least fixed point. Since the $\Box$- and $\Diamond$-modalities are absent in the combined logic PDL+MuC then we have to simulate them in some way. We will use modalities $[UNI]$ and $< UNI >$ respectively, where UNI is the program $(\cup_{a\in ACT}(a))*$ and ACT is the set of all program symbols of the formula. So we have the first new inference rule μ-Introduction (μI): $\frac{[UNI](C\rightarrow p)\rightarrow B(p)}{B(\mu p.C)}$.

The next inference rule is a PDL+MuC version of the second order equivalence $B(\nu p.C) \leftrightarrow \exists p.(\Box(p \rightarrow C) \wedge B(p))$ which holds in the combined logic PDL+MuC+SO-PDL for normal formulae and an outer the greatest fixed point. Since we have to simulate the existantional quantifier $\exists p$ by a proper formula A then we have the second new inference rule ν-Introduction (νI): $\frac{A\rightarrow C(A)\ ,\ B(A)}{B(\nu p.C)}$.

In the new inference rules UNI is the same program as above, A, B and C are normal formulae (i.e. negations may be applied to variables only), the selected fixed points are out off any other fixed point.

Lemma 10. *1. The inference rule μI is sound and invertible for the combined logic PDL+MuC.*
2. The inference rule νI is sound for the combined logic PDL+MuC.

On base of the above Lemmas the following can be proved.

Theorem 11. *1. The following expressability inequalities hold: $PDL < \mu$-formulae $\leq PDL+MuC = MuC < SO\text{-}PDL$.*
2. The axiom systems $AS + A9 + \mu I + \nu I$ is sound for $PDL+MuC$.
3. The axiom system $AS + A9 + \mu I$ is complete for μ-formulae of $PDL+MuC$.
4. The axiom system $AS + A9 + \mu I$ is incomplete for $PDL+MuC$.

Let us turn to the Gentzen-type axiomatization of PDL+MuC and extend the syntax of PDL+MuC with a countable alphabet of natural value variables, the set of programs by means of program construct as in the extended PDL and the set of formulae by means of two new constructs: for any natural value variable v, a propositional variable p and a formula A without negative instances of p let $(A(p))_v$ and $(A(p))^v$ be formulae. Semantics of this new constructions with an evaluation $E(v) \in Nat$ is the substitution of A into itself $E(v)$-times:
$(A(p))_0 = false$ and $(A(p))_{i+1} = A(\,(A(p))_i)$ for any $i \geq 0$;
$(A(p))^0 = true$ and $(A(p))^{i+1} = A(\,(A(p))^i)$ for any $i \geq 0$.
In the extended PDL+MuC we will consider the formulae without double instances of any natural value variable.

In the framework of the extended PDL+MuC notions of sequent, sequent with induction assumptions, their validity are the same as in context of the extended PDL. Gentzen type i-sequent calculus G(PDL+MuC) for the extended PDL+MuC is the extension of G(PDL) by the axiom schemata Embedding 1 and 2 and new rules G(MuC) — see the Appendix B. This new rules fall into five parts:
duality rules $((\mu - \nu \vdash), (\vdash \mu - \nu))$,
μ-introduction rules $((\mu I \vdash), (\vdash \mu I))$,
ν-introduction rules $((\nu I \vdash), (\vdash \nu I))$,
PDL-like iteration rules $((\nu \vdash), (\vdash \mu))$,
PDL-like induction rules $((\mu \vdash), (\vdash \nu))$,
which are translation into Gentzen-type form of the axiom (A9), inference rules μI, νI and a generalization of the PDL inference rules $([*] \vdash), (\vdash < * >)$ and induction inference rules $(< * > \vdash), (\vdash [*])$ respectively.

We have as a corollary from previous Lemmas and Theorems

Theorem 12. *1. $G(PDL+MuC)$ is sound, i.e. for any $PDL+MuC$ i-sequent $\Sigma \| \; \Gamma \vdash \Delta$ the following holds: if $\vdash_{G(PDL+MuC)} \; \Sigma \| \; \Gamma \vdash \Delta$ then $\models \; \Sigma \| \; \Gamma \vdash \Delta$.*
2. $G(PDL+MuC)$ is complete for μ-formulae, i.e. for any μ-formula A of $PDL+MuC$ the following holds: $\vdash_{G(PDL+MuC)} \; \emptyset \| \; \emptyset \vdash A$ iff $\models A$.

Questions.
1. Is Gentzen-type calculus G(PDL) complete for the extended PDL?
2. Are μ-formulae less expressive then PDL+MuC?
3. Is Hilbert-type calculus $AS + A9 + \mu I + \nu I$ complete for PDL+MuC?
4. Is Gentzen-type calculus G(PDL+MuC) complete for PDL+MuC?

4 Conclusion

Decidability and axiomatization are not the only algorithmic problems for program logics. An other algorithmic problem is the model-checking problem i.e. the evaluation of the validity set of a formula in a finite model. For MuC, PDL+MuC and SO-PDL this problem is decidable because the semantics of those logics can be defined constructively. But the next question arises: what about lower and upper bounds for this problem?

The model-checking problem as a mathematical problem originated as an approach to specification and verification of finite state systems. A stream of publications on applied model-checking is very wide now and can be a subject for a separate survey. We would like to point out [11] because of the importance of this paper for the verification practice. Furthermore, [11] demonstrates the following typical feature of applied computer-aided model- checking: a logic with fixed points is an internal representation of external specifications and verification is done in terms of model-checking for this logic with fixed points. The time bound for the Direct Model Checking algorithm based on the constructive semantics of MuC or PDL+MuC is exponential on the length of a formula, and it turns out that the model-checking problem for MuC is NP and co-NP problem [12] in contrast with a lot of program logics which are decidable with a one-exponential time bound as MuC itself but have a polynomial model-checking algorithm.

References

1. Fisher M.J., Ladner R.E. : Propositional dynamic logic of regular programs. J. Comput. System Sci., **18, n.2.**, (1979), 194–211
2. Harel D.: Dynamic Logic. in Handbook of Philosophical Logic, **2**, Reidel, (1984).
3. Kozen D.: Results on the Propositional Mu-Calculus. Theoretical Computer Science, **27, n.3**, (1983), 333–354.
4. Pratt V.R.: A decidable Mu-Calculus: preliminary report. 22-nd IEEE Symp. on Foundation of Computer Science, (1982), 421–427.
5. Streett R.S. Emerson E.A.: An Automata Theoretic Decision Procedure for the Propositional Mu-Calculus. Information and Computation, **81, n.3**, (1989), 249–264.
6. Walukiewicz I.: A Complete Deduction System for the μ - Calculus. Doctoral Thesis, Warsaw, (1993).
7. Bull R.A.: Cut elimination for propositional dynamic logic without *. Zeitschrift fuer Mathematische Logik und Grundlagen der Mathematik, **38**, (1992), 85–100.
8. Pliuskeviciene A.U.: Index Technique for Propositional Dynamic logic. (Russian) 10-th USSR Conf. on Math. Log., Alma-Ata, 1990, 130.
9. Shilov N.V.: Propositional Dynamic Logic with Fixed Points: Algorithmic tools for verification of Finite State Machines. Int. Symp. on Log. Found. of Comp. Sci. LFCS'92, LNCS, **620**, (1992), 452–458.
10. Shilov N.V. Program schemata vs. automata for decidability of program logics. to appear in Theoretical Computer Science, **175**, April 1997.

11. Burch J.R. Clarke E.M. McMillan K.L. Dill D.L. Hwang L.J.: Symbolic Model Checking: 10^{20} States and Beyond. Information and Computation, **98, n.2**, (1992), 142–170.

12. Emerson E.A. Jutla C.S. Sistla A.P.: On model-checking for fragments of Mu-Calculus. Int. Conf. on Computer-Aided Verification CAV'93, LNCS, **698**, (1993), 385–396.

APPENDIX A: Gentzen-type i-sequent calculus G(PDL)

Axioms are i-sequents of two types:

(Usual) $\Sigma \| \ \Gamma, A \vdash \Delta, A$

(Induction) $\Sigma \cup \{\Gamma' \vdash \Delta'\} \| \ \Gamma \vdash \Delta$
    if $\Gamma' \subseteq \Gamma, \Delta' \subseteq \Delta$

where Γ, Δ are sets of formula, Σ is a set of sequents, A is a formula.

Inference rules:

$$\frac{\Sigma \| \ \Gamma, A \vdash \Delta \quad \Sigma \cup \{\Gamma, <\alpha^v> A \vdash \Delta\} \| \ \Gamma, <\alpha><\alpha^v> A \vdash \Delta}{\Sigma \| \ \Gamma, <\alpha\star> A \vdash \Delta} (<\star>\vdash)$$

$$\frac{\Sigma \| \ \Gamma \vdash \Delta, A \quad \Sigma \cup \{\Gamma \vdash \Delta, [\alpha^v]A\} \| \ \Gamma \vdash \Delta, [\alpha][\alpha^v]A}{\Sigma \| \ \Gamma \vdash \Delta, [\alpha\star]A} (\vdash [\star])$$

$$\frac{\Sigma \| \ \Gamma, <\alpha> A \vdash \Delta \quad \Sigma \| \ \Gamma, <\beta> A \vdash \Delta}{\Sigma \| \ \Gamma, <\alpha \cup \beta> A \vdash \Delta} (<\cup>\vdash)$$

$$\frac{\Sigma \| \ \Gamma \vdash \Delta, [\alpha]A \quad \Sigma \| \ \Gamma \vdash \Delta, [\beta]A}{\Sigma \| \ \Gamma \vdash \Delta, [\alpha \cup \beta]A} (\vdash [\cup])$$

$$\frac{\Sigma \| \ \Gamma, A, B \vdash \Delta}{\Sigma \| \ \Gamma, A \wedge B \vdash \Delta}(\wedge \vdash) \qquad\qquad \frac{\Sigma \| \ \Gamma \vdash \Delta, A \quad \Sigma \| \ \Gamma \vdash \Delta, B}{\Sigma \| \ \Gamma \vdash \Delta, A \wedge B}(\vdash \wedge)$$

$$\frac{\Sigma \| \ \Gamma \vdash A, \Delta}{\Sigma \| \ \Gamma, \neg A \vdash \Delta}(\neg \vdash) \qquad\qquad \frac{\Sigma \| \ \Gamma, A \vdash \Delta}{\Sigma \| \ \Gamma \vdash \neg A, \Delta}(\vdash \neg)$$

$$\frac{\Sigma \| \ \Gamma, A \vdash \Delta \quad \Sigma \| \ \Gamma, B \vdash \Delta}{\Sigma \| \ \Gamma, A \vee B \vdash \Delta}(\vee \vdash) \qquad\qquad \frac{\Sigma \| \ \Gamma \vdash \Delta, A, B}{\Sigma \| \ \Gamma \vdash \Delta, A \vee B}(\vdash \vee)$$

$$\frac{\Sigma \| \ \Gamma, A \vdash \Delta \quad \Sigma \| \ \Gamma \vdash \Delta, B}{\Sigma \| \ \Gamma, A \to B \vdash \Delta}(\to \vdash) \qquad\qquad \frac{\Sigma \| \ \Gamma, A \vdash \Delta, B}{\Sigma \| \ \Gamma \vdash \Delta, A \to B}(\vdash \to)$$

$$\frac{\Sigma \| \ \Gamma \vdash \Delta, <\alpha><\alpha\star>A, A}{\Sigma \| \ \Gamma \vdash \Delta, <\alpha\star>A}(\vdash <\star>) \qquad\qquad \frac{\Sigma \| \ \Gamma, [\alpha][\alpha\star]A, A \vdash \Delta}{\Sigma \| \ \Gamma, [\alpha\star]A \vdash \Delta}([\star] \vdash)$$

$$\frac{\Sigma \| \ \Gamma, A, B \vdash \Delta}{\Sigma \| \ \Gamma, <A?>B \vdash \Delta}(<?>\vdash) \qquad\qquad \frac{\Sigma \| \ \Gamma \vdash \Delta, A \quad \Sigma \| \ \Gamma \vdash \Delta, B}{\Sigma \| \ \Gamma \vdash \Delta, <A?>B}(\vdash <?>)$$

$$\frac{\Sigma \| \ \Gamma \vdash \Delta, <\alpha>A, <\beta>A}{\Sigma \| \ \Gamma \vdash \Delta, <\alpha \cup \beta>A}(\vdash <\cup>) \qquad\qquad \frac{\Sigma \| \ \Gamma, [\alpha]A, [\beta]A \vdash \Delta}{\Sigma \| \ \Gamma, [\alpha \cup \beta]A \vdash \Delta}([\cup] \vdash)$$

$$\frac{\Sigma \| \ \Gamma, <\alpha><\beta>A \vdash \Delta}{\Sigma \| \ \Gamma, <\alpha;\beta>A \vdash \Delta}(<;>\vdash) \qquad\qquad \frac{\Sigma \| \ \Gamma \vdash \Delta, <\alpha><\beta>A}{\Sigma \| \ \Gamma \vdash \Delta, <\alpha;\beta>A}(\vdash <;>)$$

$$\frac{\Sigma \| \ \Gamma, [\alpha][\beta]A \vdash \Delta}{\Sigma \| \ \Gamma, [\alpha;\beta]A \vdash \Delta}([;] \vdash) \qquad\qquad \frac{\Sigma \| \ \Gamma \vdash \Delta, [\alpha][\beta]A}{\Sigma \| \ \Gamma \vdash \Delta, [\alpha;\beta]A}(\vdash [;])$$

$$\frac{\Sigma \| \ \Gamma, A \vdash \Delta \quad \Sigma \| \ \Gamma \vdash \Delta, B}{\Sigma \| \ \Gamma, [B?]A \vdash \Delta}([?] \vdash) \qquad\qquad \frac{\Sigma \| \ \Gamma, A \vdash \Delta, B}{\Sigma \| \ \Gamma \vdash \Delta, [A?]B}(\vdash [?])$$

$$\frac{\Sigma \| \ \Gamma^{[\alpha]}, A \vdash \Delta^{<\alpha>}}{\Sigma \| \ \Gamma, <\alpha>A \vdash \Delta}(<>\vdash) \qquad\qquad \frac{\Sigma \| \ \Gamma^{[\alpha]} \vdash \Delta^{<\alpha>}, A}{\Sigma \| \ \Gamma \vdash \Delta, [\alpha]A}(\vdash [\ ])$$

where Γ, Δ are sets of formulae, Σ is a set of sequents, α, β are programs, A, B are formulae, v is a new natural value variable.

APPENDIX B: Gentzen-type i-sequent calculus G(MuC)

Axioms are i-sequents of two types:
(Embedding 1) $\quad \Sigma \| \ \Gamma, (A(p))_u \vdash \Delta, (A(p))^w$
(Embedding 2) $\quad \Sigma \| \ \Gamma, (\mu p.A) \vdash \Delta, (\nu p.A)$
where Γ, Δ are sets of formula, Σ is a set of sequents, A is a formula without negative instances of a propositional variable p, u and w are natural value variables.

Inference rules:

$$\frac{\Sigma\| \; \Gamma \vdash \Delta, \nu p.\neg A(\neg p)}{\Sigma\| \; \Gamma, \mu p.A(p) \vdash \Delta}(\mu - \nu \vdash) \qquad \frac{\Sigma\| \; \Gamma, \nu p.\neg A(\neg p) \vdash \Delta}{\Sigma\| \; \Gamma \vdash \Delta, \mu p.A(p)}(\vdash \mu - \nu)$$

$$\frac{\Sigma\| \; \Gamma \vdash \Delta, (A(\mu p.A(p))), A(p)}{\Sigma\| \; \Gamma \vdash \Delta, (\mu p.A)}(\vdash \mu) \qquad \frac{\Sigma\| \; \Gamma, (A(\nu p.A(p))), A(p) \vdash \Delta}{\Sigma\| \; \Gamma, (\nu p.A(p)) \vdash \Delta}(\nu \vdash)$$

$$\frac{\Sigma \cup \{ \; \Gamma, (A(p))_v \vdash \Delta \; \}\| \; \Gamma, ((A(p))_v \vee A((A(p))_v)) \vdash \Delta}{\Sigma\| \; \Gamma, (\mu p.A(p)) \vdash \Delta}(\mu \vdash)$$

$$\frac{\Sigma \cup \{ \; \Gamma \vdash \Delta, (A(p))^v \; \}\| \; \Gamma \vdash \Delta, ((A(p))^v \wedge A((A(p))^v))}{\Sigma\| \; \Gamma \vdash \Delta, (\nu p.A(p))}(\vdash \nu)$$

$$\frac{\Sigma\| \; \Gamma, [UNI](B(p) \to p) \vdash \Delta, A(p)}{\Sigma\| \; \Gamma \vdash \Delta, A(\mu p.B(p))}(\vdash \mu I)$$

$$\frac{\Sigma\| \; \Gamma, A(p) \vdash \Delta \quad \Sigma\| \; \Gamma \vdash \Delta, [UNI](B(p) \to p)}{\Sigma\| \; \Gamma, A(\mu p.B(p)) \vdash \Delta}(\mu I \vdash)$$

$$\frac{\Sigma\| \; \Gamma, [UNI](A \to C(A)), B(A) \vdash \Delta}{\Sigma\| \; \Gamma, B(\nu p.C) \vdash \Delta}(\nu I \vdash)$$

$$\frac{\Sigma\| \; \Gamma \vdash \Delta, [UNI](A \to C(A)) \quad \Sigma\| \; \Gamma \vdash \Delta, B(A)}{\Sigma\| \; \Gamma \vdash \Delta, B(\nu p.C)}(\vdash \nu I)$$

where Γ, Δ are sets of formulae, Σ is a set of sequents, α, β are programs, A, B are formulae, v is a new natural value variable.

Quasi-Characteristic Inference Rules for Modal Logics

Vladimir V. Rybakov *

Mathematics Department, Krasnoyarsk University. Ave. Svobodnyi 79, Krasnoyarsk, 660041 Russia, *e-mail: rybakov@math.kgu.krasnoyarsk.su*

Abstract. The aim of this paper is to develop techniques characterizing quasi-characteristic inference rules for modal logics. We describe a necessary and sufficient condition for a quasi-characteristic rule to be valid on an algebra and obtain basic properties concerning derivability of quasi-characteristic rules. Using this approach we characterize all structurally complete logics with the finite model property. The main results of this paper characterize admissible quasi-characteristic inference rules for modal logics $S4$ and $K4$. We also show that the set of all quasi-characteristic inference rules admissible in the logic $S4$ have a finite basis consisting of three special rules which are precisely described.

1 Introduction, Preliminary Facts

In this paper we will investigate quasi-characteristic inference rules for modal logics. We will develop a general technique for the description and study of the family of all quasi-characteristic inference rules, but our main aim is to obtain a complete description of quasi-characteristic rules which are admissible in modal logics $S4$ and $K4$, which play an especially important role in research concerning transitive modal logics and in applications of modal logic to Computer Science. The role of inference rules in such applications consists of the simple fact that usually any atomic instruction in any procedure type programming language can be understood as an inference rule. This line is especially distinct in logical programming languages like Prolog.

We are interested in quasi-characteristic inference rules for a number of reasons. First, this notion is closely related to the algebraic concept of embedding models and structures into other ones. Second, we observe that the notion of quasi-characteristic rules is a generalization of Jankov's characteristic formulas, which played an important role in the initial research of superintuitionistic and modal logics. And third, we can using such rules easily describe structurally complete modal and superintuitionistic logics: those where the notions of admissibility and derivability rules coincide. The quasi-characteristic rules for superintuitionistic logics were invented by A. Citkin [1]. We drow on this approach to extend the notion to modal logics. In the final description of admissible quasi-characteristic rules we use the Gödel-McKinsey-Tarski translation of the G.Mints

---

* This research supported by RFBR grant 96-01-00228

rule r_M from [2] and the descriptions of n-characterizing models for modal logics from Rybakov [3], as well as certain technical results from Rybakov [4, 5]. The properties of free algebras of countable rank $\mathcal{F}_\lambda(\omega)$ from varieties $Var(\lambda)$ corresponding to logics λ, play an importante role.

For our results, we need a description of finite subdirectly irreducible pseudo-boolean and modal $K4$-algebras. With respect to pseudo-boolean algebras, it is well known that a finite pseudo-boolean algebra $\mathcal{A}$ is subdirectly irreducible iff there is greatest element ω in $\{a \mid a \in |\mathcal{A}|, a \neq \top\}$. For modal $K4$-algebras there is a similar description. Recall that we say that an element a of a modal algebra $\mathcal{A}$ is *up-stable* if $\Box a \geq a$. It is easy to show that a finite modal algebra α from $Var(K4)$ is subdirectly irreducible iff there is a greatest up-stable element ω in $(|\alpha| - \{\top\})$.

We recall the notion of characteristic formulas which goes back to Jankov who first introduced characteristic formulas for superintuitionistic logics. Let $\mathcal{A}$ be a finite subdirectly irreducible modal $K4$-algebra. As noted above, $\mathcal{A}$ has a greatest up-stable element ω. The characteristic formula for $\mathcal{A}$ is the formula $\chi(\mathcal{A})$, where

$$\chi(\mathcal{A}) := \Box\beta(\mathcal{A}) \wedge \beta(\mathcal{A}) \to p_\omega, \text{ where}$$

$$\beta(\mathcal{A}) := C(\mathcal{A}) \wedge D(\mathcal{A}) \wedge I(\mathcal{A}) \wedge N(\mathcal{A}) \wedge B(\mathcal{A}),$$

$$C(\mathcal{A}) := \bigwedge_{x,y,z \in |\mathcal{A}|, x \wedge y = z} (p_x \wedge p_y \equiv p_z),$$

$$D(\mathcal{A}) := \bigwedge_{x,y,z \in |\mathcal{A}|, x \vee y = z} (p_x \vee p_y \equiv p_z),$$

$$I(\mathcal{A}) := \bigwedge_{x,y,z \in |\mathcal{A}|, x \to y = z} (p_x \to p_y \equiv p_z),$$

$$N(\mathcal{A}) := \bigwedge_{x,y \in |\mathcal{A}|, \neg x = y} (\neg p_x \equiv p_y),$$

$$B(\mathcal{A}) := \bigwedge_{x,y \in |\mathcal{A}|, \Box x = y} (\Box p_x \equiv p_y).$$

These formulas characterize the properties of modal algebras as follows.

Lemma 1. *Let $\mathcal{A}$ be a finite subdirectly irreducible modal $K4$-algebra, also let $\mathcal{B}$ be a modal $K4$-algebra. Then $\mathcal{B} \not\models \chi(\mathcal{A})$ iff $\mathcal{A}$ is a subalgebra of a homomorphic image of $\mathcal{B}$, i.e. $\mathcal{A} \in SH(\mathcal{B})$.*

Proof. Assume that $\mathcal{A} \in SH(\mathcal{B})$, i.e. $\mathcal{A}$ is a subalgebra of $\varphi(\mathcal{B})$, where φ is a homomorphism. We take the valuation of $p_x, x \in |\mathcal{A}|$ as follows: $f(p_x)$ is a fixed element in $\varphi^{-1}(x)$. Then $\varphi(\beta(\mathcal{A})(f(p_x))) = \top$ and consequently we obtain that

$\varphi(\Box\beta(\mathcal{A})(f(p_x)) \wedge \beta(\mathcal{A})(f(p_x))) = \top$. At the same time $\varphi(f(p_\omega) = \omega \neq \top$. Therefore

$$\mathcal{B} \models \Box\beta(\mathcal{A})(f(p_x)) \wedge \beta(\mathcal{A})(f(p_x)) \nleq f(p_\omega),$$

i.e. $\mathcal{B} \nvDash \chi(\alpha)$.

Conversely, suppose $\mathcal{B} \nvDash \chi(\alpha)$, i.e. there is a valuation f of $p_x, x \in |\mathcal{A}|$ in $\mathcal{B}$ such that

$$\mathcal{B} \models \Box\beta(\mathcal{A})(f(p_x)) \wedge \beta(\mathcal{A})(f(p_x)) \nleq f(p_\omega).$$

Let $c := \Box\beta(\mathcal{A})(f(p_x)) \wedge \beta(\mathcal{A})(f(p_x))$. We take the principal filter $\nabla(c)$ generated by c. Clearly, $\nabla(c)$ is a $\Box$-filter. Then, as well known, there is the natural homomorphism φ from $\mathcal{B}$ onto the quotient-algebra $\mathcal{B}/_{\nabla(c)}$. We define the mapping g from $\mathcal{A}$ into $\mathcal{B}/_{\nabla(c)}$ as follows: $\forall x \in |\mathcal{A}|(g(x) := \varphi f(p_x))$.

First we show that g is a homomorphism. Consider, for example, the operation $\rightarrow$. Let $x, y \in |\mathcal{A}|$ and $(x \rightarrow y) = z$. Then $g(x \rightarrow y) = \varphi f(p_z)$. The element $d := (f(p_z) \equiv (f(p_x) \rightarrow f(p_y)))$ is a member of $\nabla(c)$. Therefore $\varphi(d) = \top$, i.e. $\varphi(f(p_z)) \equiv (\varphi(f(p_x)) \rightarrow \varphi(f(p_y))) = \top$. That is

$$\varphi(f(p_z)) = (\varphi(f(p_x)) \rightarrow \varphi(f(p_y))) = g(x) \rightarrow g(y).$$

That g commutes with other operations can be shown in the same way. Thus g is a homomorphism.

To show g is one-to-one mapping, suppose $g(x) = g(y)$ and $x \neq y$. Then $x \equiv y \neq \top$ and $g(x \equiv y) = \top$. Using standard arguments we can show that the set $\nabla := g^{-1}(\top)$ is a $\Box$-filter in $\mathcal{A}$, and we have $\nabla \neq \{\top\}$. Since $\mathcal{A}$ is finite, ∇ is a principal finite filter and its smallest element a is up-stable. Since $\nabla \neq \{\top\}$, we obtain $a \neq \top$. The fact that ω is greatest up-stable element of $\mathcal{A}$ among distinct with $\top$ elements entails that $a \leq \omega$. Then $g(a) \leq g(\omega) = \varphi(f(p_\omega))$. Since $c \nleq f(p_\omega)$, we infer that $\varphi(f(p_\omega)) \neq \top$ which contradicts $g(a) = \top$. $\quad\square$

The quasi-characteristic inference rules, which we introduce below, are in some sense a generalization of the characteristic formulas. For simplicity, we first define quasy-characteristic rules for the class of finite subdirectly irreducible algebras, although there is very little difference between this and the general case. Second argument for consideration of this restricted case is the following. In fact we will need only quasi-characteristic rules of finite subdirectly irreducible algebras for characterization of structurally complete modal logics with the finite model property (fmp). After giving this characterization we will return to the general case of quasi-characteristic rules. In what follows a modal logic means a modal logic extending the modal system $K4$.

We introduce the *quasi-characteristic inference rules* for finite subdirectly irreducible algebras in the following way. Suppose $\mathcal{A}$ is a certain finite subdirectly irreducible pseudo-boolean algebra or a certain finite and subdirectly-irreducible modal $K4$-algebra. We denote by ω the greatest element among those distinct from $\top$ for pseudo-boolean algebras and the greatest up-stable element among those distinct from $\top$ for modal algebras (recalling that an element a of a modal algebra is up-stable if $\Box a \geq a$).

Definition 2. Suppose $\mathcal{A}$ consists of the elements $\{a_i \mid 1 \leq i \leq n\}$. The *quasi-characteristic* inference rule for λ is the rule $r(\mathcal{A})$, where

$$r(\mathcal{A}) := \frac{\{x_a * x_b \equiv x_{a*b} \mid a, b \in \alpha\} \cup \{\circ x_a \equiv x_{\circ a} \mid a \in \alpha\}}{x_\omega},$$

where $*, \circ$ are all possible binary and unary signature operations on $\mathcal{A}$.

Note that this definition corresponds in the case of finite algebras to the general definition from Citkin [1]. We say an algebra $\mathcal{B}$ *invalidates* an inference rule r if there is a valuation of the variables from r in $\mathcal{B}$ such that all the formulas in the premise of r receive under this valuation the value T but the conclusion of r receives another value. The most important property of quasi-characteristic rules of finite algebras is described in the following theorem.

Theorem 3. *Suppose $\mathcal{A}$ is a certain finite subdirectly irreducible pseudo-boolean or modal $K4$-algebra. Then, for any algebra $\mathcal{B}$, where $\mathcal{B} \in Var(H)$ if $\mathcal{A}$ is a pseudo-boolean algebra, and $\mathcal{B} \in Var(K4)$ if $\mathcal{A}$ is a modal algebra, the inference rule $r(\mathcal{A})$ is invalid in $\mathcal{B}$ iff $\mathcal{A}$ is isomorphically embeddable in $\mathcal{B}$.*

Proof. It is clear that the inference rule $r(\mathcal{A})$ is invalid in $\mathcal{A}$ under the valuation $x_a \to a$. Therefore if $\mathcal{A}$ is embeddable in $\mathcal{B}$ then $r(\mathcal{A})$ is invalid in $\mathcal{B}$. Suppose $r(\mathcal{A})$ is invalid in $\mathcal{B}$. Then there is a valuation $x_d \to c_d \in \mathcal{B}$ of all variables x_d from $r(\mathcal{A})$ in $\mathcal{B}$ such that

$$(\forall a, b \in \mathcal{A})(c_a * c_b = c_{a*b}), \quad (\forall a \in \alpha)(\circ c_a = c_{\circ a})$$

but $c_\omega \neq T$. Then, taking into account the equalities presented above, we conclude that the mapping $h : a \to c_a$ is a homomorphism from $\mathcal{A}$ into $\mathcal{B}$. If the homomorphism h is not one-to-one then there is an element $c \in \mathcal{A}$ such that $c \neq 1$ but $h(c) = T$. If $\mathcal{A}$ is a pseudo-boolean algebra then $h(\omega) = T$, which contradicts $h(\omega) = c_\omega \neq T$. If $\mathcal{A}$ is a modal $K4$-algebra then since $\mathcal{A}$ is finite, we can assume that c is up-stable. Then $c \leq \omega$ and it follows that $h(\omega) = T$ which contradicts $h(\omega) = c_\omega \neq T$. Thus h is an isomorphism. $\square$

Using this theorem we can easily give a description of structurally complete modal and superintuitionistic logics with the fmp.

Corollary 4. *A modal logic λ with fmp extending $K4$ or a superintuitionistic logic λ with fmp is structurally complete iff any finite subdirectly irreducible algebra from the variety $Var(\lambda)$ is embeddable in the free algebra $\mathcal{F}_\lambda(\omega)$ of countable rank from $Var(\lambda)$.*

Proof. Necessity: Suppose λ is a structurally complete modal logic over $K4$ or a superintuitionistic logic. And suppose that λ has the fmp and there is a finite subdirectly irreducible algebra $\mathcal{A}$ from $Var(\lambda)$ which is not a subalgebra of $\mathcal{F}_\lambda(q)$ for all q. According to Theorem 3 this entails $r(\mathcal{A})$ is valid in the algebra $\mathcal{F}_\lambda(\omega)$; that is, $r(\mathcal{A})$ is admissible in λ. At the same time, the formula $D(r(\mathcal{A}))$ describing $r(\mathcal{A})$ is false in $\mathcal{A}$ since $\mathcal{A}$ invalidates $r(\mathcal{A})$. Thus $r(\mathcal{A})$ is not derivable in λ, a contradiction. *Sufficiency* can be shown by simple standard arguments.
$\square$

2 Main Results

Now we are going to describe the concept of quasi-characteristic rules in general, following the original definition of A. Citkin [1]. For this we need the definition of *finite presentation* for algebras, which is the well-known notion from universal algebra.

Definition 5. An algebra B is called *finitely presented* in a variety Var if the following hold. Algebra B is generated by a finite set of elements $\{a_1, ..., a_n\}$ and there is a finite list of terms $\gamma_i, \delta_i, 1 \leq i \leq m$ in the signature of A depending on variables $\{x_1, ..., x_n\}$ such that

   (a) $B \models \gamma_i(a_1, ..., a_n) = \delta_i(a_1, ..., a_n), 1 \leq i \leq m$;
   (b) for every algebra $A \in Var$, if for some $\{b_1, ..., b_n\} \subseteq |A|$
   $$A \models \gamma_i(b_1, ..., b_n) = \delta_i(b_1, ..., b_n), 1 \leq i \leq m$$
   then the mapping $a_i \to b_i$ can be extended to a homomorphism from B into A.

Suppose A is a pseudo-boolean algebra or a modal $K4$-algebra. Suppose that if A is a pseudo-boolean algebra then A has a greatest element ω among its elements which are distinct from $\top$ and if A is a modal algebra then A has a greatest up-stable element ω among its up-stable elements distinct from $\top$. Suppose that A is generated by $\{a_1, ..., a_n\}$, and the formulas $\gamma_i, \delta_i, 1 \leq i \leq m$ depending on the variables $\{x_1, ..., x_n\}$ are a finite presentation of A (in $Var(H)$ or $Var(K4)$ respectively). And suppose $\chi(x_1, ..., x_n)$ is a formula such that $\chi(a_1, ..., a_n) = \omega$.

Definition 6. The quasi-characteristic rule for A is the rule $r(A)$, where

$$r(A) := \frac{\{\gamma_i \equiv \delta_i \mid 1 \leq i \leq m\}}{\chi}.$$

It is clear that all finite algebras (of finite signature) have certain finite presentations. Also note that quasi-characteristic rules as defined above are not uniquely determined by the generating algebra A, since any given A may have distinct finite presentations and the choice of formula χ is also not uniquely determined. However, it will be shown below that all the quasi-characteristic rules of the same algebra A are pairwise equivalent in a semantic sense. But first we present the most important property of quasi-characteristic rules which is due to Citkin [1].

Theorem 7. *A quasi-characteristic rule $r(A)$ of an algebra A is invalid in an algebra B ($B \in Var(H)$ or $B \in Var(K4)$, respectively) if and only if A is isomorphically embeddable in B.*

Proof. It is clear that $r(A)$ is invalid in A under the valuation $x_i \to a_i$. Thus if A is embeddable into B then $r(A)$ is invalid in B as well. Now suppose $r(A)$ is invalid in B. Then there is a valuation $x_i \to c_i \in B$ of all the variables $x_1, ..., x_n$ from $r(A)$ in B such that

$$\gamma_i(c_1, ...c_n) = \delta_i(c_1, ...c_n), 1 \leq i \leq m,$$
$$\chi(c_1, ...c_n) \neq \top.$$

Since formulas $\gamma_i, \delta_i, 1 \le i \le m$ form a finite presentation for $\mathcal{A}$, there is a homomorphism of $\mathcal{A}$ into $\mathcal{B}$ such that $h(a_i) := c_i$. Suppose that h is not a one-to-one mapping. Then there is an element $b \ne \top$ such that $h(b) = \top$. If $\mathcal{A}$ is a pseudo-boolean algebra then $h(\omega) = \top$, which contradicts

$$h(\omega) = h(\chi(a_1, ...a_n)) = \chi(c_1, ...c_n) \ne \top.$$

If $\mathcal{A}$ is a modal $K4$-algebra then for some $b \ne \top$, $h(\Box b) = \top$ holds. If $\Box b = \top$ then b is up-stable element, i.e. $b \le \omega$, which entails $h(\omega) = \top$. If $\Box b \ne \top$ then $\Box b \le \Box\Box b$, i.e. $\Box b$ is up-stable and $\Box b \le \omega$. Thus again $h(\omega) = \top$. So in any case we have $h(\omega) = \top$. But similar to above we have $h(\omega) = h(\chi(a_1, ...a_n)) = \chi(c_1, ...c_n) \ne \top$, a contradiction. Thus h is an isomorphism into. $\qquad\square$

Corollary 8. *Let $r(\mathcal{A})$ be a quasi-characteristic rule for $\mathcal{A}$ and r_1 be a rule in the same language. Then $r(\mathcal{A})$ is a semantic corollary of r_1 (in $Var(H)$, or in $Var(K4)$, respectively) iff r_1 is invalid in $\mathcal{A}$.*

Proof. The rule $r(\mathcal{A})$ is a semantic corollary of r_1 iff for every algebra $\mathcal{B}$ (of the corresponding type), if $r(\mathcal{A})$ is invalid in $\mathcal{B}$ then r_1 is invalid in $\mathcal{B}$. By Theorem 7 the latter is equivalent to the following: for every $\mathcal{B}$, the fact that $\mathcal{A}$ is a subalgebra of $\mathcal{B}$ implies that r_1 is invalid in $\mathcal{B}$; equivalently, r_1 is disprovable in $\mathcal{A}$. $\qquad\square$

From this corollary we immediately derive the following

Corollary 9. *If r_1 and r_2 are quasi-characteristic rules of the same algebra $\mathcal{A}$ then r_1 and r_2 are equivalent in the semantic sense.*

Corollary 10. *The rule $r(\mathcal{A})$ is a semantic corollary of a family of inference rules $\mathcal{R}$ iff there is a rule $r_1 \in \mathcal{R}$ such that $r(\mathcal{A})$ is a semantic corollary of r_1.*

Proof. For suppose $\mathcal{R} \models r(\mathcal{A})$. Then since $r(\mathcal{A})$ is false in $\mathcal{A}$, there is a rule $r_1 \in \mathcal{R}$ such that r_1 is false in $\mathcal{A}$. Then by Corollary 8 $r(\mathcal{A})$ is a semantic corollary of r_1. $\qquad\square$

Corollary 11. *Let $r(\mathcal{A})$ be a quasi-characteristic rule of an algebra $\mathcal{A}$. The rule $r(\mathcal{A})$ is admissible in a logic λ, where $\lambda \subseteq \lambda(\mathcal{A})$ (and λ is a superintuitionistic logic, or is a modal logic over $K4$, according to the type of $\mathcal{A}$), iff $\mathcal{A}$ is not embeddable isomorphically in the free algebra $\mathcal{F}_\lambda(\omega)$.*

Proof. Indeed, if $r(\mathcal{A})$ is admissible in λ then we know that $r(\mathcal{A})$ is valid in $\mathcal{F}_\lambda(\omega)$. Consequently $\mathcal{A}$ is not a subalgebra of $\mathcal{F}_\lambda(\omega)$ since $r(\mathcal{A})$ is invalid in $\mathcal{A}$. Conversely, if $\mathcal{A}$ is not a subalgebra of $\mathcal{F}_\lambda(\omega)$ then by Theorem 7, $r(\mathcal{A})$ is valid in $\mathcal{F}_\lambda(\omega)$ and consequently $r(\mathcal{A})$ is admissible in λ. $\qquad\square$

Definition 12. Suppose $\mathcal{F}_1 := \langle F_1, R_1 \rangle$ and $\mathcal{F}_2 := \langle F_2, R_2 \rangle$ are certain frames. The consequent assembling of $\mathcal{F}_1$ and $\mathcal{F}_2$ is the frame $\mathcal{F}_1 \oplus \mathcal{F}_2$ which is based upon disjoint union of F_1 and F_2 and has the accessibility relation including R_1 and R_2 and the relation consisting of all pairs with a first element from F_1 and a second element from F_2.

Definition 13. A cluster C of a frame $\mathcal{F} := \langle F, R \rangle$ is a *node* if $\mathcal{F} = \mathcal{F}_1 \oplus C^{R\leq}$, where $\mathcal{F}_1$ is the frame obtained from $\mathcal{F}$ by removing all elements from $C^{R\leq}$, and the frame $C^{R\leq}$ has the accessibility relation generated from $\mathcal{F}$.

We are going to describe admissible quasi-characteristic rules for some popular modal logics. For this we need the following well known rule introduced by G. Mints [2], which is admissible but non-derivable in intuitionistic logic.

$$r_M := \frac{(x \to y) \to x \vee z}{((x \to y) \to x) \vee ((x \to y) \to z)}.$$

More precisely, we will use the Gödel-McKinsey-Tarski translation of Mint's rule:

$$T(r_M) := \frac{[\Box(\Box x \to \Box y) \to \Box x] \vee \Box z}{([\Box(\Box x \to \Box y) \to \Box x) \vee (\Box(\Box x \to \Box y) \to \Box x)] \vee \Box z}.$$

and the following two rules:

$$r_{1,1} := \frac{\Diamond x \wedge \Diamond \neg x}{y},$$

$$r_{2,2} := \frac{\alpha_1 \wedge \alpha_2 \to \Box(\alpha_1 \wedge \alpha_2 \to \Diamond(\alpha_1 \wedge \alpha_2 \wedge y) \wedge \Diamond(\alpha_1 \wedge \alpha_2 \wedge \neg y))}{\neg(\alpha_1 \wedge \alpha_2)},$$

$$\alpha_1 := \Diamond(z_1 \wedge \Box\neg z_2), \quad \alpha_2 := \Diamond(z_2 \wedge \Box\neg z_1).$$

Definition 14. Let $\mathcal{F}$ be a transitive reflexive frame. The *skeleton* of $\mathcal{F}$ is the frame obtained from $\mathcal{F}$ by contracting any cluster into a single reflexive element.

Theorem 15. *Let B be a certain finite modal subdirectly-irreducible algebra from $Var(S4)$, i.e. let B have a greatest up-stable element ω among its up-stable elements which differ from $\top$. Then the following statements are equivalent:*

> *(i) B is isomorphically embeddable in $\mathcal{F}_{S4}(\omega)$;*
> *(ii) rules $T(r_M), r_{1,1}$ and $r_{2,2}$ are valid in B;*
> *(iii) (a): The skeleton of B^+ is a rooted poset which has the form $\mathcal{F}_1 \oplus \ldots \oplus \mathcal{F}_k$, where any $\mathcal{F}_i$ is a rooted poset of width not more than 2 and $\|\mathcal{F}_i\| \leq 3$ or $\mathcal{F}_i$ is 4-element distributive lattice, or $\mathcal{F}_i$ is the two-element antichain;*
> *(b): Some cluster of B^+ maximal by the accessibility relation is a singleton;*
> *(c): For any two incomparable elements a, b from B^+, there is a single-element cluster c in B^+ which is a co-cover for a and b.*

Proof. Since $T(r_M), r_{1,1}$ and $r_{n,1}$ are admissible in $S4$, these rules are valid in $\mathcal{F}_{S4}(\omega)$, therefore (i) implies (ii).

(ii) $\Rightarrow$ (iii). Since B has the element ω, B^+ is a finite rooted reflexive transitive frame. Since $T(r_M)$ is valid in B and B^+, it follows that r_M is valid in the skeleton of B^+. Using these observations, we can show that the skeleton of B^+ has the form required in (a) of (iii). If all clusters C of B^+ maximal by the accessibility relation are proper then it is easy to see that $r_{1,1}$ is invalid in B^+ and in B, a contradiction. Thus (b) from (iii) holds. If a and b are elements incomparable in B^+ then there are co-cover clusters for a and b since B^+ has a width of not more than 2 and (a). If all these co-cover clusters C_i are proper then it is not hard to check that $r_{2,2}$ is invalid in B^+ and B, a contradiction. Thus (c) from (iii) also holds.

(iii) $\Rightarrow$ (i). It is easy to see that if B^+ has the form described in (iii) then B^+ is a p-morphic image of the frame of the n-characterizing model $Ch_{S4}(k)$ of $S4$ for some k (see Rybakov [3]). Therefore B is a subalgebra of $Ch_{S4}(k)^+$, then because $r(B)$ is invalid in B (Theorem 3), it follows that $r(B)$ is invalid in $Ch_{S4}(k)^+$. Using results and technique from Rybakov [4, 5] we can derive that $r(B)$ is invalid in $\mathcal{F}_{S4}(\omega)$. Applying Theorem 7 it follows that A is a subalgebra of $\mathcal{F}_{S4}(\omega)$ □

Using this theorem, Corollaries 8, 10 and 11 we immediately derive

Corollary 16. *A quasi-characteristic rule $r(B)$ of a finite modal algebra B is admissible in $S4$ iff $r(B)$ is a semantic consequence of rules $T(r_M), r_{1,1}$ and $r_{2,2}$. In other words, rules $T(r_M), r_{1,1}$ and $r_{2,2}$ form a basis in $S4$ for the class of all admissible in $S4$ quasi-characteristic rules of all finite subdirectly irreducible modal $S4$-algebras.*

We also can describe all finite modal $S4$-algebras such that all rules admissible in $S4$ are valid in these algebras.

Corollary 17. *All rules admissible in $S4$ are valid in a finite modal $S4$-algebra B if and only if all rules of the form $\Box\alpha \vee \Box u/\Box\beta \vee \Box u$, where u is a new variable which does not occur in α, β, and α/β is any rule from the list $T(r_M), r_{1,1}, r_{2,2}$, are valid in B*

Proof. If all admissible in $S4$ rules are valid in B then the above mentioned rules also are valid in B because $T(r_M), r_{1,1}$ and $r_{2,2}$ are admissible in $S4$, and consequently all the rules $\Box\alpha \vee \Box u/\Box\beta \vee \Box u$ are admissible in $S4$ as well (because $S4$ has the disjunction property). Conversely, suppose that all the rules $\Box\alpha \vee \Box u/\Box\beta \vee \Box u$ are valid in B. We take any maximal rooted sharp open subframe $\mathcal{F}_i$ of B^+. If some α/β would be false in $\mathcal{F}_i$ we would have $\Box\alpha \vee \Box u/\Box\beta \vee \Box u$ is false in B^+ and in B, a contradiction. Therefore all the rules α/β are valid in any $\mathcal{F}_i$ and in any $\mathcal{F}_i^+$. Then by Theorem 15, all admissible in $S4$ rules are valid in any algebra $\mathcal{F}_i^+$ and any frame $\mathcal{F}_i$, consequently all of them also are valid in the frame B^+ and in the modal algebra B. □

Now we turn to consider the situation concerning admissibility of quasi-characteristic rules in modal system $K4$.

Theorem 18. *Let B be a certain finite modal subdirectly-irreducible algebra from $Var(K4)$, i.e. let B have the greatest up-stable element ω among up-stable elements which differ from $\top$. Let $||B|| > 2$. Then B is not embeddable in $\mathcal{F}_{K4}(\omega)$ as a subalgebra.*

Proof. It is easy to see that B^+ is a rooted transitive frame and $||B|| > 1$. It is not difficult also to see that if B^+ has no two R-maximal elements one of which is reflexive and the other one is irreflexive, then B^+ is not a p-morphic image of the frame of the n-characterizing Kripke model $Ch_{K4}(k)$ for $K4$ for any k (see Rybakov [3]). In particular, B is not embeddable in $\mathcal{F}_{K4}(\omega)$. Suppose that two such R-maximal elements exist. Then again it is impossible to arrange a p-morphism from some $Ch_{K4}(k)$ onto B^+ due to the presence of irreflexive and reflexive co-covers for any two R-incomparable elements in $Ch_{K4}(k)$. Thus B^+ is not a p-morphic image of any $Ch_{K4}(k)$. In particular, B is not a subalgebra of $Ch_{K4}(k)^+$ for any k. Therefore by Theorem 7, $r(B)$ is valid in $Ch_{K4}(k)^+$. On that basis we can conclude that $r(B)$ is valid in the algebra $\mathcal{F}_{K4}(k)$ for any k. Hence $r(B)$ is valid in $\mathcal{F}_{K4}(\omega)$. Using Theorem 7 we obtain B is not a subalgebra of $\mathcal{F}_{K4}(\omega)$ □

Applying this theorem and Corollary 11, we can immediately derive

Corollary 19. *Any quasi-characteristic rule $r(B)$ for any finite modal algebra B with $||B|| > 1$ is admissible in $K4$.*

Using this corollary it is not hard to show that there are infinitely many independent quasi-characteristic rules which are admissible but not derivable in $K4$. And also we showed that the modal logic $K4$ is entirely structurally incomplete with respect to quasi-characteristic rules: any such rule is admissible but not derivable.

References

1. CITKIN A.I. On Admissible Rules of Intuitionistic Propositional Logic. *Math. USSR Sbornik*, V.31, 1977, No.2, 279 - 288.
2. MINTS G.E. Derivability of Admissible Rules. *J. of Soviet Mathematics*, V. 6, 1976, No. 4, 417 - 421.
3. RYBAKOV V.V. Criteria for Admissibility of Inference Rules. Modal and Intermediate Logics with the Branching Property. *Studia Logica*, V.53, 1994, No.2, 203-225.
4. RYBAKOV V.V. Preserving of Admissible Inference Rules in Modal Logic. In Book: *Logical Foundations of Computer Science, Lecture Notes in Computer Science*, Eds.: A.Nerode, Yu.V.Matiyasevich, V.813, 1994, Springer-Verlag, 304-316.
5. RYBAKOV V.V. Hereditarily Structurally Complete Modal Logics. *J. of Symbolic Logic*, V.60, 1995, No.1, 266-288

Provability Logic with Operations on Proofs

Tatiana Sidon *

Department of Mathematical Logic, Faculty of Mathematics and Mechanics,
Moscow State University, Moscow, RUSSIA, 119899

Abstract. We present a natural axiomatization for propositional logic with a modal operator for formal provability (Solovay, [6]) and labeled modalities for individual proofs with operations on them (Artemov, [2]). For this purpose, the language has to be extended by two new operations. The obtained system $\mathcal{MLP}$ naturally includes both Solovay's provability logic GL and Artemov's operational modal logic $\mathcal{LP}$. We prove that the system $\mathcal{MLP}$ is decidable and arithmetically complete. We also show that $\mathcal{MLP}$ realizes all operations on proofs admitting description in the modal propositional language.

1 Introduction.

The idea of interpreting the modality $\Box$ as provability is due to Gödel. He formulated the modal system $S4$ describing the behavior of provability operator. But $S4$ failed to be correct with respect to the straightforward arithmetical interpretation of $\Box F$ as an arithmetical statement "F is provable". Propositional modal logic describing properties of formal arithmetical provability was axiomatized by Solovay. In [6] he proved that the Gödel–Löb modal system GL formalizes the set of modal principles which are provable in Peano Arithmetic **PA** under this interpretation.

The Logic of Proofs was presented by S. N. Artemov in [1]. It describes properties of both the predicates "F is provable" and "p is a proof of F". In order to be able to express the proof–theorem relation, the modal propositional language is enriched with *proof variables* p_i in addition to the usual propositional variables (called *sentence variables*). Expressions of the form $[\![p]\!]F$ (where F is a formula and p is a proof variable) are formulas of this language with the intended arithmetical semantics "p is a code for the proof of F". The system $\mathcal{B}$ presented in [1] is complete with respect to the class of arithmetical interpretations based on arbitrary proof predicates.

In [2], Artemov formulates the operational modal logic of proofs $\mathcal{LP}$. $\mathcal{LP}$ describes properties of the predicate "p is a proof of F" together with computable operations on proofs. The language of $\mathcal{LP}$ is an extension of the usual propositional language by the addition of proof variables and function symbols $\times$, $+$ and $!$ for operations on proofs. Thus $\mathcal{LP}$ contains sentence variables $S_0, S_1, S_2, \ldots$ and

* Partially supported by the grant 96-01-01395 of the Russian Foundation for Basic Research

boolean constants $\top$ and $\bot$; proof variables $p_0, p_1, p_2, \ldots$ and axiom constants $a_0, a_1, a_2, \ldots$; boolean connectives and function symbols (binary $\times$, $+$ and unary $!$). Terms of the language $\mathcal{LP}$ are constructed from proof variables and axiom constants by means of function symbols. The set of formulas is generated from propositional variables by means of boolean connectives and $[\cdot](\cdot)$-constructor (i.e. if t is a term and F is a formula then $[t]F$ is a formula). In the intended semantics, terms denote proofs and formulas represent statements of **PA**. Formulas of the form $[t]F$ are interpreted as "t is a code for a proof of F". Axiom constants stand for primitive proofs of axioms, and function symbols correspond to operations on proofs defined by the following conditions:

Op1 $[t](A \to B) \to ([s]A \to [t \times s]B)$
Op2 $[t]A \to [!t][t]A$
Op3 $[t]A \to [t + s]A$, $[s]A \to [t + s]A$.

The system $\mathcal{LP}$ from [2] is proved to be complete with respect to the intended arithmetical semantics, namely $\mathcal{LP}$ axiomatizes the set of all formulas provable in **PA** (true in the standard model) under every arithmetical interpretation. It also enjoys functional completeness: $\mathcal{LP}$ realizes any operation on proofs admitting a propositional description. In [2], it is also proved that after the substitution of $\Box$ for all occurrences of $[t]$, the modal counterpart of $\mathcal{LP}$ is the Gödel system $S4$: a formula is $S4$-provable iff all boxes in it can be labeled by $\mathcal{LP}$-terms in such a way that the result is a $\mathcal{LP}$-theorem. This result allows us to give an exact meaning to the informal provability semantics of $S4$. In combination with arithmetical completeness of $S4$, it provides completeness of $S4$ with respect to this semantics.

In this paper, we formulate an operational logic of proofs $\mathcal{MLP}$ over the base of a modal propositional language. $\mathcal{MLP}$ gives a joint description of arithmetical provability which combines Solovay's proof predicate with Artemov's operations on proofs. The presence of the modal operator $\Box$ makes it possible to introduce operations on proofs that correspond to the rule

$$\mathbf{PA} \vdash \varphi \iff \mathbf{PA} \vdash Pr(\lceil \varphi \rceil).$$

These operations are $\Downarrow^{\Box}$ and $\Uparrow_{\Box}$. The former operation delivers a proof of ϕ, given a proof of $Pr(\lceil \phi \rceil)$. The later operation works in the opposite direction: given a proof of ϕ, it constructs a proof for $Pr(\lceil \phi \rceil)$. So the language of $\mathcal{MLP}$, denoted by $\mathcal{LP}_{\Box}$, is the extension of $\mathcal{LP}$ by the modal operator $\Box$ and two unary functional symbols $\Downarrow^{\Box}$ and $\Uparrow_{\Box}$. In this paper, we present a natural axiomatization for provability logic with operations on proofs. The system $\mathcal{MLP}$ formulated in this paper is proved to be complete with respect to the intended arithmetical semantics. We develop the appropriate notion of a Kripke-style model and prove the corresponding completeness theorem. $\mathcal{MLP}$ also suffices to realize all operations on proofs admitting description in the modal propositional language, thus justifying our choice of basic operations.

2 The system $\mathcal{MLP}$.

The language $\mathcal{LP}_\square$ contains sentence variables $S_0, S_1, S_2, \ldots$ and boolean constants $\top$ and $\bot$; proof variables $p_0, p_1, p_2, \ldots$ and axiom constants $a_0, a_1, a_2, \ldots$; boolean connectives $\rightarrow, \wedge, \vee$, functional symbols (binary $\times, +$ and unary $!$, $\Uparrow_\square$, $\Downarrow^\square$) and the modal operator $\square$. Terms of $\mathcal{LP}_\square$ are defined in the usual way. The set Fm of formulas includes propositional variables and boolean constants, it is closed under boolean connectives and $\square$, and if t is a term and $F \in Fm$ then $[\![t]\!]F \in Fm$. Formulas of the form $[\![t]\!]F$ are called *quasiatomic* or *q–atomic*. We denote the set of all such formulas by $Qatom$.

The axioms system for the minimal logic $\mathcal{MLP}_\emptyset$ includes axioms for the system $\mathcal{B}$ (see [1]) and operational axioms which describe operations on proofs:

axioms of $\mathcal{B}$	operational axioms
B0 all tautologies	**Op1** $[\![t]\!](A \rightarrow B) \rightarrow ([\![s]\!]A \rightarrow [\![ts]\!]B)$
B1 $\square(A \rightarrow B) \rightarrow (\square A \rightarrow \square B)$	**Op2** $[\![t]\!]A \rightarrow [\![!t]\!][\![t]\!]A$
B2 $\square(\square A \rightarrow A) \rightarrow \square A$	**Op3** $[\![t]\!]A \rightarrow [\![t+s]\!]A$
B3 $[\![t]\!]A \rightarrow A$	$\quad\;\; [\![s]\!]A \rightarrow [\![t+s]\!]A$
B4 $[\![t]\!]A \rightarrow \square[\![t]\!]A$	**Op4** $[\![t]\!]\square A \rightarrow [\![\Downarrow^\square t]\!]A$
B5 $\neg[\![t]\!]A \rightarrow \square\neg[\![t]\!]A$	**Op5** $[\![t]\!]A \rightarrow [\![\Uparrow_\square t]\!]\square A$

Rules of inference for $\mathcal{MLP}_\emptyset$:

$$\textbf{(R1)} \quad \frac{A, A \rightarrow B}{B} \qquad\qquad \textbf{(R2)} \quad \frac{A}{\square A} \qquad\qquad \textbf{(R3)} \quad \frac{\square A}{A}$$

Axiom specification AS is a finite set of formulas of the form $[\![a]\!]A$, where a is an axiom constant and A is an axiom of $\mathcal{MLP}_\emptyset$. The system $\mathcal{MLP}_{AS}$ is obtained by adding AS to $\mathcal{MLP}_\emptyset$ as new axioms. We also consider systems $\mathcal{MLP}^\omega_{AS}$ that have all theorems of $\mathcal{MLP}_{AS}$ and all formulas of the form $\square A \rightarrow A$ as axioms and the only rule of inference *modus ponens*. It is shown below that $\mathcal{MLP}_{AS}$ is decidable. So the set of axioms for $\mathcal{MLP}^\omega_{AS}$ is decidable too.

Remark. The proof of the modal completeness theorem (cf. section 5) contains the construction of a model for $\mathcal{MLP}_{AS}$, that guarantees the consistency of $\mathcal{MLP}_{AS}$, and the same for $\mathcal{MLP}^\omega_{AS}$.

Remark. Note that operator $[\![t]\!](\cdot)$ is not a modality since the rule scheme

$$\frac{A \leftrightarrow B}{[\![t]\!]A \leftrightarrow [\![t]\!]B}$$

is not admissible in $\mathcal{MLP}_{AS}$. However some analogies of modal properties still hold for it. The following lemma states a certain kind of the necessitation rule for $\mathcal{MLP}_{AS}$ and can be proved by induction on the derivation of F.

Lemma 1. *(Constructive necessitation.)*
Suppose $\mathcal{MLP}_{AS} \vdash F$. *Then* $\mathcal{MLP}_{AS'} \vdash [\![t]\!]F$ *for some term t and axiom specification AS'.*

Remark. Let $\mathcal{MLP}_{AS}^-$ denote the system that has the same axioms as $\mathcal{MLP}_{AS}$ and the sole inference rule *modus ponens*. Using lemma 1 one can show that rules (R2) and (R3) are derivable in $\mathcal{MLP}^- = \bigcup_{AS} \mathcal{MLP}_{AS}^-$, that is

$$\mathcal{MLP}^- \vdash A \iff \mathcal{MLP}^- \vdash \Box A.$$

It means that

$$\mathcal{MLP}_{AS}^- \vdash A \implies \exists AS'\, \mathcal{MLP}_{AS'}^- \vdash \Box A$$
$$\mathcal{MLP}_{AS}^- \vdash \Box A \implies \exists AS'\, \mathcal{MLP}_{AS'}^- \vdash A.$$

However in general case $AS' \neq AS$, and so $\mathcal{MLP}_{AS}^- \neq \mathcal{MLP}_{AS}$. Thus we can not omit (R2) and (R3) in the axiom system for $\mathcal{MLP}_{AS}$.

3 Arithmetical interpretation.

Let us give the precise description of arithmetical interpretation for $\mathcal{LP}_\Box$. We suppose that the language of **PA** contains ι-terms (see [4]): for every arithmetical formula φ the expression $\iota z.\varphi$ is a term. Let $\mu z.\varphi$ denote ι-term, defined by the formula $\varphi(z) \wedge \forall v < z \neg \varphi(v)$.

Definition 2. Term $\mu z.\varphi$ is called *recursive* if $\varphi(z) \wedge \forall v < z \neg \varphi(v)$ is a provably Σ_1-formula. If **PA** $\vdash \exists z \varphi(z)$, then $\mu z.\varphi$ is a *provably total* term. *Closed recursive term* is a provably total recursive term $\mu z.\varphi$, where φ does not contain any free variables other then z.

Remark. 1. Functions defined by recursive terms are partial recursive and, vise versa, every partial recursive function can be represented by such a term.

2. Suppose $\phi(x_0, \ldots, x_n, z)$ is a Σ_1-formula which is provably functional in z, namely **PA** $\vdash \phi(x_0, \ldots, x_n, z) \wedge \phi(x_0, \ldots, x_n, z') \to z = z'$. Then the term $\mu z.\phi(x_0, \ldots, x_n, z)$ is recursive.

3. For every closed recursive term $f = \mu z.F(z)$ one can calculate the number n called *the value of* f, such that **PA** $\vdash$ "$t = n$" where "$t = n$" denotes arithmetical formula $F(n) \wedge \forall v < n \neg F(v)$. Note that the function $val(x)$ that assigns the value of a closed recursive term to its Gödel number is partial recursive and thus can be represented by a recursive term.

4. Let $f_1, \ldots, f_n$ be closed recursive terms. Suppose a term $g(x_1, \ldots, x_n) = \mu z.G(x_1, \ldots, x_n, z)$ is recursive and the function represented by this term is defined at the values of $f_1, \ldots, f_n$. Then the result of substitution f_i in $g(x_1, \ldots, x_n)$ defined as $\mu z.(\exists y_1 \ldots y_n (\bigwedge_{i=1}^n "f_i = y_i" \wedge G(y_1, \ldots, y_n, z)))$ is a closed recursive term too.

5. Let $D(x)$ be a provably Δ_1-formula, and suppose f is a closed recursive term. Then the formula $D(f) = \exists y ("f = y" \wedge D(y))$ is provably Δ_1 too.

Definition 3. *Standard proof predicate* is a provably Δ_1–formula $Prf(x, y)$ with the following properties:

1. *Prf* enumerates theorems of **PA**, that is $\mathbf{PA} \vdash \varphi \iff Prf(n, \lceil \varphi \rceil)$ is true for some n.
2. The provability predicate $Pr(y) = \exists x\, Prf(x, y)$ satisfies the conditions

$$\mathbf{PA} \vdash Pr(\lceil A \rightarrow B \rceil) \rightarrow (Pr(\lceil A \rceil) \rightarrow Pr(\lceil B \rceil))$$
$$\mathbf{PA} \vdash \sigma \rightarrow Pr(\lceil \sigma \rceil) \text{ for every arithmetical statement } \sigma \in \Sigma_1.$$

Proof predicate is called *normal nondeterministic*, if for any $n \in \omega$ the set $T(n) = \{A | Prf(n, \lceil A \rceil)\}$ is finite, the function $\widetilde{T(n)} = $ "the code of $T(n)$" is total recursive and for every finite set T of **PA**–theorems there exists n such that $T \subseteq T(n)$.

The example of such a predicate is the nondeterministic variant of the standard Gödel proof predicate $PROOF(x, y)$, that is the natural Δ_1–representation of the relation "x is Gödel number of a proof containing a statement with Gödel number y".

Lemma 4. *For any normal nondeterministic proof predicate Prf there exist recursive terms* $\mathsf{e}(x, y), \mathsf{m}(x, y)$, $\mathsf{c}(x)$, $\mathsf{r}(x)$ *and* $\mathsf{n}(x)$, *such that for every closed recursive terms* $f = \mu z.F(z)$ *and* $g = \mu z.G(z)$ *and arithmetical sentences* ϕ *and* ψ *the following holds:*

1. $\mathbf{PA} \vdash Prf(f, \lceil \phi \rightarrow \psi \rceil) \rightarrow (Prf(g, \lceil \phi \rceil) \rightarrow Prf(\mathsf{m}(f, g), \lceil \psi \rceil))$
2. $\mathbf{PA} \vdash Prf(f, \lceil \phi \rceil) \vee Prf(g, \lceil \phi \rceil) \rightarrow Prf(\mathsf{e}(f, g), \lceil \phi \rceil)$
3. $\mathbf{PA} \vdash Prf(f, \lceil \phi \rceil) \rightarrow Prf(\mathsf{c}(\lceil f \rceil), \lceil Prf(f, \lceil \phi \rceil) \rceil)$
4. $\mathbf{PA} \vdash Prf(f, \lceil Pr(\lceil \phi \rceil) \rceil) \rightarrow Prf(\mathsf{r}(f), \lceil \phi \rceil)$
5. $\mathbf{PA} \vdash Prf(f, \lceil \phi \rceil) \rightarrow Prf(\mathsf{n}(f), \lceil Pr(\lceil \phi \rceil) \rceil).$

Remark. Terms $\mathsf{e}(x, y)$, $\mathsf{m}(x, y)$ and $\mathsf{c}(x)$ were constructed in [3]. According to the properties of recursive terms mentioned above in order to establish the existence of terms $\mathsf{r}(x)$ and $\mathsf{n}(x)$ we only need to produce algorithms computing values of total functions $\mathsf{r}(x)$ and $\mathsf{n}(x)$, corresponding to operations $\Downarrow^{\square}$ and $\Uparrow_{\square}$ resp. The algorithm for $\mathsf{r}(\cdot)$ given a number n recovers the finite set of arithmetical sentences $\{A_0, A_1, \ldots, A_k\}$, such that $Pr(\lceil A_i \rceil)$ occurs in the proof with the code n. In case it is empty put $\mathsf{r}(n) = 0$. Otherwise set $\mathsf{r}(n) := l$ where l is the least code of a proof, containing $A_0, A_1, \ldots, A_k$. The existence of l is provided by provability of all A_i. To calculate $\mathsf{n}(x)$ one first reconstructs the finite set $T(x) = \{\varphi \mid Prf(x, \lceil \varphi \rceil) \text{ is true}\}$. If $T(x) = \emptyset$ put $\mathsf{n}(x) = 0$. Otherwise all formulas from $\{Pr(\lceil \varphi \rceil) \mid \varphi \in T(x)\}$ are provable and the definition of a normal nondeterministic proof predicate guarantees the existence of a common proof for them. We set $\mathsf{n}(x)$ be minimal code of such a proof.

The recursive terms $\mathsf{r}(x)$ and $\mathsf{n}(x)$ (representing operations $\Downarrow^{\square}$ and $\Uparrow_{\square}$), can be easely reconstructed from the description of algorithms given above. If f is a closed recursive term and φ is an arithmetical sentences then the formulas

$$Prf(f, \lceil Pr(\lceil \varphi \rceil) \rceil) \rightarrow Prf(\mathsf{r}(f), \lceil \varphi \rceil),$$
$$Prf(f, \lceil \varphi \rceil) \rightarrow Prf(\mathsf{n}(f), \lceil Pr(\lceil \varphi \rceil) \rceil)$$

are true. By the remark on properties of recursive terms these formulas are Δ_1 whence they are provable in **PA**.

Remark. Note that operation $c(\cdot)$ is sufficiently different from the others, namely it depends on a Gödel number of a closed recursive term but not only on its value. This particularity of $c(\cdot)$ arises from the fact that proofs for $Prf(f, \lceil \varphi \rceil)$ and $Prf(g, \lceil \varphi \rceil)$, generally speaking, are different for different terms t and g with the same values. The set of Gödel numbers for closed recursive terms is not recursive but recursively enumerable, and it is the reason the function $c(\cdot)$ is not total.

Definition 5. *Arithmetical AS–interpretation* of $\mathcal{LP}_\square$ consists of a normal non-deterministic proof predicate *Prf* supplied with provably recursive terms $e(x, y)$, $m(x, y)$, $c(x)$, $r(x)$ and $n(x)$, satisfying lemma 4 and an evaluation $*$, that assigns arithmetical sentences to $\mathcal{LP}_\square$-formulas and closed recursive terms to $\mathcal{LP}_\square$-terms in such a way that

1. $\top^* = (0 = 0)$, $\perp^* = (0 = 1)$ and $*$ commutes with boolean connectives;
2. $(t \times s)^* = m(t^*, s^*)$, $(t + s)^* = e(t^*, s^*)$, $(!t)^* = c(\lceil t^* \rceil)$, $(\Downarrow^\square t)^* = r(t^*)$, $(\Uparrow_\square t)^* = n(t^*)$;
3. $(\llbracket t \rrbracket F)^* = Prf(t^*, \lceil F^* \rceil)$, $(\square F)^* = Pr(\lceil F^* \rceil)$;
4. $\mathbf{PA} \vdash A^*$ for all formulas $A \in AS$.

Theorem 6. *(Arithmetical completeness.)*

1. $\mathcal{MLP}_{AS} \vdash F \iff \mathbf{PA} \vdash F^*$ *for any AS–interpretation* $*$.
2. $\mathcal{MLP}^\omega_{AS} \vdash F \iff F^*$ *is true in the standard model of arithmetic for any AS–interpretation* $*$.

To prove soundness we first note that all axioms of $\mathcal{B}$ are arithmetically provable (cf. [1]). Interpretations of operational axioms are true Δ_1-formulas and thus they are provable in **PA**.

The completeness proof combines Solovay-like argument and the fixed-point equation defining the appropriate proof predicate and operations on proofs. We embed Kripke models for $\mathcal{MLP}_{AS}$ into formal arithmetic and need to prove some of their properties in **PA**. The notion of finitely generated model suitable for this purpose is introduced in section 5. Let us briefly sketch the completeness proof.

Proof. Suppose $\mathcal{MLP}_{AS} \nvdash A$. Modal completeness theorem provides us with a "nice" (that is finitely generated) countermodel $\mathcal{K} = (K, \prec, \models)$ for the formula A. We apply standard Solovay construction to $\mathcal{K}$ and the usual Gödel proof predicate it $\text{Proof}(x, y)$. Solovay function $h(x)$ in defined the expected way (cf. [6]). Let "$l = i$" denote natural arithmetical representation for the proposition "i *is a limit of the function* $h(x)$". The sentences "$l = i$" satisfy the following Solovay conditions (cf. [6])

1. $\mathbf{PA} \vdash$ "$0 \leq l \leq n$";
2. "$l = 0$" is true, but $\mathbf{PA} +$ "$l = i$" is consistent for all $0 \leq i \leq n$;
3. $\mathbf{PA} +$ "$l = i$" $\vdash Provable(\lceil$ "$l \neq i$" $\rceil)$ for every $i \geq 1$;
4. $\mathbf{PA} +$ "$l = i$" $\vdash \neg Provable(\lceil$ "$l \neq j$" $\rceil)$ for all $i \prec j$;

5. $\mathbf{PA} + \text{``}l = i\text{''} \vdash Provable(\lceil \text{``}l \neq j\text{''} \rceil)$ for all $j \not\vdash i$.

To define arithmetical interpretation $(Prf, *)$ we fix Gödel enumeration of $\mathcal{LP}_\square$-terms. For all sentence variables S_i, proof variables x_j and axiom constants a_j put

$$S_i^* = \bigvee_j \{ \text{``}l = j\text{''} \mid j \models S_i \} \wedge (i = i), \qquad p_j^* = \lceil p_j \rceil, \qquad a_j^* = \lceil a_j \rceil.$$

The desired proof predicate Prf and operations on proofs are constructed by a multiple fixed point equation. In what follows we suppose $*$ to be based of Prf and

$$\times^* = \mu z.M(x,y,z) \qquad +^* = \mu z.E(x,y,z) \qquad !^* = \mu z.C(x,z)$$
$$\Downarrow^{\square*} = \mu z.R(x,z) \qquad \Uparrow_\square{}^* = \mu z.N(x,z).$$

Note that $\lceil B^* \rceil$ and $\lceil t^* \rceil$ are primitive recursive in Gödel numbers of $Prf(x,y)$, $M(x,y,z)$, $\ldots$ for any $\mathcal{LP}_\square$-formula B and term t. Let $\widetilde{Tm}(x)$ denote arithmetical Δ_1-formula, representing the set of Gödel numbers of $\mathcal{LP}_\square$-terms in $\mathbf{PA}$. Here $PROOF$ is the nondeterministic Gödel proof predicate, $\oplus$ and $\otimes$ are recursive terms representing $+$ and $\times$ for it; $\Uparrow$, $\Downarrow^\blacksquare$ and $\Uparrow_\blacksquare$ are recursive terms realizing $!$, $\Downarrow^\square$ and $\Uparrow_\square$ for Prf constructed in lemma 4. From the definition of a finitely generated model it follows that the function that assigns the code of the (finite) set $I(t) = \{B \mid \mathcal{K} \models [t]B\}$ to any term t is primitive recursive. By the arithmetical fixed point argument there exists Δ_1-formula Prf and provably functional Σ_1-formulas $M(x,y,z)$, $E(x,y,z), \ldots$, such that the following fixed point equation is provable in $\mathbf{PA}$:

$$Prf(x,y) \leftrightarrow PROOF(x,y) \vee$$
$$\vee(\text{``}x = \lceil t \rceil \text{ for some term } t\text{''} \wedge y = \lceil B^* \rceil \wedge B \in I(t)\text{''});$$

$M(x,y,z) \leftrightarrow$ if $x = \lceil t \rceil$ and $y = \lceil s \rceil$, then $z = \lceil ts \rceil$;

if $x = \lceil t \rceil$ and $\neg\widetilde{Tm}(y)$,

then $z = \mu w[\bigwedge\{PROOF(w, \lceil B^* \rceil) \mid B \in I(t)\}] \otimes y$;

if $\neg\widetilde{Tm}(x)$ and $y = \lceil s \rceil$,

then $z = x \otimes \mu w[\bigwedge\{PROOF(w, \lceil B^* \rceil) \mid B \in I(s)\}]$;

$z = x \otimes y$ otherwise;

$E(x,y,z) \leftrightarrow$ if $x = \lceil t \rceil$ and $y = \lceil s \rceil$, then $z = \lceil t+s \rceil$;

if $x = \lceil t \rceil$ and $\neg\widetilde{Tm}(y)$,

then $z = \mu w[\bigwedge\{PROOF(w, \lceil B^* \rceil) \mid B \in I(t)\}] \oplus y$;

if $\neg\widetilde{Tm}(x)$ and $y = \lceil s \rceil$,

then $z = x \oplus \mu w[\bigwedge\{PROOF(w, \lceil B^* \rceil) \mid B \in I(s)\}]$;

$z = x \oplus y$ otherwise;

$C(x,z) \leftrightarrow$ if $x = \lceil \overline{t^*} \rceil$, then $z = \lceil !t \rceil$;

$z = \Uparrow x$ otherwise;

$$R(x, z) \leftrightarrow \text{if } x = \lceil t \rceil, \text{ then } z = \lceil \Downarrow^\square t \rceil;$$
$$z = \Downarrow^\bullet x \text{ otherwise};$$
$$N(x, z) \leftrightarrow \text{if } x = \lceil t \rceil, \text{ then } z = \lceil \Uparrow_\square t \rceil;$$
$$z = \Uparrow_\bullet x \text{ otherwise}.$$

The following properties of * can be easily verified

Lemma 7. *Interpretation* * *is injective on terms and formulas of* $\mathcal{LP}_\square$, *and* **PA** $\vdash t^* = \lceil t \rceil$ *for every term* t. *Moreover,* **PA** $\vdash \forall x(\neg Tm(x) \rightarrow x = val(x^*))$ *where* $val(y)$ *is a recursive term for the function computing values of closed recursive terms.*

Using this lemma, Solovay conditions and nice properties of finitely generated models one can prove that

Lemma 8. *For every formula* B *and any* $0 \le k \le n$

$$k \models B \Longrightarrow \mathbf{PA} \vdash \text{``}l = k\text{''} \rightarrow B^*$$
$$k \not\models B \Longrightarrow \mathbf{PA} \vdash \text{``}l = k\text{''} \rightarrow \neg B^*.$$

This lemma together with the arithmetization of finitely generated models enables us to prove that Prf is a normal nondeterministic proof predicate and **PA** $\vdash Pr(x) \leftrightarrow Provable(x)$. Terms interpreting functional symbols satisfy the conditions of lemma 4. To complete the proof we note that $0 \not\models \square A$, whence by lemma 8 we conclude

$$\mathbf{PA} \vdash \text{``}l = 0\text{''} \rightarrow \neg Pr(\lceil A^* \rceil).$$

Since "$l = 0$" is true we conclude $\neg Pr(\lceil A^* \rceil)$ whence **PA** $\not\vdash A^*$.

Remark. Revising the proof of Hylbert–Bernays derivability conditions (cf. [5]) we can notice that for the nondeterministic variant of Gödel proof predicate *PROOF* the terms $m(x, y)$ and $n(x)$ can be taken primitive recursive and thus provably total. Moreover, $+$ can be realised as just a concatenation of x and y and so $e(x, y)$ can be chosen provably total too. It was mentioned above that the function realizing ! is not even total. As for the term $r(x)$ representing total operation $\Downarrow^\square$ it can not be provably total either. Indeed, in case it can we have

$$\mathbf{PA} \vdash \forall x \, [PROOF(x, \lceil Provable(\lceil \varphi \rceil) \rceil) \rightarrow PROOF(r(x), \lceil \varphi \rceil)]$$

for every arithmetical sentence φ. Take $\varphi = \bot$, then

$$\mathbf{PA} \vdash PROOF(x, \lceil Provable(\lceil \bot \rceil) \rceil) \rightarrow PROOF(r(x), \lceil \bot \rceil),$$
whence $\mathbf{PA} \vdash Provable(\lceil Provable(\lceil \bot \rceil) \rceil) \rightarrow Provable(\lceil \bot \rceil).$

Then in accordance to Löb theorem **PA** $\vdash Provable(\lceil \bot \rceil)$ that contradicts the consistensy of **PA**.

The completeness result can be strengthened for the class of interpretations with provably total recursive terms representing operations $\times$, $+$ and $\Uparrow_\square$.

4 Functional completeness.

The addition of operations $\Uparrow_\square$ and $\Downarrow^\square$ to $\{\times, +, !\}$ provides functional completeness of $\mathcal{MLP}$.

Definition 9. *Positive δ-formula is any $\{\vee, \wedge\}$-combination of q-atomic formulas. Abstract propositional operation on proofs is a formula $C \to \square G$, true under every arithmetical interpretation where C is any positive δ-formula.*

Arithmetically valid formulas of the form

$$[x_1]C_1 \wedge [x_2]C_2 \wedge \ldots \wedge [x_n]C_n \to \square G,$$

that formalize the notion of the admissible inference rule are the particular cases of this definition.

All $\mathcal{LP}_\square$-operations can be described as abstract operations on proofs. For example, functions $\Downarrow^\square$ and $\Uparrow_\square$ realize respectively operations

$$[x]\square A \to \square A$$
$$\text{and } [x]A \to \square\square A.$$

Theorem 10. *(Functional completeness.)*
For any propositional operation on proofs $C \to \square G$ there exists a $\mathcal{LP}_\square$-term u and axiom specification AS, such that

$$\mathcal{MLP}_{AS} \vdash C \to [u]G.$$

The following example demonstrates the main idea of the proof.

Example 1. Let us consider proof operation $l\ddot{o}b(x)$, corresponding to Löb schema

$$[x](\square A \to A) \to \square A.$$

This formula is valid under every arithmetical interpretation. Lemma 1 provides a term t and an axiom specification AS, such that

$$\mathcal{MLP}_{AS} \vdash [t]([x](\square A \to A) \to \square A). \tag{1}$$

Then, using operational axioms and extending axiom specification if necessary we obtain

$$\mathcal{MLP}_{AS} \vdash [x](\square A \to A) \to [!x][x](\square A \to A) \quad \text{(Op2)}$$
$$\mathcal{MLP}_{AS} \vdash [x](\square A \to A) \to [t \times !x]\square A \quad \text{(Op1) together with (1)}$$
$$\mathcal{MLP}_{AS} \vdash [x](\square A \to A) \to [\Downarrow^\square (t \times !x)]A \quad \text{by (Op4).}$$

So $\Downarrow^\square (t \times !x)$ is the desired term realizing the operation *(löb)*.

5 Kripke–style semantics and decidability.

Systems $\mathcal{MLP}_{AS}$ are supplied with the appropriate Kripke–style semantics. The completeness theorem with respect to the described class of Kripke models is established and the decidability of $\mathcal{MLP}_{AS}$ for any axiom specification AS is proved via reduction of $\mathcal{MLP}_{AS}$ to the Artemov's system $\mathcal{B}$.

Definition 11. *Kripke model for* $\mathcal{MLP}_{AS}$ *is a GL-frame with forcing relation satisfying stability and q-reflexivity conditions (see [1]), operational axioms and AS. For every set of formulas Γ the model is called Γ-sound if for its root the following holds* $root \models \{\Box B \to B \mid \Box B \in Sub(\Gamma)\}$.

Lemma 12. *(Modal soundness.)*
If $\mathcal{MLP}_{AS} \vdash A$ then A is valid in all finite $\{A, AS\}$-sound models for $\mathcal{MLP}_{AS}$.

The definition of truth relation is not inductive since q-atomic formulas play the role of an infinite number of new atoms. It could provide difficulties when embedding of Kripke models into **PA** is concerned. Yet the modal completeness theorem is true for the restricted class of models with the forcing relation which is efficient in a strong sense. In what follows we consider only Kripke models for finite fragments of the language $\mathcal{LP}_\Box$ on the base of finite tree-like ordered frames.

Definition 13. Let $M \subseteq Qatom$ be an arbitrary set of q-atomic formulas. A finite set $N \supset M$ is called *operational completion* of M if the following holds

OpR1 if $[t](B \to G) \in M$ and $[s]B \in M$, then $[ts]G \in N$;
OpR2 if $[t]G \in M$ and $|s| \leq |M|$, then $[t+s]G$, $[s+t]G \in N$;
OpR3 if $[t]G \in M$, then $[!t][t]G$, $[\Uparrow_\Box t]\Box G \in N$;
OpR5 if $[t]\Box G \in M$, then $[\Downarrow^\Box t]G \in N$.

Here *operational rules* OpR1–OpR4 correspond to axioms of $\mathcal{MLP}_{AS}$. Starting from $M \subseteq Qatom$ we define the sets $M = M^0 \subseteq M^1 \subseteq M^2 \subseteq \ldots$, where M^{i+1} is the operational completion of M^i. If M is finite then all M^i are finite and the function assigning to i the code of M^i is primitive recursive. *Operation closure* of M is the set $[M]_{op} = \bigcup_i M^i$. It is the minimal set containing M and closed under operational rules.

Definition 14. Kripke model $\mathcal{K} = (K, \prec, \models)$ is *finitely generated*, if there exists a set $M \subseteq Qatom$ such that for any q-atomic formula Q

$$\mathcal{K} \models Q \iff Q \in [M]_{op}.$$

Remark. If $\mathcal{K} = (K, \prec, \models)$ is finitely generated then for every term t the set $I(t) = \{G \mid \mathcal{K} \models [t]G\}$ is finite and the function that assigns the code of $I(t)$ to every term t is primitive recursive.

The system $\mathcal{MLP}_{AS}$ is proved to be complete with respect to the Kripke–style semantics described above.

Theorem 15. *(Modal completeness.)*
If A is valid in all finitely generated $\{A, AS\}$–sound models for $\mathcal{MLP}_{AS}$ then $\mathcal{MLP}_{AS} \vdash A$.

Proof. Let us sketch the proof of this theorem. Suppose $\mathcal{MLP}_{AS} \nvdash A$. Let $F_{AS}(A)$ be the minimal (finite) set closed under subformulas and such that

1. $A \in F_{AS}(A)$ and $AS \subseteq F_{AS}(A)$;
2. if $[ts]G \in F_{AS}(A)$ and $B \to G \in Sub(A, AS)$, then $[t](B \to G) \in F_{AS}(A)$ and $[s]B \in F_{AS}(A)$;
3. if $[t+s]G \in F_{AS}(A)$, then $[t]G \in F_{AS}(A)$ and $[s]G \in F_{AS}(A)$;
4. if $[\Uparrow_\square t]\square G \in F_{AS}(A)$, then $[t]G \in F_{AS}(A)$;
5. if $[\Downarrow^\square t]G \in F_{AS}(A)$, then $[t]\square G \in F_{AS}(A)$.

To every term t, occurring in $F_{AS}(A)$, we assign a fresh proof variable x_t. The translation $(\cdot)^+$ of formulas from $F_{AS}(A)$ into the language, not containing operational symbols on proofs is defined as follows: $(\cdot)^+$ preserves atomic formulas, commutes with boolean connectives and $\square$, and $([t]B)^+ = [x_t]B^+$.

Let AX be the conjunction if all operational axioms containing q-atomic formulas from $F_{AS}(A)$. Put $A' = (AX)^+ \to A^+$. Then it is easily seen that $B \nvdash A'$. The modal completeness theorem for B (cf. [1]), guarantees the existence of a model $\mathcal{K}_0 = (K, \prec, \models_0)$ such that $a_0 \models_0 (AX)^+$ and $a_0 \nvDash_0 A^+$ for some $a_0 \in K$. For every node $a \in K$ consider the sets

$$\Gamma_a^0 = \{G \in F_{AS}(A) \mid a \models_0 G^+\}$$
$$\Delta_a^0 = \{G \in F_{AS}(A) \mid a \nvDash_0 G^+\}.$$

Let $\Gamma_a = \Gamma_a^0 \cup [\Gamma_a^0 \cap Qatom]_{op}$. By induction on the definition of operational closure and using properties of the sets Γ_a^0 and Δ_a^0 we can prove that

Lemma 16. *For any $a \in K$ the set Γ_a satisfies the following conditions*

1. $\Gamma_a \cup \Delta_a = \emptyset$;
2. *if $[t]G \in \Gamma_a$, then $G \in \Gamma_a$;*
3. $\Gamma_a \cap Qatom = \Gamma_b \cap Qatom$ for all $a, b \in K$.

Now the desired model $\mathcal{K}$ for the formula A can be constructed on the frame $(K, \prec)$ as follows: for every atomic or q-atomic G put

$$a \models G \iff G \in \Gamma_a.$$

The following lemma can be proved by induction on the length of the formula G

Lemma 17. *For every $a \in K$ and formula G*

$$G \in \Gamma_a \implies a \models G$$
$$G \in \Delta_a^0 \implies a \nvDash G.$$

From lemmas 16 and 17 it follows that the forcing relation satisfies stability and q-reflexivity conditions, operational axioms and AS. Apart from this $a_0 \nvDash A$ that proves the completeness theorem.

Examining the proof of the completeness theorem one can extract an algorithm which given a $\mathcal{LP}_\Box$-formula A and axiom specification AS constructs a formula A' such that $\mathcal{MLP}_{AS} \vdash A \iff \mathcal{B} \vdash A'$. Using this reduction and decidability of $\mathcal{B}$ (see [1]) we can conclude decidability of $\mathcal{MLP}_{AS}$.

Remark. It can be seen that $\mathcal{MLP}^\omega_{AS} \vdash A \iff A$ is valid at the root node of every $\{A, AS\}$–sound model. The system $\mathcal{MLP}^\omega_{AS}$ is decidable too.

6 Acknowledgements

I would like to thank Professor S. N. Artemov for his help at all stages of the work. I am also indebted to V. N. Krupskii for careful reading of this paper and extended discussions on the problem.

References

1. S.Artemov, "Logic of Proofs', *Annals of Pure and Applied Logic*, v.67 (1994), pp. 29–59.
2. S.Artemov, "Operational modal logic", Techn. Rep. No 95–29, Mathematical Science Institute, Cornell University, December 1995.
3. S.Artemov, "Proof Realization of Intuitionistic and Modal Logic", Techn. Rep. No 96–06, Mathematical Science Institute, Cornell University, December 1996.
4. D.Hilbert, P.Bernays, "Grundlagen der Mathematik", I, Springer–Verlag, 1968.
5. C.Smorynski, *Self-reference and modal logic*, New York, Berlin, Heidelberg, Tokio: Springer Verlag, 1985.
6. R.M.Solovay, "Provability interpretation of modal logic", *Israel Journal of Mathematics*, v.25 (1976), pp. 287–304.

Formal Verification of Logic Programs: Foundations and Implementation

Robert F. Stärk

Institute of Informatics, University of Fribourg
Rue Faucigny 2, CH–1700 Fribourg, Switzerland
Email: ⟨robert.staerk@unifr.ch⟩

Abstract. We present the theoretical foundations of LPTP, a logic program theorem prover implemented in Prolog by the author. LPTP is an interactive theorem prover in which one can prove termination and correctness properties of pure Prolog programs that contain negation and built-in predicates like is/2 and call/n. The largest program that has been verified using LPTP is 635 lines long including its specification. The full formal correctness proof is 13128 lines long (133 pages). The formal theory underlying LPTP is the inductive extension of pure Prolog programs. This is a first-order theory that contains induction principles corresponding to the definition of the predicates in the program plus appropriate axioms for built-in predicates.

1 Introduction

There are several reasons that we have implemented an interactive theorem prover for the verification of pure Prolog programs. First of all, we wanted to show that results of [8, 10, 11] about the foundations of logic programming are not only of theoretical interest. In the spirit of Apt [1] we wanted to show that the results can be extended to a rather large subset of Prolog. Secondly, we believe that if computer programs become bigger and more complex, it will be inevitable that parts of it have to be formally verified. This is one possible way to ensure that they work as they are supposed to do.

Why Prolog programs and not imperative programs? When we start to reason about imperative programs, then soon we are on the low level where the state of the system is given by the contents of the variables together with a pointer to the program code that shows where the execution is at the moment. Formulas describing relations between such states can be very complex.

The advantage of Prolog programs is their high level of abstraction. For Prolog programs the state of a computation is given by the atom that is called, i.e. it is of the form $R(t_1, \ldots, t_n)$. The predicate R corresponds to a pointer into the code of the program and the terms $t_1, \ldots, t_n$ corresponds to the contents of the registers. If we ask, for example, what can be reached from the state $R(t_1, \ldots, t_n)$, then we just have to match this atom against the clauses of the program. This shows, that for Prolog programs, formulas describing relations between states of the computation are very close to the syntax of the program.

There are other reasons in favor of Prolog. For example, Prolog is both, a specification language and a programming language. This means that one can write specifications in Prolog as well as efficient algorithms. As a consequence, correctness of Prolog programs can be reduced to equivalence of programs. One just has to show that the specification program computes the same relation as the implementation program.

Finally, there is a pragmatic reason to use Prolog. It is desirable that a theorem prover for a certain programming language is implemented in the programming language itself. For example, a theorem prover for Java programs should be implemented in Java. Since proof-checking and proof-search of our theorem prover is based on backtracking, it is natural to use a programming language that provides backtracking for free. And Prolog is such a language.

The plan of this paper is as follows. In Section 2 we define a subset of Prolog and describe a simple operational model for it. In Section 3 we introduce a first-order theory that is computationally adequate with respect to the operational model of Section 2. In Section 4 we give a short overview of LPTP. This is the logic program theorem prover based on the theoretical results of Sections 2 and 3. Section 5 finally illustrates how LPTP can be used to prove the correctness of an algorithm for inserting elements in AVL trees.

There are essential differences between this article and the first-order theory introduced in [12]. We are now working with general goals and not only with sequences of literals. This makes it possible to treat built-in predicates in a uniform and simple way. Mode assignments are no longer needed. Instead of it we have a unary predicate gr in the formal language and are now able to treat higher-order programs that use the call/n predicate.

2 Pure Prolog with negation and built-in predicates

Pure Prolog is a subset of Prolog. Which subset, however, is not always so clear. Apt, for example, uses in [1] the term "pure Prolog" for Horn clause programs when they are viewed as sequences of clauses. We use the term "pure Prolog" for a larger subset which we define below. We include negation and built-in predicates like integer/1, is/2, </2, and call/n. Even the term decomposition predicates functor/3 and arg/3 are allowed. Predicates like var/1 which tests during run-time whether a variable is bound are not included in pure Prolog. Also the predicates assert/1 and rectract/1 which modify a program during run-time are forbidden in pure Prolog. The cut operator (!) does not belong to pure Prolog, since it destroys the lifting lemma. We assume that pure Prolog performs the occurs check during unification.

Let $\mathcal{L}$ be a first-order language. The terms r, s, t of $\mathcal{L}$ are built up as usual from variables x, y, z and constants c, d using function symbols f, g. The predicate symbols of $\mathcal{L}$ are divided into user-defined and built-in predicates. If R is an n-ary predicate symbol of $\mathcal{L}$ then the expression $R(t_1, \ldots, t_n)$ is an atomic goal of $\mathcal{L}$. The atomic goal is called user-defined or built-in according to whether the predicates symbol R is user-defined or built-in. Atomic goals are denoted by

A, B. The goals of $\mathcal{L}$ are

$$E, F, G ::= \mathtt{true} \mid \mathtt{fail} \mid s = t \mid A \mid F \ \& \ G \mid F \ \mathtt{or} \ G \mid \mathtt{not} \ G \mid \mathtt{some} \ x \ G.$$

The goal $\mathtt{true}$ is the goal that always succeeds; $\mathtt{fail}$ is the goal that always fails. Equations $s = t$ are solved by unification. Conjunction ($\&$), disjunction ($\mathtt{or}$), and negation ($\mathtt{not}$) are written in Prolog as (F, G), $(F; G)$, and $\backslash + G$. The meaning of these connectives will be explained below in terms of an operational semantics and later by a transformation of goals into formulas that contain the logical connectives $\wedge$, $\vee$ and $\neg$. Conjunction and disjunction are both associated to the right. The goal $E \ \& \ F \ \& \ G$, for example, stands for $E \ \& \ (F \ \& \ G)$. The empty conjunction is identified with $\mathtt{true}$; the empty disjunction corresponds to the goal $\mathtt{fail}$. A goal of the form $G_1 \ \& \ \ldots \ \& \ G_n \ \& \ \mathtt{true}$ is called a query. It can be considered as a finite list of goals. We use $[G_1, \ldots, G_n]$ as an abbreviation for the query $G_1 \ \& \ \ldots \ \& \ G_n \ \& \ \mathtt{true}$. The existential quantifier $\mathtt{some} \ x \ G$ binds the variable x in the goal G. Existential quantification is implicit in Prolog. It is explicit in extensions of Prolog like Gödel [7] and Mercury [9].

Free and bound variables in goals are defined as usual. We use the vector notation $\mathbf{x}$ for a finite list $x_1, \ldots, x_n$. We write $G[\mathbf{x}]$ to express that all free variables of G are among the list $\mathbf{x}$; $G(\mathbf{x})$ may contain other free variables than $\mathbf{x}$. A goal or a term is called ground, if it does not contain free variables.

If A is a user-defined atomic goal and G is a goal then the expression $A :\!\!- G$ is called a clause with head A and body G. Let C be the clause

$$R(t_1[\mathbf{y}], \ldots, t_n[\mathbf{y}]) :\!\!- G[\mathbf{y}].$$

Then the *definition form* of C is defined to be the goal

$$D_C[x_1, \ldots, x_n] :\equiv \mathtt{some} \ \mathbf{y} \ (x_1 = t_1[\mathbf{y}] \ \& \ \ldots \ \& \ x_n = t_n[\mathbf{y}] \ \& \ G[\mathbf{y}]),$$

where $\mathbf{y}$ are fresh variables. The *normal form* of C is the clause

$$R(x_1, \ldots, x_n) :\!\!- D_C[x_1, \ldots, x_n].$$

A program is a finite sequence of clauses. Let P be a program and R be a user-defined predicate symbol such that the clauses for R in P are $C_1, \ldots, C_m$ (in this order). Then the *definition form* of R with respect to P is defined to be the goal

$$D_R^P[\mathbf{x}] :\equiv D_{C_1}[\mathbf{x}] \ \mathtt{or} \ \ldots \ \mathtt{or} \ D_{C_m}[\mathbf{x}].$$

The *normalized definition* of R in P is the clause $R(\mathbf{x}) :\!\!- D_R^P[\mathbf{x}]$.

Both, the definition form of a clause and the definition form of a user-defined predicate are goals. Thus, from a theoretical point of view, one could as well define a logic program to be a function that assigns to every user-defined predicate symbol R a goal $D_R^P[\mathbf{x}]$ for some distinguished variables $\mathbf{x}$.

Without general goals, a theory of built-in predicates would be rather ad-hoc, since then every built-in predicate has to be treated in a different way. Using the concept of goals, built-in predicates can be treated in a uniform way. Built-in predicates can be modeled by a set $\mathcal{D}$ of built-in atomic goals and a function $\mathcal{B}$ from $\mathcal{D}$ into the set of goals such that the following two conditions are satisfied:

(D) If $A \in \mathcal{D}$ then $A\sigma \in \mathcal{D}$ for each substitution σ.

(B) $\mathcal{B}(A\sigma) = \mathcal{B}(A)\sigma$ for each $A \in \mathcal{D}$ and each substitution σ.

The idea is that $\mathcal{D}$ contains exactly those built-in atomic goals that can be evaluated and do not report an error message because of type violations or insufficient instantiation of arguments. The goal $\mathcal{B}(A)$ is then the result of the evaluation of A. In most cases the goal $\mathcal{B}(A)$ is either the goal **true** or the goal **fail**. In other cases $\mathcal{B}(A)$ can be an equation or a conjunction of equations.

$\mathcal{D}$ and $\mathcal{B}$ can also be understood as a foreign language interface. Given an atom A from the set $\mathcal{D}$ some code in a foreign language, like for example C, is called. $\mathcal{B}(A)$ is the result of the call. In order that Prolog can use the result, it must be converted into a goal.

There is another possibility is to think of a built-in predicate R. It is given by the (possibly infinite) collection of clauses $R(\mathbf{t})$:- $\mathcal{B}(R(\mathbf{t}))$ for $R(\mathbf{t}) \in \mathcal{D}$.

Example 1. The predicates **integer/1**, **is/2**, **</2** and **call/n** satisfy conditions (D) and (B):

$\mathbf{integer}(t) \in \mathcal{D}$:$\Leftrightarrow$ t is ground.

$\mathcal{B}(\mathbf{integer}(t)) := \begin{cases} \mathbf{true}, & \text{if } t \text{ is an integer constant;} \\ \mathbf{fail}, & \text{otherwise.} \end{cases}$

$(t_1 \mathbf{\ is\ } t_2) \in \mathcal{D}$:$\Leftrightarrow$ t_2 is a ground arithmetic expression.

$\mathcal{B}(t_1 \mathbf{\ is\ } t_2) := (t_1 = n)$, where n is the value of t_2 (as an integer).

$(t_1 < t_2) \in \mathcal{D}$:$\Leftrightarrow$ t_1 and t_2 are ground arithmetic expressions.

$\mathcal{B}(t_1 < t_2) := \begin{cases} \mathbf{true}, & \text{if the value of } t_1 \text{ is less than the value of } t_2; \\ \mathbf{fail}, & \text{otherwise.} \end{cases}$

$\mathbf{call}(s, \mathbf{t}) \in \mathcal{D}$:$\Leftrightarrow$ s is a constant.

$\mathcal{B}(\mathbf{call}(s, \mathbf{t})) := s(\mathbf{t})$.

Not all of the commonly used built-in predicates can be modeled this way. The **var/1** predicate, for example, violates condition (B).

Example 2. The **var/1** predicate violates (B):

$\mathbf{var}(t) \in \mathcal{D}$:$\Leftrightarrow$ t is a term.

$\mathcal{B}(\mathbf{var}(t)) := \begin{cases} \mathbf{true}, & \text{if } t \text{ is a variable;} \\ \mathbf{fail}, & \text{otherwise.} \end{cases}$

Some multi-purpose, built-in predicates like **functor/3** have to be decomposed into their single components.

Example 3. The components of **functor/3** are **decompose/3** and **construct/3**:

$\text{decompose}(t_1, t_2, t_3) \in \mathcal{D}$ $:\Leftrightarrow t_1$ is not a variable.

$\mathcal{B}(\text{decompose}(f(r_1, \ldots, r_n), s, t)) := (s = f \ \& \ t = \bar{n}).$

$\text{construct}(t_1, t_2, t_3) \in \mathcal{D}$ $:\Leftrightarrow t_2$ is a constant, $0 \leq t_3 \leq 255.$

$\mathcal{B}(\text{construct}(t, f, \bar{n}))$ $:= \textbf{some} \ x_1, \ldots, x_n \ (t = f(x_1, \ldots, x_n)).$

$\text{arg}(t_1, t_2, t_3) \in \mathcal{D}$ $:\Leftrightarrow t_1$ is an integer, t_2 is not a variable.

$\mathcal{B}(\text{arg}(\bar{\imath}, f(s_1, \ldots, s_n), t)) := \begin{cases} t = s_i, & \text{if } 1 \leq i \leq n; \\ \textbf{fail}, & \text{otherwise.} \end{cases}$

For example, we have

1. $\mathcal{B}(\text{decompose}(f(c, d), x, y)) = (x = f \ \& \ y = 2),$
2. $\mathcal{B}(\text{construct}(x, f, 2)) = \textbf{some} \ y, z \ (x = f(y, z)),$
3. $\mathcal{B}(\text{arg}(f(c, d), 2, x)) = (x = d).$

Given a program P, the set $\mathcal{D}$ and the function $\mathcal{B}$, we can describe the evaluation of goals as a transition relation between states of a computation. States are defined in the following way:

An *environment* is a finite set of bindings $\{t_1/x_1, \ldots, t_n/x_n\}$ such that the x_i's are pairwise different variables. It is not required that $t_i \not\equiv x_i$ (cf. [4]).

A *frame* consists of a query G and an idempotent environment η. Idempotent means that if $t_i \not\equiv x_i$ then x_i does not occur in $t_1, \ldots, t_n$. Remember that a query is a list of goals.

A *frame stack* consists of a (possibly empty) sequence $\langle G_1, \eta_1; \ldots; G_n, \eta_n \rangle$ of frames. The frames G_i, η_i are alternatives, also called choice points. The query G_n together with the environment η_n is called the *topmost frame* of the stack. Capital greek letters Φ, Ψ and Θ denote finite, possibly empty, sequences of the form $G_1, \eta_1; \ldots; G_n, \eta_n$. Thus $\langle \Phi; G, \eta \rangle$ denotes a stack with topmost frame G, η.

A *state* of a computation is a finite sequence $\langle \Phi_1 \rangle \ldots \langle \Phi_n \rangle$ of frame stacks. $\langle \Phi_n \rangle$ is called the topmost stack of the state. States are denoted by the capital greek letter Σ. For a query G with free variables $x_1, \ldots, x_n$ let $init(G)$ be the state $\langle G, \{x_1/x_1, \ldots, x_n/x_n\} \rangle$. There are three kinds of final states: $yes(\eta)$, no and $error$.

Definition 1. The transition rules of the query evaluation procedure are:

1. $\Sigma \ \langle \Phi; \textbf{true} \ \& \ G, \eta \rangle \longrightarrow \Sigma \ \langle \Phi; G, \eta \rangle$
2. $\Sigma \ \langle \Phi; \textbf{fail} \ \& \ G, \eta \rangle \longrightarrow \Sigma \ \langle \Phi \rangle$
3. $\Sigma \ \langle \Phi; s = t \ \& \ G, \eta \rangle \longrightarrow \Sigma \ \langle \Phi; G, \eta\tau \rangle$ [if $\tau = mgu(s\eta, t\eta)$]
4. $\Sigma \ \langle \Phi; s = t \ \& \ G, \eta \rangle \longrightarrow \Sigma \ \langle \Phi \rangle$ [if $s\eta$ and $t\eta$ are not unifiable]
5. $\Sigma \ \langle \Phi; R(t) \ \& \ G, \eta \rangle \longrightarrow \Sigma \ \langle \Phi; D_R^P[t] \ \& \ G, \eta \rangle$ [if R is user-defined]
6. $\Sigma \ \langle \Phi; A \ \& \ G, \eta \rangle \longrightarrow \Sigma \ \langle \Phi; \mathcal{B}(A\eta) \ \& \ G, \eta \rangle$ [if A is built-in and $A\eta \in \mathcal{D}$]
7. $\Sigma \ \langle \Phi; A \ \& \ G, \eta \rangle \longrightarrow error$ [if A is built-in and $A\eta \notin \mathcal{D}$]
8. $\Sigma \ \langle \Phi; (E \ \& \ F) \ \& \ G, \eta \rangle \longrightarrow \Sigma \ \langle \Phi; E \ \& \ (F \ \& \ G), \eta \rangle$
9. $\Sigma \ \langle \Phi; (E \ \text{or} \ F) \ \& \ G, \eta \rangle \longrightarrow \Sigma \ \langle \Phi; F \ \& \ G, \eta; E \ \& \ G, \eta \rangle$
10. $\Sigma \ \langle \Phi; (E \ \text{or} \ F) \ \& \ G, \eta \rangle \longrightarrow \Sigma \ \langle \Phi; E \ \& \ G, \eta; F \ \& \ G, \eta \rangle$

11. $\Sigma \langle \Phi; (\text{some } x\, F) \,\&\, G, \eta \rangle \longrightarrow \Sigma \langle \Phi; F\{y/x\} \,\&\, G, \eta \cup \{y/y\} \rangle$ [where y is new]
12. $\Sigma \langle \Phi; (\text{not } F) \,\&\, G, \eta \rangle \longrightarrow \Sigma \langle \Phi; (\text{not } F) \,\&\, G, \eta \rangle \langle [F], \eta \rangle$ [if $F\eta$ is ground]
13. $\Sigma \langle \Phi; (\text{not } F) \,\&\, G, \eta \rangle \longrightarrow error$ [if $F\eta$ is not ground]
14. $\Sigma \langle \Phi; (\text{not } F) \,\&\, G, \eta \rangle \langle \Psi; \text{true}, \tau \rangle \longrightarrow \Sigma \langle \Phi \rangle$
15. $\Sigma \langle \Phi; (\text{not } F) \,\&\, G, \eta \rangle \langle \rangle \longrightarrow \Sigma \langle \Phi; G, \eta \rangle$
16. $\langle \Phi; \text{true}, \eta \rangle \longrightarrow yes(\eta)$
17. $\langle \rangle \longrightarrow no$

Remark. Rule 1 says that the goal **true** can be deleted. In 2, the goal **fail** starts backtracking. This means that the topmost frame of the topmost stack is popped. In 3 and 4, equations are solved by unification. If the unification is successful, it changes the current environment; if the unification fails then backtracking starts. We assume that $mgu(s, t)$ returns an idempotent most general unifier if s and t are unifiable. Rule 5 and 6 deal with atomic goals. User-defined predicates are replaced by their definition forms. Built-in predicates are replaced by their built-in definitions provided that the necessary type conditions are satisfied. Otherwise, in 7, built-in predicates report an error message. Rule 8 says that the left goal is selected in a conjunction. This corresponds to a left-most goal selection rule in standard terminology or to so-called LDNF-resolution (see [2]). Rule 9 and 10 are nondeterministic. This is the only place where nondeterminism occurs. To solve a disjunction E or F means either to solve first E and then F or to solve first F and then E. In both cases, new frames are allocated. Without rule 10 one obtains the deterministic evaluation procedure of Prolog. In 11, existential quantified variables are standardized apart. The environment is enlarged. The variable y must be chosen in such a way that is does not appear free neither in the query (**some** $x\, F$) **&** G nor in the environment η. In 12, negated goals start subcomputations. In order to process the goal **not** $F\eta$, the query $[F\eta]$ is started in a subcomputation, provided that $F\eta$ is ground. Otherwise, in 13, an error message is raised. Rule 14 and 16 deal with the cases where the query of the topmost frame is the goal **true**; rule 15 and 17 deal with the cases where the topmost stack is empty. Rule 14 says that if F succeeds then **not** F fails. Rule 15 says that if F fails then **not** F succeeds. Rule 16 corresponds to a global success and rule 17 to a global failure.

Definition 2. We say that

1. a query G *succeeds with answer* σ, if there exists a computation with initial state $init(G)$ and final state $yes(\eta)$ such that σ is the restriction of η to the variables of G;
2. a query G *succeeds with answer including* σ, if there exist substitutions τ and θ such that G succeeds with answer τ and $G\tau\theta \equiv G\sigma$;
3. a query G *fails*, if there exists a computation with initial state $init(G)$ and final state no;
4. a query G *terminates*, if *all* computations with initial state $init(G)$ are finite and do not end in *error*;
5. a query G is *safe*, if there exists no computation with initial state $init(G)$ and final state *error*.

If a query is safe then during a computation all negative goals goals are ground at the time when they are processed and all built-in atoms belong to $\mathcal{D}$ when they are called. We have defined termination in such a way that it includes safeness. If a goal terminates then it is safe. Note, that termination means universal termination. For Prolog-like systems this means that one can hit the semicolon key a finite number of times until one finally obtains the message *no more solutions*.

Practice shows that most goals are terminating in this sense (cf. [1]). Also our experience with the LPTP theorem prover supports this fact. We loose nothing if we restrict our attention to terminating goals only. However, given a program P and a goal G we have to prove that the goal G is terminating. This can be done, for example, using the method of Apt and Pedreschi in [2] by guessing a level mapping for atoms and a model of the program and showing that the program is *acceptable* with respect to the level mapping and the model. Another method which we present here is to use an appropriate first-order theory with induction. This theory is called the inductive extension of pure Prolog programs and is implemented in the interactive theorem prover LPTP.

3   The inductive extension of pure Prolog programs

The inductive extension of a logic program P is, roughly speaking, Clark's completion of a logic program (cf. [3]) plus induction along the definition of the predicates. However, there are essential differences. For instance, the inductive extension is consistent for arbitrary programs. This is not the case for Clark's completion. We can prove termination of predicates in the inductive extension. This is not possible in Clark's completion.

The inductive extension is formulated in a language $\hat{\mathcal{L}}$ which is obtained from $\mathcal{L}$ in the following way. For each predicate symbol R of $\mathcal{L}$ we take in $\hat{\mathcal{L}}$ three predicates symbols R^s, R^f and R^t of the same arity as R. The intended meaning of these predicates is that they express success, failure and termination of R. $\hat{\mathcal{L}}$ contains in addition a special unary predicate gr which expresses that an object is ground. The syntactic objects of $\hat{\mathcal{L}}$ are called formulas. They are

$$\varphi, \chi, \psi ::= \top \mid \bot \mid s = t \mid S(\mathbf{t}) \mid \varphi \wedge \psi \mid \varphi \vee \psi \mid \neg \varphi \mid \varphi \to \psi \mid \forall x\, \varphi \mid \exists x\, \varphi,$$

where S denotes any predicate symbol of $\hat{\mathcal{L}}$. We write $s \neq t$ for $\neg(s = t)$.

The meaning of formulas is given by the first-order predicate calculus of classical logic. By an $\hat{\mathcal{L}}$-theory we mean a (possibly infinite) collection T of formulas of $\hat{\mathcal{L}}$. We write $T \vdash \varphi$ to express that the formula φ can be derived from the $\hat{\mathcal{L}}$-theory T by the usual rules of predicate logic with equality.

For the declarative semantics of logic programs we need three syntactic operators $\mathbf{S}$, $\mathbf{F}$ and $\mathbf{T}$ which transform goals of the language $\mathcal{L}$ into positive formulas of $\hat{\mathcal{L}}$. The operators $\mathbf{S}$, $\mathbf{F}$ and $\mathbf{T}$ are not part of the language. They are defined notions. $\mathbf{S}\, G$ is read: G succeeds; $\mathbf{F}\, G$ is read: G fails; $\mathbf{T}\, G$ is read: G terminates and is safe. The operator have the following definitions:

$$\textbf{S true} :\equiv \top, \qquad\qquad \textbf{F true} :\equiv \bot,$$
$$\textbf{S fail} :\equiv \bot, \qquad\qquad \textbf{F fail} :\equiv \top,$$
$$\textbf{S } s = t :\equiv s = t, \qquad\qquad \textbf{F } s = t :\equiv s \neq t,$$
$$\textbf{S } R(\textbf{t}) :\equiv R^s(\textbf{t}), \qquad\qquad \textbf{F } R(\textbf{t}) :\equiv R^f(\textbf{t}),$$
$$\textbf{S}(G \,\&\, H) :\equiv \textbf{S } G \wedge \textbf{S } H, \qquad \textbf{F}(G \,\&\, H) :\equiv \textbf{F } G \vee \textbf{F } H,$$
$$\textbf{S}(G \text{ or } H) :\equiv \textbf{S } G \vee \textbf{S } H, \qquad \textbf{F}(G \text{ or } H) :\equiv \textbf{F } G \wedge \textbf{F } H,$$
$$\textbf{S some } x\, G :\equiv \exists x\, \textbf{S } G, \qquad\qquad \textbf{F some } x\, G :\equiv \forall x\, \textbf{F } G,$$
$$\textbf{S not } G :\equiv \textbf{F } G, \qquad\qquad \textbf{F not } G :\equiv \textbf{S } G,$$

$$\textbf{T true} :\equiv \top, \qquad\qquad \textbf{T}(G \,\&\, H) :\equiv \textbf{T } G \wedge (\textbf{F } G \vee \textbf{T } H),$$
$$\textbf{T fail} :\equiv \top, \qquad\qquad \textbf{T}(G \text{ or } H) :\equiv \textbf{T } G \wedge \textbf{T } H,$$
$$\textbf{T } s = t :\equiv \top, \qquad\qquad \textbf{T some } x\, G :\equiv \forall x\, \textbf{T } G,$$
$$\textbf{T } R(\textbf{t}) :\equiv R^t(\textbf{t}), \qquad\qquad \textbf{T not } G :\equiv \textbf{T } G \wedge \text{gr}(G).$$

In this definition special attention require only the cases $\textbf{T}(G \,\&\, H)$, $\textbf{T}(G \text{ or } H)$ and $\textbf{T not } G$. The other cases are as one would expect. The definition of $\textbf{T}(G \,\&\, H)$ reflects the fact that a goal $G \,\&\, H$ terminates if, and only if,

(a) G terminates, and
(b) G fails or H terminates.

The definition of $\textbf{T}(G \text{ or } H)$ shows that termination has to be understood as universal termination. The goal $G \text{ or } H$ terminates if, and only if, both branches G and H terminate.

The definition of $\textbf{T not } G$ is the essential difference between the $\textbf{T}$ operator here and the $\textbf{T}$ (resp. $\textbf{L}$) operator in [10] and [12]. There, $\textbf{T not } G$ is simply defined as $\textbf{T } G$. Here, we require in addition that G is ground using the operator gr which is defined as follows:

$$\text{gr}(\textbf{true}) :\equiv \top, \qquad\qquad \text{gr}(G \,\&\, H) :\equiv \text{gr}(G) \wedge \text{gr}(H),$$
$$\text{gr}(\textbf{fail}) :\equiv \top, \qquad\qquad \text{gr}(G \text{ or } H) :\equiv \text{gr}(G) \wedge \text{gr}(H),$$
$$\text{gr}(s = t) :\equiv \text{gr}(s) \wedge \text{gr}(t), \qquad \text{gr}(\textbf{some } x\, G) :\equiv \exists x\, \text{gr}(G),$$
$$\text{gr}(R(t_1, \ldots, t_n)) :\equiv \text{gr}(t_1) \wedge \ldots \wedge \text{gr}(t_n), \qquad \text{gr}(\textbf{not } G) :\equiv \text{gr}(G).$$

What we want is that for a goal G with free variables $x_1, \ldots, x_n$ the following is true:

$$(*) \qquad\qquad \text{gr}(G) \leftrightarrow \text{gr}(x_1) \wedge \ldots \wedge \text{gr}(x_n).$$

It is not possible to take this as a definition of $\text{gr}(G)$ directly, since then we would loose the substitution property that $(\textbf{T } G)\sigma \equiv \textbf{T}(G\sigma)$ for each substitution σ. We will see that in the inductive extension of a logic program $(*)$ will be provable.

Definition 3. The inductive extension of P, $\mathrm{IND}(P)$, comprises the following axioms:

I. The axioms of Clark's equality theory CET:

1. $f(x_1, \ldots, x_m) = f(y_1, \ldots, y_m) \to x_i = y_i$ [if f is m-ary and $1 \leq i \leq m$]
2. $f(x_1, \ldots, x_m) \neq g(y_1, \ldots, y_n)$ [if f is m-ary, g is n-ary and $f \not\equiv g$]
3. $t \neq x$ [if x occurs in t and $t \not\equiv x$]

II. Axioms for gr:

4. $\mathrm{gr}(c)$ [if c is a constant]
5. $\mathrm{gr}(x_1) \wedge \ldots \wedge \mathrm{gr}(x_m) \leftrightarrow \mathrm{gr}(f(x_1, \ldots, x_m))$ [if f is m-ary]

III. Uniqueness axioms (UNI):

6. $\neg(R^s(\mathbf{x}) \wedge R^f(\mathbf{x}))$

IV. Totality axioms (TOT):

7. $R^t(\mathbf{x}) \to R^s(\mathbf{x}) \vee R^f(\mathbf{x})$

V. Fixed point axioms for user-defined predicates R:

8. $\mathbf{S}\, D_R^P[\mathbf{x}] \leftrightarrow R^s(\mathbf{x}), \quad \mathbf{F}\, D_R^P[\mathbf{x}] \leftrightarrow R^f(\mathbf{x}), \quad \mathbf{T}\, D_R^P[\mathbf{x}] \leftrightarrow R^t(\mathbf{x})$

VI. Fixed point axioms for built-in, atomic goals $A \in \mathcal{D}$:

9. $\mathbf{S}\,\mathcal{B}(A) \leftrightarrow \mathbf{S}\,A, \quad \mathbf{F}\,\mathcal{B}(A) \leftrightarrow \mathbf{F}\,A, \quad \mathbf{T}\,\mathcal{B}(A) \leftrightarrow \mathbf{T}\,A.$

VII. True axioms for built-in predicates: We will explain below what we mean by that.

VIII. The simultaneous induction scheme for user-defined predicates: Let $R_1, \ldots, R_n$ be user-defined predicates and let $\varphi_1(\mathbf{x}_1), \ldots, \varphi_n(\mathbf{x}_n)$ be $\hat{\mathcal{L}}$ formulas such that the length of $\mathbf{x}_i$ is equal to the arity of R_i for $i = 1, \ldots, n$. Let

$$closed(\varphi_1(\mathbf{x}_1)/R_1, \ldots, \varphi_n(\mathbf{x}_n)/R_n)$$

be the formula obtained from

$$\forall \mathbf{x}_1 (\mathbf{S}\, D_{R_1}^P[\mathbf{x}_1] \to R_1^s(\mathbf{x}_1)) \wedge \ldots \wedge \forall \mathbf{x}_n (\mathbf{S}\, D_{R_n}^P[\mathbf{x}_n] \to R_n^s(\mathbf{x}_n))$$

by replacing simultaneously all occurrences of $R_i(\mathbf{t})$ by $\varphi_i(\mathbf{t})$ for $i = 1, \ldots, n$ and renaming the bound variables when necessary. Let

$$sub(\varphi_1(\mathbf{x}_1)/R_1, \ldots, \varphi_n(\mathbf{x}_n)/R_n)$$

be the formula

$$\forall \mathbf{x}_1 (R_1^s(\mathbf{x}_1) \to \varphi_1(\mathbf{x}_1)) \wedge \ldots \wedge \forall \mathbf{x}_n (R_n^s(\mathbf{x}_n) \to \varphi_n(\mathbf{x}_n)).$$

Then the simultaneous induction axiom is the formula

$$closed(\varphi_1(\mathbf{x}_1)/R_1, \ldots, \varphi_n(\mathbf{x}_n)/R_n) \to sub(\varphi_1(\mathbf{x}_1)/R_1, \ldots, \varphi_n(\mathbf{x}_n)/R_n).$$

Remark. I. Clark's equality theory CET is needed for the formalization of unification.

II. The predicate gr is used to express that a term is ground. If $\text{gr}(t)$ is provable from $\text{IND}(P)$, then t is ground. We assume that the language contains at least one constant symbol.

III. From the uniqueness axioms (UNI) one can immediately derive the principle $\neg(\mathbf{S}\,G \wedge \mathbf{F}\,G)$ for arbitrary goals G.

IV. From the totality axioms (TOT) one can derive $\mathbf{T}\,G \to \mathbf{S}\,G \vee \mathbf{F}\,G$ for each goal G.

V. The fixed point axioms for user defined-predicates express that one can read a clause both, from body to head, but also from head to body.

VI. In the fixed point axioms for built-in predicates it is important that A belongs to $\mathcal{D}$. Otherwise, $\mathcal{B}(A)$ is not defined.

VII. For example, the following axioms for built-in predicates are true:

1. $\forall x_1, x_2, y\,(\mathbf{S}\,x_1 \text{ is } y \wedge \mathbf{S}\,x_2 \text{ is } y \to x_1 = x_2)$.
2. $\forall x\,(\text{gr}(x) \leftrightarrow \mathbf{T}\,\text{integer}(x))$.
3. $\forall x(\mathbf{S}\,\text{integer}(x) \to \mathbf{F}\,x < x)$.
4. $\forall x_1, x_2, y_1, y_2\,(\mathbf{S}\,x_1 \text{ is } y_1 \wedge \mathbf{S}\,x_2 \text{ is } y_2 \to (\mathbf{S}\,x_1 < x_2 \leftrightarrow \mathbf{S}\,y_1 < y_2))$.
5. $\forall x, y, z(\mathbf{S}\,\text{integer}(x) \wedge \mathbf{S}\,\text{integer}(y) \wedge \mathbf{S}\,\text{integer}(z) \wedge \mathbf{S}\,x < y \wedge \mathbf{S}\,y < z \to \mathbf{S}\,x < z)$.

Note, that axioms like $x = 7 \leftrightarrow \mathbf{S}(x \text{ is } 3 + 4)$ are included in the fixed point axioms VI. The full version of this article [14] contains an exact definition of what it means that an axiom is true.

VIII. The simultaneous induction scheme expresses the minimality of the R^s predicates. Note, that the formulas $\mathbf{S}\,D_R^P$ are positive. Informally, the induction scheme says that one can use induction along the definition of the predicates. For example, for the append/3 and the list/1 predicate we have the following rules:

$$\frac{\forall \ell\, \varphi([], \ell, \ell) \qquad \forall x, \ell_1, \ell_2, \ell_3\,(\mathbf{S}\,\text{append}(\ell_1, \ell_2, \ell_3) \wedge \varphi(\ell_1, \ell_2, \ell_3) \to \varphi([x|\ell_1], \ell_2, [x|\ell_3]))}{\forall \ell_1, \ell_2, \ell_3\,(\mathbf{S}\,\text{append}(\ell_1, \ell_2, \ell_3) \to \varphi(\ell_1, \ell_2, \ell_3))}$$

$$\frac{\varphi([]) \qquad \forall x, \ell\,(\mathbf{S}\,\text{list}(\ell) \wedge \varphi(\ell) \to \varphi([x|\ell]))}{\forall \ell\,(\mathbf{S}\,\text{list}(\ell) \to \varphi(\ell))}$$

The predicates list/1 and append/3 have their standard definitions:

list([]).   append([], ℓ, ℓ).

list([$x|\ell$]) :- list(ℓ).   append([$x|\ell_1$], ℓ_2, [$x|\ell_3$]) :- append(ℓ_1, ℓ_2, ℓ_3).

The expression [] denotes a constant for the empty list and $[\cdot|\cdot]$ is a binary function symbol for constructing list.

The inductive extension is related to Clark's completion and Kunen's three-valued completion in the following way. Let (FIX) be the collection of the fixed-point axioms for the R^s and R^f relations. Then (CET)+(UNI)+(FIX) is equivalent to Kunen's three-valued completion of [8]. Moreover, Clark's completion of [3] can be obtained from the three-valued completion be adding the stronger totality axiom $R^s(t) \vee R^f(t)$.

Example 4. The simultaneous induction scheme is more natural for logic programs than structural induction on the Herbrand universe. Assume that the language $\mathcal{L}$ has exactly one constant symbol c and one unary function symbol f. In this case, *induction on the universe* is the scheme

$$(**) \qquad \qquad \varphi(c) \wedge \forall x \, (\varphi(x) \to \varphi(f(x))) \to \forall x \, \varphi(x).$$

Let P be the program with the two clauses $q :- r(x)$ and $r(f(x)) :- r(x)$. Using induction on the universe $(**)$ for $\varphi(x) :\equiv \mathbf{T} \, r(x)$ and the fixed point axioms

$$\forall x \, \mathbf{T} \, r(x) \leftrightarrow \mathbf{T} \, q \quad \text{and} \quad \forall y (x = f(y) \to \mathbf{T} \, r(y)) \leftrightarrow \mathbf{T} \, r(x)$$

one can easily derive $\forall x \, \mathbf{T} \, r(x)$ and hence $\mathbf{T} \, q$. But the goal q does not terminate under query evaluation. Therefore, induction on the universe is not appropriate for our purposes. We want that $\mathbf{T} \, G$ is provable if, and only if, G terminates.

Proofs of the following two theorems can be found in the full version of this paper [14]. They use ideas from [8] and [11].

Theorem 4 (Soundness). *1. If G terminates, then* IND$(P) \vdash \mathbf{T} \, G$.
2. If G succeeds with answer σ, then IND$(P) \vdash \mathbf{S} \, G\sigma$.
3. If G fails, then IND$(P) \vdash \mathbf{F} \, G$.

In the proof of this theorem the full power of the inductive extension is not used. Only CET and the directions from left to right in the fixed point axioms are needed.

Theorem 5 (Adequacy). *1. If* IND$(P) \vdash \mathbf{T} \, G$, *then G terminates.*
2. If IND$(P) \vdash \mathbf{T} \, G \wedge \mathbf{S} \, G\sigma$, *then G succeeds with answer including σ.*
3. If IND$(P) \vdash \mathbf{T} \, G \wedge \mathbf{F} \, G$, *then G fails.*

This theorem is not trivial, since the term model in which $R^t(t)$ is true iff $R(t)$ terminates is, in general, not a model of the inductive extension (cf. Example 4). Note, that the theorem implies, for example, the following existence property:

Corollary 6. *If* IND$(P) \vdash \mathbf{S}(\text{some } x \, G[x]) \wedge \mathbf{T}(\text{some } x \, G[x])$ *then there exists a term t such that the goal $G[x]$ succeeds with answer $\{t/x\}$ and* IND$(P) \vdash \mathbf{S} \, G[t]$.

It is important to note, that from the provability of $\mathbf{T} \, G$ if follows not only that all computations for G terminate but also that there are no errors in calls of built-in predicates during the computation. There is an interesting analogy between the $\mathbf{T}$ operator and the *logic of partial terms* (cf. eg. [5, 6]). In the logic of partial terms the expression $t\downarrow$ means that the functional program t terminates and that during the evaluation there are no type conflicts, i.e. the program is dynamically well-typed. The meaning of $\mathbf{T}\,G$ is similar. It means that the evaluation of the goal G terminates and that there are no error messages caused by non-ground negative goals or wrongly typed built-in atomic goals.

The next theorem is proved in [13] using standard methods like partial cut-elimination for infinitary systems and asymmetric interpretations.

Theorem 7. *Without built-in predicates,* $\mathrm{IND}(P)$ *has the same proof-theoretic strength as Peano Arithmetic.*

From the proof of this theorem in [13] one could extract a program P with a distinguished predicate symbol R such that the true formula

$$\forall x(\mathbf{S}\,\mathtt{list}(x) \to \mathbf{T}\,R(x))$$

is not provable in $\mathrm{IND}(P)$. The reason that the formula is not provable is that the computation tree for $R(\ell)$ grows too fast compared to the length of the list ℓ. In practice, however, such programs do not occur.

4 LPTP — a logic program theorem prover

In this section we give a short overview of LPTP, an interactive theorem prover which is based on the inductive extension of pure Prolog programs. LPTP is a proof refinement system that allows a user to construct formal proofs interactively. The user can generate proofs deductively from the assumptions forwards to the goal or goal directed backwards from the goal to the axioms. LPTP has the ability to search for proofs automatically. In the simplest case, LPTP just finds the name of a lemma that can be used at a certain point in a proof. In the best case, LPTP finds complete proofs. In general LPTP can complete automatically small gaps in proofs that require not more than, say, 10 steps. For the rest of the proof LPTP has to be guided by the user.

LPTP consists of 6500 lines of Prolog code. It runs in CProlog, Quintus Prolog, and SICStus Prolog under Unix. LPTP has a graphical user interface in the Gnu Emacs Editor. For example, the user can double-click on a quantifier and the whole scope of the quantifier is highlighted. LPTP generates TeX and HTML output.

The kernel of LPTP is written in exactly the fragment of Prolog that can be treated in LPTP. This means that LPTP uses no single cut. Moreover, it is possible to prove properties of LPTP within LPTP.

The largest program we have verified with LPTP is 635 lines long. It is a parser for standard ISO Prolog. The 635 lines comprise not only the implementation but also the specification of the parser. The correctness proof includes theorems like the following: if a parse tree is transformed into a token list (using **write**) and the token list is parsed back into a parse tree (using **read**), then this parse tree is identical to the original one.

The fully formalized correctness proof for the ISO Prolog parser is 13000 lines long. So we have a factor of 20 for the full verification of this example program. LPTP is able to check the whole proof (133 pages) in 99.2 seconds on a Sun SPARCstation. This speed, however, says not much about LPTP, since it is much more important how fast a user can create proofs using the system. A skilled user can generate more than 1000 lines of formal proofs in one day. Altogether we have generated 25000 lines of formal proofs with LPTP.

5 En example proof in LPTP

As an example, we now sketch a proof of an algorithm that inserts elements into AVL trees. AVL trees are ordered binary trees. They are subject to the Adelson-Velskii-Landis balance criterion: A tree is balanced iff for every node the heights of its two subtrees differ by at most 1. Our Prolog version of the algorithm is generic in two predicates $a/1$ and $r/2$. The idea is that $r/2$ is a total ordering on the set $a/1$. In fact, we only need that $a/1$ and $r/2$ satisfy the following axioms:

1. $\forall x, y, z (S\, a(x) \wedge S\, a(y) \wedge S\, a(z) \wedge S\, r(x, y) \wedge S\, r(y, z) \rightarrow S\, r(x, z))$
2. $\forall x, y (S\, a(x) \wedge S\, a(y) \rightarrow S\, r(x, y) \vee S\, r(y, x))$
3. $\forall x, y (S\, a(x) \wedge S\, a(y) \rightarrow T\, r(x, y))$
4. $\forall x (S\, a(x) \rightarrow \mathrm{gr}(x))$.

Axiom (1) says that $r/2$ is transitive on $a/1$; (2) says that $r/2$ is total on $a/1$; (3) says that $r/2$ terminates on $a/1$; (4) says that $a/1$ contains only ground terms.

The algorithm is coded as predicate $\mathrm{addavl}/3$. If x is a value of $a/1$ and t_1 is an AVL tree then $\mathrm{addavl}(x, t_1, t_2)$ inserts x into t_1 and returns the result in t_2.

The empty tree is represented as t. A tree with value x, left subtree ℓ and right subtree r is represented as $t(x, b, \ell, r)$; b is the difference of the height of r and the height of ℓ; b can be -1, 0 or 1.

For the specification of the correctness of the algorithm we need the predicates $\mathrm{avl}/1$ and $\mathrm{in}/2$. The predicate $\mathrm{avl}(t)$ expresses that (i) t is a tree with values x belonging to $a/1$; (ii) t satisfies the Adelson-Velskii-Landis balance criterion; (iii) t is ordered, i.e. in a node $t(x, b, \ell, r)$, x is an upper bound of the elements of ℓ and a lower bound of the elements of r with respect to the ordering $r/2$. The predicate $\mathrm{in}(x, t)$ expresses that the value x occurs in the tree t.

Correctness of the algorithm can be expressed by the following formulas:

1. $\forall x, t_1, t_2 (S\, a(x) \wedge S\, \mathrm{avl}(t_1) \wedge S\, \mathrm{addavl}(x, t_1, t_2) \rightarrow S\, \mathrm{avl}(t_2))$.
2. $\forall x, t_1, t_2 (S\, \mathrm{addavl}(x, t_1, t_2) \rightarrow S\, \mathrm{in}(x, t_2))$.
3. $\forall x, y, t_1, t_2 (S\, \mathrm{addavl}(x, t_1, t_2) \wedge S\, \mathrm{in}(y, t_1) \rightarrow S\, \mathrm{in}(y, t_2))$.
4. $\forall x, y, t_1, t_2 (S\, \mathrm{addavl}(x, t_1, t_2) \wedge S\, \mathrm{in}(y, t_2) \rightarrow y = x \vee S\, \mathrm{in}(y, t_1))$.
5. $\forall x, t_1, t_2 (S\, a(x) \wedge S\, \mathrm{avl}(t_1) \rightarrow T\, \mathrm{addavl}(x, t_1, t_2))$.
6. $\forall x, t_1 (S\, a(x) \wedge S\, \mathrm{avl}(t_1) \rightarrow \exists t_2 (S\, \mathrm{addavl}(x, t_1, t_2)))$.

The formulas mean the following:

1. If we insert an element into an AVL tree, then the new tree we get is also an AVL tree.
2. If we insert an element in an AVL tree, then the added element is an element of the new AVL tree.
3. If we add an element to an AVL tree containing y, then the new AVL tree also contains y.
4. If we add an element x to an AVL tree, and if the new AVL tree contains y, then y is the element x we just added, or y was already in the initial AVL tree.
5. The algorithm terminates for appropriate inputs.
6. The algorithm is complete. It can insert elements into arbitrary AVL trees.

The algorithm addavl/3 together with the predicates used in the specification is 137 lines long. The formal correctness proof of formulas 1–6 is 2903 lines long. It has been created by Patrik Fuhrer and Rene Lehmann and is part of the distribution of LPTP.

References

1. K. R. Apt. *From Logic Programming to Prolog.* International Series in Computer Science. Prentice Hall, 1996.
2. K. R. Apt and D. Pedreschi. Reasoning about termination of pure Prolog programs. *Information and Computation,* 106(1):109–157, 1993.
3. K. L. Clark. Negation as failure. In H. Gallaire and J. Minker, editors, *Logic and Data Bases,* pages 293–322. Plenum Press, New York, 1978.
4. S. K. Debray and P. Mishra. Denotational and operational semantics for Prolog. *J. of Logic Programming,* 5(1):61–91, 1988.
5. S. Feferman. Logics for termination and correctness of functional programs. In Y. N. Moschovakis, editor, *Logic from Computer Science,* pages 95–127, New York, 1992. Springer-Verlag.
6. S. Feferman. Logics for termination and correctness of functional programs, II. Logics of strength PRA. In P. Aczel, H. Simmons, and S. S. Wainer, editors, *Proof Theory,* pages 195–225. Cambridge University Press, 1992.
7. P. M. Hill and J. W. Lloyd. *The Gödel Programming Language.* MIT Press, 1994.
8. K. Kunen. Signed data dependencies in logic programs. *J. of Logic Programming,* 7(3):231–245, 1989.
9. Z. Somogyi, F. Henderson, and T. Conway. The execution algorithm of Mercury, an efficient purely declarative logic programming language. *J. of Logic Programming,* 1996. To appear.
10. R. F. Stärk. The declarative semantics of the Prolog selection rule. In *Proceedings of the Ninth Annual IEEE Symposium on Logic in Computer Science, LICS '94,* pages 252–261, Paris, France, July 1994. IEEE Computer Society Press.
11. R. F. Stärk. Input/output dependencies of normal logic programs. *J. of Logic and Computation,* 4(3):249–262, 1994.
12. R. F. Stärk. First-order theories for pure Prolog programs with negation. *Archive for Mathematical Logic,* 34(2):113–144, 1995.

13. R. F. Stärk. The finite stages of inductive definitions. In P. Hájek, editor, *GÖDEL'96. Logical Foundations of Mathematics, Computer Science and Physics — Kurt Gödel's Legacy*, pages 267–290, Brno, Czech Republic, 1996. Springer-Verlag, Lecture Notes in Logic 6.

14. R. F. Stärk. The theoretical foundations of LPTP (a logic program theorem prover). Technical report, Institute of Informatics, University of Fribourg, 1997.

LPTP is available by anonymous ftp from:
ftp://ftp-iiuf.unifr.ch/pub/dss/staerk/lptp-1.01.tar.gz
Further information on LPTP can be found at:
http://www-iiuf.unifr.ch/~staerk

Unification of Terms with Term-Indexed Variables

Igor L. Tandetnik

Moscow State University of Railway Transport (MIIT),
Obrazcova 15, 101475, Moscow, Russia

Abstract. We consider the unification problem for generalized terms. The syntax of generalized terms allows the use of indexed variables, with indexes themselves being generalized terms. This leads to an infinite set of conditional equations. We propose a reduction to the finite conditional unification proble. We prove the existence of a most general unifier (m.g.u.) for both the finite and infinite cases. An efficient (cubic in time, linear in space) algorithm for computing the m.g.u. for these problems is developed.

1 Introduction

The classic unification problem is as follows: Given a finite set of pairs of terms (L_i, R_i), $i = 1, \ldots, N$, find a most general substitution (most general unifier, m.g.u) σ for which the equalities

$$L_i \sigma = R_i \sigma, \qquad i = 1, \ldots, N \tag{1}$$

hold.

The theory of most general unifiers is developed in [1] and [2] (see also the survey in [3]). Fast unification algorithms can be found in [4] and [5]. In [6] and [7], while dealing with problems of logical description and synthesis of reference structures, the following problem was introduced: terms being unified are allowed to contain variables of the form v_p where p is again a variable. This naturally leads to the following constraint on substitution, expressing functional dependency of an indexed variable on its index:

$$p\sigma = q\sigma \Rightarrow v_p \sigma = v_q \sigma. \tag{2}$$

The algorithm given in [6] and [7] can cope only with indexed variables which have index-free variables or constants as their indexes. In this paper, we consider a more general problem, in which any term is allowed as an index, including those containing other indexed variables. This leads to the necessity of considering infinite unifiers for (1), (2), i.e. arbitrary mappings defined on the set of all terms and commuting with functional symbols. We prove the existence of an idempotent m.g.u. σ for (1), (2) and give an algorithm calculating $t\sigma$ for any term t. Exploiting the ideas of [5], we give an effective realization of this algorithm which has time complexity $O(l^3)$, where l is the total length of all terms occurring in (1) together with the term t.

2   Consistent substitutions

Let us fix the set $V_I = \{p_1, p_2, \ldots\}$ (index-free variables) and the set F of *functional symbols*.

The sets of *indexed variables* V_{II} and *terms* T are defined as the least sets satisfying the following conditions:

1. $V_I \cup V_{II} \subseteq T$.
2. If $f \in F$, $n = arity(f)$ and $t_1, \ldots, t_n \in T$ then $f(t_1, \ldots, t_n) \in T$.
3. If $t \in T$ then $v_t \in V_{II}$.

Let $V = V_I \cup V_{II}$ be the set of all variables, $Var(t)$ the set of all variables occurring in t on the first level (not in indexes) and $AVar(t)$ the set of all variables occurring in t on any depth. For $W \subseteq V$ let $W^+ = \bigcup_{x \in W} AVar(x)$.

The *substitution* is a mapping $\sigma : T \to T$ satisfying the condition

$$f(t_1, \ldots, t_n)\sigma = f(t_1\sigma, \ldots, t_n\sigma)$$

(the substitution is defined uniquely by its values on variables). Let us define $Dom(\sigma) = \{x \in V \mid x\sigma \neq x\}$, $Val(\sigma) = \bigcup_{x \in Dom(\sigma)} Var(x\sigma)$, $Var(\sigma) = Dom(\sigma) \cup Val(\sigma)$ and $AVar(\sigma) = (Var(\sigma))^+$.

The substitution σ is called *finite* when the set $Dom(\sigma)$ is finite, and *infinite* otherwise.

The substitution σ is called *W-consistent* for some set $W \subseteq V$ if for each $v_s, v_t \in W$ the equality $s\sigma = t\sigma$ implies $v_s\sigma = v_t\sigma$. The *V-consistent* substitution is called *consistent*. Note that any consistent substitution σ with $Dom(\sigma) \neq \emptyset$ is infinite.

A consistent substitution $\bar{\sigma}$ is called a consistent extension of a substitution σ if $x\bar{\sigma} = x\sigma$ for each $x \in Dom(\sigma)$.

Theorem 1. *Let σ be W-consistent substitution, where $Var(\sigma) \subseteq W$ and $W^+ = W$. Let $\bar{\sigma}$ be the substitution satisfying the conditions*

$$p\bar{\sigma} = p\sigma \quad \text{for } p \in V_I,$$

$$v_t\bar{\sigma} = \begin{cases} v_t\sigma, & \text{if } v_t \in W, \\ v_s\sigma, & \text{if } v_t \notin W \text{ and } \exists v_s \in W : s\bar{\sigma} = t\bar{\sigma}, \\ v_{t\bar{\sigma}}, & \text{if } v_t \notin W \text{ and } \neg\exists v_s \in W : s\bar{\sigma} = t\bar{\sigma}. \end{cases} \tag{3}$$

Then 1. The substitution $\bar{\sigma}$ exists and is unique.

2. The substitution $\bar{\sigma}$ is a consistent extension of σ.

3. If σ is idempotent (i.e. $\sigma^2 = \sigma$) then $\bar{\sigma}$ is also idempotent and $\sigma\bar{\sigma} = \bar{\sigma}$ (and so $s\sigma = t\sigma$ implies $s\bar{\sigma} = t\bar{\sigma}$).

Proof. **1.** We introduce function $VD(t)$ on terms as follows:

1. $VD(p) = 0$ for $p \in V_I$.
2. $VD(v_t) = 1 + VD(t)$.
3. $VD(f(t_1, \ldots, t_n)) = \max\{VD(t_1), \ldots, VD(t_n)\}$.

The first two lines of (3) uniquely define the substitution $\bar{\sigma}$ on terms t for which $Var(t) \subseteq V_I \cup W$ (this includes all terms with the property $VD(t) = 0$). Moreover, for such terms $t\bar{\sigma} = t\sigma$.

Consider a variable $v_t \notin W$. We assume that the substitution $\bar{\sigma}$ is uniquely defined on all terms r with $VD(r) < VD(v_t)$. There are three possibilities.

Case 1: there is exactly one variable $v_s \in W$ for which $s\bar{\sigma} = t\bar{\sigma}$ (both sides of this equality are defined due to inductive hypothesis). Then by (3) $v_t\bar{\sigma} = v_s\sigma$.

Case 2: there are two variables $v_r, v_s \in W$ for which $r\bar{\sigma} = s\bar{\sigma} = t\bar{\sigma}$. It follows from the condition $W^+ = W$ that $Var(r) \subseteq W$ and $Var(s) \subseteq W$, and so $r\sigma = r\bar{\sigma} = s\bar{\sigma} = s\sigma$. But the substitution σ is W-consistent which implies $v_r\sigma = v_s\sigma$. Thus $v_t\bar{\sigma}$ is uniquely defined.

Case 3: there is no variable $v_s \in W$ for which $s\bar{\sigma} = t\bar{\sigma}$. Then $v_t\bar{\sigma} = v_{t\bar{\sigma}}$ where $t\bar{\sigma}$ is uniquely defined by inductive hypothesis.

2. It is obvious that the substitution $\bar{\sigma}$ is an extension of σ i.e. $x\bar{\sigma} = x\sigma$ for each $x \in Dom(\sigma)$. It remains to be proved that $\bar{\sigma}$ is consistent.

Suppose there are two variables v_s and v_t for which $s\bar{\sigma} = t\bar{\sigma}$. We shall prove that $v_s\bar{\sigma} = v_t\bar{\sigma}$. If $v_s, v_t \in W$ then $v_s\bar{\sigma} = v_s\sigma = v_t\sigma = v_t\bar{\sigma}$. If $v_s \in W$ but $v_t \notin W$ then $v_t\bar{\sigma} = v_s\sigma = v_s\bar{\sigma}$. If $v_s, v_t \notin W$ then $v_s\bar{\sigma} = v_{s\bar{\sigma}} = v_{t\bar{\sigma}} = v_t\bar{\sigma}$. So $\bar{\sigma}$ is consistent in all three cases.

3. Let x be a variable and $\tau = x\sigma$. It follows from the condition $Var(\sigma) \subseteq W$ that $Var(\tau) \subseteq V_I \cup W$ whenever $x \in V_I \cup W$. If $x \in W$ then $x\sigma\bar{\sigma} = \tau\bar{\sigma} = \tau\sigma = x\sigma^2 = x\sigma = x\bar{\sigma}$. If $x \notin W$ then $x\sigma = x$ and the equality $x\sigma\bar{\sigma} = x\bar{\sigma}$ holds.

We now prove that $\bar{\sigma}$ is idempotent. If $x \in V_I \cup W$ then $x\bar{\sigma} = x\sigma$ and $x\bar{\sigma}^2 = x\sigma\bar{\sigma} = x\bar{\sigma}$.

Let $x = v_t \notin W$. If there is a variable $v_s \in W$ for which $s\bar{\sigma} = t\bar{\sigma}$ then $v_t\bar{\sigma}^2 = v_s\sigma\bar{\sigma} = v_s\bar{\sigma} = v_s\sigma = v_t\bar{\sigma}$. Suppose such a variable does not exist. We assume that $r\bar{\sigma}^2 = r\bar{\sigma}$ for all terms r with the property $VD(r) < VD(v_t)$. Thus $t\bar{\sigma}^2 = t\bar{\sigma}$ and there is no variable $v_s \in W$ for which $t\bar{\sigma}^2 = s\bar{\sigma}$. We have $v_t\bar{\sigma}^2 = v_{t\bar{\sigma}}\bar{\sigma} = v_{t\bar{\sigma}^2} = v_{t\bar{\sigma}} = v_t\bar{\sigma}$. $\quad\square$

The substitution $\bar{\sigma}$ from (3) is called the *standard consistent extension* of σ. Note that the conditions (3) provide a method to calculate $t\bar{\sigma}$ for any $t \in T$ whenever σ is computable (in particular, finite).

3 Unification algorithm

Consider the system of equations S of the form $L_i = R_i, i = 1, \ldots, N$. We define $Var(S) = \bigcup_{i=1}^{N}(Var(L_i) \cup Var(R_i))$ and $AVar(S) = (Var(S))^+$.

The substitution σ is called a most general (most general W-consistent, most general consistent) unifier of S, if σ is a (W-consistent, consistent) unifier of S and for any (W-consistent, consistent) unifier θ of S there exists a substitution λ such that $\theta = \sigma\lambda$ (if σ is idempotent, the last condition can be rewritten as $\theta = \sigma\theta$).

Theorem 2. *Let S be a system of equations, $W \subseteq V$, $W^+ = W$ and $Var(S) \subseteq W$. Let σ be an idempotent most general W-consistent unifier of S, $Var(\sigma) \subseteq W$. Then the standard consistent extension $\bar{\sigma}$ of σ is an idempotent most general consistent unifier of S.*

Proof. By the theorem 1 the substitution $\bar{\sigma}$ is an idempotent consistent unifier of S. We shall prove that it is also a most general consistent unifier. Let θ be some consistent unifier of S. As σ is a m.g.u. of S we have $\theta = \sigma\theta$. Thus $t\bar{\sigma}\theta = t\sigma\theta = t\theta$ for all terms t for which $Var(t) \subseteq V_I \cup W$ (this includes all terms with the property $VD(t) = 0$).

Consider a variable $v_t \notin W$. We assume that $r\bar{\sigma}\theta = r\theta$ for all terms r for which $VD(r) < VD(v_t)$. Suppose there is a variable $v_s \in W$ for which $s\bar{\sigma} = t\bar{\sigma}$. By applying the substitution θ to both sides of the last equality we have $s\bar{\sigma}\theta = t\bar{\sigma}\theta$. But $s\bar{\sigma}\theta = s\theta$ due to the fact that $Var(s) \subseteq W$ and $t\bar{\sigma}\theta = t\theta$ by the inductive hypothesis. As θ is consistent we have $v_s\theta = v_t\theta$. Thus $v_t\bar{\sigma}\theta = v_s\sigma\theta = v_s\theta = v_t\theta$.

Suppose there is no variable $v_s \in W$ for which $s\bar{\sigma} = t\bar{\sigma}$. By the inductive hypothesis $t\theta = t\bar{\sigma}\theta$ which leads to $v_t\theta = v_{t\bar{\sigma}}\theta$ because θ is consistent. Then $v_t\bar{\sigma}\theta = v_{t\bar{\sigma}}\theta = v_t\theta$. So $x\bar{\sigma}\theta = x\theta$ for any $x \in V$. $\qquad\square$

Thus in order to find the most general consistent solution $\bar{\sigma}$ of S ($\bar{\sigma}$ in most cases is infinite) it is sufficient to construct only its finite part — the most general $AVar(S)$-consistent unifier σ of S. Then the procedure (3) allows one to calculate $t\bar{\sigma}$ for any term t. Furthermore, if term t is known beforehand we can assume without loss of generality that the system S contains the equation $t = t$. In this case $t\bar{\sigma} = t\sigma$ and the application of the procedure (3) is not necessary.

Let $unify(S)$ be a unification algorithm for terms without indexed variables (e.g. algorithm from [5]). Its input is the system of equations S, and its output is a pair $(bool, \sigma_0)$ with the following properties:

1. If $bool = false$ then the system S has no solutions.
2. If $bool = true$ then σ_0 is the most general unifier of S. Moreover, σ_0 is idempotent and conservative (i.e. $Var(\sigma_0) \subseteq Var(S)$).

The conservativity property provides the finiteness of σ_0.

Consider the algorithm $unify_{ind}(S)$, accepting system S as its input and constructing a pair $(bool, \sigma)$:

1. Let $k = 0$, $S_0 = S$.
2. Let $(bool, \sigma_k) = unify(S_k)$. If $bool = false$, stop with the result $(false, \sigma_k)$.
3. Find a pair $v_s, v_t \in AVar(S)$ for which $s\sigma_k = t\sigma_k$ but $v_s\sigma_k \neq v_t\sigma_k$. If there is no such pair, stop with the result $(true, \sigma_k)$, else let $S_{k+1} = S_k \cup (v_s = v_t)$, $k = k + 1$ and go to step 2.

The theorem below states the properties of the algorithm $unify_{ind}(S)$.

Theorem 3. *1. Algorithm $unify_{ind}(S)$ always stops.*

2. If $(bool, \sigma)$ is an output of $unify_{ind}(S)$ and $bool = false$ then system S has no consistent solutions.

3. If $(bool, \sigma)$ is an output of $unify_{ind}(S)$ and $bool = true$ then σ is an idempotent most general $AVar(S)$-consistent $AVar(S)$-conservative unifier of S.

Proof. **1.** No one pair of variables can be chosen more than once during step 3, and the set $AVar(S)$ is finite.

2. Suppose there is a consistent unifier θ of the system S. θ is obviously a unifier of S_0. We assume θ is a unifier of S_k for some k so S_k is unifiable and σ_k is its idempotent m.g.u. There must exist a pair $v_s, v_t \in AVar(S)$ for which $s\sigma_k = t\sigma_k$ (otherwise the algorithm stops with $bool = true$). We have $s\theta = s\sigma_k\theta = t\sigma_k\theta = t\theta$ which implies $v_s\theta = v_t\theta$ and θ is a unifier of the system S_{k+1}. By induction the systems S_k for any k are unifiable and the algorithm cannot stop with $bool = false$.

3. Let S_n be the last system being unified by the algorithm. Note that $AVar(S_k) = AVar(S)$ for any k. As $\sigma = \sigma_n$ is the output of $unify(S_n)$ it is an idempotent $AVar(S)$-conservative m.g.u. of S_n. σ is a unifier of S because $S \subseteq S_n$. Step 3 of the algorithm ensures that σ is $AVar(S)$-consistent. We shall prove that σ is a most general $AVar(S)$-consistent unifier of S.

Let θ be some $AVar(S)$-consistent unifier of $S = S_0$. The argument similar to that of part 2 of this proof leads us to the fact that θ is a unifier of S_k for any k, particularly S_n. As σ is an idempotent m.g.u. of S_n we have $\sigma\theta = \theta$ and so σ is a most general $AVar(S)$-consistent unifier of S. $\quad\square$

Thus algorithm $unify_{ind}(S)$ obtains a finite substitution satisfying conditions of theorem 2.

The effective realizations of the algorithm $unify_{ind}(S)$ exploits a representation of terms using so-called dags (directed acyclic graphs) as in [5]. The algorithm has linear space complexity and cubic time complexity in the size of the dag representing the system of equations. The size of this dag is bound by the total length l the system S, which gives the time complexity $O(l^3)$.

References

1. J.A. Robinson. A machine oriented logic based on the resolution principle. *J. of the ACM*, 12(1), 1965, p. 23-41

2. D.E. Knuth, P.B. Bendix. Simple word problems in universal algebras. In J.Leech, editor, *Computational problems in Abstract Algebra*, Pergamon Press, Oxford, 1970

3. F. Baader and J.H. Siekmann. Unification Theory. In: *Handbook of Logic in Artificial Intelligence and Logic Programming* D.M. Gabbay, C.J. Hogger, and J.A. Robinson (Ed.), Oxford University Press, 1994

4. A. Martelli, U. Montanari. An efficient unification algorithm. *ACM Transactions on Programming Languages and Systems* 4, 258-282, 1982.

5. J. Corbin, M. Bidoit. A rehabilitation of Robinson's unification algorithm. *Information Processing* 83, R.E.A. Mason (ed.), Elsevier Science Publishers (North-Holland), p. 909-914, 1983

6. S. Artemov, V. Krupski. Referential data structures and labeled modal logic. *Lecture Notes in Computer Science*, 1994, v.13, p. 23-33

7. S. Artëmov, V. Krupski. Data storage interpretation of labeled modal logic. *Annals of Pure and Applied Logic*, v.78 (1996), pp. 57-71.

Back-Forth Equivalences for Design of Concurrent Systems *

Igor V. Tarasyuk

Institute of Informatics Systems,
Siberian Division of the Russian Academy of Sciences,
Lavrentieva ave. 6, 630090, Novosibirsk, Russia
Phone: +7 3832 35 03 60
Fax: +7 3832 32 34 94
E-mail: itar@iis.nsk.su

Abstract. The paper is devoted to the investigation of behavioural
equivalences of concurrent systems modelled by Petri nets. Back-forth
bisimulation equivalences known from the literature are supplemented
by new ones, and their relationship with basic behavioural equivalences
is examined for the whole class of Petri nets as well as for their subclass
of sequential nets. In addition, the preservation of all the equivalence
notions by refinements is examined.

1 Introduction

The notion of equivalence is central in any theory of systems. It allows to compare
systems taking into account particular aspects of their behaviour.

Petri nets became a popular formal model for design of concurrent and distributed systems. One of the main advantages of Petri nets is their ability for
structural characterization of three fundamental features of concurrent computations: causality, nondeterminism and concurrency.

In recent years, a wide range of semantic equivalences was proposed in concurrency theory. Some of them were either directly defined or transferred from
other formal models to Petri nets. The following basic notions of equivalences
for Petri nets are known from the literature (some of them were introduced by
the author in [15] to obtain the complete set of relations in interleaving/true
concurrency and linear time/branching time semantics).

- *Trace equivalences* (respect only protocols of nets functioning): interleaving
  ($\equiv_i$) [8], step ($\equiv_s$) [12], pomset ($\equiv_{pom}$) [7] and process ($\equiv_{pr}$) [15].
- *(Usual) bisimulation equivalences* (respect branching structure of nets func-
  tioning): interleaving ($\underline{\leftrightarrow}_i$) [11], step ($\underline{\leftrightarrow}_s$) [10], partial word ($\underline{\leftrightarrow}_{pw}$) [16],
  pomset ($\underline{\leftrightarrow}_{pom}$) [3] and process ($\underline{\leftrightarrow}_{pr}$) [1].

* The work is supported by Volkswagen Fund, grant I/70 564 and Russian Fund for
Basic Research, grant 96-01-01655

- *ST-bisimulation equivalences* (respect the duration of transition occurrences in nets functioning): interleaving ($\underleftrightarrow{}_{iST}$) [7], partial word ($\underleftrightarrow{}_{pwST}$) [16], pomset ($\underleftrightarrow{}_{pomST}$) [16] and process ($\underleftrightarrow{}_{prST}$) [15].
- *History preserving bisimulation equivalences* (respect the "past" or "history" of nets functioning): pomset ($\underleftrightarrow{}_{pomST}$) [14] and process ($\underleftrightarrow{}_{prST}$) [15].
- *Conflict preserving equivalences* (completely respect conflicts in nets): multi event structure ($\equiv_{mes}$) [15] and occurrence ($\equiv_{occ}$) [7].
- *Isomorphism* ($\simeq$) (i.e. coincidence of nets up to renaming of places and transitions).

Back-forth bisimulation equivalences are based on the idea that bisimulation relation do not only require systems to simulate each other behaviour in the forward direction (as usually) but also when going back in history. They are closely connected with equivalences of logics with past modalities.

These equivalence notions were initially introduced in [9] in the framework of transition systems. It was shown that back-forth variant ($\underleftrightarrow{}_{ibif}$) of interleaving bisimulation equivalence coincide with ordinary $\underleftrightarrow{}_i$.

In [4, 5, 6] the new variants of step ($\underleftrightarrow{}_{sbsf}$), partial word ($\underleftrightarrow{}_{pwbpwf}$) and pomset ($\underleftrightarrow{}_{pombpomf}$) back-forth bisimulation equivalences were defined in the framework of prime event structures and compared with usual, ST- and history preserving bisimulation equivalences. It was demonstrated that among all back-forth bisimulation equivalences only $\underleftrightarrow{}_{pombpomf}$ is preserved by refinements (it coincides with $\underleftrightarrow{}_{pomh}$ which has such a property).

In [13] the new idea of differentiating the kinds of back and forth simulations appeared (following this idea, it is possible, for example, to define step back – pomset forth bisimulation equivalence ($\underleftrightarrow{}_{sbpomf}$)). The set of all possible back-forth equivalence notions was proposed in interleaving, step, partial word and pomset semantics. Two new notions which do not coincide with known ones were proposed: step back – partial word forth ($\underleftrightarrow{}_{sbpwf}$) and step back – pomset forth ($\underleftrightarrow{}_{sbpomf}$) bisimulation equivalences. It was proved that the former is not preserved by refinements, and the question was addressed about the latter.

To choose most appropriate behavioural viewpoint on systems to be modelled, it is very important to have a complete set of equivalence notions in all semantics and understand their interrelations. This branch of research is usually called *comparative concurrency semantics*. To clarify the nature of equivalences and evaluate how they respect a concurrency, it is actual to consider also correlation of these notions on concurrency-free (sequential) nets. Treating equivalences for preservation by refinements allows one to decide which of them may be used for top-down design.

Working in the framework of Petri nets, in this paper we extend the set of back-forth equivalences from [13] by that of induced by process semantics and obtain two new notions which cannot be reduced to the known ones: step back – process forth ($\underleftrightarrow{}_{sbprf}$) and pomset back – process forth ($\underleftrightarrow{}_{pombprf}$) bisimulation equivalences. We compare all back-forth equivalences with the set of basic behavioural notions from [15] and complete the results of [6, 13].

In addition, we investigate the interrelations of all the equivalence notions on sequential nets. The merging of most of the equivalence relations in interleaving – pomset semantics is demonstrated. We prove that on sequential nets back-forth equivalences coincide with usual forth ones.

In [2], SM-refinement operator for Petri nets was proposed, which "replaces" transitions of nets by SM-nets, a special subclass of state machine nets. We treat all the considered equivalence notions for preservation by SM-refinements and establish that among back-forth relations only $\underleftrightarrow{}_{pombpomf}$ and $\underleftrightarrow{}_{prbprf}$ are preserved by SM-refinements So, we obtained the negative answer to the question from [13]: $\underleftrightarrow{}_{sbpomf}$ (and even more strict $\underleftrightarrow{}_{pombprf}$) is not preserved by refinements.

2 Basic definitions

In this section we give some basic definitions used further.

2.1 Labelled nets

Let $Act = \{a, b, \ldots\}$ be a set of *action names* or *labels*.

Definition 1. A *labelled net* is a quadruple $N = \langle P_N, T_N, F_N, l_N \rangle$, where:

- $P_N = \{p, q, \ldots\}$ is a set of *places*;
- $T_N = \{t, u, \ldots\}$ is a set of *transitions*;
- $F_N : (P_N \times T_N) \cup (T_N \times P_N) \to \mathbf{N}$ is the *flow relation* with weights ($\mathbf{N}$ denotes a set of natural numbers);
- $l_N : T_N \to Act$ is a labelling of transitions with action names.

Given labelled nets $N = \langle P_N, T_N, F_N, l_N \rangle$ and $N' = \langle P_{N'}, T_{N'}, F_{N'}, l_{N'} \rangle$. A mapping $\beta : N \to N'$ is an *isomorphism* between N and N', denoted by $\beta : N \simeq N'$, if:

1. β is a bijection such that $\beta(P_N) = P_{N'}$ and $\beta(T_N) = T_{N'}$;
2. $\forall t \in T_N \ l_N(t) = l_{N'}(\beta(t))$;
3. $\forall t \in T_N \ {}^\bullet\beta(t) = \beta({}^\bullet t)$ and $\beta(t)^\bullet = \beta(t^\bullet)$.

Labelled nets N and N' are *isomorphic*, denoted by $N \simeq N'$, if there exists an isomorphism $\beta : N \simeq N'$.

Given a labelled net N and some transition $t \in T_N$, the *precondition* and *postcondition* t, denoted by ${}^\bullet t$ and $t^\bullet$ respectively, are the multisets defined in such a way: $({}^\bullet t)(p) = F_N(p, t)$ and $(t^\bullet)(p) = F_N(t, p)$. Analogous definitions are introduced for places: $({}^\bullet p)(t) = F_N(t, p)$ and $(p^\bullet)(t) = F_N(p, t)$. Let ${}^\circ N = \{p \in P_N \mid {}^\bullet p = \emptyset\}$ is a set of *initial (input)* places of N and $N^\circ = \{p \in P_N \mid p^\bullet = \emptyset\}$ is a set of *final (output)* places of N.

A labelled net N is *acyclic*, if there exist no transitions $t_0, \ldots, t_n \in T_N$ such that $t_{i-1}^\bullet \cap {}^\bullet t_i \neq \emptyset$ $(1 \leq i \leq n)$ and $t_0 = t_n$. A labelled net N is *ordinary* if $\forall p \in P_N \ {}^\bullet p$ and $p^\bullet$ are proper sets (not multisets).

Let $N = \langle P_N, T_N, F_N, l_N \rangle$ be acyclic ordinary labelled net and $x, y \in P_N \cup T_N$. Let us introduce the following notions.

- $x \prec_N y \Leftrightarrow x F_N^+ y$, where F_N^+ is a transitive closure of F_N (*strict causal dependence* relation);
- $\downarrow_N x = \{y \in P_N \cup T_N \mid y \prec_N x\}$ (the set of *strict predecessors* of x);

A set $T \subseteq T_N$ is *left-closed* in N, if $\forall t \in T$ $(\downarrow_N t) \cap T_N \subseteq T$.

2.2 Marked nets

A *marking* of a labelled net N is a multiset $M \in \mathcal{M}(P_N)$.

Definition 2. A *marked net (net)* is a tuple $N = \langle P_N, T_N, F_N, l_N, M_N \rangle$, where $\langle P_N, T_N, F_N, l_N \rangle$ is a labelled net and $M_N \in \mathcal{M}(P_N)$ is the *initial* marking.

Let $M \in \mathcal{M}(P_N)$ be a marking of a net N. A transition $t \in T_N$ is *fireable* in M, if ${}^{\bullet}t \subseteq M$. If t is fireable in M, firing it yields a new marking $M' = M - {}^{\bullet}t + t^{\bullet}$, denoted by $M \xrightarrow{t} M'$. A marking M of a net N is *reachable*, if $M = M_N$ or there exists a reachable marking M' of N such that $M' \xrightarrow{t} M$ for some $t \in T_N$. $Mark(N)$ denotes a *set of all reachable* markings of a net N.

2.3 Partially ordered sets

Definition 3. A *labelled partially ordered set (lposet)* is a triple $\rho = \langle X, \prec, l \rangle$, where:

- $X = \{x, y, \ldots\}$ is some set;
- $\prec \subseteq X \times X$ is a strict partial order (irreflexive transitive relation) over X;
- $l : X \to Act$ is a *labelling* function.

Let $\rho = \langle X, \prec, l \rangle$ and $\rho' = \langle X', \prec', l' \rangle$ be lposets.

A mapping $\beta : X \to X'$ is a *label-preserving bijection* between ρ and ρ', denoted by $\beta : \rho \approx \rho'$, if:

1. β is a bijection;
2. $\forall x \in X$ $l(x) = l'(\beta(x))$.

We write $\rho \approx \rho'$, if there exists a label-preserving bijection $\beta : \rho \approx \rho'$.

A mapping $\beta : X \to X'$ is a *homomorphism* between ρ and ρ', denoted by $\beta : \rho \sqsubseteq \rho'$, if:

1. $\beta : \rho \approx \rho'$;
2. $\forall x, y \in X$ $x \prec y \Rightarrow \beta(x) \prec' \beta(y)$.

We write $\rho \sqsubseteq \rho'$, if there exists a homomorphism $\beta : \rho \sqsubseteq \rho'$.

A mapping $\beta : X \to X'$ is an *isomorphism* between ρ and ρ', denoted by $\beta : \rho \simeq \rho'$, if $\beta : \rho \sqsubseteq \rho'$ and $\beta^{-1} : \rho' \sqsubseteq \rho$. Lposets ρ and ρ' are *isomorphic*, denoted by $\rho \simeq \rho'$, if there exists an isomorphism $\beta : \rho \simeq \rho'$.

Definition 4. *Partially ordered multiset (pomset)* is an isomorphism class of lposets.

2.4 C-processes

Definition 5. A *causal net* is acyclic ordinary labelled net
$C = \langle P_C, T_C, F_C, l_C \rangle$, s.t:

1. $\forall r \in P_C \ |^\bullet r| \leq 1$ and $|r^\bullet| \leq 1$, i.e. places are unbranched;
2. $| \downarrow_C x| < \infty$, i.e. a set of causes is finite.

Let us note that on the basis of any causal net $C = \langle P_C, T_C, F_C, l_C \rangle$ one can
define lposet $\rho_C = \langle T_C, \prec_N \cap (T_C \times T_C), l_C \rangle$.

The fundamental property of causal nets is [1]: if C is a causal net, then
there exists an occurrence sequence $^\circ C = L_0 \overset{v_1}{\rightarrow} \cdots \overset{v_n}{\rightarrow} L_n = C^\circ$ such that
$L_i \subseteq P_C \ (0 \leq i \leq n)$, $P_C = \cup_{i=0}^n L_i$ and $T_C = \{v_1, \ldots, v_n\}$. Such a sequence is
called a *full execution* of C.

Definition 6. Given a net N and a causal net C. A mapping $\varphi : P_C \cup T_C \rightarrow$
$P_N \cup T_N$ is an *embedding* C into N, denoted by $\varphi : C \rightarrow N$, if:

1. $\varphi(P_C) \in \mathcal{M}(P_N)$ and $\varphi(T_C) \in \mathcal{M}(T_N)$, i.e. sorts are preserved;
2. $\forall v \in T_C \ l_C(v) = l_N(\varphi(v))$, i.e. labelling is preserved;
3. $\forall v \in T_C \ ^\bullet \varphi(v) = \varphi(^\bullet v)$ and $\varphi(v)^\bullet = \varphi(v^\bullet)$, i.e. flow relation is respected.

Since embeddings respect the flow relation, if $^\circ C \overset{v_1}{\rightarrow} \cdots \overset{v_n}{\rightarrow} C^\circ$ is a full
execution of C, then $M = \varphi(^\circ C) \overset{\varphi(v_1)}{\longrightarrow} \cdots \overset{\varphi(v_n)}{\longrightarrow} \varphi(C^\circ) = M'$ is an occurrence
sequence in N.

Definition 7. A *fireable in marking M C-process (process)* of a net N is a pair
$\pi = (C, \varphi)$, where C is a causal net and $\varphi : C \rightarrow N$ is an embedding such that
$M = \varphi(^\circ C)$. A *fireable in M_N process* is a *process* of N.

We write $\Pi(N, M)$ for a *set of all fireable in marking M processes of a net N*
and $\Pi(N)$ for the *set of all processes of a net N*. The *initial* process of a net N
is $\pi_N = (C_N, \varphi_N) \in \Pi(N)$, such that $T_{C_N} = \emptyset$. If $\pi \in \Pi(N, M)$, then firing of
this process transforms a marking M into $M' = M - \varphi(^\circ C) + \varphi(C^\circ) = \varphi(C^\circ)$,
denoted by $M \overset{\pi}{\rightarrow} M'$.

Let $\pi = (C, \varphi)$, $\tilde{\pi} = (\tilde{C}, \tilde{\varphi}) \in \Pi(N)$ and $\hat{\pi} = (\hat{C}, \hat{\varphi}) \in \Pi(N, \varphi(C^\circ))$.

A process $\tilde{\pi}$ is an *extension* of π *by process* $\hat{\pi}$, denoted by $\pi \overset{\hat{\pi}}{\rightarrow} \tilde{\pi}$, if $T_C \subseteq T_{\tilde{C}}$
is a left-closed set in $\tilde{C}$ and $T_{\hat{C}} = T_{\tilde{C}} \setminus T_C$.

A process $\tilde{\pi}$ is an extension of a process π *by one transition* $v \in T_{\tilde{C}}$, denoted
by $\pi \overset{v}{\rightarrow} \tilde{\pi}$, if $\pi \overset{\hat{\pi}}{\rightarrow} \tilde{\pi}$ and $T_{\hat{C}} = \{v\}$.

A process $\tilde{\pi}$ is an extension of a process π *by sequence of transitions* $\sigma =$
$v_1 \cdots v_n \in T_{\tilde{C}}^*$, denoted by $\pi \overset{\sigma}{\rightarrow} \tilde{\pi}$, if $\exists \pi_i \in \Pi(N) \ (1 \leq i \leq n) \ \pi \overset{v_1}{\rightarrow} \pi_1 \overset{v_2}{\rightarrow} \cdots \overset{v_n}{\rightarrow}$
$\pi_n = \tilde{\pi}$.

3 Back-forth bisimulation equivalences

In this section, in the framework of Petri nets, we supplement the definitions of
back-forth bisimulation equivalences [13] by the new notions induced by process
semantics and compare them with basic ones.

3.1 Definitions of back-forth bisimulation equivalences

The definitions of back-forth bisimulation equivalences are based on the following notion of sequential run.

Definition 8. A *sequential run* of a net N is a pair (π, σ), where:

- a process $\pi \in \Pi(N)$ contains the information about causal dependencies of transitions which brought to this state;
- a sequence $\sigma \in T_C^*$ such that $\pi_N \xrightarrow{\sigma} \pi$, contains the information about the order in which the transitions occur which brought to this state.

Let us denote the *set of all sequential runs* of a net N by $Runs(N)$.

The *initial* sequential run of a net N is a pair (π_N, ε), where ε is an empty sequence.

Let (π, σ), $(\tilde{\pi}, \tilde{\sigma}) \in Runs(N)$. We write $(\pi, \sigma) \xrightarrow{\hat{\pi}} (\tilde{\pi}, \tilde{\sigma})$, if $\pi \xrightarrow{\hat{\pi}} \tilde{\pi}$, $\exists \hat{\sigma} \in T_{\hat{C}}^*$ $\pi \xrightarrow{\hat{\sigma}} \tilde{\pi}$ and $\tilde{\sigma} = \sigma\hat{\sigma}$.

Definition 9. Let N and N' be some nets. A relation $\mathcal{R} \subseteq Runs(N) \times Runs(N')$ is a $\star$-*back* $\star\star$-*forth bisimulation* between N and N', $\star, \star\star \in \{interleaving, step, partial\ word, pomset, process\}$, denoted by $\mathcal{R} : N \underleftrightarrow{}_{\star b \star\star f} N'$, $\star, \star\star \in \{i, s, pw, pom, pr\}$, if:

1. $((\pi_N, \varepsilon), (\pi_{N'}, \varepsilon)) \in \mathcal{R}$.
2. $((\pi, \sigma), (\pi', \sigma')) \in \mathcal{R}$
   - (back)
     $(\tilde{\pi}, \tilde{\sigma}) \xrightarrow{\hat{\pi}} (\pi, \sigma)$,
     (a) $|T_{\hat{C}}| = 1$, if $\star = i$;
     (b) $\prec_{\hat{C}} = \emptyset$, if $\star = s$;
     $\Rightarrow \exists (\tilde{\pi}', \tilde{\sigma}') : (\tilde{\pi}', \tilde{\sigma}') \xrightarrow{\hat{\pi}'} (\pi', \sigma')$, $((\tilde{\pi}, \tilde{\sigma}), (\tilde{\pi}', \tilde{\sigma}')) \in \mathcal{R}$ and
     (a) $\rho_{\hat{C}'} \sqsubseteq \rho_{\hat{C}}$, if $\star = pw$;
     (b) $\rho_{\hat{C}} \simeq \rho_{\hat{C}'}$, if $\star \in \{i, s, pom\}$;
     (c) $\hat{C} \simeq \hat{C}'$, if $\star = pr$;
   - (forth)
     $(\pi, \sigma) \xrightarrow{\hat{\pi}} (\tilde{\pi}, \tilde{\sigma})$,
     (a) $|T_{\hat{C}}| = 1$, if $\star\star = i$;
     (b) $\prec_{\hat{C}} = \emptyset$, if $\star\star = s$;
     $\Rightarrow \exists (\tilde{\pi}', \tilde{\sigma}') : (\pi', \sigma') \xrightarrow{\hat{\pi}'} (\tilde{\pi}', \tilde{\sigma}')$, $((\tilde{\pi}, \tilde{\sigma}), (\tilde{\pi}', \tilde{\sigma}')) \in \mathcal{R}$ and
     (a) $\rho_{\hat{C}'} \sqsubseteq \rho_{\hat{C}}$, if $\star\star = pw$;
     (b) $\rho_{\hat{C}} \simeq \rho_{\hat{C}'}$, if $\star\star \in \{i, s, pom\}$;
     (c) $\hat{C} \simeq \hat{C}'$, if $\star\star = pr$.
3. As item 2, but the roles of N and N' are reversed.

Two nets N and N' $\star$-*back* $\star\star$-*forth bisimulation equivalent*, $\star, \star\star \in \{interleaving, step, partial\ word, pomset, process\}$, denoted by $N \underleftrightarrow{}_{\star b \star\star f} N'$, if $\exists \mathcal{R} : N \underleftrightarrow{}_{\star b \star\star f} N'$, $\star, \star\star \in \{i, s, pw, pom, pr\}$.

3.2 Interrelations of back-forth bisimulation equivalences

In back-forth bisimulations, it is possible to move back from a state only along the history which brought to the state. Such a determinism implies merging of some equivalences.

Proposition 10. *Let* $\star \in \{i, s, pw, pom, pr\}$. *For nets* N *and* N' *the following holds:*

1. $N \underline{\leftrightarrow}_{pwb\star f} N' \Leftrightarrow N \underline{\leftrightarrow}_{pomb\star f} N'$;
2. $N \underline{\leftrightarrow}_{\star bif} N' \Leftrightarrow N \underline{\leftrightarrow}_{\star b\star f} N'$.

Hence, interrelations of the remaining back-forth equivalences may be represented by the graph in Figure 1.

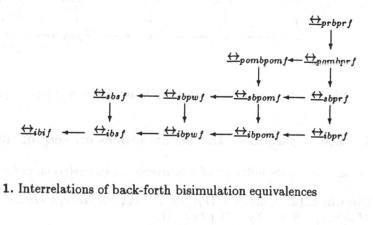

Fig. 1. Interrelations of back-forth bisimulation equivalences

3.3 Interrelations of back-forth bisimulation and basic equivalences

Let us consider how back-forth equivalences are connected with basic ones.

Proposition 11. *Let* $\star \in \{i, s, pw, pom, pr\}$, $\star\star \in \{pom, pr\}$. *For nets* N *and* N' *the following holds:*

1. $N \underline{\leftrightarrow}_{ib\star f} N' \Leftrightarrow N \underline{\leftrightarrow}_{\star} N'$;
2. $N \underline{\leftrightarrow}_{\star\star b\star\star f} N' \Leftrightarrow N \underline{\leftrightarrow}_{\star\star h} N'$;
3. $N \underline{\leftrightarrow}_{\star\star ST} N' \Rightarrow N \underline{\leftrightarrow}_{sb\star\star f} N'$.

Theorem 12. *Let* $\leftrightarrow \in \{\equiv, \underline{\leftrightarrow}, \simeq\}$ *and* $\star, \star\star \in \{i, s, pw, pom, pr, iST, pwST,$ $pomST, prST, pomh, prh, mes, occ, sbsf, sbpwf, sbpomf, sbprf, pombprf\}$. *For nets* N *and* N' *the following holds* : $N \leftrightarrow_{\star} N' \Rightarrow N \leftrightarrow_{\star\star} N'$ *iff in graph in Figure 2 there exists a directed path from* $\leftrightarrow_{\star}$ *to* $\leftrightarrow_{\star\star}$.

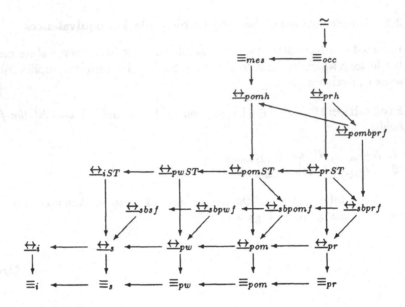

Fig. 2. Interrelations of back-forth bisimulation and basic equivalences

4 Investigation of the equivalences on sequential nets

Let us consider the influence of concurrency on interrelations of the equivalences.

Definition 13. A net $N = \langle P_N, T_N, F_N, l_N, M_N \rangle$ is *sequential*, if $\forall M \in Mark(N) \; \neg \exists t, u \in T_N : \; {}^{\bullet}t + {}^{\bullet}u \subseteq M$.

Proposition 14. *For sequential nets N and N' the following holds:*

1. $N \equiv_i N' \Leftrightarrow N \equiv_{pom} N'$;
2. $N \underline{\leftrightarrow}_i N' \Leftrightarrow N \underline{\leftrightarrow}_{pomh} N'$;
3. $N \underline{\leftrightarrow}_{pr} N' \Leftrightarrow N \underline{\leftrightarrow}_{pombprf} N'$.

Theorem 15. *Let $\leftrightarrow \in \{\equiv, \underline{\leftrightarrow}, \simeq\}$, $\star, \star\star \in \{i, pr, prST, prh, mes, occ\}$. For sequential nets N and N' the following holds: $N \leftrightarrow_\star N' \Rightarrow N \leftrightarrow_{\star\star} N'$ iff in graph in Figure 3 there exists a directed path from $\leftrightarrow_\star$ to $\leftrightarrow_{\star\star}$.*

5 Preservation of the equivalences by refinements

Let us consider which equivalences may be used for top-down design.

Definition 16. An *SM-net* is a net $D = \langle P_D, T_D, F_D, l_D, M_D \rangle$ such that:

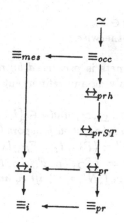

Fig. 3. The equivalences on sequential nets

1. $\exists p_{in}, p_{out} \in P_D$ such that $p_{in} \neq p_{out}$ and $^\circ D = \{p_{in}\}$, $D^\circ = \{p_{out}\}$, i.e. net D has unique input and unique output place.

2. $M_D = \{p_{in}\}$ and $\forall M \in Mark(D)$ $(p_{out} \in M \Rightarrow M = \{p_{out}\})$, i.e. at the beginning there is unique token in p_{in}, and at the end there is unique token in p_{out};

3. $p_{in}^\bullet$ and $^\bullet p_{out}$ are proper sets (not multisets), i.e. p_{in} (respectively p_{out}) represents a set of all tokens consumed (respectively produced) for any refined transition.

4. $\forall t \in T_D$ $|^\bullet t| = |t^\bullet| = 1$, i.e. each transition has exactly one input and one output place.

SM-refinement operator "replaces" all transitions with particular label of a net by SM-net.

Definition 17. Let $N = \langle P_N, T_N, F_N, l_N, M_N \rangle$ be some net, $a \in l_N(T_N)$ and $D = \langle P_D, T_D, F_D, l_D, M_D \rangle$ be SM-net. An *SM-refinement*, denoted by $ref(N, a, D)$, is (up to isomorphism) a net $\overline{N} = \langle P_{\overline{N}}, T_{\overline{N}}, F_{\overline{N}}, l_{\overline{N}}, M_{\overline{N}} \rangle$, where:

- $P_{\overline{N}} = P_N \cup \{\langle p, u \rangle \mid p \in P_D \setminus \{p_{in}, p_{out}\}, \ u \in l_N^{-1}(a)\}$;
- $T_{\overline{N}} = (T_N \setminus l_N^{-1}(a)) \cup \{\langle t, u \rangle \mid t \in T_D, \ u \in l_N^{-1}(a)\}$;
- $F_{\overline{N}}(\bar{x}, \bar{y}) = \begin{cases} F_N(\bar{x}, \bar{y}), \ \bar{x}, \bar{y} \in P_N \cup (T_N \setminus l_N^{-1}(a)); \\ F_D(x, y), \ \bar{x} = \langle x, u \rangle, \ \bar{y} = \langle y, u \rangle, \ u \in l_N^{-1}(a); \\ F_N(\bar{x}, u), \ \bar{y} = \langle y, u \rangle, \ x \in {}^\bullet u, \ u \in l_N^{-1}(a), \ y \in p_{in}^\bullet; \\ F_N(u, \bar{y}), \ \bar{x} = \langle x, u \rangle, \ y \in u^\bullet, \ u \in l_N^{-1}(a), \ x \in {}^\bullet p_{out}; \\ 0, \qquad \text{otherwise}; \end{cases}$
- $l_{\overline{N}}(\bar{u}) = \begin{cases} l_N(\bar{u}), \ \bar{u} \in T_N \setminus l_N^{-1}(a); \\ l_D(t), \ \bar{u} = \langle t, u \rangle, \ t \in T_D, \ u \in l_N^{-1}(a); \end{cases}$

$$- M_{\overline{N}}(p) = \begin{cases} M_N(p), & p \in P_N; \\ 0, & \text{otherwise.} \end{cases}$$

Some equivalence on nets is *preserved by refinements*, if equivalent nets remain equivalent after applying any refinement operator to them accordingly.

Theorem 18. *Let $\leftrightarrow \in \{\equiv, \underleftrightarrow{\ }, \simeq\}$ and $\star \in \{i, s, pw, pom, pr, iST, pwST, pomST, prST, pomh, prh, mes, occ, sbsf, sbpwf, sbpomf, sbprf, pombprf\}$. For nets $N = \langle P_N, T_N, F_N, l_N, M_N \rangle$, $N' = \langle P_{N'}, T_{N'}, F_{N'}, l_{N'}, M_{N'} \rangle$ such that $a \in l_N(T_N) \cap l_{N'}(T_{N'})$ and SM-net $D = \langle P_D, T_D, F_D, l_D, M_D \rangle$ the following holds: $N \leftrightarrow_\star N' \Rightarrow ref(N, a, D) \leftrightarrow_\star ref(N', a, D)$ iff equivalence $\leftrightarrow_\star$ is in oval in Figure 4.*

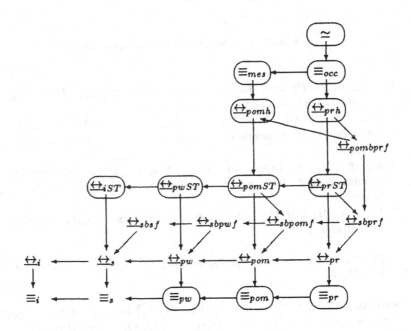

Fig. 4. Preservation of the equivalences by SM-refinements

6 Conclusion

In this paper, we examined and supplemented by new ones a group of back-forth bisimulation equivalences. We compared them with basic ones on the whole class of Petri nets as well as on their subclass of sequential nets. All the considered equivalences were treated for preservation by SM-refinements.

Further research may consist in the investigation of place bisimulation equivalences from [1] which are used for effective semantically correct reduction of nets. We intend to compare these equivalences with the ones we examined (for example, the relationship is unknown between place bisimulation equivalences and ST-, history preserving, conflict preserving and back-forth ones) and check them for preservation by refinements to establish whether they may be used for construction of multilevel concurrent systems.

Acknowledgements I would like to thank Dr. Irina B. Virbitskaite for many helpful discussions.

References

1. AUTANT C., SCHNOEBELEN PH. *Place bisimulations in Petri nets.* LNCS **616**, p. 45–61, June 1992.
2. BEST E., DEVILLERS R., KIEHN A., POMELLO L. *Concurrent bisimulations in Petri nets.* Acta Informatica **28**, p. 231–264, 1991.
3. BOUDOL G., CASTELLANI I. *On the semantics of concurrency: partial orders and transition systems.* LNCS **249**, p. 123–137, 1987.
4. CHERIEF F. *Back and forth bisimulations on prime event structures.* LNCS **605**, p.843–858, June 1992.
5. CHERIEF F. *Contributions à la sémantique du parallélisme: bisimulations pour le raffinement et le vrai parallélisme.* Ph.D. thesis, Institut National Politechnique de Grenoble, France, October 1992 (in French).
6. CHERIEF F. *Investigations of back and forth bisimulations on prime event structures.* Computers and Artificial Intelligence **11**(5), p. 481–496, 1992.
7. VAN GLABBEEK R.J., VAANDRAGER F.W. *Petri net models for algebraic theories of concurrency.* LNCS **259**, p. 224–242, 1987.
8. HOARE C.A.R. *Communicating sequential processes, on the construction of programs.* (McKeag R.M., Macnaghten A.M., eds.) Cambridge University Press, p. 229–254, 1980.
9. DE NICOLA R., MONTANARI U., VAANDRAGER F.W. *Back and forth bisimulations.* LNCS **458**, p. 152–165, 1990.
10. NIELSEN M., THIAGARAJAN P.S. *Degrees of non-determinizm and concurrency: A Petri net view.* LNCS **181**, p. 89–117, December 1984.
11. PARK D.M.R. *Concurrency and automata on infinite sequences.* LNCS **104**, p. 167–183, March 1981.
12. POMELLO L. *Some equivalence notions for concurrent systems. An overview.* LNCS **222**, p. 381–400, 1986.
13. PINCHINAT S. *Bisimulations for the semantics of reactive systems.* Ph.D. thesis, Institut National Politechnique de Grenoble, January 1993 (in French).
14. RABINOVITCH A., TRAKHTENBROT B.A. *Behaviour structures and nets.* Fundamenta Informaticae **XI**, p. 357–404, 1988.
15. TARASYUK I.V. *Equivalence notions for design of concurrent systems using Petri nets.* Hildesheimer Informatik-Bericht **4/96**, part 1, 19 p., Institut für Informatik, Universität Hildesheim, Hildesheim, Germany, January 1996.
16. VOGLER W. *Bisimulation and action refinement.* LNCS **480**, p. 309–321, 1991.

Association Nets: An Alternative Formalization of Common Thinking

G.S.Tseytin

St.Petersburg University, Russia
e-mail: tseyting@acm.org

Abstract. The development of programming as well as some studies in artificial intelligence have produced their own conceptualizations used in the representation of tasks being solved and methods applied. This paper is an attempt at logic-like representation of a programming system, based on the open data model and object-oriented programming, earlier developed and extensively used by the author for various tasks including authomatic programming and natural language understanding. This is suggested as an alternative to traditional logic systems in description of common thinking.

1   Introduction

The classic mathematical logic, after being so successful in the analysis of standard mathematical thinking, since long tends to embrace neighboring areas like computer programming and commonsense reasoning. The typical approach to commonsense reasoning is to modify the classic techniques by adding features like non-monotonicity, modality, temporal reasoning, actions, causality, fuzziness, etc. The aim of this paper is to present a different approach, not directly relying on classic techniques, but also leading to a mathematical description of some features of common thinking, though possibly not the same as in other approaches.

In fact, computer programming is pursuing the same goal, though not stating it explicitly. At the initial stages programming was too machine-oriented, based on such concepts as storage location or goto and producing entangled spaghetti-like programs that clearly called for some kind of logical verification. However, methods of logical verification didn't develop very fast and clearly lagged behind the practial needs of this rapidly expanding art. What is more, many actual applications of programming didn't yield themselves to rigorous mathematical description, and their specifications were based instead on practical experience and informal thinking. As a result, programming moved, all on its own, from machine-oriented to human-oriented concepts, with symbolic programming languages, databases, modularity and object-oriented programming as important milestones on this road. With program code represented in more humanlike terms, the need for verification with external methods decreased.

Studies in artificial intelligence also developed some forms of knowledge representation as an alternative to the classic approach. Some of those forms converged with the tendencies of programming. One can observe deep analogies between M.Minsky's theory of frames, object-oriented programming and some concepts of Prolog (seen as a programming language rather than a proof-searching technique). The central concepts of each of the three approaches, i.e., frames, classes, and predicates defined by lists of axioms, are clearly related to each other. So it seems that these concepts capture some important elements of human cognition.

One more feature long present in some AI languages is the open data model – property lists with dynamic retrieval of properties by their names with no lists of names specified in advance; another example of this is found in semantic nets. In conventional programming this was avoided until recently because deemed inefficient. But development of Internet put emphasis on interoperability of independently designed applications, which called for exactly this feature.

The system presented in this paper was originally conceived as a programming tool systematically using the open data model in various programming and AI tasks, in the first place in natural language understanding. Very soon it also acquired object-oriented features in a very permissive form. An early version of it is presented in [1], for natural language applications see [2, 3]. For a philosophic background see [4]. This paper is an attempt to present a version of this system to make it look as close as possibly to logic rather than programming, so it can be seen as a kind of alternative logic. No attempt is made to study this system mathematically; instead some examples are given to illustrate the way it can be used. Some analogies with concepts of object-oriented programming (OOP) will be mentioned later.

2 Constants and variables

We assume an unlimited supply of both constants and variables. Constants will have a concrete representation (e.g., words in a predefined alphabet) but we won't consider this representation here.

The difference between constants and variables is that each constant has its own identity and may not be equated to a different constant, while variables may be equated to each other or to constants.

A unique constant (unique variable) is a constant (a variable) created on a specific occasion and distinct from all constants (variables) similarly created on other occasions.

3 Association nets

An association net is a finite collection of formulas of the following forms $(A,B,C,$ etc. stand for constants of variables)

$$A = B$$
$$A(B) = C, \quad \text{where } A \text{ is a constant}$$
$$A \in B,$$

closed under the following inference rules

$$\frac{A = B}{A = A}$$

$$\frac{X S Y \quad S = T}{X T Y}$$

(where X, Y, S, and T are any sequences of symbols generating valid formulas in the above rule).

Some other types of formulas and related inference rules will be introduced later, but they are not regarded as part of the association net and are only needed to perform some operations. The last rule applies to those types of formulas as well.

A formula of the form

$$A = B,$$

where A and B are two distinct constants is called a contradiction.

In the sequel, unless explicitly stated otherwise, we will be considering one single association net, and will not mention it directly.

The expression

assert F,

where F is a list of one or more formulas, will mean a change in the association net resulting from addition of F and all other formulas that can be inferred.

4   Interpretation

In applications, both constants and variables are interpreted as real-world entities.

A formula $A = B$ is a statement that A and B refer to one same entity.

A formula $A(B) = C$ is interpreted as a statement about a property of the entity referred to by B; properties are distinguished by names (which can be, e.g., words of a commonly used language), A is the name of the property, and C is its value, referring to another (or possibly the same) entity. Examples:

$Husband(Alla) = Philip$
$Mother(Christine) = Alla$

A formula $A \in B$ means that the entity referred to by B, apart from any other properties it may have, can be interpreted as a finite set, and A refers to one of its elements.

The collection of formulas is meant to reflect some part of one's knowledge of the world; in applications, an essential step will be to combine two nets and establish additional relationships between them (e.g., to combine a net representing some facts with another net representing previous knowledge of the world –

"axioms"). Some restrictions stated in the next section are intended to preserve the integrity of the parts brought together.

Any changes in the association net are meant to reflect the progress of reasoning and/or decision-making. They are not interpreted as changes over time (for a time logic, special entities for moments of time and special relationships including such moments will have to be introduced).

Details of representation of constants are not considered here.

5 Principles of self-discipline

An extension of an association net N is a net obtained from N by one or more **assert** operations. If N_1 is an extension of N, an operation on N, resulting in N', is said to be *tolerant* of this extension if its application to N_1 results in an extension of N'.

All operations we are going to define shall be tolerant of the following types of extensions:

(a) **assert** *a single equality where the right member is a unique variable* (introducing a new notation);

(b) *for any constant or variable X not occurring in the net* **assert** *any number of formulas of the types $A(X) = B$, $C \in X$, and $X(A) = B$ (when X is a constant), unless this results in a contradiction* (this is an accessibility restriction: no operation shall be able to access X from these types of formulas).

The second rule requires tolerance of irrelevant information. No definition of an operation shall rely on the complete listing of the association net, except for some specially introduced primitives. This is needed because different applications can use partially overlapping data without interfering with each other, but possibly cooperating through shared data. A special case of this is the third type of added formula, $X(A) = B$, because it implies that there shall be no way of listing all properties defined for an item. Normally different applications use different, possibly overlapping, sets of properties.

Another principle of self-discipline requires that

no operation shall remove formulas from an association net except for return to a specially saved previous state in case of contradiction.

In practice, one can remove formulas from the net by reversing extensions (a) and (b) above ("garbage collection"), but, in virtue of the tolerance principle, this will not affect the results, so one can think that this operation, violating the last restriction, is never used.

These principles don't forbid to inquire whether the net contains any given formula, or to look for a formula of the form $A(B) = X$ with given A and B, or to list all formulas $X \in A$ for a given A.

As seen from these rules, there is no requirement of monotonicity (tolerance is promised only for some types of extensions but not for all). But all sources of non-monotonicity will be explicitly identified; they are two, and both related to the primitives allowed to scan the whole association net.

One of them is the test whether the association net contains a contradiction (which is not forbidden by (b)); the other is discussed in connection with procedures.

6   Terms and evaluation

Terms can be built out of constants and variables using the constructor $T_1(T_2)$, T_1 and T_2 being terms.

To evaluate a term of the form $A(B)$, where A is a constant and B a constant or a variable, one has to look for a formula $A(B) = C$; if there is no such formula extension (a) of the preceding section is used to add one. C is called the value of $A(B)$ in the given association net.

More complex terms can be evaluated by first evaluating innermost subterms. But sometimes this process will stop without a result. In evaluating a term like $A(B)(C)$ one has to evaluate $A(B)$, but if the value isn't equal to any constant there will be no result.

In the sequel terms sometimes will be used for brevity instead of their values with an implicit precondition that the values can be built.

7   Procedures

A procedure can be associated with a constant. A procedure can be applied to a constant or variable, changing the net and possibly performing some external actions (like i/o). For simplicity we'll use the constant to refer to the associated procedure without further explanations.

An application of a procedure to an entity should be construed as an attempt to ascribe to the entity a certain conceptual category represented by this procedure. If the attempt succeeds, the knowledge is extended. If it fails (i.e., produces a contradiction), one can try something else.

In OOP, the closest analogy to these procedures are classes. But here we don't generate a new instance of a class; instead, we ascribe a class to an existing object.

We introduce a new type of formula:

$$A \text{ is-a } B, \tag{1}$$

representing an intermediate step in the process of reasoning. The only way to use such a formula is to apply to A the procedure associated with B. However this transition from the formula to action can be deferred (see below).

Some procedures are primitives, not defined here (but tolerant to the extensions described above).

Other procedures are defined here in terms of properties of the constants with which they are associated. Some particular constants essential in their definition will be denoted by symbols starting with #. (The choice of representation is as arbitrary as any decisions about Gödel numbering.)

We define four types of procedures, distinguished by means of the property *#type*. The types are frame, conjunction, disjunction, and "for all". Let C be a constant, with which we associate a procedure.

Some auxiliary types of formulas and inference rules will be needed to apply procedures. One type of such formula is

$$\mathbf{Const}(X)$$

with the following inference rule:

if the association net contains no formula equating X to a constant, a contradiction is inferred from $\mathbf{Const}(X)$. (For a contradiction in such cases one fixed equality is chosen, no matter which one.)

This is the second operation requiring a complete scan of the net and thus producing non-monotonicity. Other types of auxiliary formulas and rules will be shown later.

7.1 Frames

If $\#type(C) = \#frame$ the procedure is a frame.

The application of a frame to a target consists of mapping the frame to the target and applying subprocedures of the mapped frame. More formally, the following steps are taken:

- let M be a unique constant;
- map C to X with M;
- apply subprocedures with M;
- optionally, discard M by reversing the appropriate extension.

(Mapping corresponds to frame instantiation and Prolog unification.)

Mapping. In the simplest case mapping C to X consists in building around X a copy of the structure of properties already existing around C (while some of them might have existed at X before), and possibly imposing some equalities among properties of C on the respective properties of X (this can propagate recursively).

Four sets of property names (probably overlapping and possibly empty) are needed to map a frame C:

- $\#copied(C)$, the set of properties for which new copies are built;
- $\#invariant(C)$, the set of "invariant" (not copied) properties;
- $\#inquiry(C)$, the set of properties whose values are sought from the target;
- $\#mandatory(C)$, the set of properties not marked as defaults that can be overridden by other information.

(In OOP, copied properties correspond to instance variables, and invariant properties to class variables.)

Then, one set of procedures is needed (possibly empty),

- $\#ascribe(C)$, the set of "superclasses" to be ascribed to the target in addition to C.

(This is a wide generalization of the OOP concept of derived classes.)

Formally, mapping is done with the use of some auxiliary types of formulas and inference rules. The auxiliary formulas start with Map or Map_1, and the rules are

$$\frac{Map(M,C,X)}{M(C) = X}$$

$$\frac{\begin{array}{c} Map(M,C,X) \\ S \in \#ascribe(C) \end{array}}{X \text{ is-a } S} \tag{2}$$

$$\frac{\begin{array}{c} Map(M,C,X) \\ E \in C \end{array}}{\begin{array}{c} M(E) \in X \\ Map(M,E,M(E)) \end{array}} \tag{3}$$

$$\frac{\begin{array}{c} Map(M,C,X) \\ P \in \#mandatory(C) \end{array}}{Map_1(M,C,X,P)}$$

$$\frac{\begin{array}{c} Map(M,C,X) \\ Map_1(M,C,X,P) \quad \text{doesn't produce a contradiction} \end{array}}{Map_1(M,C,X,P)} \tag{4}$$

$$\frac{\begin{array}{c} Map_1(M,C,X,P) \\ P \in \#inquiry(C) \end{array}}{\mathbf{Const}(P(X))}$$

$$\frac{\begin{array}{c} Map_1(M,C,X,P) \\ P \in \#copied(C) \end{array}}{Map(M,P(C),P(X))}$$

$$\frac{\begin{array}{c} Map_1(M,C,X,P) \\ P \in \#invariant(C) \end{array}}{P(X) = P(C)}$$

To map C onto X with M, **assert** $Map(M,C,X)$.

The rule (4) needs a special discussion. Its obvious meaning is to provide for defaults: a default can be used unless this leads to a contradiction. But if the derivation of new formulas proceeds indefinitely it can be an unsolvable problem to determine whether there ever will be a contradiction. Without expansion of "is-a" formulas (1) some obvious restrictions will guarantee that the process is finite (if X is not "accessible" from C none of the inference steps will increase the set of potential arguments for the M-mapping). If "is-a" formulas are used immediately, resulting in new applications of procedures, it is easy to build an infinite process, and even to simulate a Turing machine. If the application is deferred, there is a danger that a default conclusion accepted as not leading to a contradiction will produce a contradiction later. Thus the rule for default is not completely defined. For the time being we'll accept the interpretation that the test for contradiction is performed before expansion of any "is-a" formulas obtained during the mapping process.

Both default (non-mandatory) and "inquiry" properties of a frame generate non-monotonicity. The intuitive meaning of "inquiry" properties is that there is no point applying a procedure in the absence of reqired information.

The rule (3) could be replaced by somewhat more elaborate rules distinguishing between "copied" and "invariant" elements of C, but for simplicity it can be left as it stands now.

Calling subprocedures. With mapping done, subprocedures of the frame, if any,have to be applied.The set of subprocedures of C is specified as $\#subproc(C)$. For any subprocedure Q two properties are defined to make a call, $\#function(Q)$ and $\#argument(Q)$. It is the mapped images rather than those values that are used in the process of application.

Thus, to apply subprocedures of C with M,

- for each Q such that $Q \in \#subproc(C)$ apply $M(\#function(Q))$ to $M(\#argument(Q))$.

One could use this rule also to cover the effect of the (2) rule, but this rule is intended mainly to provide for external actions (like i/o), and it might be of interest to consider cases where this rule is not used at all.

7.2 Conjunctions and disjunctions

If $\#type(C) = \#conjunction$ (or $\#type(C) = \#disjunction$) the procedure associated with C is a conjunction (respectively, a disjunction), defined by means of a sequence of other procedures.

To apply a conjunction to a target one applies all elements of the sequence (to the same target) in succession as long as the association net doesn't contain a contradiction.

To apply a disjunction to a target one attempts to apply all elements of the sequence in succession until a successful attempt, if any. If no attempt was successful, a contradiction is asserted.

To attempt to apply a procedure (to a target) means to save the current state of the association net, then to apply the procedure, and in case of a contradiction to revert to the saved state.

For a more formal definition we will use constants $\#first$, $\#next$, $\#nil$, and $\#proc$.

To apply a conjunction C to X, and-apply $\#first(C)$ to X.

To and-apply F to X,

- if $F = \#nil$ stop;
- apply $\#proc(F)$ to X;
- if there is a contradiction, stop;
- and-apply $\#next(F)$ to X.

To apply a disjunction C to X, or-apply $\#first(C)$ to X.
To or-apply F to X,

- if $F = \#nil$ **assert** a contradiction and stop;
- save the current state of the association net;
- apply $\#proc(F)$ to X;
- if there is no contradiction stop;
- revert to the saved state;
- or-apply $\#next(F)$ to X.

7.3 For all

If $\#type(C) = \#forall$ this is a "for all" procedure. To apply C to X one has to apply $\#proc(C)$ to all Y such that $Y \in X$ (in any order).

8 Example 1: family

Consider a frame associated with the item X (in this and subsequent examples frames will be built up of constants only):

$Mother(X) = Y$
$Father(X) = Z$
$Husband(Y) = Z$
$Gender(Y) = Feminine$
$Gender(Z) = Masculine$
$Mother \in \#copied(X)$
$Mother \in \#mandatory(X)$
$Father \in \#copied(X)$
$Husband \in \#copied(Y)$
$Husband \in \#mandatory(Y)$
$Gender \in \#invariant(Y)$
$Gender \in \#mandatory(Y)$
$Gender \in \#invariant(Z)$
$Gender \in \#mandatory(Y)$

Try apply X to $Christine$ of the section 4. X will map to $Christine$, Y to $Alla$, and Z to $Philip$. Since $Father$ is not in $\#mandatory(X)$ the conclusion

$Father(Christine) = Philip$

will not be made in the presence of incompatible information.

9 Example 2: simulation

Consider computer simulation of a group of two populations which we'll call unicorns and griffins (this will relieve us of the need of giving realistic relationships and data).

Unicorns **is-a** *Population*
Griffins **is-a** *Population*

We build a model of population that will allow to build arithmetic relationships between the number of the population, its fertility (birth rate) and death reate; the death rate will consist of the basic death rate to which something can be added for reasons not shown in this model. Auxiliary procedures for *PlainValue*, *Sum*, and *Product* will be defined later. We'll assume for simplicity that all copied and invariant properties are also mandatory, and will not write out the respective formulas.

$\#type(Population) = \#frame$

$Number(Population) = vNumber$
$Number \in \#copied(Population)$
$PlainValue \in \#ascribe(vNumber)$

$Fertility(Population) = vFertility$
$Fertility \in \#copied(Population)$

$BasicDeathRate(Population) = vBasicDeathRate$
$BasicDeathRate \in \#copied(Population)$

$DeathRate(Population) = vDeathRate$
$DeathRate \in \#copied(Population)$
$Sum \in \#ascribe(vDeathRate)$
$vBasicDeathRate \in vDeathRate$

Before we continue let us apply this to one of the populations, just to see how it works. Newly introduced unique variables will be prefixed with an underscore.

$Number(Unicorns) = \_uNumber$
$\_uNumber$ **is-a** *PlainValue*
$Fertility(Unicorns) = \_uFertility$
$BasicDeathRate(Unicorns) = \_uBasicDeathRate$
$DeathRate(Unicorns) = \_uDeathRate$
$\_uDeathRate$ **is-a** *Sum*
$\_uBasicDeathRate \in \_uDeathRate$

Continue the population model with the construction of the differential equation for the values. The left member of the equation for a value is its derivative with respect to time. Thus we only have to build the right member using sums, products and a special value for -1, represented here as *MinusOne*.

$RightMember(vNumber) = vRightMember$

$RightMember \in \#copied(vNumber)$
$Product \in \#ascribe(vRightMember)$
$vNumber \in vRightMember$
$vBirthMinusDeath \in vRightMember$
$Sum \in \#ascribe(vBirthMinusDeath)$
$vFertility \in vBirthMinusDeath$
$vMinusDeath \in vBirthMinusDeath$
$Product \in \#ascribe(vMinusDeath)$
$MinusOne \in vMinusDeath$
$vDeathRate \in vMinusDeath$

The simplified rule (3) of 7.1 will result in unnecessary copying of $MinusOne$, but this is harmless and optionally can be eliminated with a simple trick.

Let griffins prey on unicorns. We define a preying relationship $PrGU$ which will be an instance of a special model Pr, by which both the fertility and the death rate of the predator depend on the availability of prey, and the basic death rate of the prey is further increased by the predator.

$PrGU$ **is-a** Pr
$Predator(PrGU) = Griffins$
$Prey(PrGU) = Unicorns$

The Pr frame might look like this.

$\#type(Pr) = \#frame$
$Predator(Pr) = vPredator$
$Predator \in \#copied(Pr)$
$Prey(Pr) = vPrey$
$Prey \in \#copied(Pr)$
$Population \in \#ascribe(vPredator)$
$Population \in \#ascribe(vPrey)$

(The last two formulas are redundant in this example.)

Now introduce an additional member in the prey's death rate (defined as a sum).

$vPreyDeath \in DeathRate(vPrey)$
$Product \in \#ascribe(vPreyDeath)$
$Number(vPredator) \in vPreyDeath$
$vPreyFactor \in vPreyDeath$

So we assume the additional death rate to be proportional to the number of the predator. We need now to link $vPreyDeath$ and $vPreyFactor$ to the model to make sure the values can be accessed by other relationships and to be used in mapping.

$PreyFactor(Pr) = vPreyFactor$
$PreyFactor \in \#copied(Pr)$
$PreyDeath(Pr) = vPreyDeath$
$PreyDeath \in \#copied(Pr)$

To secure copying, *vPreyDeath* could be also linked to *vPrey* instead of *Pr*; but this would lead to a confusion if the same population suffered from two different predators. On the other hand, *vPreyFactor* being a member of *vPreyDeath* set, it would be copied without the link, but the link will be needed to provide a numerical value.

Now we define fertility and death rate of the predator as depending on the number of prey.

$Fertility(vPredator) = vPredFertility$
$Product \in \#ascribe(vPredFertility)$
$Number(vPrey) \in vPredFertility$
$vFertilityFactor \in vPredFertility$
$FertilityFactor(Pr) = vFertilityFactor$
$FertilityFactor \in \#copied(Pr)$

$vDeathDecrease \in DeathRate(vPredator)$
$Product \in \#ascribe(vDeathDecrease)$
$Number(vPrey) \in vDeathDecrease$
$vDeathDecreaseFactor \in vDeathDecrease$
$MinusOne \in vDeathDecrease$
$DeathDecrease(Pr) = vDeathDecrease$
$DeathDecrease \in \#copied(Pr)$
$DeathDecreaseFactor(Pr) = vDeathDecreaseFactor$
$DeathDecreaseFactor \in \#copied(Pr)$

The next thing to do is to provide values for some parameters and to print out the differential equations. We'll only discuss printing of expressions in the right members of the differential equations, because printing is an example of an action other than just inference, and we will use steps not defined in the preceding exposition.

Some expressions have been described as sums or products; others have to be described as plain values. Each of them will have a "method" for printing. A plain value will have its print representation, a constant (number or symbolic name).

$\#type(PlainValue) = \#frame$
$Representation \in \#inquiry(PlainValue)$
$PrintingMethod \in \#invariant(PlainValue)$
$PrintingMethod(PlainValue) = PrintPlain$

PrintPlain is a procedure that will be applied to an argument, which will be an instance of *PlainValue*; the procedure must find and print *Representation* of its argument.

For sums and products, other printing procedures will be defined. We consider only sums because the other case is absolutely analogous.

$\#type(Sum) = \#frame$
$PrintingMethod \in \#invariant(Sum)$
$PrintingMethod(Sum) = PrintSum$

PrintSum will print the addends in succession, but it has also to keep track of whether any addends were printed before the current addend. We'll assume this is done by providing a special "object" (in the standard sense of OOP) called "sum separator", with a constructor that sets it to the state of "nothing printed", a "separate" method which makes it print a left parenthesis at the first call and a plus sign at subsequent calls, changing the state to "something printed", and a "finalize" method that prints either a right parenthesis, or zero (if nothing was printed).

Thus *PrintSum* will be a conjunction (7.2) of three smaller procedures:

- build a sum separator,
- print addends,
- finalize the separator.

The second step will, in its turn, be a "for all" procedure (7.3). Its internal procedure (*#proc* property) will be applied to all addends, and will be a conjunction of two steps: a call to the "separate" method of the separator and printing of the addend. We consider the latter in more detail because at this moment we don't know what kind of expression is the addend. The procedure *PrintExpression* will be defined as follows.

$$\#type(PrintExpression) = \#frame$$
$$PrintingMethod(PrintExpression) = vMethod$$
$$PrintingMethod \in \#copied(PrintExpression)$$
$$PrintingMethod \in \#inquiry(PrintExpression)$$
$$Printsubproc \in \#subproc(PrintExpression)$$
$$\#function(Printsubproc) = vMethod$$
$$\#argument(Printsubproc) = PrintExpression$$

Let us see how *PrintExpression* applies to an expression X. We start with building a mapping function, M.

$$M(PrintExpression) = X$$
$$M(vMethod) = PrintingMethod(X)$$

If *PrintingMethod*(X) was not previously defined as a constant, we get a contradiction due to the "inquiry" formula. Otherwise we get the specific printing method defined for this X, which might be *PrintPlain*, *PrintSum*, *PrintProduct*, or something else. To invoke this method, we call subprocedures of *PrintExpression*. The only existing subprocedure requires that $M(vMethod)$ be applied to $M(PrintExpression)$, i.e., *PrintingMethod*(X) to X.

Now to obtain the system of differential equations we still need to describe some parameters as plain values and to provide print representations. Then we get something like this.

```
d(uNumber)/dt=(uNumber*(uFertility+((-1)*(uBasicDeathRate+
(gNumber*pPreyFactor)))));
    d(gNumber)/dt=(gNumber*((uNumber*gFertilityFactor)+
((-1)*(gBasicDeathRate+((-1)*uNumber*gDeathDecreaseFactor)))));
```

Let us summarize what has been done in this example. We use two sources of knowledge. One is information about specific populations and their relationships, it is just a few lines (including the information about numeric values or symbolic names of some parameters, not given explicitly in this example). The other source is the supply of models. It should be probably expanded by introducing models for other types of propagation, dying, preying, migration, etc. – we don't impose any particular pattern of numerical relationships. But different models for different populations will be able to interact properly in a single system producing a set of simulating equations, provided that we use the same names (like *Number* or *DeathRate*) for the concepts shared by different models. Few programming tools provide this flexibility, except possibly for latest object-oriented systems.

References

1. Cejtin, G.S.: Programmirovanie na associativnyq setjaq. In *EhVM v proektirovanii i proizvodstve, vyp. 2, pod red. G.V.Orlovskogo*, 16-48. "Mashinostroenie", Leningr. otdelenie, Leningrad, 1985 (G.S.Tseytin. Programming with association nets. English version at
   http://gamma.niimm.spb.su/~tseytin/netslong.html)
2. Zheleznjakov, M.M., Nevleva, T.N., Novickaja, I.M., Smirnova, L.N., Cejtin, G.S.: Opyt postroenija modeli "tekst ⇒ dejstviteljnostj" s ispoljzovaniem associativnyq setej. In *Mashinnyj fond russkogo jazyka: predproektnye issledovanija*, 140-167. Institut russkogo jazyka, Moskva, 1988 (M.M.Zheleznyakov, T.N.Nevleva, I.M.Novitskaya, L.N.Smirnova, G.S.Tseytin. An attempt of building a "text ⇒ reality" model using association nets. English version at http://gamma.niimm.spb.su/~tseytin/article.html)
3. Tseytin, G.S.: On some mechanisms of representation of meaning in natural languages. The Prague Bulletin of Mathematical Linguistics, **65-66** (1996) 5-12 (See also at http://gamma.niimm.spb.su/~tseytin/mytalke.html)
4. Tseytin, G.S.: From Logicism to Proceduralism (an autobiographical account). In Algorithms in Modern Mathematics and Computer Science, Lecture Notes in Computer Science, **122** (1981) 390-396

Simulating η-expansions with β-reductions in the Second-Order Polymorphic λ-calculus

Hongwei Xi

Department of Mathematical Sciences
Carnegie Mellon University
Pittsburgh, PA 15213, USA

email: hwxi+@cs.cmu.edu

Abstract. We introduce an approach to simulating η-expansions with β-reductions in the second-order polymorphic λ-calculus. This generalizes the work of Di Cosmo and Delia Kesner which simulates η-expansions with β-reductions in simply typed settings, positively solving the conjecture on whether the simulation technique can be extended to polymorphic settings. We then present a *modular* proof that the second-order polymorphic λ-calculus with an expansive version of η-reduction is strong normalizing and confluent. The simulation is also promising to provide modular proofs showing that other rewriting systems are also strongly normalizing after expanded with certain versions of η-expansion.

1 Introduction and Related Work

η-conversion presents an approach to studying extensional equalities for λ-terms. Given an η-equality $\lambda x.M(x) =_\eta M$, where x has no free occurrences in M; one can either say $\lambda x.M(x)$ η-contracts ($\rightarrow_\eta$) to M, or M η-expands ($\rightarrow_{\eta_\bullet}$) to $\lambda x.M(x)$; the former is usually adopted as a rewrite rule since every λ-term can then be η-contracted to an η-normal form. This strategy leads to a confluent untyped λ-calculus $\lambda\beta\eta$, in which every λ-term M has a β-normal form if and only if M has a $\beta\eta$-normal. Carrying η-contraction into the simply typed λ-calculus $\lambda^\rightarrow\beta\eta$, one can prove that $\lambda^\rightarrow\beta\eta$ is confluent and strongly normalizing. However, problems occur when $\lambda^\rightarrow\beta\eta$ is augmented with a unit type $\mathbf{T}$ representing the terminal object of a cartesian closed category, and an extensional rule

$$M \rightarrow_\mathbf{T} *,$$

where M is of type $\mathbf{T}$ and $*$ is the only element in $\mathbf{T}$. The confluence breaks down as shown in the following well-known example [16]:

$$\lambda x.f(x) \rightarrow_\eta f \quad \text{and} \quad \lambda x.f(x) \rightarrow_\mathbf{T} \lambda x.*,$$

where f is of type $\mathbf{T} \rightarrow \mathbf{T}$ and x of type $\mathbf{T}$. Note that both f and $\lambda x.*$ cannot be contracted further. This is a serious drawback since it can easily occur when one adds η-contraction to algebraic rewriting systems. One immediate remedy is to allow $N \to_{\mathbf{T} \to \mathbf{T}} \lambda x.*$ for every term N of type $\mathbf{T} \to \mathbf{T}$. However, this method can produce an unwieldy system since infinitely many rules have to be included in order to handle terms of other similar types.

The use of η-expansion as a rewrite rule was suggested in [18]. Mints [17] presented a proof for the confluence and (weak) normalization of a simply typed λ-calculus with surjective pairing and η-expansion. A flaw in Mints's proof was later corrected in [6]. Y. Akama [1] proved the confluence and strong normalization of the above system, which is also given by C.B. Jay and N. Ghani [15]. Di Cosmo and Kesner discovered a translation $(\cdot)^{\circ}$ which simulates η-expansion with β-reduction in a simply typed setting [8]. An application of this translation can also be found in [7].

The following example shows that applying η-expansion unconditionally easily leads to infinite reduction sequence.

$$x^{A \to B} \to_{\eta_*} \lambda y^A . x^{A \to B}(y^A) \to_{\eta_*} \lambda y^A . (\lambda y^A . x^{A \to B}(y^A))(y^A) \to_{\eta_*} \cdots$$

However, this can be remedied by applying $\to_{\eta_*}$ with restrictions; namely, λ-abstraction cannot be η-expanded, nor can terms which are applied to other terms. Systems with such restricted η-expansions have been receiving attentions of a growing number of researchers as shown in the reference.

The translation discovered by Di Cosmo and Kesner provides a modular approach to proving that many rewriting systems are still strong normalizing and confluent after augmented with restricted versions of η-expansion [7]. This is a very useful proof technique in the first-order settings. Unfortunately, their translation cannot be applied to polymorphic settings directly. Although they have proven that the second-order polymorphic λ-calculus with surjective pairing and a version of η-expansion is strongly normalizing and confluent, their proof is not modular. Compared with the corresponding results in simply typed settings, this is less satisfactory.

We will generalize the translation of Di Cosmo and Kesner to the second-order polymorphic λ-calculus. We then present a modular proof, showing the second-order polymorphic λ-calculus with a version of η-expansion is strongly normalizing and confluent. This new translation is also promising to be a powerful technique dealing with η-expansions in other polymorphic settings [5].

Extensional polymorphism is studied in [5, 9, 10, 13]. A modular proof for the strong normalisation and confluence of $\lambda\beta\beta_2\eta_*$ (in our notation) is given in [10], which is rather long and complicated. Since $\lambda\beta\beta_2\eta_*$ is $\lambda\beta\beta_2\eta_*\eta_*^2$ excluding $\to_{\eta_*^2}$, the result simply follows from our main result (Corollary 9). We point out that proofs for our main result have been already presented in [9] and [13], but those proofs are not modular. They are established upon the well-known *reducibility candidates* method due to Tait [21] and Girard [14].

It is interesting to mention that the author learned the use η-expansion while studying the equivalence between strong and weak normalizations in various typed λ-calculi. Schwichtenberg [20] used η-expansion to translate simply typed λ-terms into simply typed λI-terms so that β-reduction sequences from the former can be simulated by those from the latter. This can yield a bound for the lengths of β-reduction sequence from simply typed terms. Details can also be founded in [22].

2 Preliminaries

The second-order polymorphic typed λ-calculus $\lambda 2$ is originally introduced in [14] and [19], where Church typing is involved. We now give a slightly different formulation of $\lambda 2$.

$$\text{Types } A, B ::= a \mid b \mid X \mid A \to B \mid \forall X.A$$
$$\text{Terms } M, N ::= x^A \mid (\lambda x^A.M) \mid M(N) \mid (\Lambda X.M) \mid M(A)$$

We use a, b for base types, X for type variables, A, B for types, M, N for terms and x^A, y^A for variables of type A. We write $[B/X]A$ for the type obtained by substituting B for every free occurrence of X in A. We often write $\lambda x^A.M$ for $(\lambda x^A.M)$ and $\Lambda X.M$ for $(\Lambda X.M)$, enhancing clarity.

As usual, let $FV(M)$ be the set of free (term) variables in term M and $FTV(A)$ be the set of free type variables in type A. We also need the following notions: $FTV(M)$ is the set of free type variables in term M and $FTFV(M)$ is the set of free type variables in the types of free variables in M.

$$FTV(x^A) = FV(A) \qquad\qquad FTV(M(N)) = FTV(M) \cup FTV(N)$$
$$FTV(\lambda x^A.M) = FTV(A) \cup FTV(M) \quad FTV(\Lambda X.M) = FTV(M) \backslash \{X\}$$
$$FTV(M(A)) = FTV(A) \cup FTV(M) \quad FTFV(M) = \bigcup \{FTV(A) : x^A \in FV(M)\}$$

We assume the familiarity of the reader with free and bound occurrences of variables. Term equalities are modulo α-conversions, which are often taken implicitly. The followings are typing rules for $\lambda 2$.

$$\frac{}{\vdash x^A : A}(var)$$

$$\frac{\vdash M : B}{\vdash (\lambda x^A.M) : A \to B}(\to I) \qquad \frac{\vdash M : A \to B \qquad \vdash N : A}{\vdash M(N) : B}(\to E)$$

$$\frac{\vdash M : A}{\vdash (\Lambda X.M) : \forall X.A}(\forall I)^* \qquad \frac{\vdash M : \forall X.A}{\vdash M(B) : [B/X]A}(\forall E)$$

Note that the rule $\forall I$ can be applied only if $X \notin FTFV(M)$. A term M is a $\lambda 2$-term of type A if $\vdash M : A$ is derivable.

An advantage of this formulation is that we can define a function τ which maps terms to their types without bothering their typing derivations.

$$\tau(x^A) = A \qquad\qquad \tau(\lambda x^A.M) = A \to \tau(M)$$
$$\tau(M(N)) = B, \text{ if } \tau(M) = A \to B \text{ and } \tau(N) = A;$$
$$\tau(\Lambda X.M) = \forall X.\tau(M) \quad \tau(M(B)) = [B/X]A, \text{ if } \tau(M) = \forall X.A$$

Since η-expansions needs the guidance of types, this formulation of $\lambda 2$ brings a great deal of convenience in our following development. If we use typing rules with contexts in the formulation of $\lambda2$, we have to carry contexts around in order to compute the types of open terms, complicating our following presentation *significantly*.

Proposition 1. *For every term M, M is a $\lambda2$-term of type A if and only if $\tau(M) = A$.*

Proof. A structural induction on M yields the result. $\qquad\qquad\qquad\qquad\square$

Given a $\lambda2$-term M of type A; for every λ-term of type B, $[N/x^B]M$ is the term obtained by substituting N for every free occurrence of x^B in M; for every type B, $[B/X]M$ is the term obtained by substituting B for every free occurrence of X in M; it can be readily verified that $\tau([N/x^B]M) = A$ and $\tau([B/X]M) = [B/X]A$. We now present the reduction rules for the second order extensional λ-calculus $\lambda\beta\beta_2\eta_*\eta_*^2$.

- (β-contraction) $(\lambda x^A.M)(N) \overset{\beta}{\to} [N/x^A]M$
- (β^2-contraction) $(\Lambda X.M)(A) \overset{\beta^2}{\to} [A/X]M$
- (η-expansion) $M \overset{\eta}{\to} \lambda x^A.M(x^A)$, if $x^A \notin \mathrm{FV}(M)$ and $\tau(M) = A \to B$ for some B and M is not a λ-abstraction
- (η^2-expansion) $M \overset{\eta^2}{\to} \Lambda X.M(X)$, if $X \notin \mathrm{FTV}(M)$ and $\tau(M) = \forall X.B$ for some B and M is not a Λ-abstraction

β-reduction $(\to_\beta)$ is given below, and β^2-reduction $(\to_{\beta^2})$ can be given accordingly.

$$\frac{M_1 \overset{\beta}{\to} M_2}{M_1 \to_\beta M_2} \qquad \frac{M_1 \to_\beta M_2}{\lambda x^A.M_1 \to_\beta \lambda x^A.M_2} \qquad \frac{M_1 \to_\beta M_2}{M_1(N) \to_\beta M_2(N)}$$

$$\frac{M_1 \to_\beta M_2}{N(M_1) \to_\beta N(M_2)} \qquad \frac{M_1 \to_\beta M_2}{\Lambda X.M_1 \to_\beta \Lambda X.M_2} \qquad \frac{M_1 \to_\beta M_2}{M_1(A) \to_\beta M_2(A)}$$

η-reduction $(\to_{\eta_*})$ and η^2-reduction $(\to_{\eta_*^2})$ can also be defined accordingly with the following restrictions, respectively.

$$\frac{M_1 \to_{\eta_*} M_2 (\text{except } M_1 \overset{\eta}{\to} M_2)}{M_1(N) \to_{\eta_*} M_2(N)} \qquad \frac{M_1 \to_{\eta_*^2} M_2 (\text{except } M_1 \overset{\eta^2}{\to} M_2)}{M_1(A) \to_{\eta_*^2} M_2(A)}$$

Now we define reduction $\underset{\beta\beta^2\eta_*\eta_*^2}{\longrightarrow}$ and $\underset{\beta\beta^2}{\longrightarrow}$ as follows.

$$\underset{\beta\beta^2\eta_*\eta_*^2}{\longrightarrow} = \to_\beta \cup \to_{\beta^2} \cup \to_{\eta_*} \cup \to_{\eta_*^2} \quad \text{and} \quad \underset{\beta\beta^2}{\longrightarrow} = \to_\beta \cup \to_{\beta^2}.$$

For any decorated reduction notation of $\to$ in this paper, the corresponding decorated reduction notations of $\twoheadrightarrow$ and $\twoheadrightarrow^+$ stand for some (possibly zero) and a positive number of steps of such a reduction, respectively. We shall use *subject reduction* property of $\lambda2$ implicitly in our following presentation. We assume the familiarity of the reader with this property.

3 Translation and Simulation

For each type variable X, we assign to it a *fresh* term variable $v^{X \to X}$ of type $X \to X$, which will *only* be used for the following translation purpose.

$$|b| = b \qquad\qquad\qquad |X| = X$$

$$|A \to B| = |A| \to |B| \qquad |\forall X.A| = \forall X.(X \to X) \to |A|$$

$$\Delta_b = \lambda x^b.x^b \qquad\qquad \Delta_X = v^{X \to X}$$

$$\Delta_{A \to B} = \lambda x^{|A \to B|}.\lambda y^{|A|}.\Delta_B(x^{|A \to B|}(\Delta_A(y^{|A|})))$$

$$\Delta_{\forall X.A} = \lambda x^{|\forall X.A|}.\Lambda X.\lambda v^{X \to X}.\Delta_A(x^{|\forall X.A|}(X)(v^{X \to X}))$$

Notice $\Delta_{a \to b}(x^{a \to b}) \twoheadrightarrow_\beta (\lambda y^a.x^{a \to b}(y^a))$, which is the $\to_{\eta_\bullet}$-normal form of $x^{a \to b}$. In general, for M of type A, $\Delta_A(M)$ $\beta\beta^2$-reduces a $\to_{\eta_\bullet \eta_\bullet^2}$-normal form which closely relates to the $\to_{\eta_\bullet \cdot \eta_\bullet^2}$-normal form of M as shown below. These Δ's are called expansors, which set up the machinery to simulate η-expansions with β-reductions.

Proposition 2. *Given types A and B, the following hold.*

1. $[|B|/X]|A| = |[B/X]A|$.
2. $\tau(\Delta_A) = |A| \to |A|$.
3. $[\Delta_B/v^{|B| \to |B|}][|B|/X]\Delta_A = \Delta_{[B/X]A}$.

Proof. An structural induction on A yields the results. $\qquad\qquad\qquad\square$

We use $\Delta_A^0(M)$ for M and $\Delta_A^{n+1}(M)$ for $\Delta_A(\Delta_A^n(M))$ when $n \geq 0$.

Definition 3. We now define versions of $|M|$ and $|M|^+$ inductively.

$$|x^A| = \Delta_A^n(x^{|A|}) \qquad\qquad\qquad |x^A|^+ = \Delta_A(|x^A|)$$

$$|(\lambda x^A.M)| = \Delta_{A \to \tau(M)}^n(\lambda x^{|A|}.|M|^+) \qquad |(\lambda x^A.M)|^+ = |(\lambda x^A.M)|$$

$$|M(N)| = \Delta_{\tau(M(N))}^n(|M|(|N|^+)) \qquad |M(N)|^+ = \Delta_{\tau(M(N))}(|M(N)|)$$

$$|(\Lambda X.M)| = \Delta_{\forall X.\tau(M)}^n(\Lambda X.\lambda v^{X \to X}.|M|^+) \quad |(\Lambda X.M)|^+ = |(\Lambda X.M)|$$

$$|M(B)| = \Delta_{\tau(M(B))}^n(|M|(|B|)(\Delta_B)) \qquad |M(B)|^+ = \Delta_{\tau(M(B))}(|M(B)|)$$

Note that n can be any nonnegative integers here. Hence, for every $\lambda 2$-term, $|M|$ and $|M|^+$ have infinitely many different versions.

We shall try to simulate a $\underset{\beta\beta^2\eta_\bullet \eta_\bullet^2}{\longrightarrow}$-reduction sequence from M with some $\underset{\beta\beta^2}{\longrightarrow}$- reduction sequence from $|M|^+$. The need of versions can be explained as follows. Given $M \to_\beta N$, we need to have $|M|^+ \twoheadrightarrow_\beta |N|^+$ in order to make the simulation work. Suppose we translate $(\lambda x^A.x^A)(M)$ to $(\lambda x^A.\Delta_A(x^A))(\Delta_A(M))$, which β- reduces to $\Delta_A^2(M)$. Note that $\Delta_A^2(M)$ can not be β-reduced to $\Delta_A(M)$, though they may be β-equivalent. This forces us to adopt that $\Delta_A^2(M)$ is also a version of $|M|^+$ since the simulation would break down otherwise.

Examples Given $M = x^{a \to b}(y^a)$, where a, b are base types; $\Delta_b^2(x^{a \to b}(\Delta_a(y^a)))$ is a version of $|M|$ and a version of $|M|^+$; $\Delta_{a \to b}(x^{a \to b})(\Delta_a^2(y^a))$ is a version of $|M|$, but not a version of $|M|^+$. Let $N = \Lambda X.(\lambda x^X.x^X)$ and $A = a \to b$; note $N(A)$ is in $\eta_* \eta_*^2$-normal form; we have $N(A) \to_{\beta^2} N_1 = (\lambda x^A.x^A)$ but N_1 is not in η_*-normal form; we translate $N(A)$ to

$$(\Lambda X.\lambda v^{X \to X}.(\lambda x^X.v^{X \to X}(x^X)))(A)(\Delta_A);$$

this translation $\beta\beta^2$-reduces to $(\lambda x^A.\Delta_A(x^A))$, which is a version of $|N_1|^+$; then η-expansion from N_1 can be simulated. The use of $v^{X \to X}$ is like an *expansor candidate* which will be instantiated with some expansor when X is instantiated.

We often use $|M|$ ($|M|^+$) for a *version* of $|M|$ ($|M|^+$) in our following writing. This convention brings a great deal of convenience and turns out to be adequately clear, esp. with the help of contexts. For instance, $|M| \underset{\beta\beta^2}{\twoheadrightarrow} |N|$ means that *every* version of $|M|$ $\beta\beta^2$-reduces to *some* version of $|N|$.

If the reader is interested in the proof details for the following results, please see [23].

Proposition 4. *Given $\lambda 2$-term M of type A and N of type B, we have the following.*

1. $|M|$ *and* $|M|^+$ *are $\lambda 2$-terms of type* $|A|$.
2. $[|N|/x^{|B|}]|M|$ *is a version of* $|[N/x^B]M|$, *and*
   $[|N|/x^{|B|}]|M|^+$ *is a version of* $|[N/x^B]M|^+$.
3. $[\Delta_B/v^{|B| \to |B|}][|B|/X]|M|$ *is a version of* $|[B/X]M|$, *and*
   $[\Delta_B/v^{|B| \to |B|}][|B|/X]|M|^+$ *is a version of* $|[B/X]M|^+$.

Proof. By a structural induction on M. □

Proposition 5. *We have the following.*

1. *For every $\lambda 2$-term $\lambda x^A.M$ of type $A \to B$, $|\lambda x^A.M| \twoheadrightarrow_\beta \lambda x^{|A|}.|M|^+$.*
2. *For every $\lambda 2$-term $\Lambda X.M$ of type $\forall X.A$, $|\Lambda X.M| \underset{\beta\beta^2}{\twoheadrightarrow} \Lambda X.\lambda v^{X \to X}.|M|^+$.*

Proof. We may assume

$$|\lambda x^A.M| = \Delta_{A \to B}^n(\lambda x^A.|M|^+) \quad \text{and} \quad |\Lambda X.M| = \Delta_{\forall X.A}^n(\Lambda X.\lambda v^{X \to X}.|M|^+)$$

for some n. The proof proceeds by induction on n.

- $n = 0$. (1) and (2) hold by definition.
- $n = k + 1$ for some $k \geq 0$. Then by induction hypothesis, we have

$$
\begin{aligned}
\Delta_{A \to B}^{k+1}(|\lambda x^A.M|) &= \Delta_{A \to B}(\Delta_{A \to B}^k(|\lambda x^A.M|)) \\
&\twoheadrightarrow_\beta \Delta_{A \to B}(\lambda x^{|A|}.|M|^+), \text{ by induction hypothesis} \\
&\to_\beta \lambda y^{|A|}.\Delta_B((\lambda x^{|A|}.|M|^+)(\Delta_A(y^{|A|}))) \\
&\to_\beta \lambda y^{|A|}.\Delta_B([\Delta_A(y^{|A|})/x^{|A|}]|M|^+) \\
&= \lambda y^{|A|}.|[y^A/x^A]M|^+, \text{ by Proposition 4 (2)} \\
&= \lambda x^{|A|}.|M|^+, \alpha\text{-conversion}
\end{aligned}
$$

Also, we have

$$
\begin{aligned}
\Delta_{\forall X.A}^{k+1}(|AX.M|) &= \Delta_{\forall X.A}(\Delta_{\forall X.A}^{k}(|AX.M|)) \\
&\xrightarrow[\beta\beta^2]{} \Delta_{\forall X.A}(AX.\lambda v^{X\to X}.|M|^+), \text{ by induction hypothesis} \\
&\to_\beta AX.\lambda v^{X\to X}.\Delta_A((AX.\lambda v^{X\to X}.|M|^+)(X)(v^{X\to X})) \\
&\xrightarrow[\beta\beta^2]{} AX.\lambda v^{X\to X}.|M|^+.
\end{aligned}
$$

Hence, we are done.

$\square$

The next lemma will be used to establish our main simulation result.

Lemma 6. *(Main Lemma) We have the followings.*

1. *Given* $R = (\lambda x^A.M)(N)$ *and* $M^* = [N/x]M$, $|R| \twoheadrightarrow_\beta^+ |M^*|$.
2. *Given* $R = (AX.M)(B)$ *and* $M^* = [B/X]M$, $|R| \xrightarrow[\beta\beta^2]{}^+ |M^*|$.
3. *Given* $M \xrightarrow{\eta} \lambda x^A.M(x^A)$, $|M|^+ \twoheadrightarrow_\beta |\lambda x^{|A|}.M(x)|^+$.
4. *Given* $M \xrightarrow{\eta^2} AX.M(X)$, $|M|^+ \xrightarrow[\beta\beta^2]{} |AX.M(X)|^+$.

Proof. For (1), we have

$$
\begin{aligned}
|R| &= \Delta_{\tau(R)}^n(|\lambda x^A.M|(|N|^+)), \text{ by definition} \\
&\twoheadrightarrow_\beta \Delta_{\tau(R)}^n((\lambda x^{|A|}.|M|^+)(|N|^+)), \text{ by Proposition 5 (1)} \\
&\to_\beta \Delta_{\tau(R)}^n([|N|^+/x^{|A|}]|M|^+) \\
&= \Delta_{\tau(R)}^n(|M^*|^+), \text{ by Proposition 4 (2)} \\
&= |M^*|^+, \text{ since } \tau(R) = \tau(M^*)
\end{aligned}
$$

For (2), we have

$$
\begin{aligned}
|R| &= \Delta_{\tau(R)}^n(|AX.M|(|B|)(\Delta_B)), \text{ by definition} \\
&\xrightarrow[\beta\beta^2]{} \Delta_{\tau(R)}^n((AX.\lambda v^{X\to X}.|M|^+)(|B|)(\Delta_B)), \text{ by Proposition 5 (2)} \\
&\xrightarrow[\beta\beta^2]{}^+ \Delta_{\tau(R)}^n([\Delta_B/v^{|B|\to|B|}][|B|/X]|M|^+) \\
&= \Delta_{\tau(R)}^n(|M^*|^+), \text{ by Proposition 4 (3)} \\
&= |M^*|^+, \text{ since } \tau(R) = \tau(M^*)
\end{aligned}
$$

For (3), we have

$$
\begin{aligned}
|M|^+ &= \Delta_{\tau(M)}(|M|), \text{ since } M \text{ is not a } \lambda\text{-abstraction} \\
&\twoheadrightarrow_\beta \lambda y^{|A|}.\Delta_B(|M|(\Delta_A(y^{|A|}))), \\
&\quad \text{by Proposition 5 (1), where } A \to B = \tau(M) \\
&= \lambda y^{|A|}.|M(y)|^+, \text{ by definition} \\
&= \lambda x^{|A|}.|M(x)|^+, \text{ by } \alpha\text{-conversion} \\
&= |\lambda x^{|A|}.M(x)|^+, \text{ by definition}
\end{aligned}
$$

For (4), we have

$$|M|^+ = \Delta_{\tau(M)}(|M|), \text{ since } M \text{ is not a } \Lambda\text{-abstraction}$$
$$\underset{\beta\beta^2}{\twoheadrightarrow} \Lambda X.\lambda v^{X \to X}.\Delta_A(|M|(X)(v^{X \to X})),$$

$$\text{by Proposition 5 (1), where } \forall X.A = \tau(M)$$
$$= \Lambda X.\lambda v^{X \to X}.|M(X)|^+, \text{ by definition}$$
$$= |\Lambda X.M(X)|^+, \text{ by definition}$$

$\square$

Note that $\underset{\beta\beta^2\eta_\bullet\eta_\bullet^2}{\longrightarrow}$ is defined as $\to_\beta \cup \to_{\beta^2} \cup \to_{\eta_\bullet} \cup \to_{\eta_\bullet^2}$ and $\underset{\beta\beta^2}{\longrightarrow}$ as $\to_\beta$ $\cup \to_{\beta^2}$. We now state and prove our main simulation result.

Theorem 7. *(Simulation)* $M \underset{\beta\beta^2\eta_\bullet\eta_\bullet^2}{\longrightarrow} M'$ *implies* $|M|^+ \underset{\beta\beta^2}{\twoheadrightarrow}{}^+ |M'|^+$.

Proof. We proceed by a structural induction on M to show the followings simultaneously.

1. $M \underset{\beta\beta^2\eta_\bullet\eta_\bullet^2}{\longrightarrow} M'$ implies $|M| \underset{\beta\beta^2}{\twoheadrightarrow}{}^+ |M'|$ unless $M \overset{\eta_\bullet}{\to} M'$ or $M \overset{\eta_\bullet^2}{\to} M'$.
2. $M \underset{\beta\beta^2\eta_\bullet\eta_\bullet^2}{\longrightarrow} M'$ implies $|M|^+ \underset{\beta\beta^2}{\twoheadrightarrow}{}^+ |M'|^+$.

Note $x^{a \to b} \overset{\eta_\bullet}{\to} (\lambda y^a.x^{a \to b}(y^a))$, but $x^{a \to b}$, a version of $|x^{a \to b}|$, is in β-normal form. This explains why the restriction on (1) is necessary. We now do a case analysis on the structure of M.

- $M \overset{\beta}{\to} M'$, $M \overset{\beta^2}{\to} M'$, $M \overset{\eta_\bullet}{\to} M'$ or $M \overset{\eta_\bullet^2}{\to} M'$. This follows from Lemma 6.
- $M = M_1(N)$ and $M' = M_1'(N)$ and $M_1 \underset{\beta\beta^2\eta_\bullet\eta_\bullet^2}{\longrightarrow} M_1'$. Then $M_1 \overset{\eta_\bullet}{\to} M_1'$ cannot hold. By induction hypothesis on (1), $|M_1| \underset{\beta\beta^2}{\twoheadrightarrow}{}^+ |M_1'|$. Hence,

$$|M| = \Delta_{\tau(M)}^n(|M_1|(|N|^+)) \underset{\beta\beta^2}{\twoheadrightarrow}{}^+ \Delta_{\tau(M)}^n(|M_1'|(|N|^+)) = |M'|.$$

Therefore (1) holds. (2) follows immediately.
- $M = N(M_1)$ and $M' = N(M_1')$ and $M_1 \underset{\beta\beta^2\eta_\bullet\eta_\bullet^2}{\longrightarrow} M_1'$. By induction hypothesis on (2), $|M_1|^+ \underset{\beta\beta^2}{\twoheadrightarrow}{}^+ |M_1'|^+$. Hence,

$$|M| = \Delta_{\tau(M)}^n(|N|(|M_1|^+)) \underset{\beta\beta^2}{\twoheadrightarrow}{}^+ \Delta_{\tau(M')}^n(|N|(|M_2|^+)) = |M'|.$$

Therefore (1) holds. (2) follows immediately.
- $M = M_1(B)$ and $M' = M_1'(B)$. Then $M_1 \overset{\eta_\bullet^2}{\to} M_1'$ cannot hold. By induction hypothesis on (1), $|M_1| \underset{\beta\beta^2}{\twoheadrightarrow}{}^+ |M_1'|$. This implies that (1) and (2) hold.
- $M = \lambda x^A.M_1 \underset{\beta\beta^2\eta_\bullet\eta_\bullet^2}{\longrightarrow} \lambda x^A.M_1' = M'$ or $M = \Lambda X.M_1 \underset{\beta\beta^2\eta_\bullet\eta_\bullet^2}{\longrightarrow} \Lambda X.M_1' = M'$. (1) and (2) follow from induction hypothesis.

$\square$

4 $\underset{\beta\beta^2\eta_*\eta_*^2}{\longrightarrow}$ is strongly normalizing and confluent

Given some reduction $\to$, we write $\lambda 2 \models \mathcal{SN}(\to)$ meaning that $\to$ is strongly normalizing in $\lambda 2$, i.e., there exist no infinite $\to$-reduction sequences starting from $\lambda 2$-terms.

Theorem 8. $\lambda 2 \models \mathcal{SN}(\underset{\beta\beta^2\eta_*\eta_*^2}{\longrightarrow})$ if and only if $\lambda 2 \models \mathcal{SN}(\underset{\beta\beta^2}{\longrightarrow})$.

Proof. If there exists an infinite $\underset{\beta\beta^2\eta_*\eta_*^2}{\longrightarrow}$-reduction sequence

$$M_1 \underset{\beta\beta^2\eta_*\eta_*^2}{\longrightarrow} M_2 \underset{\beta\beta^2\eta_*\eta_*^2}{\longrightarrow} M_3 \underset{\beta\beta^2\eta_*\eta_*^2}{\longrightarrow} \cdots,$$

then by Theorem 7 we have the following corresponding sequence

$$|M_1|^+ \underset{\beta\beta^2}{\twoheadrightarrow}{}^+ |M_2|^+ \underset{\beta\beta^2}{\twoheadrightarrow}{}^+ |M_3|^+ \underset{\beta\beta^2}{\twoheadrightarrow}{}^+ \cdots.$$

Therefore, $\lambda 2 \models \mathcal{SN}(\underset{\beta\beta^2}{\longrightarrow})$ implies $\lambda 2 \models \mathcal{SN}(\underset{\beta\beta^2\eta_*\eta_*^2}{\longrightarrow})$. The other direction is trivial. □

Notice that this is a proof which can be formulated in the first-order Peano arithmetic.

Corollary 9. *Every $\lambda 2$-term is strongly $\underset{\beta\beta^2\eta_*\eta_*^2}{\longrightarrow}$-normalizing and confluent.*

Proof. Since it is well-known that $\lambda 2 \models \mathcal{SN}(\underset{\beta\beta^2}{\longrightarrow})$, $\lambda 2 \models \mathcal{SN}(\underset{\beta\beta^2\eta_*\eta_*^2}{\longrightarrow})$ follows immediately. It can be verified that $\underset{\beta\beta^2\eta_*\eta_*^2}{\longrightarrow}$ is weakly confluent [8]. Therefore, $\underset{\beta\beta^2\eta_*\eta_*^2}{\longrightarrow}$ is confluent by Newman's Lemma.

Roberto Di Cosmo and Delia Kesner have also proven Corollary 9 in [9] with a method involving *reducibility candidates* due to Tait [21] and Girard [14]. Hence, their proof is not modular. Also a modular proof for the strong normalization and confluence of a subsystem of $\lambda\beta\beta^2\eta_*\eta_*^2$ is presented in [10], but the proof — in the author's opinion — is very much involved and can hardly be scaled to other more complicated systems such as $\lambda\beta\beta^2\eta_*\eta_*^2$.

5 Conclusions and Future Work

We have demonstrated how to simulate η-expansion with β-reduction in the second-order polymorphic λ-calculus $\lambda 2$. This yields a proof of the equivalence between $\lambda 2 \models \mathcal{SN}(\underset{\beta\beta^2}{\longrightarrow})$ and $\lambda 2 \models \mathcal{SN}(\underset{\beta\beta^2\eta_*\eta_*^2}{\longrightarrow})$, which can be formulated in the first-order Peano arithmetic. A clean modular proof of $\lambda 2 \models \mathcal{SN}(\underset{\beta\beta^2\eta_*\eta_*^2}{\longrightarrow})$

follows immediately. We intend to investigate the effects of augmenting first-order algebraic rewriting systems with $\xrightarrow[\beta\beta^2\eta_*\eta_*^2]{}$. Improvements on [5] seems to be immediate. Also we shall study the use of our technique in more complicated λ-calculi such as the high-order polymorphic λ-calculus $\lambda\omega$ and the construction of calculus λC. Some current work on combining algebraic term rewriting systems with $\lambda\beta\beta^2\eta_*\eta_*^2$ can be found through the pointer below.

http://www.cs.cmu.edu/ hwxi/papers/TRS.ps

6 Acknowledgement

I thank Frank Pfenning, Peter Andrews and Richard Statman for their support and for providing me a nice work environment. I also thank Roberto Di Cosmo for his comments on a draft of the paper.

References

1. Y. Akama (1993), On Mints' reduction for ccc-calculus. In *Typed lambda-calculi and applications*, vol. 664 of LNCS, pp 1-12.
2. H.P. Barendregt (1984), The Lambda Calculus: Its Syntax and Semantics, *North-Holland publishing company, Amsterdam.*
3. H.P. Barendregt (1992), Lambda calculi with types, *Handbook of Logic in Computer Science edited by S. Abramsky, Dov M. Gabbay and T.S.E. Maibaum*, Clarendon Press, Oxford, pp. 117-414.
4. Val Breazu-Tannen and Jean Gallier (1991), Polymorphic rewriting conserves strong normalization, *Theoretic Computer Sicence*, vol. 83, pp 3-28.
5. Val Breazu-Tannen and Jean Gallier (1994), Polymorphic rewriting preserves algebraic confluence, *Information and Computation*, vol. 114(1), pp. 1-29.
6. Djordje Cubric (1992), On free ccc, *Manuscripts.*
7. R. Di Cosmo and D. Kesner (1994), Combining the first order algebraic rewriting systems, recursion and extensional lambda calculi. In *Serge Abiteboul and Eli Shamir, editors, International Conference on Automata, Languages and Programming*, vol. 820 of LNCS, pp. 462-472.
8. R. Di Cosmo and D. Kesner (1994), Simulating expansions without expansions. *Mathematical Structures in Computer Science*, vol. 4, pp. 1-48.
9. R. Di Cosmo and D. Kesner (1995), Rewriting with polymorphic extensional lambda-calculus. In *Proceedings of Computer Science Logic '95*, vol. 1092 of Lecture Notes in Computer Science, pages 215-232.
10. R. Di Cosmo and A. Piperno (1995), Expanding Extensional Polymorphism, In *Proceedings of Typed Lambda-Calculi and Applications*, vol. 902 of LNCS, pp. 139-153.
11. Daniel J. Dougherty (1993), Some lambda calculi with categorical sums and products. In *Proceedings of the 5th International Conference on Rewriting Techniques and Applications.*
12. N. Ghani (1995), $\beta\eta$-equality for coproducts. In *Typed lambda-calculi and applications*, vol. 902 of LNCS, pp. 171-185.
13. N. Ghani (1996), Eta Expansions in System F, *Manuscripts.*

14. J.-Y. Girard (1972), Interprétation fonctionnelle et élimination des coupures de l'arithmétique d'ordre supérieur, *Thèse de doctorat d'etat, Université Paris VII*.
15. C.B. Jay and N. Ghani (1996), The virtues of eta-expansion, *Journal of Functional Programming*, vol. 5(2), pp. 135-154.
16. J. Lambek and P.J. Scott (1986), *An introduction to higher order categorical logic*, Cambridge University Press.
17. G.E. Mints (1979), Theory of categories and theory of proofs (I). In *Urgent Question of Logic and the Methodology of Science* [In Russian], Kiev.
18. D. Prawitz (1971), Ideas and results of proof theory, Proceedings of the 2nd Scandinavia logic symposium, *editor J.E. Fenstad, North-Holland Publishing Company, Amsterdam*.
19. J. Reynolds (1974), Towards a theory of type structure, *Colloquium sur la Progrmmation*, vol. 19 of LNCS, pp. 408-423.
20. H. Schwichtenberg (1991), An upper bound for reduction sequences in the typed lambda-calculus, *Archive for Mathematical Logic*, 30:405-408.
21. W. Tait (1967), Intensional Interpretations of functionals of finite type I, *J. symbolic logic 32*, pp. 198-212.
22. H. Xi (1996), Upper bounds for standardizations and an application, *Research Report 96-189, Department of Mathematical Sciences, Carnegie Mellon University, Pittsburgh*.
23. H. Xi (1996), Simulating eta-expansions with beta-reductions in the second-order polymorphic lambda-Calculus, *Research Report, Department of Mathematical Sciences, Carnegie Mellon University, Pittsburgh*. Available through pointer: http://www.cs.cmu.edu/~hwxi/papers/EtaSim.ps

Logical Schemes for First Order Theories

Rostislav E. Yavorsky *

Moscow State University, Moscow, 119899, RUSSIA

Abstract. The logic $\mathcal{L}(T)$ of an arbitrary first order theory T is the set of predicate formulas provable in T under every interpretation into the language of T. We prove that if T is an arithmetically correct theory in the language of arithmetic, or T is the theory of fields, the theory of rings, or an inessential extension of the theory of groups, then $\mathcal{L}(T)$ coincides with the predicate calculus PC. We also study inclusion relations and decidability for the logics of Presburger's arithmetic of addition, Skolem's arithmetic of multiplication and other decidable theories.

1 Introduction

According to the Gödel completeness theorem for the predicate calculus PC, the set of formulas true in all models under every interpretation coincides with PC. On the other hand, if we restrict the class of considered models and interpretations, the corresponding class of logical laws will increase. For example, there are well known formulas true in all finite models but falsified in an infinite one. So the logic of the class of all finite models (we denote it by FIN) is a proper extension of PC. Following this idea, we define the logic $\mathcal{L}(T)$ of a first order theory T as the set of all predicate formulas provable in T under every interpretation into the language of T. Thus, the logic $\mathcal{L}(T)$ is an extension of PC which describes all universal logical schemes for a theory T.

Properties of $\mathcal{L}(T)$ for constructive arithmetical theories were considered in [3]. In this paper we study the logic $\mathcal{L}(T)$ for distinct classical first order theories. It was shown by V. A. Vardanyan [7] that for every degree of unsolvability d there is an extension T of the Peano arithmetic PA such that degree of unsolvability for T and $\mathcal{L}(T)$ equals d. In Section 3.1 we prove that for all arithmetically correct theories in the full arithmetical language, as well as the theories of rings and of fields, and some extensions of theory of groups, $\mathcal{L}(T)$ coincides with PC. On the other hand, for wide class of classical first order theories, $\mathcal{L}(T)$ turns out to be proper extension of PC.

First, for every decidable theory T, the logic $\mathcal{L}(T)$ is co-r.e., so $\mathcal{L}(T) \neq PC$. In Section 3.2 we prove that for the decidable arithmetical subtheories: Skolem's arithmetic of multiplication Sko, Presburger's arithmetic of addition Pre, and the theory of discrete linear order with minimal element DO, the logics are distinct and $\mathcal{L}(Sko) \subset \mathcal{L}(Pre) \subset \mathcal{L}(DO)$.

* The research described in this publication was made possible in part by the Russian Foundation for Basic Research (project 96-01-01395).

Second, $\mathcal{L}(T) \neq PC$ for expressively weak theories. As noted above, the poorer the class of models of T and the weaker the expressive power of its language, the greater the logic $\mathcal{L}(T)$. It is clear that if a theory T is complete under some class of finite models, then $FIN \subseteq \mathcal{L}(T)$. It is shown in Section 3.3 that for axiomatic theories of finite fields, rings, groups and other algebraic systems, $\mathcal{L}(T) = FIN$. For the theory of equality Eq, the logic $\mathcal{L}(Eq)$ is proper extension of FIN. Moreover, in Section 3.4 we prove that $\mathcal{L}(Eq)$ is decidable, but can not be axiomatized by any set of schemes with restricted arity. The same holds for the theory of dense linear order without minimal and maximal elements.

Finally, we show that there are infinitely many distinct decidable logics and that the logic ClP of singletons turns out to be the maximal element in the class of all predicate logics, i.e. for arbitrary first order theory T, one has $\mathcal{L}(T) \subseteq ClP$. We also give an example of undecidable theory with a decidable logic.

2 Definition and basic properties

Let PC denotes the Predicate Calculus in a language containing infinitely many n-ary predicate symbols, for every natural n, but containing no function symbols, no constants and no symbol for equality. Formulas in this language are called *predicate formulas*. Let T be an arbitrary first order theory. A predicate formula A is called to be T-*valid* [3] if for every interpretation f into the language of T, the formula $f(A)$ turns out to be provable in T. The set of all T-valid formulae is called the *logic of* T. We will denote it $\mathcal{L}(T)$.

It follows from the definition that for every theory T, the logic $\mathcal{L}(T)$ extends the predicate calculus PC. Note that $\mathcal{L}(T)$ satisfies the monotonicity condition:

Lemma 1. *If T_1 and T_2 are first order theories in the same language, and $T_1 \subset T_2$, then $\mathcal{L}(T_1) \subseteq \mathcal{L}(T_2)$.*

Proof. It follows immediately from the definition of the $\mathcal{L}(T)$.   □

Let FIN denotes the set of all predicate formulas true in all finite models. The following lemma provides us with a necessary and sufficient condition for the inclusion $\mathcal{L}(T) \subseteq FIN$. Since FIN is essentially undecidable, the undecidability of $\mathcal{L}(T)$ then follows.

Let LA_T^1 denotes the Lindenbaum Algebra with one variable of a theory T. We say, that LA_T^1 is model-infinite if for each natural number n, there is a model $\mathcal{M}$ of the theory T in which n distinct sets of elements can be defined, i.e. there are formulas with one free variable $A_1(x)$, $A_2(x)$, ..., $A_n(x)$ in the language of T such that sets $M_i = \{a \in \mathcal{M} \mid \mathcal{M} \models A_i(a)\}$ are all distinct.

Lemma 2. *Let T be a first order theory. Then*

$$\mathcal{L}(T) \subseteq FIN \quad \text{iff} \quad LA_T^1 \text{ is model-infinite.}$$

Proof. ($\Leftarrow$) Let φ be a predicate formula such that $\varphi \notin FIN$ and let $\mathcal{K}$ be a finite model in which φ is falsified. We have to form an interpretation f such that $T \nvdash f(\varphi)$. Let $\mathcal{K} = \{1, 2, \ldots, k\}$ for some natural number k. It follows from the model-infiniteness of T that there is a model $\mathcal{M}$ and formulas $A_1(x)$, $A_2(x)$, $\ldots$, $A_k(x)$ in the language of T such that $M_i \neq \emptyset$ for every i, $M_i \cap M_j = \emptyset$ for $i \neq j$ and $\bigcup_{i=1}^{N} M_i = \mathcal{M}$.

For every predicate letter $P(x_1, \ldots, x_n)$ we define

$$f(P(x_1, \ldots, x_n)) \rightleftharpoons \bigvee_{\{(i_1, \ldots, i_k) | \mathcal{K} \models P(i_1, \ldots, i_n)\}} A_{i_1}(x_1) \& \cdots \& A_{i_n}(x_n).$$

Roughly speaking, we simulate $\mathcal{K}$ in $\mathcal{M}$. Straightforward induction on the complexity of formulas shows that for every closed predicate formula ψ one has

$$\mathcal{K} \models \psi \Leftrightarrow \mathcal{M} \models f(\psi).$$

So $\mathcal{M} \nvDash f(\varphi)$ and $T \nvdash f(\varphi)$.

($\Rightarrow$) Suppose now that LA_T^1 is not model-infinite and in every model of T one can define not more than s different sets, for some natural s. Let $P_1(x)$, $P_2(x), \ldots, P_{s+1}(x)$ be $(s+1)$ different one-variable predicate symbols. Let also

$$\varphi_s \rightleftharpoons \bigvee_{1 \leq i < j \leq s+1} \forall x (P_i(x) \leftrightarrow P_j(x)).$$

It is clear that for any interpretation f, the formula $f(\varphi_s)$ is true in all models of T, hence provable in T, so $\varphi_s \in \mathcal{L}(T)$, but $\varphi_s \notin FIN$. $\square$

Our next goal is to describe other sufficient condition for the inclusion $\mathcal{L}(T_2) \subseteq \mathcal{L}(T_1)$. Without loss of generality, we assume that T_1 is formulated in a language with no function symbols or constants.

Definition 3. A model M_1 of the theory T_1 is said to be relatively definable in a model M_2 of the theory T_2 if there is a formula $\delta(x)$ in the language of T_2, which describes in M_2 some non-empty set M_2', and for every predicate symbol P in the language of T_1, there is a formula P^I in the language of T_2 with the same free variables, such that M_2' considered as a model of T_1 turns out to be isomorphic to M_1.

Definition 4. The class of models of a theory T_1 is said to be relatively definable in the class of models of a theory T_2, if for every model M_1 of T_1 there exists a model M_2 of T_2 such that M_1 is relatively definable in M_2.

Lemma 5. *If at least one constant is definable in the theory T_1 and the class of all models of T_1 is relatively definable in the class of all models of T_2, then $\mathcal{L}(T_2) \subseteq \mathcal{L}(T_1)$.*

Proof. Let $A \notin \mathcal{L}(T_1)$. This means that there is an interpretation f of the predicate language into the language of T_1, such that $T_1 \nvdash f(A)$. So $f(A)$ is falsified in a model M_1 of the theory T_1. According to the premise of the theorem, there is a model M_2 of the theory T_2 such that M_1 is relatively definable in M_2.

We will construct a translation g of T_1-formulas into the language of T_2 such that for any formula B the following property will hold:

$$M_1 \models B \Leftrightarrow M_2 \models g(B). \tag{$*$}$$

Let a constant c be defined in the theory T_1. Let $C(x)$ be the corresponding T_1-formula, and $C^I(x)$ its I-image, i.e. the result of substituting every predicate symbol P by the corresponding formula P^I in the language of T_2. We consider an equivalence relation on M_2 given by the formula:

$$x \approx y \rightleftharpoons x = y \vee \neg\delta(x) \& C^I(y) \vee \neg\delta(y) \& C^I(x) \vee \neg\delta(x) \& \neg\delta(y).$$

This equivalence relation coincides with equality on the elements of M_2', while all other elements of M_2 are equivalent to the I-image of the constant c.

We define the translation g in the following way:

$$g(P(x_1, \ldots, x_k)) \rightleftharpoons \exists z_1, \ldots, z_k (\delta(z_1) \& \ldots \& \delta(z_k) \& \\ x_1 \approx z_1 \& \ldots \& x_k \approx z_k \& P^I(x_1, \ldots, x_k)).$$

An easy induction on the complexity of formulas shows that for every formula B in the language of T_1, we have

$$M_2 \models g(B) \Leftrightarrow M_2' \models B.$$

The property $(*)$ follows from the isomorphism between M_1 and M_2'.

To complete the proof, note that the the composition $g \circ f$ is an interpretation of the predicate language into the language of T_2, such that $M_2 \nvDash g(f(A))$, so $T_2 \nvdash g(f(A))$. Thus, $A \notin \mathcal{L}(T_2)$. $\qquad\square$

3 $\mathcal{L}(T)$ for different theories

We now study properties of $\mathcal{L}(T)$ for different classical first order theories.

3.1 Expressively strong theories

Here we show, that for expressively strong theories $\mathcal{L}(T)$ coincides with PC.

Theorem 6. *If the language of a theory T includes the arithmetical one and all arithmetical consequences of T are true in the standard model, then $\mathcal{L}(T) = PC$.*

Proof. The inclusion $PC \subseteq \mathcal{L}(T)$, as noted above, holds for every first order theory. The converse one follows immediately from the Gilbert-Bernays theorem [2], which states that for every formula A not provable in PC there exists an arithmetical interpretation f, such that $f(A)$ is falsified in the standard model of arithmetic. So if $A \notin PC$, then $T \nvdash f(A)$ and $A \notin \mathcal{L}(T)$. Hence, $\mathcal{L}(T) \subseteq PC$. $\quad\square$

It follows from Theorem 6 that for Peano Arithmetic PA and for True Arithmetic TA, the logic $\mathcal{L}(T)$ coincides with PC. On the other hand, for Presburger's arithmetic of addition and Skolem's arithmetic of multiplication, it is not true: their languages are weaker then the full arithmetical language.

Corollary 7. *If the standard model of arithmetic is relatively defined in some model of a theory T then $\mathcal{L}(T) = PC$.*

Proof. It follows from the Theorem 6 and Lemma 1. □

Using the well known results about the relative interpretability of TA into the field of rationals [5], the ring of integer numbers and some extensions of the theory of groups [6], we obtain the following:

Corollary 8. *The logic of the theory of fields coincides with PC.*

Corollary 9. *The logic of the theory of rings coincides with PC.*

Corollary 10. *Let G' denote the theory of groups in the language, enriched with one individual constant, then $\mathcal{L}(G') = PC$.*

3.2 Decidable arithmetical subtheories

Let Pre stand for Presburger's arithmetic of addition, let Sko be Skolem's arithmetic of multiplication and let DO be the theory of discrete order (the order type of the natural numbers). The undecidability of $\mathcal{L}(Sko)$, $\mathcal{L}(Pre)$ and $\mathcal{L}(DO)$ follows from Lemma 2, because FIN is essentially undecidable, so every subtheory of FIN is also undecidable.

Theorem 11. $PC \subset \mathcal{L}(Sko) \subset \mathcal{L}(Pre) \subset \mathcal{L}(DO) \subset FIN$

Proof. 1. $PC \subset \mathcal{L}(Sko)$. This inclusion holds for every first order theory. Here, the inclusion is strict because $\mathcal{L}(Sko)$ is co-r.e.(since Sko is decidable) while PC is not.

2. $\mathcal{L}(Sko) \subset \mathcal{L}(Pre)$ follows from Lemma 5 because Pre is relatively interpretable in Sko. The idea is to consider the set $\{2^k \mid k \in N\}$ and then define $(x = n)^* \rightleftharpoons (x = 2^n)$ and $(x + y)^* \rightleftharpoons (x \cdot y)$. To prove that these logics do not coincide, consider the following predicate formula φ:

$$\forall u R(u, u) \& \forall u v w (R(u, v) \& R(v, w) \rightarrow R(u, w))$$
$$\&\ \forall u v \exists w (R(w, u) \& R(w, v) \& \forall s (R(s, u) \& R(s, v) \rightarrow R(s, w)))$$
$$\&\ \forall u v \exists w (R(u, w) \& R(v, w) \& \forall s (R(u, s) \& R(v, s) \rightarrow R(w, s)))$$
$$\&\ \forall u \exists v (R(u, v) \& \neg R(v, u))$$
$$\&\ \forall u \exists v (\neg R(u, v) \& \neg R(v, u)).$$

We claim that $\neg\varphi \in \mathcal{L}(Pre)$ but $\neg\varphi \notin \mathcal{L}(Sko)$.

Suppose that under some interpretation f, the formula $f(\varphi)$ is true. It is known that every subset of natural numbers defined in Pre is a union of finite number of arithmetical progressions [1]. Exploring this fact, one can show that the relation $f(R(x, y))$ with the properties described in φ could not be defined in *Pre*, so $\neg\varphi \in \mathcal{L}(Pre)$. To see that $\neg\varphi \notin \mathcal{L}(Sko)$, let g be an interpretation into the language of *Sko* such that

$$g(\varphi) \rightleftharpoons \exists z(x \cdot z = y \,\&\, x \neq 1).$$

It is clear that $Sko \vdash g(\varphi)$, hence $\neg\varphi \notin \mathcal{L}(Sko)$.

3. $\mathcal{L}(Pre) \subset \mathcal{L}(DO)$. This inclusion follows from Lemma 5. Indeed, the relation $x \leq y$ is definable in *Pre* by the formula $\exists z(x + z = y)$. To prove that these logics do not coincide, observe that every set definable in *DO* is either finite or co-finite [4], while for *Pre* this property does not hold. So for the following formula φ, we have $\varphi \in \mathcal{L}(DO)$ and $\varphi \notin \mathcal{L}(Pre)$:

$$\forall u R(u, u) \,\&\, \forall uvw[R(u, v) \& R(v, w) \to R(u, w)]$$
$$\&\, \forall uv[R(u, v) \vee R(v, u)] \,\&\, \forall u \exists v[R(u, v) \& \neg R(v, u)]$$
$$\to \exists u[\forall v(R(u, v) \to P(v)) \vee \forall v(R(u, v) \to \neg P(v))]$$

If the antecedent of φ is true under some interpretation f, then $f(R(x, y))$ represents a discrete order without maximal elements. In this case, the succedent of φ will always be true in *DO*. On the other hand, if $f(R(u, v))$ is $(u \leq v)$ and $f(P(u))$ is $\exists v(v + v = u)$ then $Pre \not\vdash f(\varphi)$.

4. $\mathcal{L}(DO) \subset FIN$. This inclusion follows immediately from Lemma 2. As an example of formula in *FIN* but not in $\mathcal{L}(DO)$ we can take the negation of the antecedent of φ from 3. above. It is clear that an infinite discrete order cannot be defined in any finite model. □

3.3 Theories of finite models

Theorem 12. *Let T denote one of the following theories:*

a. *the logic of unary predicates PC1;*
b. *the axiomatic theory of finite cyclic groups;*
c. *the axiomatic theory of finite fields.*

Then $\mathcal{L}(T) = FIN$.

Proof. All of these theories are complete under the class of their finite models, so $FIN \subseteq \mathcal{L}(T)$. The converse inclusion follows immediately from Lemma 2. The only thing we need is to ensure that in each case, LA_T^1 is model-infinite.

a. Given a natural number n, consider a model $\mathcal{M} = \{1, 2, \ldots, n\}$. We define $\mathcal{M} \models P_i(a) \rightleftharpoons a = i$, for $i \leq n$. It is clear that formulas $P_1(x), \ldots, P_n(x)$ define n different sets in $\mathcal{M}$.

b. Let $\mathcal{M}$ be a cyclic group of degree $d = p_1 \cdot \ldots \cdot p_n$, where $p_1, \ldots, p_n$ are fixed distinct prime numbers. Let $A_i(x) \rightleftharpoons (x^{p_i} = e)$ for $i \leq n$. The sets $M_1, \ldots, M_n$ defined in $\mathcal{M}$ by these formulas are surely different.

c. Taking $\mathcal{M}$ to be a large enough finite field, we can define $A_i(x) \rightleftharpoons (x = 1 + \ldots + 1)$ i times. □

Similar reasoning can be applied to other classes of finite algebraic systems. Applying Lemma 2 to this theorem we obtain

Corollary 13. *For the axiomatic theories of finite groups, finite abelian groups and finite rings the logic $\mathcal{L}(T)$ coincides with FIN.*

3.4 Theories with decidable logics

Let Eq denotes the theory of equality.

Theorem 14. *The logic $\mathcal{L}(Eq)$ of the theory of equality is decidable.*

Proof. It is known that Eq admits quantifier elimination and every formula in the language of Eq is provably equivalent to a boolean combination of atoms and formulas σ_n, where

$$\sigma_n \rightleftharpoons \forall x_1 \ldots x_n \exists y (x_1 \neq y \,\&\, \cdots \,\&\, x_n \neq y).$$

An interpretation f_a of the predicate language into the language of Eq will be called atomic if for every predicate symbol R, the formula $f_a(R)$ is quantifier free, i.e. $f_a(R)$ is a Boolean combination of atoms. We will show that for every formula φ not in $\mathcal{L}(Eq)$, there is an atomic interpretation f_a such that $Eq \not\vdash f_a(\varphi)$.

Let $\varphi \notin \mathcal{L}(Eq)$. Then by the definition of $\mathcal{L}(Eq)$, there is an interpretation f such that $Eq \not\vdash f(\varphi)$, so $f(\varphi)$ is falsified in some model $\mathcal{M}$ of the Eq. On the other hand, for every natural n, the formula σ_n contains no free variables, so it is equivalent in $\mathcal{M}$ to the false or the true. We define an atomic interpretation f_a as follows: for every predicate symbol R we eliminate quantifiers in $f(R)$ and then replace every formula σ_n with $x = x$ or $x \neq x$, depending on its truth value in $\mathcal{M}$. It is easy to see that for every predicate formula ψ

$$\mathcal{M} \models f(\psi) \Leftrightarrow \mathcal{M} \models f_a(\psi).$$

Therefore $Eq \not\vdash f_a(\varphi)$.

Thus a predicate formula φ belongs to the $\mathcal{L}(Eq)$ iff $Eq \vdash f_a(\varphi)$ for every atomic interpretation f_a. The decidability of $\mathcal{L}(Eq)$ now follows from the decidability of Eq and the finiteness of the set of atomic interpretations for every predicate formula. $\qquad\Box$

The following result makes impossible any finite axiomatization for $\mathcal{L}(Eq)$ [8].

Proposition 15. *The logic $\mathcal{L}(Eq)$ cannot be axiomatized by any set of schemes with restricted arity.*

The same is true for the logic of the theory of dense linear order without minimal and maximal elements [8].

Proposition 16. *The logic of the theory of dense linear order without minimal and maximal elements is decidable but can not be axiomatized by any set of schemes with restricted arity.*

4 Concluding remarks

In conclusion, we note some general facts about the logics of theories.

First, there are infinitely many decidable logics of theories. An inspection of the proof of Theorems 15 and 14 shows that the set of decidable extensions of Eq by different combinations of σ_n formulas gives us an infinite sequence of decidable logics, all extensions of $\mathcal{L}(Eq)$.

Second, the decidability of T is not necessary for the decidability of the logic $\mathcal{L}(T)$. Indeed, let X be an arbitrary undecidable set of natural numbers. Consider the theory T_X in a language containing infinitely many unary predicate symbols $P_i(x)$ with the following axioms:

A1. $\forall u P_k(u)$, $k \in X$;
A2. $\forall u P_i(u) \vee \forall u \neg P_i(u)$, $i \in N$.

The undecidability of T_X follows from the undecidability of X and the following equivalence:

$$T_X \vdash \forall u P_k(u) \iff k \in X.$$

On the other hand one can show that, independent of X, the logic $\mathcal{L}(T_X)$ coincides with ClP — the set of predicate formulas true in all one-element models under every interpretation. ClP can be easily reduced to propositional calculus, so it is decidable. In addition, it has a good axiomatization [8].

Proposition 17. *ClP is axiomatized over the predicate calculus PC by the following scheme:*

$$\forall x y R(x, y) \vee \forall x y \neg R(x, y),$$

The logic ClP has one more interesting property — it is maximal in the class of all logics.

Theorem 18. *Let T be an arbitrary first order theory. Then*

$$\mathcal{L}(T) \subseteq ClP.$$

Proof. Let $\varphi \notin ClP$ and let M be a one element model such that $M \not\models \varphi$. It is clear that every predicate symbol is either always true, or always false in M. We define an interpretation f into the language of T as follows:

$$f(R(x_1, \ldots, x_n)) \rightleftharpoons \begin{cases} \top, \text{ if } M \models \forall x_1 \ldots x_n R(x_1, \ldots, x_n) \\ \bot, \text{ if } M \models \forall x_1 \ldots x_n \neg R(x_1, \ldots, x_n). \end{cases}$$

It is clear that for every formula ψ, $M \models \psi \Leftrightarrow T \vdash f(\psi)$, so $T \vdash f(\neg\varphi)$ and $\varphi \notin \mathcal{L}(T)$. $\square$

Acknowledgements

I would like to thank Professor S. N. Artemov, who has advised me on this work and given me many helpful comments.

References

1. G.S.Boolos, R.S.Jeffrey. Computability and Logic. Cambridge University Press, 1989.
2. D.Hilbert, P.Bernays, "Grundlagen der Mathematik", I, Springer–Verlag, 1968.
3. V.E.Plisko. Konstruktivnaja formalizatsiya teoremy Tennenbauma i ee primenenie. Mat. zametki, 48 (1994), 4, pp. 108–118 (in Russian).
4. M.O.Rabin. Decidable Theories, in Handbook of Mathematical Logic. J. Barwise (ed.). North-Holland Publishing Company, 1977.
5. J. Robinson. Definability and decision problem in arithmetic. Journal of Symbolic Logic, 14 (1949), no. 2, pp. 98–114.
6. A. Tarski. Undecidable theories. By A. Tarski. In collab. with A. Mostovski and R.M. Robinson. Amsterdam, North-Holland publ., 1953.
7. V.A. Vardanyan. Predikatnaya logika dokazuemosti bez dokazuemosti. (in Russian)
8. R.E.Yavorsky. Razreshimye logiki pervogo poryadka. Fundamentalnaja i Prikladnaja Matemetika, 1997 (in Russian, to appear).
9. R.E.Yavorsky. Predikatnye logiki razreshymykh fragmentov arifmetiki. Vestnik Moskov. Univ. Ser. 1. Mat. Mekh. 1997 (in Russian, to appear).

Verification of PLTL Formulae
by Means of Monotone Disjunctive
Normal Forms

Vladimir Zakharov

Department of Computational Mathematics and Cybernetics,
Moscow State University, Moscow, Russia

Abstract. This paper offers a satisfiability checking algorithm for the
future fragment of Propositional Linear Temporal Logic. The algorithm
combines the automata theoretic approach to the verification of temporal
formulae and the symbolic computation technique in terms of monotone
boolean disjunctive normal forms. The algorithm is given as follows. A
temporal formula φ to be verified is related with a finite automaton $\mathcal{A}_\varphi$
so that φ has a temporal model iff $\mathcal{A}_\varphi$ admits at least one successful
run. The behavior of $\mathcal{A}_\varphi$ is specified by means of boolean functions and
the depth-first search of a successful run of $\mathcal{A}_\varphi$ is described in terms of
transformation of monotone disjunctive normal forms.

1 Introduction

Since the seminal paper of Pnueli [1] Temporal Logics are extensively exploited in
computer science as a useful tool for the specification and verification of reactive
programming systems. In most cases a behavior and desired properties of reactive
system may be expressed in terms of temporal formulae while the processes of
a system and the set of its computations may be considered as a finite-state
transition diagram or a relational structure. This approach allows to reduce
the problem of program verification to the model-checking and satisfiability-
checking problems for the appropriate Temporal Logic. It was proved (see [2]
for survey) that formula verification problems are decidable for many Temporal
Logics, such as Propositional Linear Time Logics (PLTL) and Computation Tree
Logics (CTL). But in practice the temporal specifications of real programs may
have a very large size. Therefore the practical application of Temporal Logics to
the verification of reactive systems needs in efficient checking procedures.

In this paper the satisfiability problem for the future fragment of PLTL is
considered. This problem is decidable since PLTL has a finite axiomatization and
a finite model property. The latter maintains a tableau technique for satisfiability
checking in PLTL (see [3] for survey). A further development of tableau-based
decision procedures was initialized in [4]. A finite tableau for PLTL formula φ can
be viewed as a transition system of a finite automaton $\mathcal{A}_\varphi$ on infinite words. The
satisfiability of PLTL formula is thus reduced to the emptiness problem of the
corresponding finite automaton. The automata-theoretic approach has proven
to be an efficient technique for linear-time satisfiability checking. It was applied

to the future fragment of PLTL in [4] and extended to the full PLTL in [5]. The complete description of the efficient model-checking and satisfiability-checking procedures was offered in [6]. It is important, that the automata-theoretic framework reveals the combinatoric nature of the satisfiability problem. In fact, a transition system of a finite automaton $\mathcal{A}_\varphi$ may be viewed as a finite directed graph. Then the satisfiability of φ means that $\mathcal{A}_\varphi$ contains a strongly connected component which justify all eventualities of φ. To reveal the strongly connected components one should either compute a transitive closure of the graph [7], or apply a depth-first search algorithm [8]. However, both algorithms require an explicit representation (enumeration) of all reachable states of the transition system. Since, in general, $\mathcal{A}_\varphi$ may have a size $O(exp(|\varphi|))$, the enumerative automata-theoretic approach suffers from the state explosion difficulties.

To overcome this obstacle the symbolic computation technique was developed in [9]. The main idea of this method is looking as follows. An adjacency matrix of a transition system may be viewed as a tableau of some boolean function f. Since in most practical cases the transition system of an automaton $\mathcal{A}_\varphi$ has a regular structure, the characteristic function f may be expressed by means of symbolic constructions (boolean formulae, polynomials, logical networks, decision diagrams, etc.) of a relatively small size. Making symbolic manipulations with these expressions one may derive the desired properties of the transition system in real time with a plausible storage consumption. In [9, 10] this approach was applied to the verification problems for Branching-Time Propositional μ-calculus and CTL. Later in [11] the symbolic model-checking and satisfiability-checking algorithms for full PLTL were presented.

This paper offers a satisfiability checking algorithm for the future fragment of PLTL. The algorithm combines the automata theoretic approach to the verification of temporal formulae [5, 6] and the symbolic computation technique in terms of monotone disjunctive normal forms. The main differences of the present paper over [11] are:

1. The approach presented here is based on a depth-first search algorithm while the decision procedures in [11] relies on the transitive closure algorithm. In some cases a depth-first search of a strongly connected component is more efficient then a computation of the transitive closure of boolean matrix.

2. Decision procedure presented below deals mostly with monotone boolean functions. The monotone boolean functions have some nice properties (the unique minimal disjunctive normal form, a simple logical network realization) which essentially simplify their symbolic transformation.

3. We consider only the future fragment of PLTL, while [11] covers the full PLTL. But this restriction gives us some advantages. In the process of depth-first search of a model for a given PLTL formula the algorithm does not compute the proper valuations of the propositional variables at each state of a model but merely determines if an appropriate valuation exists or not.

2 Temporal Logic PLTL

We deal with the future fragment of Propositional Linear Temporal Logic (PLTL) language with sintax and semantics as defined in [2]. To simplify the presentation only the following operators are considered:

- Boolean connectives: $\neg$, $\wedge$, $\vee$.
- Temporal operators: $\bigcirc$ – Next Time, U – Until, $\tilde{U}$ – Dual Until.

Let AP be a set of *atomic propositions*. The set of formulae of PLTL is generated by the following rules:

- each atomic proposition p is a PLTL formula;
- if p and q are PLTL formulae then $\neg p$, $p \wedge q$, $p \vee q$, $\bigcirc p$, pUq and $p\tilde{U}q$ are PLTL formulas.

The other connectives and temporal operators may be introduced as abbreviations in the usual way. If PLTL formula p is one of the form qUr, then it is called an *eventuality* formula. We say that PLTL formula p is in a *positive normal form* iff negations are applied only to the atomic propositions.

The semantics of PLTL is provided by a *linear temporal structure*, $M = \langle S, L \rangle$, where $S = \{s_0, s_1, \ldots\}$ is an infinite sequence of *states* and $L : S \to 2^{AP}$ is a *valuation* which maps each state to a set of atomic propositions true in this state. We write $M, s_i \models p$ to indicate that a formula p holds at a state s_i of a structure M. When M is understood the indication of structure is omitted. For a given structure M the relation $\models$ is inductively defined as follows:

- For an atomic proposition $p \in AP$, $s_i \models p$ iff $p \in L(s_i)$.
- $s_i \models \neg p$ iff it is not the case $s_i \models p$;
- $s_i \models p \vee q$ iff $s_i \models p$ or $s_i \models q$;
- $s_i \models p \wedge q$ iff $s_i \models p$ and $s_i \models q$;
- $s_i \models \bigcirc p$ iff $s_{i+1} \models p$;
- $s_i \models pUq$ iff there exists $n \geq i$, such that $s_n \models q$, and for all $i \leq m < n$, $s_m \models p$;
- $s_i \models p\tilde{U}q$ iff for every $n \geq i$, such that $s_n \models \neg q$, there exists $i \leq m < n$, such that $s_m \models p$.

We say that a structure M *satisfies* (is a *model* of) a PLTL formula p iff $M, s_0 \models p$. A formula p is called a *satisfiable* iff it has a model. *PLTL satisfiability problem* is to determine if a given PLTL formula has a model or not.

It is easy to see that every PLTL formula p can be translated using the duality of boolean connectives and temporal operators into a positive normal form q which has the same set of models as p. Therefore in what follows only the PLTL positive normal forms are considered.

3 PLTL Satisfiability via Boolean Functions

The automata-theoretic approach to the analysis of temporal logic formulae is based on the assumption that each PLTL formula φ specifies the behavior of some finite automaton $\mathcal{A}_\varphi$ on infinite words. Hence, to prove that a given formula φ is satisfiable it is sufficient to check that an automaton $\mathcal{A}_\varphi$ is nonempty, i.e. it admits some successful run. Relating PLTL formulae with the finite automata one can reduce the satisfiability problem in PLTL to known automata theory nonemptiness problem. But since the number of states of an automaton $\mathcal{A}_\varphi$ may be exponential on the length of φ, the straightforward realization of such automaton by a transition system leads to the state-explosion difficulties. That is why it is reasonable to use an intermediate language to specify the behavior of finite automata. This language should have less complicated semantics then PLTL, but at the same time it should be rather suitable for the realization and analysis of the automata specifications. In this paper we take for this purpose the language of monotone boolean functions. On the one hand, the boolean functions are closely related with the PLTL formulae and the finite automata. With a given PLTL formula φ we associate a finite set of boolean functions $S(\varphi)$ so that both φ and $S(\varphi)$ specifies the behavior of the same automaton. Thus, PLTL satisfiability problem can be reduced to the analysis of some special properties of $S(\varphi)$. On the other hand, there are many efficient methods of boolean function realization (disjunctive normal forms, Reed-Muller polynomials, oriented binary decision diagrams, contact circuits, etc.) to avoid the negative effects of the state explosion in practice.

Definition 1. The *closure of a formula* p, $Cl(p)$, is the set of PLTL formulae inductively defined as follows:

- If $p \in AP$ then $Cl(p) = \{p\}$.
- If p is one of $\neg q$ or $\bigcirc q$, then $Cl(p) = Cl(q) \cup \{p\}$.
- If p is one of $q \wedge r$, $q \vee r$, qUr or $q\tilde{U}r$, then $Cl(p) = Cl(q) \cup Cl(q) \cup \{p\}$.

The elements of $Cl(p)$ are called the *subformulae* of p. We denote by $TCl(p)$ the subset of $Cl(p)$ which contains all subformulae of the form $\bigcirc q$, qUr, or $q\tilde{U}r$.

Let φ be a PLTL formula such that $TCl(\varphi) = \{q_1, \ldots, q_m\}$, and let $AP(\varphi) = \{p_1, \ldots, p_n\}$ be the set of all atomic propositions in $Cl(\varphi)$.

We introduce two finite sets of boolean variables $X_\varphi = \{x_1, \ldots, x_n\}$ and $Y_\varphi = \{y_1, \ldots, y_m\}$. The elements of X_φ are called the propositional variables. Each propositional variable $x_i \in X_\varphi$ is assumed to be associated with an atomic proposition $p_i \in AP(\varphi)$. The elements of Y_φ are called the temporal variables. We associate each temporal variable $y_j \in Y_\varphi$ with a subformula $q_j \in TCl(\varphi)$. If q_j is an eventuality subformula then the corresponding temporal variable y_j is called an eventuality variable. We denote by E_φ the set of all eventuality variables associated with the eventuality subformulae of φ.

For each subformula $\psi \in Cl(\varphi)$, we define a boolean function f_ψ over the set of boolean variables $X_\varphi \cup Y_\varphi$, using the following rules:

- If $\psi = p_i$ (or $\psi = \neg p_i$), where $p_i \in AP(\varphi)$, then $f_p = x_i$ (or $f_p = \neg x_i$).
- If $\psi = p \vee r$ (or $\psi = p \wedge r$), then $f_\psi = f_p \vee f_r$ (or $f_\psi = f_p \wedge f_r$).
- If $\psi = q_j$, where $q_j = \bigcirc p$ is in $TCl(\varphi)$, then $f_\psi = y_j$.
- If $\psi = q_j$, where $q_j = pUr$ (or $q_j = p\tilde{U}r$) is in $TCl(\varphi)$, then $f_\psi = f_r \vee (f_p \wedge y_j)$ (or $f_\psi = f_r \wedge (f_p \vee y_j)$).

For each temporal variable y associated with a subformula ψ, we denote by f_y a boolean function which is f_p in the case that $\psi = \bigcirc p$, and f_ψ in the case that ψ is one of the form pUq, or $p\tilde{U}q$. Since by the convention above φ is in positive normal form, all functions f_y are monotone on the variables Y_φ.

For every boolean function g over the set of variables $X_\varphi \cup Y_\varphi$, we denote by $\exists \overline{x} g$ a projection of g on the temporal variables Y_φ, defined by the quantified boolean formula $\exists x_1 \exists x_2 \ldots \exists x_n g$. Clearly, a function $\exists \overline{x} f_p$ corresponding to a PLTL formula $p \in Cl(\varphi)$, is also a monotone function of the temporal variables.

Definition 2. We call every subset K of Y_φ a *monotone conjunction*, and denote by $f(K)$ a boolean function such that $f(K)$ is a constant 1 in the case $K = \emptyset$, and $f(K) = \bigwedge_{y \in K} y$ otherwise.

Definition 3. We call every subset D of 2^{Y_φ} a *monotone disjunctive normal form (d.n.f.)*, and denote by $f(D)$ a boolean function such that $f(D)$ is a constant 0 in the case $D = \emptyset$, and $f(D) = \bigvee_{K \in D} f(K)$ otherwise.

In what follows we deal, as a rule, only with the monotone boolean functions and the adjective "monotone" is often omitted.

Definition 4. A conjunction K'' is called a *refinement* of a conjunction K' iff $K' \subseteq K''$.

Definition 5. We say that a d.n.f. D' *covers* a d.n.f. D'' iff each conjunction K'' in D'' is a refinement of some conjunction K' in D'. If D' covers D'' but not vice versa then D' *strictly covers* D''.

Definition 6. A monotone d.n.f. D is called *irredundant* iff for every $K \in D$ a d.n.f. $D - \{K\}$ covers D.

It is well known (see [12]) that each monotone boolean function g over the set of variables Y_φ can be realized by the unique irredundant monotone d.n.f. D such that $f(D) = g$; we denote this irredundant d.n.f. D by $D(g)$.

We associate with each PLTL formula φ whose satisfiability we wish to check a finite set of boolean functions $S(\varphi) = \{f_\varphi\} \cup \{f_y : y \in Y_\varphi\}$. Then PLTL satisfiability problem can be defined in terms of monotone boolean functions and the properties of $S(\varphi)$.

Definition 7. We say that *a conjunction K_2 follows by a conjunction K_1 (w.r.t. the system $S(\varphi)$)*, and denote this relation by $K_1 \prec K_2$, iff either $K_1 = K_2 = \emptyset$, or $K_1 = \{y_{i_1}, \ldots, y_{i_k}\}$ and $D(\exists \overline{x} \bigwedge_{y \in K_1} f_y)$ covers $\{K_2\}$.

Definition 8. A finite sequence $T = K_1, \ldots, K_N$, $N \geq 0$, of conjunctions is called a *pre-structure (for $S(\varphi)$)* iff it meets the following conditions:
(1) $D(\exists \overline{x} f_\varphi)$ covers $\{K_1\}$;
(2) $K_i \prec K_{i+1}$, for every i, $1 \leq i < N$.

If in addition to these requirements T contains a conjunction K_j, $1 \leq j \leq N$, such that

(3) $K_N \prec K_j$, and $E_\varphi \cap \bigcap_{i=j}^{N} K_i = \emptyset$,

then a prestructure T is called a *pseudo-structure (for $S(\varphi)$)*.

By means of this definition we relate the system of boolean functions $S(\varphi)$ with a nondeterministic finite automaton $\mathcal{A}_\varphi = \langle 2^{Y_\varphi}, \prec, D_0, F \rangle$ on the 1-letter infinite words. A set of all conjunctions 2^{Y_φ} appears to be a set of states of $\mathcal{A}_\varphi$; $\prec$ corresponds to a transition relation; a set of initial states D_0 is distinguished by the specification $\exists \overline{x} f_\varphi$. A run of $\mathcal{A}_\varphi$ is defined in the usual way by means of the rules (1), (2) above. An infinite run is successful iff all eventuality assertions are justified infinitely often according to the rule (3). This rule manifests the acceptance condition F of $\mathcal{A}_\varphi$.

Theorem 1. *Let φ be a PLTL formula. Then φ is satisfiable iff $S(\varphi)$ has a pseudo-structure.*

Proof. $\Rightarrow$ Suppose φ is a satisfiable formula. Let $M = \langle S, L \rangle$ be a model of φ. We may assume without loss of generality that M is a periodic structure, i.e. for some $l \geq 0$, $m \geq 0$ an equality $L(s_i) = L(s_{i+m})$ holds for every $i \geq l$. Then for every $i \geq 0$ define a conjunction $K_i \subseteq Y_\varphi$ and a substitution $\sigma_i : X_\varphi \to \{0,1\}$ by taking

• $y \in K_i$ iff y is associated with a temporal subformula $\psi \in TCl(\varphi)$ so that either $\psi = \bigcirc \psi$ and $s_i \models \psi$, or $\psi = pUq$, $s_i \models \psi$ and $s_i \not\models q$, or $\psi = p\tilde{U}q$, $s_i \models \psi$ and $s_i \not\models p$;

• $\sigma_i(x) = 1$ iff x is associated with an atomic proposition p and $p \in L(s_i)$.

By induction on the construction of $\psi \in Cl(\varphi)$ we prove for every state $s_i \in S$ that

$$\text{if } s_i \models \psi \text{ then } D(f_\psi \sigma_i) \text{ covers } K_i; \tag{1}$$

$$\text{if } \psi \in TCl(\varphi) \text{ is associated with } y \in K_i \text{ then } D(f_y \sigma_{i+1}) \text{ covers } K_{i+1}. \tag{2}$$

The basis of induction is obvious, since in the case that $\psi = p$ or $\psi = \neg p$, where p is an atomic proposition associated with boolean variable x, $s_i \models \psi$ implies $D(f_\psi \sigma_i) = \{\emptyset\}$.

Suppose $\psi = pUq$ is a temporal subformula of φ associated with temporal variable y, and $s_i \models \psi$. Then $D(f_\psi) = D(f_y) = D(f_q \vee (y \wedge f_p))$, and $s_{i+n} \models q$ for some n, $n \geq 0$. By induction on n we prove (1), (2) for ψ. If $n = 0$ then $y \notin K_i$ and, by the induction hypothesis, $D(f_q \sigma_i)$ covers K_i. Hence, $D(f_\psi \sigma_i)$ covers K_i also. Suppose that (1), (2) hold for every n, $n \leq j$, and $s_i \not\models q$. Then $y \in K_i$, $s_i \models p$, and $s_{i+1} \models \psi$. The outer induction hypothesis implies $D((y \wedge f_p)\sigma_i)$ covers K_i, and, hence, (1) is satisfied. The inner induction hypothesis implies $D(f_\psi \sigma_{i+1})$ covers K_{i+1}, and, hence, (2) is satisfied also.

The other cases of ψ are considered analogously.

¿From the above it follows that the sequence of conjunctions $K_1, K_2, \ldots$ is a prestructure for $S(\varphi)$ and $K_l = K_{l+m}$. One should notice that for every eventuality variable $y \in E_\varphi$ there is at least one conjunction K_j, $l \leq j \leq l+m$, such that $y \notin K_j$. Therefore, $T = K_1, \ldots, K_l, \ldots, K_{l+m-1}$ is a pseudo-structure for $S(\varphi)$.

$\Leftarrow$ Suppose $S(\varphi)$ has a pseudo-structure $T = K_1, \ldots, K_N$ such that $K_N \prec K_n$ for some n, $1 \leq n \leq N$. Let $\sigma_i : X_\varphi \to \{0, 1\}$, $0 \leq i \leq N$, be a sequence of substitutions such that $D(f_\varphi \sigma_0)$ covers K_1, $D(\bigwedge_{y \in K_i} f_y \sigma_i)$ covers K_{i+1} for every i, $1 \leq i < N$, and $D(\bigwedge_{y \in K_N} f_y \sigma_N)$ covers K_n according to the definition of pseudo-structure. Unfolding the cycles $K_n, \ldots, K_N$ and $\sigma_n, \ldots, \sigma_N$, we get two infinite sequences $T' = K_1', K_2', \ldots$ and $\Sigma = \sigma_1', \sigma_2' \ldots$ of conjunctions and substitutions such that $K_i' = K_i$, $\sigma_i' = \sigma_i$ for i, $1 \leq i \leq N$, and $K_j' = K_{j-(N-n+1)}'$, $\sigma_j' = \sigma_{j-(N-n+1)}$ for j, $j > N$. Define a structure $M = \langle S, L \rangle$ by taking for every atomic proposition $p \in AP(\varphi)$ associated with boolean variable x and for every state $s_i \in S$

$$p \in L(s_i) \text{ iff } x\sigma_i' = 1$$

By the induction on the construction of $\psi \in Cl(\varphi)$ it is easy to prove for every state $s_i \in S$ that

$$\text{if } D(f_\psi \sigma_i') \text{ covers } K_i' \text{ then } s_i \models \psi$$

If $\psi = p$ (or $\psi = \neg p$), and p is an atomic proposition associated with boolean variable x, then $D(f_\psi \sigma_i')$ covers K_i' iff $D(f_\psi \sigma_i') = \emptyset$ and, hence, $p \in L(s_i)$ (or $p \notin L(s_i)$, respectively).

Suppose $\psi = \bigcirc p$ is a temporal subformula of φ associated with temporal variable y, and $D(f_\psi \sigma_i')$ covers K_i'. Then $f_\psi = y$, $f_y = f_p$ and so, $y \in K_i'$. Clearly, $D(f_p \sigma_{i+1}')$ covers K_{i+1}', since $K_i' \prec K_{i+1}'$, and so, by the induction hypothesis, $s_{i+1} \models p$. Thus $s_i \models \psi$.

The other cases are considered analogously. $\square$

4 PLTL Satisfiability Checking Algorithm

In this section we offer a PLTL satisfiability checking algorithm which verify the PLTL formulae treating the monotone conjunctions and d.n.f. Let $S(\varphi) = \{f_\varphi\} \cup \{f_1, \ldots, f_m\}$ be a system of boolean functions associated with the PLTL formula φ whose satisfiability we wish to verify. In what follows we assume that each temporal variable $y_i \in Y_\varphi$ refers to the function f_i of this system.

In addition to the basic set-theoretic operations (union, intersection and subtraction) we introduce three new operations (*reduction*, *fusion* and *filtration*) and a binary relation (*connectedness*) on the set of monotone d.n.f.

(1) **Reduction**

$D \downarrow = \{K : D \text{ strictly covers } D - \{K\}\}$.

(2) Fusion

$D_1 + D_2 = (D_1 \cup D_2) \downarrow.$

(3) Filtration

$D_1 | D_2 = (D_1 + D_2) - D_2.$

(4) Connectedness

$K \rightsquigarrow D$ iff there exists a conjunction K' in D which is a refinement of K.

PLTL satisfiability checking algorithm is given as follows. For a given system of boolean functions $S(\varphi)$ it attempts to construct a pseudo-structure. At each stage t of computation it deals with a prestructure T for $S(\varphi)$ and try to extend T to a pseudo-structure selecting one of the feasible alternatives. If no alternatives are available then an algorithm performs a backtracking and try the alternatives which were put aside at the previous stages.

The principal data structures the algorithm treats at each stage t of computation are:

- a list of d.n.f. $T_t = [H_0, \ldots, H_n]$ associated with a current prestructure;
- a list of d.n.f. $A_t = [D_0, D_1, \ldots, D_n]$ referring to the available alternatives;
- a d.n.f. G_t of the useless alternatives ("garbage collector").

A triple $\langle T_t, A_t, G_t \rangle$ is called a *state of computation* at the stage $t + 1$.

PLTL satisfiability checking algorithm
MONOTONE CHECKING

INITIAL STAGE $(t = 0)$:
$T_0 := [\{\emptyset\}]; A_0 := [D(\exists \overline{x} f_\varphi)]; G_0 := \emptyset.$

Suppose that $T_{t-1} = [H_0, H_1, \ldots, H_n], A_{t-1} = [D_0, D_1, \ldots, D_n].$

SEARCHING STAGE $(t > 0)$:
if $D_n = \emptyset$ then *BACKTRACKING* else *DEVELOPMENT*.

BACKTRACKING:
if $n = 0$ then *STOP("failure")*
else
> do
>> $T_t := [H_0, H_1, \ldots, H_{n-1}]; A_t := [D_0, D_1, \ldots, D_{n-1}]; G_t := G_{t-1} + H_n;$
>> goto *SEARCHING STAGE* $(t + 1)$.
> od

DEVELOPMENT:
Choose an arbitrary conjunction $K = \{y_{i_1}, \ldots, y_{i_k}\} \in D_n;$
$D := D(\exists \overline{x} \bigwedge_{j=1}^{k} f_{i_j}) | G_{t-1};$
if $\exists i \, (K \rightsquigarrow H_i)$ then *COMPLETION* else *EXTENSION*;

COMPLETION:
$m := min\{i : 1 \le i \le n, K \rightsquigarrow H_i\};$
if $(E_\varphi \cap \bigcap_{j=m}^{n} \bigcap_{K' \in H_j} K' = \emptyset)$ then *STOP("success")*
else

do

$$H_m := \{K\} \cup \bigcup_{j=m}^{n} H_j; \; D_m := (D + \bigcup_{j=m}^{n} D_j) - H_m;$$

$$T_t := [H_0, H_1, \ldots, H_m]; \; A_t := [D_0|D, D_1|D, \ldots, D_{m-1}|D, D_m];$$

$$G_t := G_{t-1}; \; \underline{goto} \; SEARCHING \; STAGE \; (t+1)$$

od.

EXTENSION:

do

$$T_t := [H_0, H_1, \ldots, H_n, \{K\}]; \; A_t := [D_0|D, \ldots, D_{n-1}|D, (D_n - \{K\})|D, D];$$

$$G_t := G_{t-1}; \; \underline{goto} \; SEARCHING \; STAGE \; (t+1).$$

od.

Theorem 2. Termination. *For every PLTL formula φ the MONOTONE CHECKING applied to $S(\varphi)$ always terminates.*

Proof. (Sketch) Suppose $\langle T_t, A_t, G_t \rangle$ and $\langle T_{t+1}, A_{t+1}, G_{t+1} \rangle$ are the states of computation at the stages $t+1$ and $t+2$, $t \geq 0$, such that, $T_t = [H_0, H_1, \ldots, H_n]$, $T_{t+1} = [H'_0, H'_1, \ldots, H'_m]$. ¿From the description of MONOTONE CHECKING it follows that

either $m = n - 1, G_{t+1} = G_t + H_n$, and $H'_i = H_i$ for each i, $1 \leq i \leq m$,

or $m < n, G_{t+1} = G_t, H'_m \subset \bigcup_{j=m}^{n} H_j$ and $H'_i = H_i$ for each i, $1 \leq i \leq m$,

or $m = n+1, G_{t+1} = G_t, H_m = \{K\}$, and $H'_i = H_i, K \notin H_i$ for each $i, 1 \leq i \leq n$.

Then it is easy to prove by induction on t that the following assertions hold

(1) $H_i \neq \emptyset$ for each i, $1 \leq i \leq n$;

(2) if $K' \in H_i, K'' \in H_j, 1 \leq i < j \leq n$, then $K'' \not\subseteq K'$;

(3) if $K' \in H_i, 1 \leq i \leq n$, and $K'' \in G_t$, then $K'' \not\subseteq K'$.

Thus we obtain that either G_{t+1} strictly covers G_t, or $G_{t+1} = G_t$ and $\bigcup_{i=1}^{n} H_i \subset \bigcup_{i=1}^{m} H'_i$. So, the verification of $S(\varphi)$ by MONOTONE CHECKING always termi-

nates in $O(2^N)$ stages where N is the number of temporal variables in $S(\varphi)$.

Theorem 3. Correctness. *If for a given PLTL formula φ MONOTONE CHECKING terminates successfully then φ is satisfiable.*

Proof. (Sketch) Assume that $H = \{K_1, \ldots, K_n\}$ is a monotone d.n.f. over the temporal variables Y_φ. We say that a conjunction K_j is accessible from a conjunction K_i in H and denote this relation by $K_i \prec^*_H K_j$ iff either $K_i = K_j$, or there is a subset $\{K_{i_1}, \ldots, K_{i_m}\}$ of H such that $K_i \prec K_{i_1} \prec, \ldots, \prec K_{i_m} \prec K_j$.

Suppose $\langle T_t, A_t, G_t \rangle$ is a state of computation at some stage $t+1$, $t \geq 0$, such that $T_t = [H_0, H_1, \ldots, H_n]$. Then it may be proved by induction on t that the following assertions hold for each i, $1 \leq i \leq n$

(1) If $K_1 \in H_i$ and $K_2 \in H_i$ then $K_1 \prec^*_{H_i} K_2$ for each $i, 1 \leq i \leq n$;

(2) If $K_1 \in H_{i-1}$ and $K_2 \in H_i$ then $K_1 \prec^*_{H_{i-1}} K' \prec K'' \prec^*_{H_i} K_2$ for some $K' \in H_{i-1}$, $K'' \in H_i$.

It follows from the above that for each i, $1 \le i \le n$, there exist a d.n.f. H'_i, $H'_i \subseteq H_i$, such that for some permutations $P(H'_i)$, $P(H_i)$ of H'_i and H_i, a sequence

$$P(H_1), P(H'_1), \ldots, P(H_{n-1}), P(H'_{n-1}), P(H_n)$$

is a prestructure for $S(\varphi)$. So, the successful termination of MONOTONE CHECKING means that this prestructure is a pseudo-structure for $S(\varphi)$. □

Proposition 1. *Assume that* $T = K_1, \ldots, K_n, \ldots, K_m$ *is a prestructure for* $S(\varphi)$, *a conjunction* K' *follows by both* K_n *and* K_m, *and a prestructure* $K_1, \ldots, K_n, K'$ *may be extended to a pseudo-structure for* $S(\varphi)$. *Then a prestructure* $K_1, \ldots, K_n, \ldots, K_m, K'$ *may be also extended to a pseudo-structure for* $S(\varphi)$.

Theorem 4. Completeness. *If for a given PLTL formula* φ *MONOTONE CHECKING terminates with failure then* φ *is unsatisfiable.*

Proof. (Sketch) We say that a conjunction K is *useless for* $S(\varphi)$ iff every pseudo-structure for $S(\varphi)$ (if any) contains no refinements of K.

Suppose that $\langle T_t, A_t, G_t \rangle$ is a state of computation at some stage $t+1$, $t \ge 0$, such that $T_t = [H_0, H_1, \ldots, H_n]$, $A_t = [D_0, D_1, \ldots, D_n]$. Then the following assertions may be proved by induction on t

(1) if $K \in G_t$ then K is useless for $S(\varphi)$;

(2) if K' in H_n and some conjunction K'' follows for K' w.r.t. $S(\varphi)$, then K'' is a refinement of some conjunction K in $H_n \cup D_n \cup G_t$.

Taking in account proposition 1 and applying once more induction on t one may prove the following statement

(3) if $S(\varphi)$ has a pseudo-structure then at each stage of computation $\langle T_t, A_t, G_t \rangle$, $T_t = [H_0, H_1, \ldots, H_n]$, $A_t = [D_0, D_1, \ldots, D_n]$ there exists m, $0 \le m \le n$, such that for some subsets $H'_i \subseteq H_i$, $1 \le i \le m$ and a conjunction $K \in D_m \cup D_m$ a sequence of permutations

$$P(H_1), \ P(H'_1), \ldots, P(H_{n-1}), \ P(H'_{n-1}), \ P(H_n), \ K$$

is a prefix of some pseudo-structure for $S(\varphi)$.

It immediately follows that MONOTONE CHECKING makes out the existence of pseudo-structure for $S(\varphi)$ in the case that φ is satisfiable. □

5 Conclusions

MONOTONE CHECKING can be easily adapted to the symbolic manipulations. Claim 1 guarantees that at each stage t the sets of d.n.f. T_t, A_t and G_t have no common conjunctions. Therefore the whole data structure for a state of computation of MONOTONE CHECKING may be encode by means of the unique multi-pole oriented binary decision diagram (OBDD) [13] or contact (switching) circuit [14]. Each operation over monotone d.n.f. may be implemented following the style used in [13]. In fact, the most of these operations affected only on the bounded fragments of the structure and may be executed in parallel. We are currently working on an implementation of the algorithm.

References

1. Pnueli A.: The temporal logics of programs. Proc. 18th Ann. IEEE Symp. on Foundations of Computer Science, (1977) 46–57
2. Emerson E.A.: Temporal and modal logic. Handbook of theoretical computer science, Ed. by J.van Leeuwen, Elsevier Science Publishers, (1990) 997–1072
3. Wolper P.: The tableau method for Temporal Logic: an overview. Logique et Anal., 28 (1985) 119–136
4. Vardi M., Wolper P.: Automata theoretic techniques for modal logics of programs. Proc. 16th Ann. ACM Symp. on Theory of Computing, (1984) 446–456
5. Lichtenstain O., Pnueli A., Zuck L.: The glory of the past. Lecture Notes in Computer Science, 193 (1985) 196–218
6. Kesten Y., Manna Z., McGuire H., Pnueli A.: A decision algorithm for full propositional temporal logic. Lecture Notes in Computer Science, 697 (1993) 97–109
7. Tarjan R.E.: Depth first search and linear graph algorithms, SIAM J. Comput., 1, N 2, (1972) 146–160
8. Aho A.V., Hopcroft J.E., Ullman J.D.: The design and analysis of computer algorithms. Addison-Weslay P.C. (1976)
9. Burch J.R., Clarke E.M., McMillan K.L., Dill D.L., Hwang H.: Symbolic model checking: 10^{20} states and beyond. Information and Computation, 98, N 2, (1992) 142–170
10. Clarke E.M., Grumberg O., Hamaguchi K.: Another look at LTL model checking. Lecture Notes in Computer Science, 818 (1994) 415–427
11. Kesten Y., Pnueli A., Raviv L.: Model checking of Linear TL, using OBDD. Manuscript
12. Quine W.V.: On cores and prime implicants of truth functions. Amer. Math. Monthly. 62, N 9, (1959)
13. Bryant R.E.: Graph-based algorithms for boolean function manipulation, IEEE Trans. Comput. 8 (1986)
14. Povarov G.N.: Mathematical theory of $(1, k)$–pole contact circuit synthesis. Doklady Akad. Nauk SSSR. 100, N 5, (1955) 909–912.

Springer
and the
environment

At Springer we firmly believe that an international science publisher has a special obligation to the environment, and our corporate policies consistently reflect this conviction.

We also expect our business partners – paper mills, printers, packaging manufacturers, etc. – to commit themselves to using materials and production processes that do not harm the environment. The paper in this book is made from low- or no-chlorine pulp and is acid free, in conformance with international standards for paper permanency.

List of Authors

Lecture Notes in Computer Science

For information about Vols. 1–1156

please contact your bookseller or Springer-Verlag